Developing Life Thinking Skills:

From Grade School to Business Professionals

Steve Munkeby

Marti Elliott

Title: Developing Life Thinking Skills: From Grade School to Business Professionals

Authors: Dr. Steve Munkeby and Marti Elliott

NextGen Publications, LLC

Website: https://SteveMunkeby.com

Email: drstevemunkeby@gmail.com

Phone: 470-302-4626

ISBNs:

Paperback: 978-1-968702-50-2

eBook: 978-1-968702-51-9

First Edition: October 2025

Printed in the United States of America

Cover Design: NextGen Publications, LLC

Interior Formatting: NextGen Publications, LLC

For more books and free resources, visit https://SteveMunkeby.com

Table of Contents

Foreword

The ability to think effectively is one of the most profound and essential life skills an individual can develop. Beyond the confines of rote memorization and mechanical problem-solving, the art of thinking is a dynamic, deliberate, and reflective process—one that empowers individuals to approach challenges with clarity, creativity, and wisdom. It is both a skill and a discipline, requiring intellectual rigor, emotional intelligence, and innovative reasoning to navigate the complexities of an ever-evolving world.

True thinking is not merely about acquiring information but about developing the capacity to analyze, synthesize, and apply knowledge in meaningful ways. It involves curiosity, adaptability, and resilience; the ability to question assumptions, recognize patterns, and construct well-reasoned arguments. Whether in personal decision-making, academic pursuits, or professional problem-solving, cultivating strong thinking skills serves as a foundational pillar for lifelong success.

In this context, *Developing Life Thinking Skills: From Grade School to Business Professionals* explores the transformative journey of cognitive growth across various life stages. This work delves into the progression of thinking abilities from childhood to adulthood, addressing the evolving cognitive and emotional needs of learners at every stage. By mapping the stages of intellectual development, it highlights the critical transitions from basic comprehension to advanced reasoning, strategic decision-making, and executive-level problem-solving.

At the heart of this framework is Progressive Cognitive Structuring (PCS), a scaffolded approach that ensures the gradual and systematic enhancement of thinking skills. PCS fosters cognitive development by aligning learning strategies with the increasing complexity of real-world challenges. It begins with foundational cognitive abilities in early education—such as pattern recognition, logical reasoning, and conceptual understanding—and advances toward sophisticated analytical, strategic, and adaptive thinking techniques essential in professional and leadership settings.

While this book is substantial in length—spanning nearly 750 pages—it is designed to be highly practical, not theoretical or overwhelming. Readers should know that only about 10% of the book (roughly 80 pages) is devoted to explaining the PCS framework and introducing its foundational techniques for cultivating life thinking skills. The remaining 90%—nearly 670 pages—is a rich repository of hands-on, stage-specific teaching strategies.

These strategies are clearly organized by five developmental stages: Grade School, Middle School, High School, College/Technical School, and Business Professionals.

This structure makes the book more of a reference guide and toolkit than a dense theoretical manual. Whether you are an educator, parent, corporate trainer, or lifelong learner, you can turn directly to the sections that apply to your audience or situation. It is a book designed for use, not just for reading.

By emphasizing structured cognitive growth across life stages, this work equips individuals with the intellectual agility needed to excel in an ever-changing world. Whether a young student learning to think critically, a graduate preparing for the workforce, or a seasoned business professional navigating complex decision-making, this approach provides the tools necessary to thrive in both personal and professional spheres.

Through the integration of theory, practical application, and real-world case studies, this book serves as both a guide and a blueprint for cultivating a lifelong habit of effective thinking. It underscores the idea that thinking is not a passive skill but an active, evolving process—one that, when nurtured, becomes the key to unlocking personal growth, innovation, and leadership in any domain.

Introduction

The art of thinking is one of the most intricate and powerful human faculties, enabling individuals to process information, solve problems, make decisions, and innovate. Thinking is not merely a mechanical function but a deeply intellectual, emotional, and creative endeavor that allows us to interpret the world, communicate ideas, and adapt to new circumstances. At its core, thinking involves several interconnected dimensions:

- **Analytical Thinking** – The ability to break down complex ideas and problems into smaller, more manageable components, identifying relationships, patterns, and logical structures (Paul & Elder, 2019).
- **Critical Thinking** – The disciplined practice of evaluating information objectively, questioning assumptions, and distinguishing between fact, opinion, and bias (Facione, 2020).
- **Creative Thinking** – The capacity to generate novel ideas, approach problems from unconventional angles, and synthesize diverse concepts into innovative solutions (Sawyer, 2012).
- **Reflective Thinking** – The process of examining one's thoughts, beliefs, and experiences to gain deeper insights and refine one's understanding (Dewey, 1933).
- **Practical Thinking** – The application of knowledge and reasoning to real-world situations, ensuring that abstract ideas translate into meaningful action (Sternberg, 1997).

A truly developed thinker seamlessly integrates these elements, moving fluidly between logic and intuition, analysis and creativity, skepticism, and open-mindedness. The art of thinking is both a skill and a discipline—one that demands continuous refinement, intellectual curiosity, and the courage to challenge one's own assumptions.

The Decline of Thought: Current Weaknesses and Problems

Despite the power and potential of human cognition, modern society faces a growing crisis in effective thinking. Various cultural, educational, and technological shifts have contributed to a decline in deep, disciplined, and independent thought. The most pressing weaknesses include:

1. The Overreliance on Information Over Understanding

The digital age has flooded society with an overwhelming amount of information, but sheer access to knowledge does not equate to understanding. Many individuals mistake data accumulation for intelligence, consuming endless streams of content without processing it critically (Carr, 2010). This leads to:

- **Surface-Level Thinking** – People skim information rather than engage in deep analysis (Wolf, 2018).
- **Lack of Contextual Awareness** – Facts are absorbed in isolation, without connecting them to broader frameworks of knowledge (Kahneman, 2011).
- **Diminished Retention and Application** – Rapid consumption prevents meaningful internalization, making it difficult to apply knowledge effectively (Newport, 2016).

2. The Decline of Critical Thinking Skills

The ability to evaluate information logically and skeptically is essential in an age of misinformation, propaganda, and ideological echo chambers. However, several factors contribute to a weakening of critical thinking abilities:

- **Cognitive Bias Reinforcement** – Many individuals seek information that confirms their existing beliefs rather than challenging their perspectives (Tversky & Kahneman, 1974).
- **Emotional Over Reasoned Discourse** – Emotional reactions increasingly replace reasoned argumentation, leading to polarization and knee-jerk decision-making (Haidt, 2012).
- **Misuse of Logical Fallacies** – Poor reasoning skills result in widespread reliance on faulty logic, such as ad hominem attacks, strawman arguments, and false dichotomies (Weston, 2018).

3. The Erosion of Deep Focus and Patience

Modern lifestyles promote short attention spans, leading to an inability to sustain deep thinking. The causes of this trend include:

- **Instant Gratification Culture** – Social media, fast content, and on-demand entertainment discourage patience and long-term intellectual investment (Twenge, 2017).
- **Multitasking and Distraction** – Constant notifications, digital interruptions, and information overload fragment cognitive focus, reducing the ability to engage in prolonged reflection (Gazzaley & Rosen, 2016).
- **Lack of Intellectual Endurance** – The effort required for deep, analytical thinking is often perceived as unnecessary or too demanding (Carr, 2010).

4. The Weakening of Independent Thought

The ability to think autonomously and form original conclusions is increasingly under threat due to:

- **Groupthink and Social Conformity** – Fear of social rejection discourages individuals from expressing independent or controversial ideas (Janis, 1972).
- **Algorithmic Influence** – Search engines, social media feeds, and AI-driven content curation shape perspectives by reinforcing preferred narratives rather than exposing individuals to diverse viewpoints (Pariser, 2011).

- **Overreliance on Experts Without Personal Evaluation** – While expertise is valuable, blindly accepting authority without questioning can lead to passive dependence rather than active reasoning (Nichols, 2017).

5. The Crisis of Applied Thinking

While academic and theoretical thinking are emphasized in education, real-world application often falls short. This is evident in:

- **Education Focused on Memorization Over Problem-Solving** – Many students are trained to recall facts rather than engage in deep problem-solving or adaptive reasoning (Robinson, 2011).
- **Workplace Deficiencies in Strategic Thinking** – Employees may possess technical knowledge but struggle with decision-making, adaptability, and problem resolution (Sternberg, 1997).
- **Failure to Integrate Thinking with Action** – Many individuals can articulate ideas but fail to translate them into meaningful action due to fear, indecision, or lack of real-world experience (Dweck, 2006).

Restoring the Art of Thinking

To counteract these weaknesses, a deliberate effort must be made to cultivate a more rigorous and engaged thinking culture. This requires:

- **Educational Reform** – Shifting away from rote memorization and emphasizing deep inquiry, debate, and problem-solving (Robinson, 2011).
- **Mindfulness and Focus Training** – Encouraging practices that strengthen attention, patience, and deep concentration (Newport, 2016).
- **Promotion of Intellectual Independence** – Teaching individuals how to challenge assumptions, embrace uncertainty, and construct their own well-reasoned conclusions (Facione, 2020).
- **Integration of Theory and Practice** – Ensuring that thinking is not just an academic exercise but a tool for navigating real-life challenges effectively (Sternberg, 1997).

Despite its rich history and numerous contributions, the art of thinking also faces significant challenges in the modern era. These limitations hinder its potential and highlight areas where further development is essential:

- **Overemphasis on Specialization:** Education and professional training often focus narrowly on domain-specific skills, leading to siloed thinking and a lack of interdisciplinary integration (Schön, 1983).

- **Cognitive Overload:** The sheer volume of information available in the digital age overwhelms cognitive capacities, making it difficult to discern what is relevant and essential (Sweller, 1988).

- **Bias and Heuristics:** Fast, intuitive thinking—while efficient—is prone to errors, such as confirmation bias, availability heuristics, and overconfidence, which distort decision-making (Kahneman, 2011).
- **Neglect of Emotional Integration:** Many frameworks prioritize logical and analytical thinking while neglecting emotional intelligence, empathy, and interpersonal dynamics, which are crucial for effective collaboration and leadership (Goleman, 1995).
- **Stifled Creativity:** Traditional education and workplace cultures often undervalue creative thinking, emphasizing conformity and routine over innovation and exploration (Robinson, 2011).
- **Insufficient Focus on Practical Application:** Theories of thinking are frequently taught in abstract terms without clear pathways to real-world implementation, leaving individuals ill-equipped to apply their skills effectively (Brown et al., 2014).
- **Resistance to Reflection:** In a fast-paced, results-driven world, there is little emphasis on introspection and adaptive thinking, which are necessary for continuous learning and personal growth (Argyris & Schön, 1978).
- **Limited Systems Perspective:** Complex, interconnected challenges—such as climate change or organizational dynamics—require systems thinking, yet it is rarely emphasized in traditional curricula (Meadows, 2008).

Addressing the Weaknesses and Challenges Through Progressive Cognitive Structuring

Recognizing these weaknesses, **"Developing Life Thinking Skills: From Grade School to Business Professionals"** begins where these weaknesses and challenges leave off. By leveraging the historical insights of thinking pioneers and addressing the current limitations, this book introduces a transformative approach: **Progressive Cognitive Structuring (PCS)**. The PCS framework is designed to:

- **Integrate Thinking Dimensions:** Combine cognitive, emotional, and creative thinking to develop a balanced and multidimensional skillset.
- **Promote Interdisciplinary Learning:** Foster the ability to draw connections across disciplines, breaking down silos in both education and professional environments.
- **Encourage Reflection and Adaptation:** Embed habits of mindfulness and reflection to enable adaptive responses to evolving challenges.
- **Bridge Theory and Practice:** Provide practical, actionable strategies to apply thinking skills in real-world contexts, from classrooms to boardrooms.
- **Cultivate Systems Thinking:** Emphasize the importance of understanding interconnected systems to tackle complex, multifaceted problems.

Through these strategies, Progressive Cognitive Structuring aims to equip individuals with the tools they need to navigate the complexities of the modern world, fostering lifelong growth and success.

The Structure of This Book

The journey of developing life thinking skills begins with foundational lessons in childhood, progresses through the refinement of critical and creative thinking in adolescence, and culminates in the mastery of strategic and reflective thinking in adulthood. By addressing current weaknesses and building on the strengths of historical contributions, this book provides a blueprint for cultivating the art of thinking across all stages of life while exploring:

- The foundational principles of the art of thinking.
- The role of strategic simplification in problem-solving.
- Detailed learning strategies for each developmental stage.
- Practical applications for integrating thinking skills into personal and professional contexts.

In a world characterized by rapid change, complexity, and uncertainty, the ability to think effectively has never been more critical. Teaching life thinking skills ensures that individuals are equipped to:

- Adapt to new challenges and environments.
- Collaborate across disciplines and cultures.
- Lead with vision and empathy.
- Innovate in the face of adversity.

By following this structured approach, educators, trainers, and learners can foster a new generation of critical thinkers, creative problem-solvers, and visionary leaders and create a foundation for lifelong learning and success.

The ultimate aim of developing life thinking skills is to cultivate a culture of lifelong cognitive excellence. By systematically nurturing these skills from an early age, learners become equipped not only to meet immediate challenges but also to evolve and grow as thinkers throughout their lives. The journey begins with foundational lessons that encourage curiosity and observation, collaborative thinking, and goal setting in childhood, progresses through the exploration of critical thinking, decision making, and data analysis and interpretation in adolescence, and culminates in the mastery of visionary thinking, continuous improvement, and global perspectives in adulthood.

This book is designed to be a transformative resource, a blueprint for integrating the art and science of thinking into every stage of development. By doing so, we empower individuals to harness the full potential of their minds, create innovative solutions, and lead lives of purpose and impact.

The Historical Overview of the Art of Thinking

The art of thinking has shaped human progress for millennia, evolving through philosophy, science, and psychology to refine how we reason, solve problems, and make decisions. From the structured logic of ancient philosophers to the cognitive models of modern psychology, each era has contributed new insights into how we process information and navigate complexity.

This journey through the history of thinking reveals not only its intellectual evolution but also its practical applications—how shifting perspectives have influenced leadership, innovation, and everyday problem-solving. By examining the development of thought across time, we can identify the key contributors who shaped its trajectory, the core principles they emphasized, and the ways in which their ideas continue to inform contemporary thinking. This historical perspective also sheds light on the field's current state, highlighting both its advancements and the persistent challenges that limit our ability to think more effectively in an increasingly complex world.

The Evolution of Thinking: From Ancient Wisdom to Modern Complexity

The art of thinking has evolved through centuries of philosophical inquiry, scientific discovery, and cognitive exploration. Each era has contributed to foundational principles that continue to shape how we reason, solve problems, and adapt to new challenges. From the dialectical methods of ancient Greece to the cognitive models of artificial intelligence, the progression of thought reflects humanity's ongoing quest to understand and refine the way we think.

1. The Ancient Foundations of Thinking

- Socrates (470-399 BCE)
    - **Contribution:** Socratic Method.
    - **Emphasis:** The use of questioning to stimulate critical thinking and expose underlying assumptions.
    - Socrates encouraged individuals to challenge conventional wisdom and seek deeper truths through dialectical reasoning, laying the foundation for critical thinking (Brickhouse & Smith, 2000; Vlastos, 1991).

- Plato (428-348 BCE)
    - **Contribution:** Theory of Forms and Rational Thought.
    - **Emphasis:** Abstract thinking and the pursuit of ideal truths.
    - Plato highlighted the importance of reasoning to understand concepts beyond the physical world, emphasizing the role of structured thought (Fine, 1999; Plato, c. 380 BCE).

- Aristotle (384-322 BCE)
    - **Contribution:** Formal Logic.
    - **Emphasis:** Developing systematic approaches to reasoning through syllogisms and categorization.
    - Aristotle's works formed the basis for logical reasoning, which became a cornerstone of intellectual thought for centuries (Aristotle, c. 350 BCE; Smith, 1997).

2. **Medieval and Renaissance Contributions**

- Thomas Aquinas (1225-1274)
    - **Contribution:** Scholasticism.
    - **Emphasis:** Integrating logic with theology.
    - Aquinas combined Aristotelian logic with religious doctrine, demonstrating how structured thinking could explore abstract and metaphysical ideas (Aquinas, 1274; Copleston, 1955).

- Francis Bacon (1561-1626)
    - **Contribution:** Empiricism and the Scientific Method.
    - **Emphasis:** Observation and inductive reasoning as tools for discovery.
    - Bacon emphasized evidence-based reasoning, which paved the way for modern science and problem-solving methodologies (Bacon, 1620; Urbach, 1987).

- René Descartes (1596-1650)
    - **Contribution:** Cartesian Doubt and Deductive Reasoning.
    - **Emphasis:** Methodical doubt as a starting point for certainty.
    - Descartes' famous phrase, *"Cogito, ergo sum"* (I think, therefore I am), placed thinking at the center of human existence and knowledge (Cottingham, 1992; Descartes, 1637).

3. **The Enlightenment and Rational Thought**

- Immanuel Kant (1724-1804)
    - **Contribution:** Critique of Pure Reason.
    - **Emphasis:** The interplay between sensory experience and rational thought.
    - Kant argued for the integration of empirical observation with innate cognitive structures, a precursor to modern cognitive science (Gardner, 1999; Kant, 1781).

- John Stuart Mill (1806-1873)
    - **Contribution:** Utilitarian Logic
    - **Emphasis:** Logical reasoning for ethical and societal progress.
    - Mill highlighted the importance of critical thinking for evaluating consequences and fostering collective well-being (Mill, 1863; Skorupski, 1989).

4. **Modern Foundations of the Art of Thinking**

- Charles Sanders Peirce (1839-1914)
    - **Contribution:** Pragmatism and Abductive Reasoning.
    - **Emphasis:** Thinking as a process of inquiry that adapts to context.
    - Peirce introduced the concept of abductive reasoning, expanding beyond deduction and induction (Misak, 2004; Peirce, 1877).
- Jean Piaget (1896-1980)
    - **Contribution:** Cognitive Development Theory.
    - **Emphasis:** The stages of cognitive growth from childhood to adulthood.
    - Piaget's work laid the foundation for understanding how thinking evolves with age and experience (Flavell, 1963; Piaget, 1952).
- Edward de Bono (1933-2021)
    - **Contribution:** Lateral Thinking.
    - **Emphasis:** Creative approaches to problem-solving by challenging linear thought patterns.
    - De Bono advocated for creativity as an essential complement to logical reasoning (De Bono, 1967, 1992).

5. **The Contemporary Era: Psychology, AI, and Systems Thinking**

- Daniel Kahneman (b. 1934)
    - **Contribution:** Behavioral Economics and Cognitive Biases.
    - **Emphasis:** The dual-system model of thinking (fast, intuitive thinking vs. slow, deliberate thinking).
    - Kahneman's work highlighted the limitations of human cognition and the need for reflective practices to counter biases (Kahneman, 2011; Tversky & Kahneman, 1974).
- Peter Senge (b. 1947)
    - **Contribution:** Systems Thinking.
    - **Emphasis:** Viewing organizations and problems as interconnected systems.
    - Senge's work applied holistic thinking to leadership and organizational dynamics (Senge, 2006).
- Artificial Intelligence Research (1950s-Present)
    - **Contribution:** Machine Learning and Cognitive Models.
    - **Emphasis:** Mimicking human thought processes through algorithms and AI systems.
    - AI has introduced tools for enhancing human thinking but also highlighted gaps in human adaptability and emotional intelligence (LeCun et al., 2015; Russell & Norvig, 2020; Turing, 1950).

Lessons from the Evolution of Thinking

The progression of thought from Socrates to artificial intelligence reflects a continuous refinement of how we process information, solve problems, and make decisions. Each era has introduced new methodologies, from dialectical reasoning and logic to empirical observation, cognitive development, and systems thinking. These contributions have shaped the modern understanding of thought, providing tools to navigate increasingly complex challenges.

In today's world, the integration of **logical reasoning, creativity, systems thinking, and AI-driven insights** is more important than ever. As we face rapid technological advancements, information overload, and ethical dilemmas, the lessons from history remind us that **effective thinking is both an art and a science**—a balance between structured analysis and adaptability. By drawing on the wisdom of past thinkers and the innovations of the present, we can cultivate a more deliberate, strategic, and holistic approach to thinking that meets the needs of today's dynamic world.

Today's Understanding of the Art of Thinking

The **Art of Thinking** in the modern era encompasses **deliberate, reflective, and strategic thought processes** that help individuals understand problems, generate ideas, and make well-informed decisions. It is a multidimensional skill set that integrates logic, creativity, emotional intelligence, and adaptability. As the world grows more complex, the ability to think effectively and critically has become indispensable for navigating personal, professional, and societal challenges.

Thinking is both an art—infused with creativity, intuition, and flexibility—and a discipline—anchored in structure, logic, and evidence-based reasoning. It consists of several interrelated components:

- **Cognitive Skills** – Logical analysis, pattern recognition, and critical evaluation.
- **Emotional Intelligence** – Empathy, self-regulation, and interpersonal understanding.
- **Habits of Mind** – Perseverance, curiosity, and open-mindedness.

By developing these skills, individuals enhance their ability to:

- Solve complex problems.
- Make informed and balanced decisions.
- Innovate and generate novel ideas.
- Navigate interpersonal and societal dynamics effectively.

The following sections explore the key elements of today's Art of Thinking, their significance, and their relevance in the modern world.

---

1. Critical Thinking

Critical thinking involves systematic evaluation of information to reach well-reasoned conclusions. It requires analyzing arguments, identifying biases, and seeking evidence-based solutions.

- **Analyzing Arguments** – Breaking down statements into premises and conclusions to assess their validity and soundness (Paul & Elder, 2019).
- **Recognizing Biases** – Identifying and mitigating personal or external biases that may distort judgment (Kahneman, 2011).
- **Seeking Evidence** – Demanding empirical or logical support before accepting claims as true (Facione, 2011).
- **Problem Solving** – Evaluating multiple solutions and their implications before acting (Ennis, 2018).

Why It Matters Today

With information overload and widespread misinformation, critical thinking is crucial for distinguishing fact from fiction, making rational choices, and questioning assumptions. It is a key competency for professionals, leaders, and informed citizens in an age dominated by rapid communication and complex global challenges.

---

2. Creative Thinking

Creativity is essential for problem-solving, innovation, and adaptability. It enables individuals to generate novel and valuable ideas and approach challenges from new perspectives.

- **Divergent Thinking** – Exploring multiple possible solutions before narrowing down choices (Guilford, 1967).
- **Connecting Ideas** – Finding relationships between seemingly unrelated concepts to generate fresh insights (Runco & Jaeger, 2012).
- **Risk-Taking** – Being open to uncertainty and unconventional ideas without fear of failure (De Bono, 1992).

Why It Matters Today

In a world driven by innovation and technological advancements, creative thinking fosters groundbreaking discoveries, artistic expression, and strategic adaptability. It is particularly valuable in entrepreneurship, research, and leadership, where novel approaches often lead to disruptive solutions.

---

3. Reflective Thinking

Reflective thinking encourages self-awareness, introspection, and learning from experience to refine decision-making.

- **Self-Awareness** – Recognizing one's thought processes, emotions, and biases (Dewey, 1933).
- **Learning from Mistakes** – Reviewing past decisions and their consequences to improve future choices (Schön, 1983).
- **Mindfulness** – Being present and engaged to fully understand a situation before reacting (Epstein, 1999).

Why It Matters Today

In high-pressure environments, reflective thinking helps individuals avoid impulsive decisions, build resilience, and cultivate wisdom. It is vital for leaders, educators, and lifelong learners who must continuously adapt and improve.

---

4. Systems Thinking

Systems thinking involves seeing the bigger picture and understanding how different elements interact within a broader framework.

- **Holistic Perspective** – Recognizing interdependencies and interactions in complex systems (Senge, 1990).
- **Feedback Loops** – Understanding how actions lead to chains of consequences (Meadows, 2008).
- **Long-Term Thinking** – Evaluating the future impact of current decisions rather than focusing solely on short-term gains (Arnold & Wade, 2015).

Why It Matters Today

Systems thinking is essential for solving large-scale challenges such as climate change, economic policy, business strategy, and global health crises. It enables individuals and organizations to anticipate unintended consequences and make more informed, sustainable decisions.

---

5. Emotional Intelligence

Thinking effectively requires managing emotions and understanding how they influence cognition.

- **Empathy** – Considering others' perspectives and emotions in decision-making (Salovey & Mayer, 1990).
- **Emotional Regulation** – Staying composed under pressure to maintain clarity of thought (Goleman, 1995).
- **Interpersonal Skills** – Using emotional awareness to navigate social and professional relationships (Petrides & Furnham, 2001).

Why It Matters Today

Emotional intelligence is critical for effective leadership, teamwork, conflict resolution, and mental well-being. It helps individuals build strong relationships and navigate complex social interactions.

---

6. Logical and Analytical Thinking

Logical and analytical thinking form the foundation of structured reasoning and informed decision-making. These skills allow individuals to evaluate information, detect inconsistencies, and derive conclusions based on sound principles of logic and analysis.

- **Reasoning Skills** – Applying deductive and inductive reasoning to solve problems (Weston, 2018). Deductive reasoning moves from general premises to specific conclusions, while inductive reasoning draws general conclusions from specific observations.

- **Data Interpretation** – Analyzing quantitative and qualitative data to extract meaningful insights and inform decisions (Toulmin, 1958).
- **Pattern Recognition** – Identifying recurring structures, trends, or anomalies that can improve forecasting and strategic planning (Gigerenzer & Brighton, 2009).
- **Logical Consistency** – Structuring arguments coherently and avoiding logical fallacies to ensure sound reasoning (Aristotle, c. 350 BCE).

Why It Matters Today

In an era of big data, artificial intelligence, and rapidly evolving information landscapes, logical and analytical thinking are critical for problem-solving, policymaking, and business strategy. These skills enable individuals to distill complex data into actionable insights, reduce cognitive biases, and enhance critical decision-making in fields ranging from finance to medicine to artificial intelligence.

---

7. Meta-Thinking (Thinking About Thinking)

Meta-thinking, or metacognition, is the ability to reflect on and regulate one's own thought processes. It involves recognizing cognitive biases, evaluating reasoning patterns, and adjusting strategies to improve clarity and effectiveness.

- **Metacognition** – Monitoring how we think, learn, and process information, then adjusting these processes for improvement (Flavell, 1979).
- **Cognitive Flexibility** – Shifting mental frameworks based on new information or changing contexts (Kahneman, 2011).
- **Heuristic Awareness** – Recognizing mental shortcuts (heuristics) that can lead to cognitive biases, such as overgeneralization or confirmation bias (Dunning, 2011).
- **Self-Questioning** – Engaging in internal dialogues to challenge assumptions, ask "why" and "how," and ensure deeper understanding (Paul & Elder, 2019).

Why It Matters Today

As misinformation, cognitive biases, and rapid decision-making pressures increase, individuals with strong meta-thinking skills are better equipped to analyze their own biases, refine their decision-making strategies, and foster intellectual humility. This skill is particularly valuable for leaders, educators, researchers, and professionals navigating complex, evolving industries.

---

8. Cultivating Intellectual Virtues

Intellectual virtues support effective thinking by fostering habits that promote curiosity, resilience, and ethical reasoning. These virtues are not innate but can be developed through deliberate effort and practice.

- **Curiosity** – Maintaining an inquisitive mindset, seeking out new information, and engaging in lifelong learning (Roberts & Wood, 2007).

- **Humility** – Acknowledging one's own cognitive limitations and being open to new perspectives and corrections (Baehr, 2013).
- **Perseverance** – Persisting in solving complex problems despite challenges or failures (Dweck, 2006).
- **Intellectual Courage** – Willingness to question deeply held beliefs, challenge authority when necessary, and engage in difficult discussions (Zagzebski, 1996).
- **Fair-Mindedness** – Striving to evaluate opposing viewpoints objectively rather than favoring one's own perspective (Paul & Elder, 2019).

Why It Matters Today

In an era of polarized opinions, rapid technological change, and ethical dilemmas, intellectual virtues ensure that individuals think with depth, fairness, and adaptability. They help professionals, educators, and policymakers make ethical decisions, engage in constructive debates, and continuously refine their understanding of complex issues.

---

9. Practical Application

Theoretical knowledge is valuable, but true mastery of the Art of Thinking comes from applying these principles in real-world situations. This requires active engagement, strategic decision-making, and adaptive learning.

- **Decision-Making** – Using structured thinking frameworks such as cost-benefit analysis, ethical reasoning, and risk assessment to make sound choices (Heath & Heath, 2013).
- **Innovation** – Applying both critical and creative thinking to develop new solutions in business, technology, and everyday life (Simon, 1996).
- **Conflict Resolution** – Leveraging logical reasoning, emotional intelligence, and negotiation skills to de-escalate conflicts and find equitable solutions (Fisher et al., 2011).
- **Strategic Thinking** – Anticipating long-term implications of actions and optimizing solutions for sustained success (Rumelt, 2011).

Why It Matters Today

With increasing complexity in global markets, leadership roles, and social dynamics, the ability to apply thinking skills to practical challenges is a key differentiator between effective and ineffective decision-makers. Those who excel in real-world applications can turn knowledge into action, drive meaningful changes, and lead innovation.

---

10. Habit Formation

Becoming an expert thinker requires consistent practice, habit formation, and continuous refinement of cognitive skills. Mental sharpness, like physical fitness, is built through repeated effort and disciplined engagement.

- **Daily Reflection** – Setting aside time for critical self-assessment, learning from experiences, and refining thought processes (James, 1890).
- **Lifelong Learning** – Seeking new knowledge, cross-disciplinary insights, and diverse perspectives to expand thinking capabilities (Duhigg, 2012).
- **Mental Exercises** – Engaging in activities such as puzzles, debates, philosophical discussions, and journaling to enhance cognitive flexibility (Ericsson, 2006).
- **Mindfulness Practices** – Using techniques such as meditation, focused breathing, and deliberate pausing to improve concentration and clarity (Epstein, 1999).

Why It Matters Today

In a fast-paced world with short attention spans, digital distractions, and constant information influx, forming structured mental habits ensures that individuals develop resilience, retain intellectual sharpness, and continue evolving in their cognitive abilities. Habit formation supports long-term success, deep thinking, and adaptability in both personal and professional life.

What We Know About the Modern Art of Thinking

Today, the Art of Thinking is a balance between logic, creativity, emotional intelligence, and adaptability. As the world becomes more interconnected and complex, those who refine their ability to think critically, reflectively, and strategically will be better equipped to lead, innovate, and solve global challenges. By integrating structured reasoning, intellectual virtues, and real-world application, individuals can:

- Make **informed, ethical decisions** in complex situations.
- Adapt to rapid technological and social changes.
- Engage in deep, meaningful learning and self-improvement.
- Cultivate resilience, problem-solving skills, and strategic foresight.

Mastering the Art of Thinking is not merely an intellectual exercise, it is a lifelong pursuit that empowers individuals to navigate uncertainty, drive progress, and contribute meaningfully to society.

Strategic Thinking: The Essential Next Step in the Art of Thinking

In today's fast-paced and interconnected world, the art of thinking faces significant challenges that hinder its full potential. While individuals and organizations have embraced critical thinking, creativity, and problem-solving as key cognitive skills, strategic thinking remains an underdeveloped area that must be prioritized and strengthened (Mintzberg, 1994).

Strategic thinking is a higher-order cognitive skill that allows individuals and organizations to anticipate the future, set clear goals, and align short-term actions with long-term objectives (Liedtka, 1998). It requires a multidisciplinary mindset that combines logic, creativity, foresight, emotional intelligence, and adaptability (Goldman et al., 2015). Unlike traditional decision-making processes that focus on immediate problem-solving, strategic thinking is about navigating complexity, managing uncertainty, and developing sustainable solutions (Gavetti & Rivkin, 2007).

By integrating strategic thinking into the art of thinking, individuals can enhance their ability to innovate, lead, and solve problems with greater clarity and effectiveness. This expansion is crucial for business leaders, policymakers, educators, and anyone striving to make a meaningful impact in an increasingly volatile and uncertain world.

What Is Strategic Thinking?

Strategic thinking is the process of envisioning the future, setting goals, and making decisions that align short-term actions with long-term objectives (Liedtka, 1998). It is both a mindset and a methodology that combines big-picture vision with detailed execution, emphasizing adaptability, foresight, and collaboration (Goldman et al., 2015). Unlike conventional problem-solving, which focuses on immediate challenges, strategic thinking involves:

- **Analyzing complex situations** to identify underlying patterns and relationships.
- **Anticipating outcomes** to prepare for potential challenges and opportunities.
- **Developing innovative yet pragmatic solutions** that are both forward-thinking and actionable.
- Balancing logical reasoning, creative problem-solving, and emotional intelligence to optimize decision-making (Schoemaker, 1995).

By incorporating systems thinking, iterative learning, and collaborative leadership, strategic thinking becomes an essential cognitive tool for navigating complexity and driving sustainable success (Gavetti & Rivkin, 2007).

Why Strategic Thinking Is the Next Step to Be Addressed in the Art of Thinking

While traditional cognitive skills such as critical thinking, creativity, and problem-solving are essential, they often operate in isolation without a unifying framework. Strategic thinking provides this missing link, integrating these cognitive skills into a comprehensive approach that enhances decision-making, leadership, and innovation (Goldman et al., 2015).

1. Holistic Integration

- Strategic thinking bridges the gap between logical reasoning, creative problem-solving, and emotional intelligence, integrating them into a cohesive, goal-oriented approach (Senge, 1990).
- While the art of thinking focuses on individual cognitive skills, strategic thinking elevates these skills to tackle complex, interconnected challenges that require a multidisciplinary perspective (Goldman et al., 2015).

2. Vision and Execution

- Strategic thinking bridges the gap between abstract vision and practical implementation (Mintzberg, 1994).
- This duality allows individuals to conceptualize solutions while also ensuring they are realistically actionable (Liedtka, 1998).
- By balancing long-term planning with short-term execution, strategic thinkers create sustainable and impactful solutions (Gavetti & Rivkin, 2007).

3. Adaptability

- In an uncertain and rapidly evolving world, strategic thinking enables individuals to anticipate change, adjust plans dynamically, and remain resilient in the face of uncertainty (Schoemaker, 1995).
- Unlike rigid decision-making models, strategic thinking embraces iteration and flexibility, making it a natural evolution of reflective and critical thinking (Gavetti & Rivkin, 2007).
- This adaptability is crucial for leaders, entrepreneurs, and policymakers who must navigate shifting economic, social, and technological landscapes.

4. Collaborative Application

- While traditional thinking methods often focus on individual cognitive development, strategic thinking extends into group dynamics, fostering collaboration and shared decision-making (Hughes & Beatty, 2005).
- Effective strategic thinking requires empathy, communication, and the ability to integrate diverse perspectives to generate inclusive, long-term solutions (Goldman et al., 2015).
- Organizations and teams that embrace strategic thinking develop more innovative solutions, as they leverage collective intelligence rather than relying solely on individual insight.

5. **Practical Impact**

    - Strategic thinking is action-oriented, ensuring that abstract thought and theoretical understanding translate into tangible results (Liedtka, 1998).
    - It drives innovation in business strategy, policy-making, and personal goal-setting, helping individuals and organizations achieve sustainable success (Senge, 1990).
    - By focusing on long-term value creation, strategic thinking ensures that decisions are not just reactive but proactive and forward-looking.

Key Features of Strategic Thinking

To develop effective strategic thinking, individuals must cultivate the following core capabilities:

1. **Foresight**

    - The ability to anticipate future trends, challenges, and opportunities based on current patterns (Schoemaker, 1995).
    - Involves scenario planning and risk assessment to prepare for multiple potential outcomes.

2. **Systems Thinking**

    - Understanding how individual components interact within a larger system and recognizing cause-and-effect relationships (Senge, 1990).
    - Helps individuals identify leverage points where small actions can lead to significant impact.

3. **Prioritization**

    - The ability to distinguish between urgent and important tasks, focusing on what matters most to achieve long-term goals (Hughes & Beatty, 2005).
    - Prevents cognitive overload and decision fatigue, ensuring efficient use of resources and effort.

4. **Iterative Improvement**

    - Strategic thinking is not a one-time process, it requires continuous refinement based on feedback and results (Gavetti & Rivkin, 2007).
    - Encourages learning from failure and adjusting strategies to enhance effectiveness over time.

5. **Collaboration**

    - Harnessing collective intelligence to co-create solutions that are more innovative and sustainable (Goldman et al., 2015).
    - Requires strong communication skills, active listening, and the ability to integrate diverse viewpoints.

Elevating the Art of Thinking with Strategic Thinking

The Art of Thinking has long emphasized critical thinking, creativity, and logical reasoning as fundamental cognitive skills. However, in today's complex and fast-changing world, strategic thinking is the essential next step in cognitive development. It integrates multiple thinking disciplines, enhances adaptability, fosters collaboration, and ensures real-world impact.

By cultivating foresight, systems thinking, prioritization, iterative learning, and collaboration, individuals can move beyond reactive decision-making and embrace proactive, goal-oriented strategies that drive long-term success.

Strategic thinking is not just for business leaders and policymakers, it is a skill that anyone can develop to make better decisions, solve problems creatively, and navigate the complexities of modern life with confidence and clarity.

The Theory of Tic-Tac-Toe

The theory of the Tic-Tac-Toe game serves as an apt metaphor for **strategic thinking**. On the surface, Tic-Tac-Toe appears to be a simple game with straightforward rules. However, mastery of the game reveals deeper principles of foresight, pattern recognition, and iterative improvement. These principles exemplify the dual nature of strategic thinking:

1. **Simplicity**: Understanding the fundamental rules and objectives.
2. **Depth**: Anticipating moves, recognizing patterns, and adapting strategies.

When applied to real-world problems, this duality—referred to as **strategic simplification**—becomes essential (Eisenhardt & Sull, 2001). Strategic simplification involves thinking both big and small simultaneously by integrating visionary foresight (big thinking) with tactical precision (small thinking) (Rumelt, 2011). This dual approach ensures that overarching goals remain aligned with actionable steps, creating a dynamic and balanced pathway to success (Lafley & Martin, 2013).

The Principles of Strategic Simplification

To achieve strategic simplification, individuals must develop a mindset and skillset that balances big-picture vision with detailed execution (Rumelt, 2011). This involves:

1. **Adopting a Systems Perspective:** Viewing problems as interconnected systems with multiple components and feedback loops (Senge, 1990).
2. **Breaking Goals into Actionable Steps:** Translating broad objectives into specific, manageable tasks to bridge vision with execution (Lafley & Martin, 2013).
3. **Practicing Adaptability and Iterative Learning:** Remaining flexible and continuously refining strategies based on feedback and outcomes (Eisenhardt & Sull, 2001).
4. **Prioritizing Simplicity and Collaboration:** Stripping away unnecessary complexity and fostering teamwork to address challenges effectively (Christensen, 1997).

Incorporating the theory of Tic-Tac-Toe into the Art of Thinking emphasizes **strategy, pattern recognition, adaptability, and simplicity** within complex decision-making. Tic-Tac-Toe, despite its simplicity, is a game that illustrates fundamental principles of logic, optimization, and foresight. If these principles were emphasized within the art of thinking, the result would be a structured yet creative approach to solving problems, managing conflicts, and strategizing in life. Here is a detailed breakdown of how the art of thinking changes when the theory of Tic-Tac-Toe is incorporated:

1. **Emphasis on Strategic Thinking**
   - **Anticipation:** Tic-Tac-Toe requires players to think ahead and anticipate their opponent's moves. This translates into the art of thinking by emphasizing the ability to predict outcomes and prepare contingencies.
   - **Optimal Decision-Making:** Every move in Tic-Tac-Toe is about maximizing chances to win while minimizing the opponent's opportunities. Similarly, thinking becomes a balance of optimizing outcomes while managing risks.
2. **Pattern Recognition**
   - **Identifying Trends:** In Tic-Tac-Toe, recognizing patterns—such as rows, columns, or diagonals—before they fully form is crucial. This skill, applied to thinking, emphasizes recognizing emerging trends, connections, and anomalies in broader contexts.
   - **Preventive Action:** The ability to spot a potential "win" for an opponent teaches proactive intervention, a critical skill in problem-solving and conflict resolution.
3. **Simplicity in Complexity**
   - **Reducing Overwhelm:** Tic-Tac-Toe simplifies strategy to a manageable set of rules and a finite grid, showing that even complex systems can be broken into smaller, more understandable parts. This aligns with the art of thinking by encouraging a focus on simplicity and clarity, even in intricate situations.
   - **Essence of the Problem:** Thinking through a Tic-Tac-Toe lens prioritizes identifying the essence of any issue—what matters most—before acting.
4. **Zero-Sum Game Perspective**
   - **Win-Loss Dynamics:** Tic-Tac-Toe is a zero-sum game, meaning one player's gain is another's loss. If this perspective is emphasized, the art of thinking might lean toward competitive strategies, analyzing situations where resources are limited, or goals are mutually exclusive.
   - **Collaboration Potential:** However, recognizing that Tic-Tac-Toe often ends in a draw when both players are competent also underscores the importance of collaboration or compromise when the game is balanced.
5. **Learning Through Repetition**
   - **Mastery Through Iteration:** Tic-Tac-Toe teaches that repeated practice and analysis of past games lead to mastery. Similarly, the art of thinking incorporates iterative learning—analyzing outcomes, learning from mistakes, and refining strategies over time.
6. **Adaptability and Resource Management**
   - **Dynamic Thinking:** In Tic-Tac-Toe, each move changes the possibilities of subsequent moves, requiring constant adaptation. Thinking becomes more dynamic, emphasizing the ability to revise plans based on new information or changing conditions.
   - **Efficient Use of Resources:** The game limits the number of moves and space on the board, teaching the importance of making the most of available resources.

7. **Game-Theoretical Applications**

   - **Decision Trees:** Tic-Tac-Toe is a simple example of a decision tree, where every choice branches into different possibilities. The art of thinking would place greater emphasis on mapping out possible scenarios and selecting paths based on desired outcomes.
   - **Minimax Strategies:** The game teaches minimax theory—minimizing potential losses while maximizing potential gains. This principle can refine risk management and strategic planning in complex decisions.

8. **Resilience and Acceptance of Draws**

   - **The Inevitability of Stalemates:** Tic-Tac-Toe often ends in a draw, teaching resilience and acceptance of situations where no perfect solution exists. Thinking becomes more realistic, focusing on making the best possible moves even when a win is unlikely.
   - **Iterative Problem-Solving:** Recognizing that a single "game" might not resolve an issue encourages iterative thinking—continuing to seek solutions through repeated efforts.

9. **Ethical and Philosophical Implications**

   - **The Value of Fair Play:** Tic-Tac-Toe relies on adherence to rules and fairness. In the art of thinking, this translates to integrity and respect for ethical considerations.
   - **Balancing Competition and Cooperation:** While the game emphasizes competition, the inevitability of "draws" teaches the value of coexistence and mutual benefit in some scenarios.

10. **Childlike Curiosity and Playfulness**

    - **Encouraging Exploration**: As a simple, universal game, Tic-Tac-Toe reminds us of the joy of exploration and the power of accessible tools. Thinking becomes less about overcomplication and more about returning to foundational principles with a sense of curiosity.

When the theory of Tic-Tac-Toe is incorporated into the Art of Thinking, **the Art of Thinking transforms into a structured, strategic, and adaptive system** that balances simplicity with depth. It teaches foresight, pattern recognition, and the importance of iterative improvement while acknowledging the inevitability of limitations and stalemates. This perspective fosters both practical decision-making and philosophical understanding, blending logic with creativity in a way that enhances life's "game."

The Art of Strategic Simplicity: A New Paradigm for Thinking

The **Art of Thinking** is a constantly evolving discipline, shaped by our ability to process, analyze, and refine the way we solve problems. When we incorporate the **theory of Tic-Tac-Toe** into this framework, **The Art of Thinking** transforms into **"The Art of Strategic Simplicity."** This evolution emphasizes that even the most complex problems often have simple, strategic solutions when approached with clarity, structure, and foresight.

At first glance, Tic-Tac-Toe appears to be an elementary game, often dismissed as a basic exercise in logic. However, beneath its simplicity lies a powerful model of strategic thinking—one that embodies anticipation, pattern recognition, and efficiency. Every move within the game is dictated by deliberate choices, forward-thinking calculations, and an understanding of cause-and-effect relationships. Success is not about complexity but about the mastery of fundamental principles and the ability to streamline decisions toward an optimal outcome.

This principle applies beyond the game itself. In leadership, business, problem-solving, and personal decision-making, strategic simplicity is the key to success. Often, we fall into the trap of overcomplicating our approaches, adding unnecessary layers of analysis, or getting lost in extraneous details. However, true expertise lies in the ability to distill complexity into its most essential components, creating clear, actionable strategies that lead to effective solutions.

What is The Art of Strategic Simplicity?

The **Art of Strategic Simplicity** is a mindset and methodology that encourages individuals to navigate complexity with structured clarity. It is about stripping away unnecessary distractions, redundant processes, and cognitive overload, allowing for efficient yet powerful decision-making.

This concept does not imply that difficult challenges should be reduced to simplistic solutions—rather, it asserts that every problem has an underlying structure, and by identifying the core patterns, one can find the most effective path forward.

Just as a master Tic-Tac-Toe player learns to anticipate an opponent's moves, recognize winning patterns, and focus only on the most impactful decisions, a strategic thinker must be able to:

- Break down complex problems into fundamental components.
- Recognize key patterns and prioritize the most critical elements.
- Eliminate unnecessary complexity to focus on the most effective solutions.
- Act with efficiency, foresight, and adaptability.

This approach mirrors the principles found in lean decision-making, effective leadership, and strategic problem-solving, where the ability to identify the simplest yet most impactful course of action determines success.

Tic-Tac-Toe as a Blueprint for Strategic Simplicity

Tic-Tac-Toe serves as a metaphor for intelligent simplicity because it operates on three foundational principles that translate into real-world applications:

1. Pattern Recognition and Predictive Thinking

- In Tic-Tac-Toe, success comes from recognizing patterns—knowing where to place an "X" or "O" before the opponent does.
- In real-world decision-making, recognizing trends, behaviors, and underlying structures allows us to anticipate outcomes and take proactive steps.
- **Example:** A successful entrepreneur can recognize emerging market trends before competitors, much like a Tic-Tac-Toe player foreseeing a winning move before the opponent.

2. Efficiency and Eliminating Unnecessary Complexity

- In Tic-Tac-Toe, unnecessary moves lead to missed opportunities or wasted turns. The best players focus only on the most effective plays.
- In problem-solving, focusing on what truly matters prevents over-analysis, decision fatigue, and inefficiency.
- **Example:** In business strategy, leaders who focus on a clear vision and essential priorities outperform those who are bogged down by irrelevant details and excessive micromanagement.

3. Strategic Adaptability and Real-Time Adjustments

- A good Tic-Tac-Toe player adjusts their strategy based on their opponent's moves, demonstrating agility and responsiveness.
- In real-world situations, unexpected challenges require rapid adaptation, making it essential to have a flexible yet structured approach.
- **Example:** A military strategist, a CEO, or an athlete must adjust tactics dynamically while staying focused on the ultimate goal.

Why Strategic Simplicity Matters in the Modern World

In today's world of information overload, rapid technological change, and increasing complexity, the ability to simplify without oversimplifying is a critical skill. The most effective leaders, decision-makers, and problem-solvers are those who can cut through complexity and focus on what truly matters.

1. Decision-Making in High-Stakes Environments

- Complexity leads to hesitation—too much information can cause paralysis by analysis (Kahneman, 2011).
- Strategic simplicity ensures that decisions are made efficiently, without losing sight of long-term objectives.

2. **Leadership & Organizational Strategy**

   - The best leaders focus on core priorities rather than getting distracted by superfluous tasks and initiatives (Collins, 2001).
   - Simplifying internal processes leads to greater efficiency, faster execution, and higher productivity (Drucker, 1993).

3. **Problem-Solving & Innovation**

   - The most successful innovations arise from solving real problems with clear, elegant solutions rather than overcomplicating technology or design (Maeda, 2006).
   - Breakthrough discoveries often come from rethinking problems in their simplest terms (Einstein, 1934).

4. **Reducing Cognitive Overload**

   - In an age of constant notifications, endless choices, and digital distractions, people suffer from decision fatigue (Newport, 2019).
   - Strategic simplicity promotes focus, clarity, and mental efficiency, helping individuals avoid mental exhaustion.

Thinking Both Big and Small: The Shift Toward Strategic Simplicity

Achieving strategic simplicity requires a delicate balance between thinking big and thinking small. This duality—**the ability to envision the bigger picture while executing with precision on the details that matter**—is the cornerstone of effective decision-making, leadership, and problem-solving. Most people tend to gravitate toward one extreme or the other:

- **Big thinkers** are visionaries who dream of long-term impact but may struggle with execution.
- **Small thinkers** excel at details and incremental steps but may lose sight of overarching goals.

However, true strategic mastery comes from integrating both perspectives simultaneously. This requires a fundamental shift in how we approach problems, decisions, and goals, a shift that embraces both broad, high-level vision and granular, detail-driven execution.

Why Thinking Both Big and Small is Essential

The most successful leaders, strategists, and problem-solvers know how to zoom in and zoom out, shifting between macro and micro perspectives seamlessly. This ability fosters clarity, adaptability, and long-term success in a rapidly changing world.

1. **Aligning Vision with Execution**

   - A big-picture vision is meaningless without the small, tactical steps required to implement it.

- Strategic simplicity ensures that high-level aspirations are translated into concrete, actionable plans (Collins, 2001).
- **Example:** A CEO who envisions global expansion must also understand local markets, operational logistics, and customer behaviors to make it happen.

2. **Breaking Down Complexity into Actionable Steps**

   - Big problems often seem overwhelming until they are deconstructed into manageable components (Maeda, 2006).
   - Strategic simplicity involves identifying core priorities and eliminating unnecessary complexity to streamline execution.
   - **Example:** A scientific breakthrough often starts with a broad research goal but requires years of detailed experiments and iterative refinements before achieving success.

3. **Adaptability in Decision-Making**

   - Thinking big helps anticipate long-term industry shifts, while thinking small enables real-time adjustments based on immediate feedback (Mintzberg, 1994).
   - The ability to move fluidly between these perspectives is critical for leaders, entrepreneurs, and strategists.
   - **Example:** A company launching a new product must consider big-picture branding and market positioning while also refining specific product features, pricing strategies, and distribution logistics.

4. **Balancing Innovation with Feasibility**

   - Groundbreaking innovations require bold, forward-thinking ideas—but they must be realistic and executable (Liedtka, 1998).
   - The most successful innovators think expansively about possibilities while simultaneously working within constraints to turn ideas into reality.
   - **Example:** Space exploration requires ambitious, long-term vision, but every mission depends on precise engineering, testing, and calculated risk assessments.

How to Shift Toward Strategic Simplicity

Shifting toward **strategic simplicity** requires a **mindset shift**—a redefinition of **how we frame problems, set goals, and execute solutions.** Here's how individuals and organizations can cultivate this approach:

1. **Establish a Clear Vision While Defining Tactical Milestones**

   - Start with a compelling "big idea" or goal, but immediately break it down into achievable, measurable steps.
   - Define key performance indicators (KPIs) that align long-term objectives with short-term progress markers (Drucker, 1993).
   - **Example:** A nonprofit aiming to eradicate hunger must focus on specific initiatives such as local food banks, supply chain improvements, and policy advocacy.

2. **Develop Systems Thinking to Connect Macro and Micro Perspectives**

   - Recognize how small decisions affect the larger system and vice versa (Senge, 1990).
   - Map out cause-and-effect relationships to ensure alignment between immediate actions and long-term impact (Meadows, 2008).
   - **Example:** In healthcare reform, a single policy change can have ripple effects on providers, insurers, and patient care, requiring a holistic approach to decision-making.

3. **Prioritize Ruthlessly to Avoid Cognitive Overload**

   - Not all details are equally important—effective strategic thinking means knowing what not to focus on (Newport, 2019).
   - Apply the Pareto Principle (80/20 Rule) to identify which 20% of efforts will drive 80% of results.
   - **Example:** A startup with limited resources must prioritize its most valuable customers instead of trying to capture the entire market at once.

4. **Encourage a Culture of Iterative Learning and Feedback**

   - The most effective strategies are not rigid, they evolve based on real-world data and insights (Gavetti & Rivkin, 2007).
   - Adopt an experiment-and-adapt mindset, ensuring that small adjustments lead to long-term success.
   - **Example:** Tech companies like Google and Tesla release beta versions, gather feedback, and refine their products through multiple iterations before mass adoption.

5. **Simplify Communication Without Oversimplifying**

   - **Clarity is Key:** Big ideas must be conveyed in ways that are accessible and actionable (Heath & Heath, 2007).
   - Avoid jargon and focus on essential insights that drive action.
   - **Example:** The best leaders can articulate a complex vision in a single sentence, ensuring that teams across all levels understand and align with strategic objectives.

Further Examples of Shifting Toward Strategic Simplicity

1. **Adopt a Systems Mindset**

   - **Change Needed**: Transition from linear thinking (cause-effect) to systems thinking.
   - **Description**:
     - **Big Thinking:** Understand how different parts of a system interact, recognizing interdependencies and emergent properties.
     - **Small Thinking:** Focus on optimizing specific components of the system to ensure they contribute effectively to the whole.
   - **Example**: When managing a business, think of it as an ecosystem where marketing, sales, operations, and customer service interact. Strategize at the systems level while fine-tuning individual processes.

---

2. **Define Clear Goals (Big) and Milestones (Small)**
   - **Change Needed**: Move from vague aspirations to well-defined objectives with measurable steps.
   - **Description**:
     - **Big Thinking:** Envision long-term success and articulate a compelling vision.
     - **Small Thinking:** Break the vision into manageable milestones and actionable tasks.
   - **Example**: If your goal is to develop a new product, the big picture is delivering value to the market. Small thinking involves setting specific development phases like prototyping, testing, and marketing.

---

3. **Practice Contextual Zooming**
   - **Change Needed**: Cultivate the habit of zooming in and out of contexts.
   - **Description**:
     - **Big Thinking:** Maintain a high-level perspective to avoid getting lost in details.
     - **Small Thinking:** Dive into the specifics when necessary, ensuring critical details are managed without losing sight of the big picture.
   - **Example**: A city planner balances a grand vision for urban growth with the technicalities of zoning laws and infrastructure needs.

---

4. **Embrace Iterative Thinking**
   - **Change Needed**: Shift from static plans to iterative and flexible strategies.
   - **Description**:
     - **Big Thinking:** Develop adaptable strategies that account for uncertainty and change.
     - **Small Thinking:** Use small-scale experiments or pilots to evaluate ideas and refine them.
   - **Example**: Entrepreneurs often start with a broad business idea (big) and refine their product through prototypes and customer feedback (small).

---

5. **Balance Risk and Reward**
   - **Change Needed**: Replace fear-based decision-making with calculated risk-taking.
   - **Description**:
     - **Big Thinking:** Assess long-term risks and rewards of decisions on a macro scale.
     - **Small Thinking:** Mitigate immediate risks by focusing on controllable factors and implementing safeguards.

- **Example**: When investing, think about portfolio diversification (big) and evaluate individual stock performance (small).

---

6. **Cultivate Focused Curiosity**

- **Change Needed**: Replace passive consumption of information with targeted exploration.
- **Description**:
    - **Big Thinking:** Stay curious about overarching trends and global patterns.
    - **Small Thinking:** Dive deeply into niche topics or specific issues to gain expertise.
- **Example**: A scientist works on climate change (big) but focuses their research on specific phenomena like ocean acidification (small).

---

7. **Prioritize Simplicity**

- **Change Needed**: Avoid overcomplication by distilling complexity into core principles.
- **Description**:
    - **Big Thinking:** Simplify overarching strategies to focus on what truly matters.
    - **Small Thinking:** Simplify individual processes by eliminating inefficiencies and redundancies.
- **Example**: A software developer creates an app with a user-friendly interface (big) and streamlined coding practices (small).

---

8. **Foster Collaboration**

- **Change Needed**: Move from individualistic thinking to collective problem-solving.
- **Description**:
    - **Big Thinking:** Align diverse stakeholders around a shared vision.
    - **Small Thinking:** Leverage the expertise of team members for specific tasks.
- **Example**: A project manager collaborates with a team to align a company's sustainability goals (big) and delegates tasks to develop reusable materials (small).

---

9. **Develop Metacognition (Thinking About Thinking)**

- **Change Needed**: From reactive thinking to intentional and reflective thought processes.
- **Description**:
    - **Big Thinking:** Regularly evaluate how your thought processes align with your goals.

- ◦ **Small Thinking:** Pay attention to cognitive biases and refine decision-making at the micro level.

- **Example**: A leader reflects on their strategic vision (big) while analyzing how their daily decisions impact progress (small).

---

10. Balance Long-Term Vision with Immediate Action

- **Change Needed**: Replace procrastination or short-termism with disciplined progress toward long-term goals.
- **Description**:
  - **Big Thinking:** Keep your eye on the future while being flexible about the path to get there.
  - **Small Thinking:** Act decisively in the present, focusing on one step at a time.
- **Example**: A student working toward a doctorate envisions their future dissertation (big) while completing coursework and research one semester at a time (small).

The Power of Thinking Both Big and Small

In a world that often glorifies either overcomplicated strategies or overly simplistic solutions, the real key to success lies in **strategic simplicity, the discipline of thinking big while mastering the small**. It requires the vision to see what is possible while ensuring practical execution at every stage of the journey. We must develop thc ability to:

- Combine the long-term goals with short-term execution.
- Break down complexity into manageable steps.
- Balance innovation with feasibility.
- Adapt to shifting circumstances while maintaining strategic clarity.

By making these changes, we develop a more dynamic, balanced, and impactful way of thinking, one that harmonizes the grand scope of "big thinking" with the focused detail of "small thinking" to navigate complexity effectively. To achieve strategic simplification, a structured progression of thinking skills must be taught systematically—starting with foundational cognitive abilities in grade school and advancing to complex problem-solving and decision-making in professional settings.

Each level must be built upon the previous one, using a scaffolded instructional technique that emphasizes growth in both "big thinking" (strategic vision) and "small thinking" (attention to detail). This approach is formally structured as **"Progressive Cognitive Structuring (PCS)."**

Progressive Cognitive Structuring: A Framework for Life Thinking Skills Development

Progressive Cognitive Structuring (PCS) is an instructional framework designed to systematically develop and refine cognitive abilities across different stages of education and professional life. It ensures a structured progression from fundamental cognitive skills in childhood to advanced strategic thinking and decision-making in professional settings. PCS emphasizes both "big thinking" (strategic vision) and "small thinking" (attention to detail) to cultivate a balanced and adaptable cognitive approach (Anderson & Krathwohl, 2001; Bloom, 1956; Guilford, 1967; Partnership for 21st Century Learning - P21, 2007; Piaget, 1952; Sternberg, 1985; Vygotsky, 1978). PCS spans **five developmental stages**, each building upon the previous:

1. **Grade School:** Foundational skills such as pattern recognition, logical reasoning, and simple problem-solving.
2. **Middle School:** Development of critical thinking, creative exploration, and early systems thinking.
3. **High School:** Introduction to abstract reasoning, strategic planning, and reflective thinking.
4. **College/Technical School:** Integration of theoretical knowledge with practical application, emphasizing interdisciplinary problem-solving and adaptability.
5. **Business Professionals:** Mastery of strategic decision-making, leadership, and managing uncertainty in complex environments.

PCS ensures a smooth transition from structured academic thinking to professional expertise, making it a valuable framework for education, business training, and leadership development.

Why PCS is Needed

PCS is necessary to bridge cognitive development gaps, ensuring individuals develop well-rounded, adaptable, and practical thinking skills. The need for PCS arises from several key educational and professional challenges:

1. **Fragmented Thinking in Education and Training**
    - Traditional education often focuses on subject-specific learning without integrating broader cognitive skills.
    - Students may excel in memorization and test-taking but struggle with real-world problem-solving and decision-making.

  - PCS addresses this by providing a structured, interdisciplinary approach to cognitive development.

2. **The Growing Complexity of Problems in the Real World**
   - Modern challenges—technological advancements, global interconnectedness, and rapid change—demand complex problem-solving, adaptability, and interdisciplinary thinking.
   - PCS prepares individuals to navigate uncertainty, synthesize information from multiple disciplines, and apply knowledge in varied contexts.

3. **Bridging the Gap Between Academic Learning and Workforce Expectations**
   - Employers seek critical thinkers, problem-solvers, and adaptable leaders rather than just individuals with technical knowledge.
   - PCS ensures individuals are ready to transition from structured learning environments to unstructured professional challenges.

4. **Avoiding the Pitfall of One-Dimensional Thinking**
   - Many traditional frameworks emphasize only one type of thinking (e.g., critical thinking or creativity) without balancing them with other necessary skills.
   - PCS ensures a synergistic development of multiple cognitive skills—analytical, creative, reflective, precise, and collaborative—creating a holistic cognitive approach.

5. **Supporting Lifelong Cognitive Growth**
   - Learning should not stop after formal education; cognitive skills must be continuously refined for professional and personal growth.
   - PCS provides a progressive structure that supports learning from childhood through professional life.

Where PCS Came From: Theoretical and Practical Origins

PCS is rooted in cognitive science, educational psychology, and strategic thinking models. It draws from several well-established theories and frameworks, including:

1. **Bloom's Taxonomy (1956, Revised 2001)**
   - PCS follows a **progressive cognitive hierarchy** similar to Bloom's:
     - Remember → Understand → Apply → Analyze → Evaluate → Create.
   - Each PCS stage builds on the prior one, ensuring a logical progression from basic to complex thinking skills.

2. **Triarchic Theory of Intelligence (Sternberg, 1985)**
   - PCS incorporates Sternberg's three types of intelligence:
     - Analytical Intelligence → Precision Thinking.

    - Creative Intelligence → Innovative Thinking.
    - Practical Intelligence → Integrative and Adaptive Thinking.
  - This ensures PCS addresses diverse cognitive abilities rather than just one mode of intelligence.

3. **Constructivist Learning Theory (Piaget, 1936; Vygotsky, 1978)**
   - PCS supports the idea that learning is an active process where knowledge is constructed rather than passively absorbed.
   - Cognitive development is stage-based and evolves through progressively more complex mental structures.

4. **Divergent vs. Convergent Thinking (Guilford, 1967)**
   - PCS balances both styles of thinking:
     - **Divergent Thinking** (creative exploration, open-ended problem-solving) → Innovative Thinking.
     - **Convergent Thinking** (structured, evidence-based reasoning) → Precision Thinking.

5. **21st Century Skills Framework (Partnership for 21st Century Learning - P21, 2007)**
   - PCS aligns with modern workforce needs by integrating:
     - **Critical thinking** → Integrative & Precision Thinking.
     - **Creativity** → Innovative Thinking.
     - **Collaboration** → Collaborative & Contextual Thinking.
     - **Adaptability and Communication** → Reflective & Adaptive Thinking.

6. **Practical Insights from Business and Leadership Development**
   - PCS mirrors cognitive progression seen in leadership training programs.
   - Business schools and executive training programs emphasize strategic thinking, decision-making, and adaptability, which align with PCS's highest levels.

PCS is a structured, research-backed framework designed to guide cognitive development from foundational skills to advanced problem-solving and leadership. It addresses educational and workforce gaps by ensuring individuals develop a balanced set of cognitive skills that allow them to think critically, adapt flexibly, create innovatively, reason precisely, and collaborate effectively.

By integrating theories from educational psychology, intelligence research, and real-world problem-solving, PCS serves as a comprehensive approach to fostering cognitive excellence across a lifetime.

PCS Development Stages vs. the Five Synergistic Thinking Skills

To ensure no confusion where PCS is mistakenly considered a "final solution," PCS avoids siloed learning by advancing cognitive development around five synergistic groupings of thinking skills:

1. **Integrative Thinking** – Includes systems thinking, critical thinking, and meta-thinking.
2. **Innovative Thinking** – Incorporates creative exploration, practical application, and habit formation.
3. **Reflective & Adaptive Thinking** – Comprises introspection, emotional intelligence, and adaptability.
4. **Precision Thinking** – Embodies logical analysis, evidence-based reasoning, and intellectual rigor.
5. **Collaborative & Contextual Thinking** – Encompasses teamwork, empathy, and contextual awareness.

Understanding the Development of the Five Synergistic Thinking Skills

While the PCS incorporates the learning strategies needed to develop the learning skills at each of the designated stages, additional training and application are necessary to fully obtain and master the Five Synergistic Thinking Skills. These advanced cognitive skills do not develop in isolation; rather, they require targeted training, real-world application, and continuous learning beyond standard education.

- The **PCS developmental stages provide the foundation** by strengthening cognitive abilities, but they do not independently cultivate fully developed synergistic thinking skills.
- The Five Synergistic Thinking Skills—**Integrative, Innovative, Reflective & Adaptive, Precision, and Collaborative & Contextual Thinking**—are higher-order cognitive abilities that necessitate deliberate practice and exposure to complex problem-solving scenarios.
- For example:
    - **Precision Thinking** (logical analysis, evidence-based reasoning) may begin developing in grade school. However, without advanced training in data analysis, critical evaluation, or structured argumentation, it remains incomplete.
    - **Reflective & Adaptive Thinking** (emotional intelligence, adaptability) may be introduced in high school, but without leadership training, real-world experience, or crisis management exposure, it cannot be fully realized.

Thus, while the **PCS framework lays the groundwork**, mastery of the Five Synergistic Thinking Skills requires a combination of structured learning, professional development, and practical application.

The Role of PCS Developmental Stages in Five Synergistic Thinking Skills Acquisition

The **PCS developmental stages are essential** for building the foundational cognitive skills that eventually support the Five Synergistic Thinking Skills. These stages function as **progressive stepping-stones**, strengthening and refining fundamental cognitive abilities in preparation for advanced thinking.

- If a learner **skips** or does not complete certain PCS stages, they may experience gaps in their cognitive skillset that hinder their ability to develop synergistic thinking skills effectively.
- Examples:
    - A lack of structured reasoning skills from middle school can result in struggles with **Integrative Thinking** later on, as systems thinking and logical analysis were not reinforced.
    - If a professional has never been exposed to adaptability training or contextual awareness exercises, they will struggle to master **Collaborative & Contextual Thinking** in leadership roles.
- While alternative learning methods—such as **self-directed learning, accelerated training, or experiential learning**—can support skill development, they cannot replace the foundational cognitive abilities established through the PCS framework.

The Difference Between PCS Development Stages and the Five Synergistic Thinking Skills

- The Five Synergistic Thinking Skills require additional training beyond PCS development.
- The PCS developmental stages are essential for laying the foundation of cognitive development.
- Without completing the PCS stages, gaps in cognitive development may hinder full mastery of the Five Synergistic Thinking Skills.

Ultimately, developing these advanced cognitive abilities demands a combination of structured learning, real-world application, and ongoing professional growth. These Five Synergistic Thinking Skills, considered beyond the Five Developmental Stages of PCS emphasized in this book, are briefly discussed further in Appendix A.

Big Thinking and Small Thinking in Progressive Cognitive Structuring (PCS)

Progressive Cognitive Structuring (PCS) is a framework that organizes and optimizes cognitive processes for effective decision-making, problem-solving, and strategic execution. Within PCS, **big thinking** and **small thinking** operate in tandem to enable individuals to navigate complex systems while executing precise, structured actions. These two cognitive modes are **not opposites but complementary**, forming a dynamic interplay essential for cognitive growth and performance.

Definitions of Big Thinking and Small Thinking

---

1. Big Thinking

Big thinking is the cognitive process that addresses macro-level vision, abstract reasoning, and systemic understanding. Big thinking is expansive, strategic, and integrative, allowing individuals to recognize patterns, set ambitious goals, and conceptualize overarching frameworks.

Characteristics of Big Thinking

- **Vision-Oriented**: Defines long-term objectives and grand strategies.
- **Abstract and Holistic**: Recognizes interconnected systems, relationships, and high-level patterns.
- **Innovative**: Encourages disruptive ideas, paradigm shifts, and revolutionary changes.
- **Risk-Tolerant**: Accepts uncertainty and explores untested possibilities.

Examples of Big Thinking in PCS

- Designing a lifelong cognitive development model that spans across different life and career stages.
- Establishing the five synergistic thinking skills that integrate different cognitive approaches.
- Developing a new interdisciplinary problem-solving method that combines multiple knowledge domains.

---

2. Small Thinking

Small thinking is the cognitive process that addresses micro-level execution, precision, and structured implementation. Small thinking is focused, detail-oriented, and methodical, ensuring that grand visions are translated into actionable steps with measurable results.

Characteristics of Small Thinking

- **Execution-Oriented**: Breaks down large goals into practical steps and achievable milestones.
- **Detail-Focused**: Ensures precision, accuracy, and logical consistency in processes.
- **Incremental and Analytical**: Implements, evaluates, refines, and improves solutions progressively.
- **Risk-Averse**: Minimizes uncertainty by validating methods before scaling.

Examples of Small Thinking in PCS

- Creating structured lesson plans and training modules for different levels of cognitive skill development.
- Designing daily exercises that reinforce specific thinking skills through practice.
- Implementing step-by-step frameworks for applying PCS in business decision-making, education, or personal development.

Why Are Big Thinking and Small Thinking Used in PCS?

PCS emphasizes structured cognitive growth through the deliberate interplay of different thinking approaches. Big thinking and small thinking are fundamental because they:

- **Bridge Theory and Practice**: Big thinking develops broad conceptual models, while small thinking translates them into executable processes.
- **Ensure Both Innovation and Precision**: Innovation alone is not enough; small thinking ensures ideas are realistic and implementable.
- **Optimize Problem-Solving**: Complex problems require both big-picture analysis and meticulous execution for effective solutions.
- **Facilitate Continuous Cognitive Development**: Big thinking expands cognitive boundaries, while small thinking sharpens critical details.
- **Support Structured Learning and Adaptation**: PCS enables individuals to adapt their thinking skills dynamically across different scenarios.

Functions of Big Thinking and Small Thinking in PCS

---

1. Big Thinking's Role in PCS

- **Sets the Vision**: Defines overarching cognitive strategies and high-level goals.
- **Creates Context**: Establishes how smaller components fit into a broader framework.
- **Drives Innovation**: Develops groundbreaking concepts and reimagines conventional approaches.
- **Guides Leadership**: Provides strategic direction for organizations, education, and decision-making.

2. **Small Thinking's Role in PCS**
   - **Executes the Details**: Translates visionary goals into structured, repeatable actions.
   - **Ensures Precision**: Reduces errors through meticulous validation and analysis.
   - **Enhances Feasibility**: Evaluates and refines ideas to ensure practical applicability.
   - **Supports Practitioners**: Helps individuals apply skills effectively in real-world situations.

How Big Thinking and Small Thinking Function Together to Form the Five Developmental Stages

PCS incorporates the developmental thinking skills, each requiring a balance of big thinking and small thinking to be effective. Here is how they integrate:

1. **Systemic Thinking (Seeing the Whole and the Parts)**
   - **Big Thinking**: Understands large-scale systems and their interconnections.
   - **Small Thinking**: Analyzes individual components for precision and accuracy.
   - **Example**: Designing an organization-wide leadership framework (big thinking) while developing individual leadership training programs (small thinking).
2. **Analytical Thinking (Breaking Down Complexity)**
   - **Big Thinking**: Identifies broad categories and conceptual groupings.
   - **Small Thinking**: Examines specific variables, data points, and causal relationships.
   - **Example**: Developing a multi-step research methodology (big thinking) and refining each step's data collection methods (small thinking).
3. **Strategic Thinking (Aligning Goals with Actions)**
   - **Big Thinking**: Establishes long-term goals and strategic direction.
   - **Small Thinking**: Maps out tactical plans and execution timelines.
   - **Example**: Creating a corporate growth strategy (big thinking) and setting quarterly implementation milestones (small thinking).
4. **Critical Thinking (Validating and Refining Ideas)**
   - **Big Thinking**: Evaluates arguments in a broad context, considering multiple perspectives.
   - **Small Thinking**: Scrutinizes details, logic, and supporting evidence.
   - **Example**: Assessing the ethical implications of AI (big thinking) while conducting rigorous bias analysis on AI datasets (small thinking).
5. **Adaptive Thinking (Adjusting to Changing Conditions)**
   - **Big Thinking**: Anticipates large-scale shifts and emerging trends.
   - **Small Thinking**: Adjusts specific processes in response to immediate changes.
   - **Example**: Developing a lifelong learning model (big thinking) and updating training modules based on new technological advancements (small thinking).

Big thinking and small thinking are essential cognitive processes within PCS, **working together to create a structured yet adaptable cognitive framework**.

- **Big thinking provides the why and what**—the vision, strategy, and system-wide perspective.
- **Small thinking provides the how**—the detailed execution, precision, and practical implementation.

By integrating these two cognitive modes, PCS enables individuals to **navigate complexity effectively, achieve tangible results, and continuously evolve their cognitive capabilities**. This balance ensures that individuals and organizations can **innovate while maintaining practical, structured approaches** to problem-solving and decision-making.

The Interplay of Big Thinking and Small Thinking in PCS

Within **PCS**, **big thinking** and **small thinking** occur together as part of a **dynamic interplay** that allows individuals to navigate complexity while executing precise actions. Big thinking and small thinking are not opposites but rather complementary modes of thought. Here is how they function together:

- Big Thinking Sets the Vision, Small Thinking Executes the Details
    - **Big Thinking**: Establishes overarching frameworks, strategies, and transformative goals (e.g., designing the five synergistic thinking groupings or developing a lifelong cognitive growth model).
    - **Small Thinking**: Focuses on the implementation, step-by-step processes, and detailed refinements that make the vision actionable (e.g., creating lesson plans, daily exercises, or training modules for each stage).
- Big Thinking Provides Context, Small Thinking Ensures Precision
    - **Big Thinking**: Helps individuals see patterns, connections, and the broader system, ensuring that problem-solving is holistic and adaptive.
    - **Small Thinking**: Ensures logical accuracy, detailed analysis, and rigorous validation, preventing errors and refining insights within the broader framework.
- Big Thinking Enables Innovation, Small Thinking Ensures Feasibility
    - **Big Thinking**: Generates disruptive ideas, reimagines systems, and explores new paradigms (e.g., proposing a new interdisciplinary approach to thinking skills).
    - **Small Thinking**: Focuses on practicality, incremental improvements, and feasibility testing to refine and implement these innovations effectively.
- Big Thinking Supports Leadership, Small Thinking Supports Execution
    - **Big Thinking**: Guides leaders, strategists, and educators in shaping a vision for cognitive development and decision-making at a macro level.
    - **Small Thinking**: Supports individual learners, employees, and practitioners in applying specific skills and structured techniques in their daily work.

Example in PCS Application

- **Big Thinking**: Designing a cognitive development framework that spans from grade school to business leadership.
- **Small Thinking**: Developing specific exercises, assessments, and feedback loops that allow individuals to apply and refine their thinking skills within each stage.

Big thinking and small thinking **must work together within PCS**:

- **Big thinking** provides the why and what—the vision, strategy, and system-wide perspective.
- **Small thinking** provides the how—the detailed execution, precision, and practical implementation.

When combined, they create a comprehensive cognitive structure that ensures both innovation and effectiveness, allowing individuals to navigate complexity while achieving tangible results.

Examples of Big Thinking in PCS with Teaching Strategies

Within PCS, big thinking sets the vision, small thinking ensures that vision is actionable, measurable, and applied effectively while small thinking refers to the detailed, precise, and execution-focused aspects that support the broader cognitive framework.

1. **Developing a Universal Cognitive Growth Model**

- Designing a framework that supports cognitive development from early education to professional leadership.
- Teaching Strategies:
  - **Framework-Based Learning Progressions:** Create structured cognitive development maps that define key skills for each stage of growth (e.g., elementary school to executive leadership).
  - **Longitudinal Cognitive Growth Studies:** Design multi-year assessments to track how individuals develop thinking skills over time, refining the model based on real-world data.
  - **Tiered Learning Systems:** Establish progressive levels of cognitive training that build on one another, ensuring continuity in skill development.
  - **Cross-Industry Cognitive Application Models:** Develop case studies demonstrating how cognitive skills transfer across disciplines, helping learners see the broad application of these skills.
  - **AI-Driven Cognitive Growth Maps:** Integrate AI tools that track user progress, recommend personalized learning paths, and adapt training based on strengths and weaknesses.

---

2. **Establishing a Multidisciplinary Thinking Integration System**

- Creating an approach that synthesizes analytical, strategic, systemic, critical, and adaptive thinking into a unified methodology.
- Teaching Strategies:
  - **Cognitive Fusion Frameworks:** Develop structured models that blend analytical, strategic, systemic, critical, and adaptive thinking into one cohesive system.

- **Interdisciplinary Case Studies:** Present complex scenarios requiring learners to apply multiple thinking styles simultaneously (e.g., engineering ethics, business strategy, and public policy).
- **Comparative Thinking Skill Evaluations:** Have students analyze problems using different cognitive approaches, then compare and contrast their effectiveness.
- **Cross-Domain Learning Simulations:** Design exercises where students must solve problems that span multiple fields (e.g., merging economic and environmental considerations in urban planning).
- **Expert-Led Multidisciplinary Workshops:** Invite professionals from different industries to demonstrate how they integrate multiple cognitive strategies in their work.

---

3. Designing a Future-Oriented Decision-Making Framework

- Constructing a model that prepares individuals to make effective decisions in rapidly evolving technological and social landscapes.
- Teaching Strategies:
  - **Predictive Analysis Exercises:** Train learners to analyze current trends and anticipate future developments using data-driven forecasting techniques.
  - **Scenario-Based Strategic Planning:** Have students work through different future scenarios and devise action plans for each possible outcome.
  - **Simulated Crisis Decision-Making:** Create high-stakes problem-solving simulations that force students to think long-term under uncertainty.
  - **AI-Augmented Decision Labs:** Use AI modeling to evaluate different decision pathways and their potential consequences in real-world contexts.
  - **Reverse-Engineering Case Studies:** Analyze past decisions of leaders, innovators, or policymakers to determine how their long-term strategies succeeded or failed.

---

4. Revolutionizing Learning Through Cognitive AI Tools

- Proposing AI-driven personalized learning paths for enhancing individual cognitive strengths.
- Teaching Strategies:
  - **AI-Driven Personalized Learning Plans:** Develop adaptive cognitive training programs that adjust based on learner performance.
  - **AI-Powered Thought Mapping:** Use AI tools to visualize thought processes and reveal areas of cognitive bias or gaps in reasoning.
  - **Automated Feedback Mechanisms:** Implement AI tutors that provide real-time suggestions for improving problem-solving approaches.
  - **Simulated AI-Enhanced Thinking Challenges:** Design exercises where students interact with AI systems to evaluate different decision-making models.
  - **Ethical AI Debates:** Engage students in discussions about the role of AI in shaping human thought and decision-making.

5. **Developing Global Cognitive Competency Standards**

- Establishing universal benchmarks for cognitive skills applicable across industries and educational systems.
- Teaching Strategies:
    - **International Benchmarking Studies:** Determine how different cultures approach cognitive development and integrate the best practices into a universal standard.
    - **Comparative Education Models:** Study how various educational systems teach cognitive skills and develop adaptable frameworks that accommodate cultural differences.
    - **Cross-Cultural Cognitive Skill Testing:** Implement standardized global assessments to measure cognitive proficiency in diverse populations.
    - **Universal Thinking Certification Programs:** Design credentialing systems that recognize and validate cognitive expertise across industries.
    - **World-Class Cognitive Conferences:** Organize global summits where experts refine and evolve cognitive competency standards.

---

6. **Creating a Leadership Development Matrix Based on PCS**

- Mapping out cognitive competencies required at different levels of leadership and designing a structured learning progression.
- Teaching Strategies:
    - **Cognitive Leadership Stages:** Develop a structured matrix that outlines the cognitive skills required at different leadership levels.
    - **Leadership Mindset Training Modules:** Train leaders to shift between different cognitive styles depending on the challenges they face.
    - **Strategic Leadership Simulations:** Design role-playing scenarios where emerging leaders must make high-stakes decisions based on PCS principles.
    - **Executive-Level Cognitive Assessments:** Create testing mechanisms that evaluate an individual's ability to process complex leadership challenges.
    - **Mentorship-Based Cognitive Development:** Pair junior professionals with experienced mentors who model PCS-based leadership thinking.

---

7. **Formulating a PCS-Based Organizational Innovation Strategy**

- Constructing a corporate strategy that fosters continuous innovation through structured cognitive methodologies.
- Teaching Strategies:
    - **Organizational Cognitive Health Assessments:** Evaluate a company's current thinking culture and identify areas for improvement.

  - **Innovation-Driven Cognitive Workshops:** Develop training sessions that enhance employees' ability to generate, refine, and implement innovative ideas.
  - **PCS-Based Change Management Plans:** Use structured thinking frameworks to help organizations navigate transformations efficiently.
  - **Cross-Industry Innovation Competitions:** Encourage collaboration between industries to develop novel solutions based on PCS principles.
  - **Cognitive KPIs for Innovation Tracking:** Establish measurable indicators to assess how well PCS-driven thinking is enhancing innovation within a company.

---

8. Implementing a Cross-Cultural Cognitive Skill Transfer Program

- Designing a system for sharing cognitive approaches across different cultural and professional environments.
- Teaching Strategies:
  - **International Cognitive Skill Exchange Programs:** Develop partnerships between schools and businesses worldwide to facilitate the sharing of thinking strategies.
  - **Diverse Problem-Solving Panels:** Organize multicultural think tanks that tackle global issues using different cognitive approaches.
  - **Cultural Cognitive Adaptation Training:** Teach individuals how different cultural perspectives shape cognitive processes and decision-making.
  - **Global Problem-Solving Challenges:** Assign real-world issues and require learners to incorporate multiple cultural viewpoints in their solutions.
  - **AI-Assisted Language and Cognition Studies:** Use machine learning to analyze how cognitive skills vary across languages and regions.

---

9. Defining a Cognitive Growth Trajectory for Problem-Solving Excellence

- Creating a tiered system that gradually enhances an individual's ability to tackle increasingly complex problems.
- Teaching Strategies:
  - **Stepwise Problem-Solving Competency Models:** Develop structured training that teaches individuals how to progress from novice to expert problem solver.
  - **Tiered Analytical Challenges:** Offer increasingly complex problem-solving exercises to enhance cognitive resilience.
  - **Cognitive Growth Reflection Frameworks:** Encourage students to analyze past problem-solving attempts and track their progress.
  - **Personalized Problem-Solving Journals:** Have learners document their problem-solving methods and refine their strategies over time.
  - **Industry-Specific PCS Application Models:** Develop problem-solving frameworks tailored to business, engineering, medicine, and other professional fields.

---

10. Integrating PCS into Lifelong Learning and Career Progression

- Establishing a structured approach that enables individuals to continuously refine and expand their cognitive skills throughout life.
- Teaching Strategies:
  - **Career-Specific Cognitive Development Plans:** Help professionals map out their cognitive skill growth based on long-term career goals.
  - **Self-Paced PCS Certification Programs:** Design online courses that allow individuals to build thinking skills at their own pace.
  - **Longitudinal Cognitive Coaching Systems:** Offer mentorship programs where professionals receive ongoing guidance in developing cognitive expertise.
  - **AI-Based Career Pathway Recommendations:** Use machine learning to analyze an individual's strengths and suggest cognitive skill-building strategies.
  - **Personalized Metacognition Training:** Teach professionals how to self-assess and refine their cognitive strategies throughout their careers.

Big thinking in PCS establishes the **visi**on, strategic framework, and overarching cognitive structures that shape an individual's long-term cognitive development. Through these detailed teaching strategies, learners can systematically develop their thinking abilities, ensuring they are well-prepared to adapt, innovate, and lead in an ever-changing world.

Examples of Small Thinking in PCS with Detailed Teaching Strategies

Big thinking in PCS sets the overarching vision and conceptual frameworks, while small thinking ensures those frameworks are actionable, structured, and applied with precision. Together, they form a comprehensive approach to cognitive skill development, enabling individuals to navigate complexity while executing effectively.

1. Structuring Learning Modules for Different Age Groups

- Developing separate curricula for elementary, high school, and adult learners, focusing on stage-appropriate cognitive skill development.
- Teaching Strategies:
  - **Age-Specific Cognitive Development Tracks:** Design lesson plans that align with cognitive abilities at different life stages. For example, elementary students may focus on basic pattern recognition, middle school students on logical sequencing, and high school students on strategic decision-making.
  - **Scaffolded Curriculum Design:** Introduce fundamental skills early and progressively increase complexity. For instance, teaching simple problem-solving methods before introducing advanced analytical frameworks.
  - **Custom Learning Tracks:** Provide multiple pathways for learners based on skill level. For example, differentiated instruction techniques ensure advanced students remain challenged while struggling learners receive additional support.

- **Interactive Age-Appropriate Activities:** Use visual storytelling, games, and hands-on projects for younger learners, while incorporating case studies and simulations for older students and professionals.
- **Transition Planning Between Levels:** Develop structured transition plans to ensure smooth progression from one level of cognitive development to the next, such as introducing preparatory lessons before moving to more advanced thinking exercises.

---

2. Designing Individual Thinking Exercises

- Crafting problem sets and puzzles that specifically target analytical or strategic thinking at different difficulty levels.
- Teaching Strategies:
  - **Personalized Cognitive Exercises:** Design skill-specific activities that cater to individual needs, such as logic puzzles for analytical thinking or real-world scenarios for strategic decision-making.
  - **Incremental Problem-Solving Tasks:** Start with simple problems and increase difficulty, ensuring learners build confidence before tackling more complex challenges.
  - **Self-Guided Reflection Logs:** Encourage students to document their thought processes and evaluate their problem-solving approaches, helping them identify areas for improvement.
  - **Peer Collaboration and Review:** Assign group-based exercises where students solve problems together, discuss different approaches, and provide constructive feedback.
  - **Timed Challenge Drills:** Use time-constrained exercises to simulate real-world decision-making pressure, helping students develop efficiency in their thinking processes.

---

3. Refining Assessment Metrics for Cognitive Growth

- Creating precise rubrics that measure progress in areas such as abstract reasoning, decision-making, and adaptability.
- Teaching Strategies:
  - **Detailed Cognitive Skill Rubrics:** Develop structured evaluation criteria that measure logical reasoning, adaptability, and strategic thinking, ensuring assessments are objective and measurable.
  - **Competency-Based Evaluations:** Use practical tasks and real-world case studies instead of traditional exams to assess cognitive skill application.
  - **Formative vs. Summative Assessments:** Implement ongoing assessments (formative) that guide cognitive growth, alongside final assessments (summative) that measure overall skill mastery.

- **Automated and AI-Assisted Feedback Tools:** Utilize digital assessment platforms that provide real-time feedback and personalized improvement suggestions.
- **Self and Peer-Assessment Models:** Encourage learners to evaluate their own performance using structured reflection frameworks and peer review systems.

---

4. Implementing Step-by-Step Decision-Making Guides

- Developing detailed frameworks for breaking down complex problems into manageable steps for structured problem-solving.
- Teaching Strategies:
  - **Structured Decision Trees:** Teach students how to break down choices using flowcharts, helping them visualize potential outcomes before making decisions.
  - **Guided Case Study Analysis:** Assign real-world business, military, or ethical dilemmas where students must apply step-by-step decision-making strategies.
  - **Scenario-Based Training:** Provide hypothetical situations where students work through a structured approach to solve problems, ensuring they follow a logical process.
  - **Comparative Decision-Making Models:** Introduce different decision frameworks (e.g., cost-benefit analysis, SWOT (strengths, weaknesses, opportunities, threats) analysis) and have students evaluate which is most effective for a given problem.
  - **Reflection and Iteration Practices:** Encourage students to revisit past decisions, analyze what worked and what did not, and refine their approach for future scenarios.

---

5. Teaching Specific Logical Evaluation Techniques

- Providing learners with structured methods to identify biases, logical fallacies, and weak arguments in reasoning.
- Teaching Strategies:
  - **Logical Fallacy Identification:** Provide examples of flawed arguments and have students pinpoint logical errors and suggest corrections.
  - **Structured Argument Mapping:** Teach students how to break down and visualize arguments using logic diagrams, improving their analytical reasoning.
  - **Socratic Questioning Methods:** Use structured questioning techniques to push students to analyze assumptions, identify biases, and refine their arguments.
  - **Comparative Logical Analysis:** Present two conflicting viewpoints and have students determine which is stronger based on evidence and reasoning.
  - **Critical Thinking Debates:** Organize structured debates where students must construct logical arguments, defend their stance, and critique opposing viewpoints.

---

6. **Creating Cognitive Skill Application Drills**

- Designing daily exercises that reinforce systematic thinking, critical questioning, and rapid adaptability.
- Teaching Strategies:
    - **Daily Cognitive Warm-Ups:** Start each class or training session with a short exercise (e.g., logic puzzles, quick decision-making tasks) to engage cognitive processes.
    - **Timed Analytical Challenges:** Provide time-sensitive problem-solving drills to encourage efficiency and adaptability in decision-making.
    - **Reverse Engineering Exercises:** Give students a solution and have them work backward to determine the original problem or logical sequence.
    - **Cross-Disciplinary Cognitive Integration:** Design drills that require students to use multiple cognitive skills (e.g., strategic + analytical thinking) in a single exercise.
    - **Repetitive Skill Reinforcement:** Use gamified repetition techniques to help students internalize cognitive frameworks, improving retention and application.

---

7. **Developing Micro-Learning Sessions for Professionals**

- Constructing short, targeted learning modules that teach thinking strategies in digestible, real-world applications.
- Teaching Strategies:
    - **10-Minute Skill Refreshers:** Create quick, targeted lessons that professionals can complete between tasks to reinforce key thinking strategies.
    - **Case Study Quick-Reads:** Provide condensed case studies highlighting cognitive skills applied in business, leadership, or problem-solving.
    - **On-the-Go Learning Modules:** Develop mobile-friendly lessons that professionals can engage with during commutes or short breaks.
    - **Practical Real-World Challenges:** Assign quick decision-making tasks that directly relate to workplace scenarios.
    - **Self-Paced Learning Paths:** Allow learners to progress through micro-learning modules at their own speed while tracking skill development.

---

8. **Crafting PCS-Based Interview and Hiring Assessments**

- Designing evaluation tools for organizations to assess job candidates based on their cognitive skill application.
- Teaching Strategies:
    - **Behavioral Interview Questions:** Develop interview prompts that assess cognitive flexibility, strategic foresight, and problem-solving skills.
    - **Situational Judgment Tests:** Create scenario-based assessments where candidates must choose the best course of action based on given criteria.

- **On-the-Spot Analytical Exercises:** Have candidates complete a short logic or problem-solving challenge to demonstrate their cognitive abilities.
- **Structured Reflection Interviews:** Ask candidates to describe past decision-making experiences using PCS-based frameworks.
- **Custom Role-Specific Cognitive Assessments:** Tailor cognitive evaluations to job roles, ensuring relevance and effectiveness in hiring decisions.

---

9. Ensuring Consistency in Cognitive Training Across Teams

- Developing standardized training materials that align all team members with PCS-based cognitive strategies.
- Teaching Strategies:
  - **Company-Wide Cognitive Training Standards:** Develop standardized learning materials to ensure all employees receive the same cognitive training.
  - **Team-Based Thinking Challenges:** Assign problem-solving tasks that require teams to collaborate and apply cognitive strategies.
  - **Regular Cognitive Benchmark Assessments:** Use periodic evaluations to measure team-wide cognitive progress and adjust training as needed.
  - **Simulation-Based Team Training:** Provide interactive learning environments where teams must apply PCS principles in real-world business or crisis scenarios.
  - **Cross-Departmental Cognitive Integration:** Conduct workshops where different departments share problem-solving techniques to foster interdisciplinary thinking.

---

10. Customizing PCS Implementation Based on Individual Needs

- Using feedback loops and data-driven refinement to personalize cognitive training programs for different learning styles.
- Teaching Strategies:
  - **Personalized Learning Analytics:** Use AI-driven data analysis to customize training based on an individual's strengths and weaknesses.
  - **Adaptive Learning Modules:** Provide self-paced courses that adjust difficulty based on user performance.
  - **One-on-One Cognitive Coaching:** Offer personalized mentoring sessions focused on PCS skill application.
  - **Competency-Based Learning Paths:** Allow learners to choose their cognitive development trajectory based on career goals and interests.
  - **Reflective Learning Techniques:** Encourage the use of learning journals or self-assessment reports to track individual progress and refine cognitive strategies.

Detailed small thinking strategies ensure that big thinking concepts in PCS are effectively translated into actionable, measurable, and structured learning processes. By combining vision with execution, PCS provides a comprehensive framework for lifelong cognitive skill development.

Scaffolded Cognitive Development in PCS: Structuring Big and Small Skills for Growth

Using **PCS (Progressive Cognitive Structuring)**, we can determine **what big skills** and **what small skills** should be taught at each level by following a **scaffolded approach** that ensures cognitive growth over time. The structure follows a **progression from foundational to advanced cognitive abilities**, ensuring that each stage builds upon the previous one.

Step 1: Understanding the Progression of Thinking Development Across Each Stage of PCS

Each level (i.e., Grade School, Middle School, High School, College/Technical School, and Business Professionals) requires different types of cognitive skills, both big thinking skills (broad cognitive strategies) and small thinking skills (step-by-step execution).

General Progression of the Five Developmental Stages of PCS Across Levels:

1. **Grade School** – Foundational thinking skills (basic patterns, logic, and problem-solving).
2. **Middle School** – Intermediate thinking skills (critical thinking, creativity, and basic systems thinking).
3. **High School** – Advanced thinking skills (abstract reasoning, strategic planning, and reflective thinking).
4. **College/Technical School** – Applied thinking skills (integrating theory and practice, interdisciplinary problem-solving, and adaptability).
5. **Business Professionals** – Mastery-level thinking skills (strategic decision-making, leadership thinking, and high-stakes problem-solving).

Step 2: Identifying Big Thinking Skills vs. Small Thinking Skills at Each Level of PCS

For each level of PCS shown above, we now define the big thinking skills (broad conceptual frameworks) and small thinking skills (specific, practical exercises that build those big skills).

Grade School: Foundational Cognitive Abilities

Big Thinking Skills (Concepts & Strategies)

- **Pattern Recognition** (recognizing repeated structures in numbers, nature, and language).
- **Logical Reasoning** (understanding cause and effect, sequencing, and basic inferences).

- **Simple Problem-Solving** (breaking down problems, trial and error, basic decision-making).

Small Thinking Skills (Step-by-Step Execution)

- Matching and sorting objects by pattern.
- If-then reasoning exercises with simple cause and effect.
- Step-by-step problem-solving tasks (puzzles, simple games).

Middle School: Intermediate Cognitive Abilities

Big Thinking Skills (Concepts & Strategies)

- **Critical Thinking** (analyzing multiple viewpoints, questioning assumptions).
- **Creative Exploration** (brainstorming solutions, testing ideas, learning from failure).
- **Basic Systems Thinking** (understanding how things interact in ecosystems, society, and technology).

Small Thinking Skills (Step-by-Step Execution)

- Learning how to ask deeper questions about topics.
- Practicing structured brainstorming techniques (mind maps, lists).
- Mapping out cause-effect relationships in different subjects.

High School: Advanced Cognitive Abilities

Big Thinking Skills (Concepts & Strategies)

- **Abstract Reasoning** (solving non-linear problems, making predictions, understanding philosophy).
- **Strategic Planning** (analyzing options, predicting consequences, decision trees).
- **Reflective Thinking** (self-awareness, metacognition, ethical reasoning).

Small Thinking Skills (Step-by-Step Execution)

- Practicing debate and argumentation with structured reasoning.
- Creating detailed project plans with alternative pathways.
- Writing reflections on learning experiences and decision-making.

College/Technical School: Applied Thinking Abilities

Big Thinking Skills (Concepts & Strategies)

- **Interdisciplinary Problem-Solving** (combining knowledge from multiple fields).
- **Adaptive Thinking** (adjusting strategies in uncertain or changing environments).
- **Theoretical Application** (applying abstract concepts to real-world scenarios).

Small Thinking Skills (Step-by-Step Execution)

- Case study analysis with multiple variables.
- Working through complex simulations or scenario-based learning.
- Applying theory through research projects and experiments.

Business Professionals: Mastery-Level Cognitive Abilities

Big Thinking Skills (Concepts & Strategies)

- **Strategic Decision-Making** (weighing trade-offs, forecasting future impacts).
- **Leadership Thinking** (influencing, motivating, guiding teams).
- **High-Stakes Problem-Solving** (solving crises, navigating uncertainty).

Small Thinking Skills (Step-by-Step Execution)

- Using decision matrices for evaluating complex choices.
- Practicing leadership scenarios with role-playing.
- Conducting risk assessments and mitigation planning.

Step 3: Aligning the Development Stages with Learning Goals to Finalize Big and Small Skills

At this stage, the initial big skills identified for each level must now be explicitly aligned with the learning goals that define what must be taught at each stage. This process transforms the preliminary set of big and small skills into the finalized skills framework, ensuring that each cognitive ability builds upon the previous stage in a scaffolded manner.

Using **PCS (Progressive Cognitive Structuring)**, we determine what big skills and what small skills should be taught at each level by following a structured, hierarchical approach that promotes cognitive development over time. The learning goals at each stage define the necessary big skills, while the associated small skills provide the foundational steps required for mastery.

Step 4: Applying PCS to Curriculum and Learning Design

To implement PCS effectively, consider the following:

- Assess the student's current level of cognitive development.
- Introduce big thinking skills as overarching learning goals.
- Use small thinking skills to provide concrete, step-by-step ways to build those big skills.
- Ensure that each level builds upon previous ones, reinforcing and advancing cognitive abilities.
- Align learning strategies with PCS-based scaffolding, ensuring smooth cognitive progression.

Step 5: Transitioning from PCS Framework to Teaching Strategies

This final step marks a critical transition from understanding the PCS framework to what must be taught and the big thinking skills and small thinking skills that must be paired to create a clear roadmap for cognitive development at each stage.

Teaching strategies must now reflect the scaffolded nature of PCS by providing practical ways to integrate big and small skills into lesson plans, training modules, and real-world applications.

By bridging the gap between conceptual cognitive structures and practical teaching applications, PCS ensures that learners evolve from foundational thinkers to high-level decision-makers, fully equipped for cognitive adaptability in real-world scenarios.

Instructional Framework: Progressive Cognitive Structuring with Big vs. Small Thinking Skill Derivations

The development of cognitive abilities follows a structured progression, where each stage builds upon the previous one. To effectively teach and develop these abilities, it is essential to align learning strategies with the cognitive skills that must be taught at each level. The **Progressive Cognitive Structuring (PCS) Framework** provides a structured approach to this alignment by distinguishing between **big thinking skills** (broad cognitive strategies) and **small thinking skills** (specific step-by-step exercises that develop these larger abilities).

Each of the five developmental stages—Grade School, Middle School, High School, College/Technical School, and Business Professionals—requires a unique set of skills tailored to the cognitive growth expected at that level. The **big thinking skills** represent overarching abilities that learners should develop, while the **small thinking skills** serve as the practical steps required to build and reinforce those abilities.

The **big and small thinking skills** outlined in this framework have been derived from an analysis of cognitive progression, instructional best practices, and real-world applications. They reflect the fundamental skills required for problem-solving, reasoning, and decision-making at each stage of learning and professional development. By pairing these skills in a scaffolded manner, we can establish **effective teaching strategies** that ensure cognitive growth and adaptability over time.

Grade School Level (Foundations of Thinking)

What Must Be Taught:

1. **Curiosity and Observation:**
    - Encourage questioning and exploration.
    - Teach how to observe patterns in everyday activities (e.g., shapes, sequences).
2. **Basic Problem-Solving:**
    - Introduce simple cause-and-effect relationships.
    - Practice solving age-appropriate puzzles and games (e.g., Tic-Tac-Toe).
3. **Collaborative Thinking:**
    - Foster teamwork through group projects.
    - Teach listening and sharing ideas.

Building Blocks:

- Recognize patterns and relationships.
- Understand the importance of order and sequence.
- Develop basic communication skills for collaboration.

Instructional Techniques:

- **Play-Based Learning**: Use games and activities to teach problem-solving and collaboration.
- **Guided Discovery**: Provide opportunities for students to explore answers rather than directly providing them.

What are the Big Thinking Skills vs. Small Thinking Skills for the Grade School Level?

1. Curiosity and Observation

Big Thinking Skills (Concepts & Strategies)

- **Asking Why & How:** Encouraging students to ask deep questions about their surroundings.
- **Pattern Recognition:** Identifying repeated structures in numbers, nature, and language.
- **Conceptual Understanding:** Seeing connections between observed patterns and larger ideas (e.g., "Why do the seasons change?").

Small Thinking Skills (Step-by-Step Execution)

- **Matching & Sorting:** Grouping objects by color, shape, or function.
- **Step-by-Step Exploration:** Learning to carefully observe and record findings (e.g., counting similar objects in nature).
- **Following Guided Observations:** Completing teacher-led activities that highlight patterns (e.g., "Find all the circles in the classroom").

---

2. Basic Problem-Solving

Big Thinking Skills (Concepts & Strategies)

- **Logical Reasoning:** Understanding cause and effect relationships (e.g., "What happens if you do not water a plant?").
- **Strategic Thinking:** Choosing the best move in simple games like **Tic-Tac-Toe**.
- **Abstract Thinking:** Understanding that different solutions can solve the same problem.

Small Thinking Skills (Step-by-Step Execution)

- **Trial and Error:** Testing different puzzle solutions to see what works.
- **If-Then Reasoning:** Practicing "If I do X, Y will happen" with simple tasks.
- Breaking Down a Problem: Completing a puzzle by focusing on one piece at a time.

3. Collaborative Thinking

Big Thinking Skills (Concepts & Strategies)

- **Understanding Teamwork:** Learning that working together can solve problems faster.
- **Idea Sharing & Communication:** Knowing when to speak, listen, and contribute effectively.
- **Conflict Resolution:** Recognizing that different opinions exist and how to work through them.

Small Thinking Skills (Step-by-Step Execution)

- **Turn-Taking & Listening:** Practicing how to wait for one's turn in group activities.
- **Following Group Rules:** Completing a shared task using a step-by-step approach.
- **Hands-on Collaboration:** Collaborating with a partner to build something, with clear individual steps (e.g., "You place the first block, I place the second").

Summary: Big Skills vs. Small Skills Across All Areas for Grade School Level

Category	Big Thinking Skills (Concepts & Strategies)	Small Thinking Skills (Step-by-Step Execution)
Curiosity & Observation	Asking "why," recognizing patterns, conceptual thinking	Matching, sorting, following guided observations
Basic Problem-Solving	Logical reasoning, strategic thinking, abstract thinking	Trial and error, if-then reasoning, breaking problems into steps
Collaborative Thinking	Understanding teamwork, communication, conflict resolution	Turn-taking, following group rules, hands-on collaboration

How This Helps Grade School Learning

- **Big Thinking Skills** help students learn how to recognize patterns, think through problems, and create smart strategies.
- **Small Thinking Skills** help students use those big ideas to solve everyday problems.
- **Instructional Techniques (Play-Based Learning and Guided Discovery)** use games and activities to teach problem-solving and collaboration as well as providing opportunities for students to explore answers rather than directly providing them.

Middle School Level (Structuring Thought)

What Must Be Taught:

1. **Logical Thinking:**
   - Introduce basic logic (e.g., "if-then" reasoning).
   - Teach students to evaluate evidence and draw conclusions.
2. **Goal Setting:**
   - Encourage students to set small, achievable goals and track progress.
3. **Systems Thinking Introduction:**
   - Show how parts of a system (e.g., ecosystems, communities) interact.
4. **Critical Thinking:**
   - Develop skills for questioning assumptions and analyzing arguments.

Building Blocks:

- Link cause-and-effect with broader outcomes.
- Develop initial self-awareness of thinking processes (metacognition).
- Understand how small actions impact larger systems.

Instructional Techniques:

- **Socratic Method**: Use guided questioning to promote critical thinking.
- **Scenario-Based Learning**: Provide real-world examples to analyze and discuss.

What are the Big Thinking Skills vs. Small Thinking Skills for the Middle School Level?

1. Logical Thinking

Big Thinking Skills (Concepts & Strategies)

- **If-Then Reasoning:** Understanding how logical conditions work in different contexts (e.g., "If I do not study, then I will not perform well on the test").
- **Evaluating Evidence:** Learning to differentiate facts from opinions and identify reliable sources.
- **Drawing Conclusions:** Using logical steps to make well-reasoned decisions rather than relying on assumptions.

Small Thinking Skills (Step-by-Step Execution)

- **Breaking Down Arguments:** Identifying premises, evidence, and conclusions in a structured way.
- **Applying Logical Rules:** Solving logic puzzles, practicing syllogisms, and simple coding exercises.
- **Organizing Information:** Using graphic organizers to categorize evidence in debates or essays.

2. Goal Setting

Big Thinking Skills (Concepts & Strategies)

- **Long-Term Vision:** Understanding how setting smaller goals contributes to larger life achievements.
- **Strategic Thinking:** Developing a roadmap for achieving personal or academic goals.
- **Self-Monitoring & Adaptation:** Evaluating progress and making necessary adjustments.

Small Thinking Skills (Step-by-Step Execution)

- **Breaking Goals into Actionable Steps:** Learning how to deconstruct a big goal into smaller, achievable milestones.
- **Tracking Progress:** Using checklists, habit trackers, or goal charts to measure success.
- **Time Management:** Practicing prioritization and scheduling tasks effectively.

3. Systems Thinking Introduction

Big Thinking Skills (Concepts & Strategies)

- **Understanding Interconnections:** Recognizing how different parts of a system (e.g., ecosystems, economies, communities) influence each other.
- **Cause-and-Effect on a Large Scale:** Analyzing how small changes in a system impact the whole (e.g., pollution affecting marine life).
- **Feedback Loops & Complexity:** Understanding how actions create cycles (e.g., supply and demand in economics).

Small Thinking Skills (Step-by-Step Execution)

- **Mapping Systems Visually:** Drawing diagrams, mind maps, or flowcharts to represent interconnections.
- **Step-by-Step System Analysis:** Breaking down how a system works in small, sequential steps.
- **Testing Small Changes:** Running simulations or role-playing different scenarios to see how variables affect a system.

4. Critical Thinking

Big Thinking Skills (Concepts & Strategies)

- **Questioning Assumptions:** Encouraging students to ask, "How do we know this is true?" and seek deeper answers.
- **Analyzing Arguments:** Learning how to spot bias, fallacies, and weak reasoning.
- **Weighing Different Perspectives:** Understanding that different viewpoints exist and considering multiple sides of an issue.

Small Thinking Skills (Step-by-Step Execution)

- **Identifying Key Claims & Evidence:** Learning how to break down an argument into main points and supporting details.
- **Comparing Sources:** Practicing how to evaluate different pieces of evidence for reliability.
- **Structured Argumentation:** Using frameworks like the Claim-Evidence-Reasoning (CER) method to organize thoughts logically.

Summary: Big Skills vs. Small Skills Across All Areas for Middle School Level

Category	Big Thinking Skills (Concepts & Strategies)	Small Thinking Skills (Step-by-Step Execution)
Logical Thinking	If-then reasoning, evaluating evidence, drawing conclusions	Breaking down arguments, applying logic, organizing information
Goal Setting	Long-term vision, strategic thinking, self-monitoring	Breaking goals into steps, tracking progress, time management
Systems Thinking	Understanding interconnections, large-scale cause-and-effect, feedback loops	Mapping systems, step-by-step system analysis, testing small changes
Critical Thinking	Questioning assumptions, analyzing arguments, weighing perspectives	Identifying claims & evidence, comparing sources, structured argumentation

---

How This Supports Middle School Learning

- Big Thinking Skills develop higher-level reasoning, pattern recognition, and strategy formation.
- Small Thinking Skills ensure students can apply those concepts in real-world problem-solving.
- **Instructional Techniques (Socratic Method & Scenario-Based Learning)** encourage both big and small thinking by pushing students to ask questions and analyze situations logically.

High School Level (Advanced Analytical Thinking)

What Must Be Taught:

1. **Strategic Planning:**
   - Teach how to create a plan with long-term goals and short-term steps.
2. **Decision-Making:**
   - Explore techniques for weighing pros and cons and evaluating risks.
3. **Creative Problem-Solving:**
   - Encourage brainstorming and lateral thinking.
4. **Adaptability:**
   - Teach how to revise plans based on new information.

Building Blocks:

- Develop the ability to think both abstractly (big picture) and concretely (specific details).
- Strengthen metacognitive skills to assess one's own thinking process.
- Build resilience and flexibility in the face of uncertainty.

Instructional Techniques:

- **Project-Based Learning:** Engage students in complex, multi-step projects.
- **Case Studies:** Analyze historical or hypothetical situations to apply strategic thinking.

What are the Big Thinking Skills vs. Small Thinking Skills for the High School Level?

1. Strategic Planning

Big Thinking Skills (Concepts & Strategies)

- **Long-Term Vision:** Understanding how small steps contribute to long-term success (e.g., college/career planning).
- **Prioritization & Resource Management:** Identifying the most critical steps in a plan and allocating time/resources effectively.
- **Contingency Thinking:** Developing backup plans in case the original plan fails.

Small Thinking Skills (Step-by-Step Execution)

- **Breaking Down Plans:** Creating detailed action steps to achieve a goal.
- **Tracking & Adjusting Progress:** Using planners, Gantt charts, or checklists to stay on course.
- **Time Management:** Scheduling tasks efficiently to maximize productivity.

---

2. Decision-Making

Big Thinking Skills (Concepts & Strategies)

- **Weighing Pros & Cons:** Analyzing benefits and drawbacks before making a choice.
- **Risk Assessment:** Evaluating potential consequences and uncertainties in decisions.
- **Logical and Ethical Considerations:** Considering logical, emotional, and ethical factors in choices.

Small Thinking Skills (Step-by-Step Execution)

- **Structured Decision Frameworks:** Applying decision-making models like SWOT analysis or the Decision Matrix.
- **Data Gathering & Comparison:** Collecting relevant information to compare different options.
- **Justifying Decisions:** Writing or presenting rational explanations for choices.

---

3. Creative Problem-Solving

Big Thinking Skills (Concepts & Strategies)

- **Divergent Thinking:** Generating multiple solutions to a problem through brainstorming.
- **Lateral Thinking:** Making unexpected connections between ideas to find innovative solutions.
- **Pattern Recognition in Problems:** Identifying underlying structures in complex issues.

Small Thinking Skills (Step-by-Step Execution)

- **Applying Problem-Solving Techniques:** Using structured methods like **mind mapping**, **Six Thinking Hats** (De Bono, 1985), or **SCAMPER** (Substitute, Combine, Adapt, Modify, Put to another use, Eliminate, Reverse) (Eberle, 1996).
- **Experimenting with Ideas:** Testing different solutions in a controlled manner.
- **Refining Solutions:** Iterating on an idea based on feedback.

---

4. Adaptability

Big Thinking Skills (Concepts & Strategies)

- **Flexibility in Thinking:** Recognizing when a change in approach is necessary.
- **Resilience & Growth Mindset:** Learning from failures rather than seeing them as roadblocks.
- **Abstract & Concrete Thinking Balance:** Switching between big-picture strategies and fine details.

Small Thinking Skills (Step-by-Step Execution)

- **Revising & Iterating Plans:** Adjusting strategies based on new information or setbacks.
- **Scenario Planning:** Practicing "what if" exercises to anticipate changes.
- **Stress Testing Decisions:** Evaluating how robust a plan is by considering worst-case scenarios.

Summary: Big Skills vs. Small Skills Across All Areas for High School Level

Category	Big Thinking Skills (Concepts & Strategies)	Small Thinking Skills (Step-by-Step Execution)
Strategic Planning	Long-term vision, prioritization, contingency thinking	Breaking down plans, tracking progress, time management
Decision-Making	Weighing pros & cons, risk assessment, logical/ethical considerations	Decision frameworks, data gathering, justifying choices
Creative Problem-Solving	Divergent & lateral thinking, pattern recognition in problems	Applying techniques, experimenting, refining solutions
Adaptability	Flexible thinking, resilience, balancing abstract vs. concrete thinking	Revising plans, scenario planning, stress testing decisions

---

How This Supports High School Learning

- Big Thinking Skills develop abstract reasoning, long-term strategy, and higher-order analysis.
- Small Thinking Skills ensure students can execute those strategies efficiently in real-world settings.
- **Instructional Techniques (Project-Based Learning & Case Studies)** naturally integrate both big and small thinking by requiring students to analyze, plan, execute, and adapt in real-world scenarios.

College/Technical School Level (Integration and Application)

What Must Be Taught:

1. **Systems Thinking Mastery:**
   - Delve deeply into interconnected systems (e.g., economics, ecosystems, organizations).
2. **Ethical Decision-Making:**
   - Introduce frameworks for evaluating decisions' moral and societal impacts.
3. **Leadership Thinking:**
   - Teach how to align team efforts with strategic goals.
4. **Data Analysis and Interpretation:**
   - Provide skills to interpret data and integrate it into strategic decisions.

Building Blocks:

- Achieve full integration of big and small thinking for complex problem-solving.
- Cultivate the ability to synthesize information from multiple disciplines.
- Develop strong analytical and reflective thinking capabilities.

Instructional Techniques:

- **Capstone Projects**: Require students to design and execute comprehensive projects that demonstrate strategic simplification.
- **Interdisciplinary Learning**: Combine insights from multiple fields to approach problems holistically.

What are the Big Thinking Skills vs. Small Thinking Skills for the College/Technical School Level?

1. Systems Thinking Mastery

Big Thinking Skills (Concepts & Strategies)

- **Understanding Interconnected Systems:** Recognizing how components interact in fields like economics, ecosystems, and organizational structures.
- **Identifying Systemic Patterns:** Recognizing feedback loops, delays, and cascading effects.
- **Holistic Problem-Solving:** Considering multiple perspectives when analyzing complex systems.

Small Thinking Skills (Step-by-Step Execution)

- **Mapping System Interactions:** Using flowcharts, causal loop diagrams, and network models to visualize systems.
- **Simulating System Changes:** Running models to predict the effects of small changes on an entire system.

- **Applying Systems Thinking Tools:** Using methodologies like stock-and-flow diagrams and leverage point identification.

---

2. Ethical Decision-Making

Big Thinking Skills (Concepts & Strategies)

- **Applying Ethical Frameworks:** Evaluating moral and societal implications using models like utilitarianism, deontology, and virtue ethics.
- **Assessing Stakeholder Impact:** Understanding how decisions affect individuals, organizations, and society at large.
- **Weighing Competing Values:** Balancing business goals, social responsibility, and personal ethics.

Small Thinking Skills (Step-by-Step Execution)

- **Breaking Down Ethical Dilemmas:** Identifying key stakeholders, conflicts, and possible resolutions.
- **Applying Case Study Analysis:** Evaluating real-world ethical situations and extracting lessons.
- **Justifying Ethical Decisions:** Forming structured arguments using established ethical reasoning models.

---

3. Leadership Thinking

Big Thinking Skills (Concepts & Strategies)

- **Visionary & Strategic Alignment:** Understanding how to align team goals with broader organizational objectives.
- **Influence & Decision-Making:** Developing the ability to inspire, guide, and persuade others.
- **Situational Adaptability:** Adjusting leadership style based on team needs, crises, or strategic shifts.

Small Thinking Skills (Step-by-Step Execution)

- **Delegating & Prioritizing Tasks:** Breaking down objectives into actionable steps for team members.
- **Facilitating Team Communication:** Leading discussions, resolving conflicts, and ensuring clear messaging.
- **Tracking Leadership Effectiveness:** Using performance metrics and feedback loops to assess and improve leadership impact.

---

4. Data Analysis and Interpretation

Big Thinking Skills (Concepts & Strategies)

- **Extracting Meaning from Data:** Understanding how to derive insights from raw numbers and trends.
- **Integrating Data into Decision-Making:** Using analytics to guide strategic choices.
- **Cross-Disciplinary Synthesis:** Combining data from multiple fields (e.g., economics + sociology) for holistic analysis.

Small Thinking Skills (Step-by-Step Execution)

- **Using Analytical Tools:** Applying Excel, Python, R, or business intelligence software for data processing.
- **Performing Statistical Analysis:** Calculating measures such as mean, median, correlation, and regression analysis.
- **Visualizing Data:** Creating dashboards, charts, and reports to present findings effectively.

Summary: Big Skills vs. Small Skills Across All Areas for College/Technical School Level

Category	Big Thinking Skills (Concepts & Strategies)	Small Thinking Skills (Step-by-Step Execution)
Systems Thinking Mastery	Understanding interconnections, systemic patterns, holistic problem-solving	Mapping systems, running simulations, applying system tools
Ethical Decision-Making	Applying ethical frameworks, assessing stakeholder impact, weighing competing values	Breaking down dilemmas, case study analysis, justifying ethical choices
Leadership Thinking	Strategic alignment, influence & decision-making, situational adaptability	Delegation, team communication, tracking leadership effectiveness
Data Analysis & Interpretation	Extracting meaning, integrating data into strategy, cross-disciplinary synthesis	Using tools, performing statistical analysis, visualizing data

---

How This Supports College/Technical Learning

- **Big Thinking Skills** ensure students can analyze, synthesize, and strategize at a high level.
- **Small Thinking Skills** provide the ability to execute and apply these strategies in real-world contexts.

- **Instructional Techniques (Capstone Projects & Interdisciplinary Learning)** require students to integrate both **big** and **small** thinking, applying knowledge from multiple disciplines.

Business Professional Level (Strategic Execution)

What Must Be Taught:

1. **Visionary Thinking:**
   - Teach how to create and communicate a strategic vision that inspires others.
2. **Tactical Execution:**
   - Break down strategic goals into actionable steps and ensure their implementation.
3. **Continuous Improvement:**
   - Emphasize iterative thinking and the importance of learning from feedback.
4. **Global Perspective:**
   - Introduce considerations of global trends, markets, and cultural dynamics.

Building Blocks:

- Master the balance of strategic vision with operational precision.
- Enhance emotional intelligence to lead effectively.
- Refine adaptability to thrive in dynamic, uncertain environments.

Instructional Techniques:

- **Simulations**: Use organizational simulations to mimic real-world challenges.
- **Mentorship Programs**: Pair learners with experienced professionals to gain practical insights.
- **Executive Training Workshops**: Provide hands-on sessions focusing on leadership and decision-making.

What are the Big Thinking Skills vs. Small Thinking Skills for the Business Professional Level?

1. Visionary Thinking

Big Thinking Skills (Concepts & Strategies)

- **Strategic Visioning:** Developing a long-term vision that aligns with business goals and market trends.
- **Inspirational Leadership:** Communicating a vision in a way that motivates and unifies teams.

- **Foresight & Innovation:** Anticipating industry shifts and proactively shaping strategies.

Small Thinking Skills (Step-by-Step Execution)

- **Translating Vision into Action:** Breaking down broad strategic goals into clear operational tasks.
- **Crafting Persuasive Messaging:** Writing and delivering effective vision statements and presentations.
- **Stakeholder Alignment:** Ensuring that different teams understand and commit to the vision.

---

2. Tactical Execution

Big Thinking Skills (Concepts & Strategies)

- **Operationalizing Strategy:** Designing structured processes to move from idea to implementation.
- **Resource Allocation:** Understanding how to distribute time, talent, and capital effectively.
- **Risk Management:** Identifying potential roadblocks and developing contingency plans.

Small Thinking Skills (Step-by-Step Execution)

- **Project Planning & Task Management:** Using frameworks like OKRs, KPIs, and Gantt charts to track progress.
- **Delegation & Accountability:** Assigning tasks and ensuring follow-through.
- **Process Optimization:** Identifying inefficiencies and refining workflows for better performance.

---

3. Continuous Improvement

Big Thinking Skills (Concepts & Strategies)

- **Iterative Thinking:** Viewing business processes as cycles of improvement rather than static plans.
- **Growth Mindset & Adaptability:** Embracing change and learning from failures.
- **Data-Driven Decision-Making:** Using metrics and feedback to refine strategies.

Small Thinking Skills (Step-by-Step Execution)

- **Implementing Feedback Loops:** Collecting and applying customer, employee, and market feedback.
- **Conducting Retrospectives:** Running structured reviews (e.g., post-mortems, lessons learned sessions).

- **Performance Monitoring:** Tracking trends using dashboards and key performance indicators.

---

4. Global Perspective

Big Thinking Skills (Concepts & Strategies)

- **Cross-Cultural Competence:** Understanding and respecting international business norms and cultural differences.
- **Global Market Analysis:** Evaluating economic, political, and technological trends in different regions.
- **International Strategy Development:** Adapting business models to different markets and regulatory environments.

Small Thinking Skills (Step-by-Step Execution)

- **Navigating International Business Regulations:** Applying compliance frameworks (e.g., GDPR, trade laws).
- **Global Communication Skills:** Using effective negotiation and collaboration techniques across cultures.
- **Monitoring Global Trends:** Keeping up with emerging market shifts and economic reports.

Summary: Big Skills vs. Small Skills Across All Areas for Business Professional Level

Category	Big Thinking Skills (Concepts & Strategies)	Small Thinking Skills (Step-by-Step Execution)
Visionary Thinking	Strategic visioning, inspirational leadership, foresight, and innovation	Translating vision into action, persuasive messaging, stakeholder alignment
Tactical Execution	Operationalizing strategy, resource allocation, risk management	Project planning, delegation, process optimization
Continuous Improvement	Iterative thinking, growth mindset, data-driven decisions	Implementing feedback loops, conducting retrospectives, performance monitoring
Global Perspective	Cross-cultural competence, global market analysis, international strategy development	Navigating regulations, global communication, monitoring trends

---

How This Supports Business Professional Learning

- Big Thinking Skills ensure leaders can strategize, innovate, and inspire change.
- Small Thinking Skills enable professionals to execute strategies with precision and accountability.
- Instructional Techniques (Simulations, Mentorship, Executive Training) integrate real-world applications with structured learning.

Key Elements of Progressive Cognitive Structuring Leading to Teaching Strategies

The instructional technique for **Progressive Cognitive Structuring** ensures a **step-by-step development of thinking skills** from the grade school level to the business professional level. By focusing on foundational skills early and progressively incorporating complexity, learners are equipped to manage both the **big picture** and the **details** in any context, achieving true **strategic simplification** in their personal and professional lives.

1. **Sequential Learning:**
    - Each level builds on prior knowledge and skills, ensuring a continuous development of thinking capabilities.
2. **Scaffolded Instruction:**
    - Support gradually decreases as learners become more competent, promoting independence.
3. **Experiential Learning:**
    - Real-world applications are emphasized at every stage, ensuring skills are relevant and practical.

Example Teaching Strategies for Big vs. Small Thinking Skill Pairings of Curiosity and Observation at the Grade School Level

Here are 90 unique teaching strategy examples for the big thinking and small thinking skill pairings, each described in detail to align with **Big Thinking Skills** (Concepts & Strategies) and **Small Thinking Skills** (Step-by-Step Execution) for the **Curiosity and Observation Skill** taught at the grade school level.

Curiosity and Observation

Big Thinking Skills (Concepts & Strategies)

- **Asking Why & How:** Encouraging students to ask deep questions about their surroundings.
- **Pattern Recognition:** Identifying repeated structures in numbers, nature, and language.
- **Conceptual Understanding:** Seeing connections between observed patterns and larger ideas (e.g., "Why do the seasons change?").

Small Thinking Skills (Step-by-Step Execution)

- **Matching & Sorting:** Grouping objects by color, shape, or function.
- **Step-by-Step Exploration:** Learning to carefully observe and record findings (e.g., counting similar objects in nature).
- **Following Guided Observations:** Completing teacher-led activities that highlight patterns (e.g., "Find all the circles in the classroom").

Curiosity and Observation – (Big Skill) Asking Why & How vs. (Small Skill) Matching & Sorting

---

Pairing 1: Floating vs. Sinking – Exploring Material Properties

Big Skill: Asking "Why" & "How" – Encouraging students to ask deep questions about their surroundings.

Small Skill: Matching & Sorting – Grouping objects by color, shape, or function.

Teaching Technique: Hands-on experiments.

Actions: Provide students with different materials and ask them to group them based on shared properties while questioning why and how certain materials are alike.

Example: Students test various materials (wood, metal, plastic) in water and sort them into "float" or "sink" groups, then discuss why some materials float while others sink.

Pairing 2: Nature's Patterns – Sorting Objects by Physical Traits

Big Skill: Asking "Why" & "How" – Encouraging inquiry into relationships between objects.

Small Skill: Matching & Sorting – Grouping based on physical characteristics.

Teaching Technique: Guided exploration.

Actions: Take students on a classroom or outdoor exploration where they collect and sort objects based on similarities, discussing why some objects belong together.

Example: On a nature walk, students gather leaves and sort them by size, then ask why some trees produce small leaves while others have large ones.

Pairing 3: Tools & Functionality – Matching Objects to Their Uses

Big Skill: Asking "Why" & "How" – Encouraging students to think critically about object functionality.

Small Skill: Matching & Sorting – Grouping by function.

Teaching Technique: Play-based learning.

Actions: Set up a matching activity where students pair objects with their intended use, explaining why each object is suited for its function.

Example: Students match tools (hammer, screwdriver, paintbrush) with their purpose and discuss why each tool is designed the way it is.

Pairing 4: Observing Nature – Identifying and Categorizing Patterns

Big Skill: Asking "Why" & "How" – Exploring patterns in the natural world.

Small Skill: Matching & Sorting – Grouping based on observable patterns.

Teaching Technique: Observation journals.

Actions: Have students keep a journal documenting different patterns they observe in nature and sort them into categories, asking why these patterns exist.

Example: Students document cloud shapes over a week, sorting them into cirrus, cumulus, and stratus, then discuss how and why clouds form.

Pairing 5: Cause and Effect – Classifying Objects Based on Reactions

Big Skill: Asking "Why" & "How" – Understanding cause-and-effect relationships.

Small Skill: Matching & Sorting – Categorizing objects by reaction or behavior.

Teaching Technique: Problem-solving exercises.

Actions: Present students with a scenario where they must sort objects based on what happens when they interact with certain conditions (e.g., heat, water, pressure).

Example: Students test objects under heat (metal spoon, plastic spoon, wooden spoon) and sort them into "changes shape" or "stays the same," then discuss why different materials react differently.

Pairing 6: What Is It Made Of? – Investigating Material Types

Big Skill: Asking "Why" & "How" – Encouraging students to investigate how things are made.

Small Skill: Matching & Sorting – Grouping by material type.

Teaching Technique: Role-playing.

Actions: Students take on the role of scientists or engineers and sort everyday items based on what they are made of, questioning why certain materials are used.

Example: Students sort clothing into "cotton," "wool," and "synthetic" and discuss why each fabric is used for different purposes (e.g., wool for warmth, cotton for breathability).

Pairing 7: Artistic Expression – Classifying Art Styles and Elements

Big Skill: Asking "Why" & "How" – Encouraging students to compare and contrast similarities and differences.

Small Skill: Matching & Sorting – Classifying based on shared and differing attributes.

Teaching Technique: Art-based learning.

Actions: Provide students with various art pieces and have them sort them into categories, discussing why certain artistic styles or elements match.

Example: Students sort paintings by color schemes (warm vs. cool tones) and discuss how color affects mood in art.

Pairing 8: Dissolving vs. Not Dissolving – Exploring Chemical Reactions

Big Skill: Asking "Why" & "How" – Investigating scientific properties.

Small Skill: Matching & Sorting – Classifying based on chemical or physical reactions.

Teaching Technique: Sensory exploration.

Actions: Have students test different substances (sugar, salt, baking soda) in water and sort them into "dissolves" and "does not dissolve" groups, explaining why some dissolve while others do not.

Example: Students mix different powders with water and predict which ones will dissolve, then explain why sugar dissolves while flour does not.

Pairing 9: Animal Classification – Grouping by Habitat and Diet

Big Skill: Asking "Why" & "How" – Understanding ecosystems and biodiversity.

Small Skill: Matching & Sorting – Grouping animals based on habitat or diet.

Teaching Technique: Collaborative group work.

Actions: Assign groups different animal species and have them sort them into categories based on diet (herbivore, carnivore, omnivore), then discuss why certain animals eat what they do.

Example: Students sort animals into "meat-eaters" and "plant-eaters" and explain why a lion eats meat while a deer eats plants.

Pairing 10: Cultural Traditions – Matching Artifacts to Their Origins

Big Skill: Asking "Why" & "How" – Encouraging exploration of cultural diversity.

Small Skill: Matching & Sorting – Grouping based on cultural symbols or traditions.

Teaching Technique: Cultural exploration.

Actions: Have students research and match cultural artifacts, clothing, or traditions to their country of origin, discussing how and why different cultures developed these traditions.

Example: Students match traditional Japanese kimonos, Indian saris, and Scottish kilts to their respective cultures and discuss why different regions have unique clothing styles.

Curiosity and Observation – (Big Skill) Asking Why & How vs. (Small Skill) Step-by-Step Exploration

---

Pairing 1: Observing Plant Growth Over Time

Big Skill: Asking "Why" & "How" – Encouraging students to ask deep questions about their surroundings.

Small Skill: Step-by-Step Exploration – Learning to carefully observe and record findings (e.g., counting similar objects in nature).

Teaching Technique: Hands-on experiments.

Actions: Guide students through an experiment where they observe and record step-by-step changes over time, questioning why and how changes occur.

Example: Students plant seeds and track their growth daily, noting differences in height and leaf count, then ask why some plants grow faster than others.

Pairing 2: Discovering Patterns in Nature

Big Skill: Asking "Why" & "How" – Encouraging curiosity about natural patterns.

Small Skill: Step-by-Step Exploration – Observing and recording recurring patterns in the environment.

Teaching Technique: Guided exploration.

Actions: Take students on a nature walk where they document repeating patterns in tree rings, leaves, or animal tracks, asking why these patterns exist.

Example: Students count the rings on tree stumps and ask why older trees have more rings than younger ones.

Pairing 3: Solving Mysteries with the Scientific Method

Big Skill: Asking "Why" & "How" – Encouraging students to investigate the scientific method.

Small Skill: Step-by-Step Exploration – Conducting structured observations and documenting findings.

Teaching Technique: Problem-solving exercises.

Actions: Assign students a mystery to solve using step-by-step clues, encouraging them to ask why each clue leads to the next.

Example: Students follow a sequence of footprints in the schoolyard, documenting and analyzing them to determine which animal or shoe made them.

Pairing 4: Tracking Weather Changes

Big Skill: Asking "Why" & "How" – Developing inquiry-based thinking.

Small Skill: Step-by-Step Exploration – Recording data in a structured format to find patterns.

Teaching Technique: Observation journals.

Actions: Have students maintain a daily observation journal of weather patterns, noting temperature, cloud cover, and precipitation while questioning why weather changes occur.

Example: Students observe that rain often follows cloudy days and ask why cloud formations predict upcoming storms.

Pairing 5: Investigating Motion and Energy

Big Skill: Asking "Why" & "How" – Investigating real-world cause-and-effect relationships.

Small Skill: Step-by-Step Exploration – Tracking changes over time and documenting observations.

Teaching Technique: Play-based learning.

Actions: Design an interactive activity where students measure and record the distance toy cars travel based on different ramp angles, asking why certain angles result in longer or shorter distances.

Example: Students test a car on ramps with different slopes and discuss why a steeper ramp makes the car go farther.

Pairing 6: Observing Ecosystem Diversity

Big Skill: Asking "Why" & "How" – Encouraging systematic thinking.

Small Skill: Step-by-Step Exploration – Counting and categorizing objects through direct observation.

Teaching Technique: Role-playing.

Actions: Have students act as field researchers, recording the number of insects, birds, or plants in a defined area and questioning why certain species appear more frequently than others.

Example: Students count bees visiting flowers in a garden and ask why some flowers attract more bees than others.

Pairing 7: Testing Engineering Design Principles

Big Skill: Asking "Why" & "How" – Examining human-made structures and engineering.

Small Skill: Step-by-Step Exploration – Observing and analyzing the effects of design choices.

Teaching Technique: Hands-on building activities.

Actions: Guide students in constructing paper bridges of different shapes and test how much weight they hold, documenting their observations and asking why some shapes are stronger.

Example: Students build triangle and square-based bridges, then ask why triangles provide better structural support.

Pairing 8: Analyzing Historical Artifacts

Big Skill: Asking "Why" & "How" – Encouraging historical inquiry.

Small Skill: Step-by-Step Exploration – Observing and recording artifacts to uncover historical insights.

Teaching Technique: Collaborative group work.

Actions: Provide historical artifacts or images and have students analyze and document details over time, asking why certain materials or symbols were used in the past.

Example: Students examine old coins, recording their engravings, and ask why rulers or national symbols were included on the coins.

Pairing 9: Documenting Biological Transformations

Big Skill: Asking "Why" & "How" – Exploring biological changes.

Small Skill: Step-by-Step Exploration – Observing and tracking living organisms.

Teaching Technique: Sensory exploration.

Actions: Have students observe and document butterfly metamorphosis step by step, questioning how and why changes occur at each stage.

Example: Students watch caterpillars form chrysalises and emerge as butterflies, recording the timeline and asking why metamorphosis is necessary for survival.

Pairing 10: Measuring Environmental Impact

Big Skill: Asking "Why" & "How" – Understanding environmental impact.

Small Skill: Step-by-Step Exploration – Measuring and documenting pollution or human impact.

Teaching Technique: Real-world applications.

Actions: Guide students in observing and recording litter around the school over a week, then discuss why certain areas accumulate more trash and how it affects the environment.

Example: Students tally the number of plastic vs. paper items in the schoolyard and ask why plastic waste is more common and takes longer to decompose.

Curiosity and Observation – (Big Skill) Pattern Recognition vs. (Small Skill) Following Guided Observations

---

Pairing 1: Identifying Patterns in Crystal Structures

Big Skill: Asking "Why" & "How" – Encouraging students to ask deep questions about their surroundings.

Small Skill: Following Guided Observations – Completing teacher-led activities that highlight patterns (e.g., "Find all the circles in the classroom").

Teaching Technique: Hands-on experiments.

Actions: Guide students through an experiment where they follow step-by-step teacher-led observations to identify patterns, then ask why and how these patterns occur.

Example: Students use magnifying glasses to observe salt and sugar crystals, finding similarities in shape and asking why both substances form cubic structures.

Pairing 2: Recognizing Repeating Elements in Everyday Life

Big Skill: Asking "Why" & "How" – Encouraging curiosity about patterns in everyday life.

Small Skill: Following Guided Observations – Identifying repeated structures in a controlled setting.

Teaching Technique: Guided exploration.

Actions: Lead students through a guided walk around the classroom or school, asking them to find repeating colors, shapes, or objects while questioning why those repetitions exist.

Example: Students find that all stop signs are red octagons and ask why that shape and color were chosen for traffic safety.

Pairing 3: Observing Geometric Structures in Nature

Big Skill: Asking "Why" & "How" – Encouraging students to think critically about structures in nature.

Small Skill: Following Guided Observations – Completing structured observation activities.

Teaching Technique: Problem-solving exercises.

Actions: Present students with a pattern-based problem, guiding them through structured observations to analyze similarities and differences before answering why and how they occur.

Example: Students observe honeycombs and turtle shells, identifying hexagonal patterns and asking why this shape appears in nature so often.

Pairing 4: Recording Observations for Pattern Recognition

Big Skill: Asking "Why" & "How" – Encouraging inquiry-based thinking.

Small Skill: Following Guided Observations – Noticing and documenting key features in objects.

Teaching Technique: Observation journals.

Actions: Assign students a guided observation task where they record specific characteristics of objects (e.g., counting windows in buildings) and then ask why they are designed that way.

Example: Students count the number of petals on different flowers and ask why certain flowers always have the same number of petals.

Pairing 5: Discovering Patterns in Object Design

Big Skill: Asking "Why" & "How" – Investigating real-world patterns.

Small Skill: Following Guided Observations – Observing and categorizing repeating elements.

Teaching Technique: Play-based learning.

Actions: Engage students in a game where they find all objects of a certain shape or function in the classroom, then ask why those objects share the same design.

Example: Students identify all rectangles in the classroom (books, doors, desks) and discuss why rectangles are commonly used in construction.

Pairing 6: Acting as Scientists to Observe Animal Features

Big Skill: Asking "Why" & "How" – Encouraging students to investigate relationships between objects.

Small Skill: Following Guided Observations – Completing activities that require pattern recognition.

Teaching Technique: Role-playing.

Actions: Have students pretend to be scientists following an observation checklist, analyzing and questioning repeating patterns in different settings.

Example: Students act as animal researchers, following guided observations to count spots on different ladybugs and asking why some have more spots than others.

Pairing 7: Identifying Patterns in Art and Design

Big Skill: Asking "Why" & "How" – Examining human-made design choices.

Small Skill: Following Guided Observations – Noticing and explaining design patterns.

Teaching Technique: Art-based learning.

Actions: Provide students with famous paintings or designs and guide them in identifying repeated patterns, then discuss why those patterns were used.

Example: Students observe Van Gogh's *Starry Night* and note the swirling patterns, then ask why he used that technique in his artwork.

Pairing 8: Detecting Rhythmic Patterns in Sounds

Big Skill: Asking "Why" & "How" – Investigating sound and rhythm patterns.

Small Skill: Following Guided Observations – Recognizing auditory patterns.

Teaching Technique: Sensory exploration.

Actions: Lead students in an activity where they listen for repeating sound patterns in music or nature, then ask why and how those patterns form.

Example: Students listen to bird calls and identify repeated sequences, asking why some birds repeat sounds while others sing varied tunes.

Pairing 9: Finding Mathematical Patterns in Architecture

Big Skill: Asking "Why" & "How" – Exploring mathematical structures in the environment.

Small Skill: Following Guided Observations – Finding numerical patterns in real-world examples.

Teaching Technique: Collaborative group work.

Actions: Assign groups different number-based observation tasks (e.g., looking for sets of three in architecture) and have them discuss why that number is used.

Example: Students count the number of steps on staircases in different buildings and ask why they often come in sets of even numbers.

Pairing 10: Analyzing Repeating Cultural Symbols

Big Skill: Asking "Why" & "How" – Investigating cultural and historical patterns.

Small Skill: Following Guided Observations – Observing and recording design elements from different cultures.

Teaching Technique: Cultural exploration.

Actions: Provide students with images of historical buildings or artifacts, guiding them to identify repeating symbols and discussing why those patterns exist.

Example: Students examine Egyptian pyramids and Mayan step-pyramids, asking why different civilizations independently used triangular structures for monumental buildings.

Curiosity and Observation – (Big Skill) Pattern Recognition vs. (Small Skill) Matching & Sorting

---

Pairing 1: Investigating Snowflake Symmetry

Big Skill: Pattern Recognition – Identifying repeated structures in numbers, nature, and language.

Small Skill: Matching & Sorting – Grouping objects by color, shape, or function.

Teaching Technique: Hands-on experiments.

Actions: Guide students through an experiment where they identify and sort objects with similar repeating structures, discussing why those patterns exist.

Example: Students analyze the structure of snowflakes under a magnifying glass, sorting them by similarities and discussing why snowflakes always form six-sided patterns.

Pairing 2: Exploring Leaf Patterns in Nature

Big Skill: Pattern Recognition – Observing repeating structures in the environment.

Small Skill: Matching & Sorting – Grouping items based on shared features.

Teaching Technique: Guided exploration.

Actions: Take students outside to find and categorize natural objects that follow patterns, encouraging them to ask why certain patterns appear.

Example: Students find different types of leaves and sort them into categories based on vein patterns, questioning why some have a symmetrical design while others do not.

Pairing 3: Recognizing Number Sequences and Math Patterns

Big Skill: Pattern Recognition – Identifying recurring sequences in numbers and language.

Small Skill: Matching & Sorting – Grouping numbers or words by shared characteristics.

Teaching Technique: Problem-solving exercises.

Actions: Provide students with a set of number sequences and guide them to sort numbers based on repeating patterns, discussing why those patterns exist in math.

Example: Students sort numbers into groups based on odd/even properties and Fibonacci sequences, then explore why Fibonacci numbers appear in nature.

Pairing 4: Identifying Rhythmic Patterns in Music

Big Skill: Pattern Recognition – Finding repeating structures in music and rhythm.

Small Skill: Matching & Sorting – Identifying and grouping similar musical elements.

Teaching Technique: Sensory exploration.

Actions: Play different rhythms and have students match beats that follow the same pattern, discussing why music relies on repetition.

Example: Students listen to drumming sequences and sort them into "fast," "slow," and "syncopated" rhythms, discussing why different genres use distinct patterns.

Pairing 5: Exploring Symmetry in Art and Design

Big Skill: Pattern Recognition – Observing symmetry in shapes and objects.

Small Skill: Matching & Sorting – Grouping objects based on symmetrical properties.

Teaching Technique: Art-based learning.

Actions: Guide students in creating symmetrical art projects, sorting images based on symmetry type and discussing why symmetry appears in design.

Example: Students fold paper to create inkblot symmetry art, then match their designs to real-world objects that share similar symmetrical structures.

Pairing 6: Tracking Cyclical Patterns in Nature

Big Skill: Pattern Recognition – Recognizing cycles in seasons, life processes, and history.

Small Skill: Matching & Sorting – Grouping events or objects based on cyclical behavior.

Teaching Technique: Observation journals.

Actions: Have students document and sort observations over time, tracking patterns in the environment.

Example: Students keep a weather journal, recording sunny, rainy, and cloudy days, then sort their findings into seasonal patterns, asking why weather changes occur in cycles.

Pairing 7: Identifying Patterns in Animal Behavior

Big Skill: Pattern Recognition – Identifying patterns in animal behavior and adaptations.

Small Skill: Matching & Sorting – Categorizing animals based on similar traits.

Teaching Technique: Role-playing.

Actions: Students pretend to be wildlife researchers, grouping animals based on observed patterns such as camouflage or migration habits.

Example: Students match animals that hibernate vs. migrate, then discuss why different species developed these survival strategies.

Pairing 8: Recognizing Cultural Patterns in Art & Symbols

Big Skill: Pattern Recognition – Recognizing repetition in cultural designs and symbols.

Small Skill: Matching & Sorting – Grouping artifacts based on shared artistic or symbolic features.

Teaching Technique: Cultural exploration.

Actions: Provide students with images of traditional patterns from different cultures and have them match and sort designs based on common features.

Example: Students compare geometric patterns from Islamic tile art and Native American textiles, questioning why repeating shapes are common in cultural expression.

Pairing 9: Identifying Storytelling Patterns Across Cultures

Big Skill: Pattern Recognition – Identifying repeated story structures and themes.

Small Skill: Matching & Sorting – Sorting books or stories based on shared narrative patterns.

Teaching Technique: Collaborative group work.

Actions: Assign groups different fables or fairy tales, guiding them to recognize and categorize recurring themes or morals.

Example: Students match classic fairy tales that share the "hero's journey" pattern and discuss why similar story arcs appear across cultures.

Pairing 10: Recognizing Structural Patterns in Architecture

Big Skill: Pattern Recognition – Finding repeated structures in architecture and city planning.

Small Skill: Matching & Sorting – Grouping buildings or layouts based on structural similarities.

Teaching Technique: Real-world applications.

Actions: Guide students to examine photos of famous buildings, sorting them by repeating patterns (e.g., arches, domes, columns) and discussing why these structures are used.

Example: Students analyze bridges with triangular supports vs. rectangular ones, discussing why engineers prefer triangles for stability.

Curiosity and Observation – (Big Skill) Pattern Recognition vs. (Small Skill) Step-by-Step Exploration

---

Pairing 1: Investigating Chemical Reactions Over Time

Big Skill: Pattern Recognition – Identifying repeated structures in numbers, nature, and language.

Small Skill: Step-by-Step Exploration – Learning to carefully observe and record findings (e.g., counting similar objects in nature).

Teaching Technique: Hands-on experiments.

Actions: Guide students in an experiment where they observe patterns in chemical reactions over time, recording their findings and analyzing emerging trends.

Example: Students drop different substances (sugar, salt, baking soda) into water, documenting how each dissolves step by step, and noting why some dissolve faster than others.

Pairing 2: Recognizing Fibonacci Sequences in Nature

Big Skill: Pattern Recognition – Observing recurring patterns in nature.

Small Skill: Step-by-Step Exploration – Tracking natural elements and recording similarities.

Teaching Technique: Guided exploration.

Actions: Take students outside to count and record repeated structures in plants or animals, leading them to analyze why these structures appear.

Example: Students count petals on different flowers, noticing that many have sets of 3, 5, or 8, and discuss why Fibonacci sequences appear in nature.

Pairing 3: Tracking Repeated Number Sequences in Math

Big Skill: Pattern Recognition – Identifying cycles and repeated sequences in math.

Small Skill: Step-by-Step Exploration – Solving problems in a structured way while documenting steps.

Teaching Technique: Problem-solving exercises.

Actions: Guide students through number sequences where they track repeating patterns and explore why they occur.

Example: Students document the multiplication table for 3 and observe the pattern in the "ones" place (3, 6, 9, 2, 5, 8, 1, 4, 7, 0).

Pairing 4: Exploring Symmetry in Art and Nature

Big Skill: Pattern Recognition – Recognizing symmetry in structures.

Small Skill: Step-by-Step Exploration – Analyzing and recording mirror-image reflections.

Teaching Technique: Art-based learning.

Actions: Have students create symmetrical drawings, recording how each shape corresponds to its mirror image and discussing why symmetry appears in design.

Example: Students fold paper in half, paint one side, and press it to create symmetrical butterfly prints, then analyze how each side matches.

Pairing 5: Discovering Rhythmic Patterns in Music

Big Skill: Pattern Recognition – Discovering rhythmic patterns in music.

Small Skill: Step-by-Step Exploration – Recording beats and tempo to identify patterns.

Teaching Technique: Sensory exploration.

Actions: Play different rhythms and have students document repeating patterns, analyzing why certain beats repeat in different types of music.

Example: Students clap along to a drumbeat, marking where patterns repeat and discussing why most songs follow a predictable structure.

Pairing 6: Identifying Patterns in Weather Cycles

Big Skill: Pattern Recognition – Finding patterns in weather and climate.

Small Skill: Step-by-Step Exploration – Recording daily observations and analyzing trends.

Teaching Technique: Observation journals.

Actions: Have students record weather conditions daily for a month and identify repeating trends, questioning why seasons affect temperature and precipitation.

Example: Students observe that mornings are often cooler than afternoons and investigate why the sun's position affects temperature.

Pairing 7: Tracking Life Cycles of Plants and Animals

Big Skill: Pattern Recognition – Examining life cycles of plants and animals.

Small Skill: Step-by-Step Exploration – Observing and tracking biological changes over time.

Teaching Technique: Role-playing.

Actions: Students pretend to be scientists tracking an animal's life stages, recording step-by-step observations.

Example: Students raise caterpillars, documenting each stage from larva to butterfly, then discuss why metamorphosis follows a set sequence.

Pairing 8: Identifying Patterns in Cultural Artifacts

Big Skill: Pattern Recognition – Finding repeated symbols and motifs in cultural artifacts.

Small Skill: Step-by-Step Exploration – Documenting details about different objects.

Teaching Technique: Cultural exploration.

Actions: Provide students with images of textiles, pottery, or architecture, guiding them to identify and record repeating patterns.

Example: Students examine Native American woven blankets, counting how often geometric shapes appear and discussing why those patterns hold cultural significance.

Pairing 9: Recognizing Historical Patterns in Events

Big Skill: Pattern Recognition – Observing historical trends.

Small Skill: Step-by-Step Exploration – Identifying and recording repeating events in history.

Teaching Technique: Collaborative group work.

Actions: Have students analyze historical timelines, tracking repeated themes or conflicts, and discussing why history often follows cycles.

Example: Students compare different economic recessions, noting similar causes and effects, then ask why financial cycles repeat over time.

Pairing 10: Identifying Architectural Patterns in Design

Big Skill: Pattern Recognition – Identifying repeating structures in architecture.

Small Skill: Step-by-Step Exploration – Noticing and recording patterns in design.

Teaching Technique: Real-world applications.

Actions: Take students on a virtual or real field trip to observe repeating patterns in buildings and record their observations.

Example: Students analyze famous bridges, recording repeating use of triangles for support, and discuss why triangles provide strength in engineering.

Curiosity and Observation – (Big Skill) Pattern Recognition vs. (Small Skill) Following Guided Observations

---

Pairing 1: Observing Chemical Reaction Patterns

Big Skill: Pattern Recognition – Identifying repeated structures in numbers, nature, and language.

Small Skill: Following Guided Observations – Completing teacher-led activities that highlight patterns (e.g., "Find all the circles in the classroom").

Teaching Technique: Hands-on experiments.

Actions: Guide students through an experiment where they observe and identify patterns in chemical reactions under teacher guidance.

Example: Students mix vinegar and baking soda with different amounts of each ingredient, observing and recording how reaction size follows a predictable pattern.

Pairing 2: Finding Spiral Patterns in Nature

Big Skill: Pattern Recognition – Observing repeated patterns in natural environments.

Small Skill: Following Guided Observations – Searching for and categorizing repeated shapes in the surroundings.

Teaching Technique: Guided exploration.

Actions: Lead students on an observation walk where they follow teacher instructions to find repeating natural patterns in trees, flowers, and rocks.

Example: Students identify and record spiral patterns in pinecones, shells, and flower petals, questioning why this pattern appears in different species.

Pairing 3: Recognizing Repeated Number Sequences

Big Skill: Pattern Recognition – Discovering patterns in numbers and mathematical sequences.

Small Skill: Following Guided Observations – Finding repeating numerical sequences with teacher guidance.

Teaching Technique: Problem-solving exercises.

Actions: Provide students with a guided number sequence activity where they analyze repeating patterns in multiplication, division, and Fibonacci numbers.

Example: Students complete a number grid, circling multiples of 2, 3, and 5, then observe and discuss overlapping patterns.

Pairing 4: Observing Symmetry in Objects and Designs

Big Skill: Pattern Recognition – Recognizing symmetry in objects and designs.

Small Skill: Following Guided Observations – Identifying symmetrical objects in a structured activity.

Teaching Technique: Art-based learning.

Actions: Have students follow a teacher-led exercise where they fold paper to create symmetrical shapes, then find real-world examples of symmetry.

Example: Students fold and cut paper snowflakes, then compare them to images of real snowflakes, noting how nature creates symmetrical patterns.

Pairing 5: Identifying Repeating Sounds in Music

Big Skill: Pattern Recognition – Finding repeated sounds and rhythms in music.

Small Skill: Following Guided Observations – Identifying and marking repeated beats in a structured listening exercise.

Teaching Technique: Sensory exploration.

Actions: Guide students through a musical exercise where they listen to and clap along with repeated rhythmic patterns, recording their observations.

Example: Students listen to a drumbeat, tapping when they recognize the repeated pattern, then discuss why certain beats repeat in different musical styles.

Pairing 6: Tracking Temperature Changes in Weather

Big Skill: Pattern Recognition – Identifying cycles in weather and climate.

Small Skill: Following Guided Observations – Tracking temperature and weather conditions under teacher guidance.

Teaching Technique: Observation journals.

Actions: Have students document daily weather conditions and follow teacher prompts to recognize seasonal patterns.

Example: Students track the temperature for two weeks, highlighting rising and falling trends, then discuss why mornings are cooler than afternoons.

Pairing 7: Recognizing Repeating Animal Behaviors

Big Skill: Pattern Recognition – Recognizing repeated behaviors in animals.

Small Skill: Following Guided Observations – Noticing and recording patterns in animal movement and interaction.

Teaching Technique: Role-playing.

Actions: Guide students through an activity where they observe animal videos and document repeated behaviors such as migration, hibernation, or camouflage.

Example: Students watch a time-lapse video of bird migration and note repeated flight formations, discussing why certain birds fly in V-shapes.

Pairing 8: Identifying Repeated Cultural Symbols

Big Skill: Pattern Recognition – Identifying repeating symbols in cultural designs.

Small Skill: Following Guided Observations – Finding and categorizing patterns in different cultural artworks.

Teaching Technique: Cultural exploration.

Actions: Provide students with guided worksheets featuring traditional textiles and architecture, asking them to locate and describe repeating patterns.

Example: Students analyze African, Islamic, and Native American geometric patterns, discussing why cultures use symmetry and repetition in their designs.

Pairing 9: Recognizing Common Themes in Stories

Big Skill: Pattern Recognition – Recognizing common themes and structures in literature.

Small Skill: Following Guided Observations – Identifying and categorizing repeated themes in stories.

Teaching Technique: Collaborative group work.

Actions: Assign groups different folk tales or fables and guide them to find repeating narrative structures such as the "rule of three" or "hero's journey."

Example: Students compare *The Three Little Pigs* and *Goldilocks and the Three Bears*, discussing why stories often feature patterns of three.

Pairing 10: Observing Architectural Patterns in Design

Big Skill: Pattern Recognition – Identifying repeated structures in architecture.

Small Skill: Following Guided Observations – Locating and classifying building patterns in a guided activity.

Teaching Technique: Real-world applications.

Actions: Take students on a virtual or real field trip to identify repeated architectural features such as columns, arches, or domes, discussing their purpose.

Example: Students examine Roman aqueducts and modern bridges, identifying repeating arch structures and discussing why arches provide strong support.

Curiosity and Observation – (Big Skill) Conceptual Understanding vs. (Small Skill) Matching & Sorting

---

Pairing 1: Understanding How Materials React to Heat

Big Skill: Conceptual Understanding – Seeing connections between observed patterns and larger ideas (e.g., "Why do the seasons change?").

Small Skill: Matching & Sorting – Grouping objects by color, shape, or function.

Teaching Technique: Hands-on experiments.

Actions: Guide students through an experiment where they categorize and match materials based on how they respond to heat, leading to a discussion on why different materials behave differently.

Example: Students test different materials (metal, plastic, wood) in sunlight, sorting them into "warms quickly" and "stays cool" groups, then discuss why metals heat up faster.

Pairing 2: Identifying Seasonal Changes in Leaves

Big Skill: Conceptual Understanding – Identifying connections between natural cycles and real-world changes.

Small Skill: Matching & Sorting – Sorting objects that follow predictable cycles.

Teaching Technique: Guided exploration.

Actions: Take students outside to collect and sort leaves based on seasonal changes, then discuss why some trees lose their leaves while others stay green.

Example: Students group leaves by color (green, yellow, red, brown) and analyze why color changes occur as seasons shift.

Pairing 3: Classifying Living and Non-Living Things

Big Skill: Conceptual Understanding – Recognizing how objects and organisms are classified.

Small Skill: Matching & Sorting – Categorizing living and non-living things.

Teaching Technique: Problem-solving exercises.

Actions: Provide students with a mixed set of objects and guide them in sorting based on whether they are living, once living, or non-living, explaining the reasoning behind the classifications.

Example: Students sort leaves, rocks, and insects, then discuss why some objects are alive and others are not, leading to the concept of life cycles.

Pairing 4: Understanding Structure & Function in Nature

Big Skill: Conceptual Understanding – Understanding how structure relates to function.

Small Skill: Matching & Sorting – Grouping tools or body parts based on function.

Teaching Technique: Role-playing.

Actions: Have students act as scientists sorting different tools or body structures based on function, then discuss why each object is designed the way it is.

Example: Students group different bird beaks by shape (hooked, long, flat) and analyze why some birds need sharp beaks for tearing while others need long beaks for probing flowers.

Pairing 5: Exploring How Materials Interact with Water

Big Skill: Conceptual Understanding – Exploring how different materials interact with the environment.

Small Skill: Matching & Sorting – Categorizing objects by how they react to water.

Teaching Technique: Observation journals.

Actions: Students test different materials (paper, plastic, sponge, metal) in water and record their observations, sorting materials into "absorbs water" and "repels water" categories.

Example: Students drop various objects into a bowl of water, documenting which ones soak up water and discussing why some materials are waterproof.

Pairing 6: Understanding Animal Adaptations & Habitats

Big Skill: Conceptual Understanding – Connecting animal adaptations to survival.

Small Skill: Matching & Sorting – Grouping animals by environment or diet.

Teaching Technique: Sensory exploration.

Actions: Guide students in sorting animals into different habitats and discussing why certain adaptations help them survive in those environments.

Example: Students classify animals into desert, rainforest, and polar regions, then analyze why thick fur benefits polar bears while webbed feet help ducks swim.

Pairing 7: Sorting Recyclable vs. Non-Recyclable Materials

Big Skill: Conceptual Understanding – Identifying the relationships between human activities and environmental impact.

Small Skill: Matching & Sorting – Sorting objects by recyclability.

Teaching Technique: Collaborative group work.

Actions: Have students sort classroom waste into recyclables, compostable materials, and trash, leading to a discussion on sustainability and why certain materials can be reused.

Example: Students separate items into "plastic recycling," "paper recycling," and "non-recyclable," discussing why plastics need special recycling processes.

Pairing 8: Recognizing Cultural Similarities Through Artifacts

Big Skill: Conceptual Understanding – Exploring similarities and differences in cultural traditions.

Small Skill: Matching & Sorting – Grouping artifacts by cultural origin.

Teaching Technique: Cultural exploration.

Actions: Provide students with images of traditional clothing, food, or celebrations from different cultures, guiding them to sort them into regional groups and analyze common themes.

Example: Students match lantern festivals to China, Diwali lamps to India, and Christmas lights to Western countries, discussing why lights are a shared symbol across cultures.

Pairing 9: Understanding Shapes in Architecture

Big Skill: Conceptual Understanding – Understanding why objects are shaped a certain way.

Small Skill: Matching & Sorting – Sorting geometric shapes based on real-world uses.

Teaching Technique: Art-based learning.

Actions: Guide students in identifying and sorting geometric shapes in everyday objects, leading to a discussion on why certain shapes are more common in structures.

Example: Students sort images of bridges, buildings, and tools into groups based on triangles, squares, and circles, discussing why triangles are used for support in architecture.

Pairing 10: Exploring Renewable vs. Non-Renewable Energy

Big Skill: Conceptual Understanding – Recognizing energy sources and their uses.

Small Skill: Matching & Sorting – Categorizing energy sources by renewable vs. nonrenewable.

Teaching Technique: Real-world applications.

Actions: Guide students through an activity where they classify different energy sources, then discuss why some are more sustainable than others.

Example: Students sort solar panels, wind turbines, coal, and gasoline into "renewable" and "nonrenewable" groups and discuss why solar, and wind energy are better long-term solutions.

Curiosity and Observation – (Big Skill) Conceptual Understanding vs. (Small Skill) Step-by-Step Exploration

---

Pairing 1: Exploring Temperature Patterns Throughout the Day

Big Skill: Conceptual Understanding – Seeing connections between observed patterns and larger ideas (e.g., "Why do the seasons change?").

Small Skill: Step-by-Step Exploration – Learning to carefully observe and record findings (e.g., counting similar objects in nature).

Teaching Technique: Hands-on experiments.

Actions: Guide students through an experiment where they track temperature changes throughout the day and analyze the relationship between sunlight and temperature.

Example: Students measure the temperature every hour and record patterns, discussing why mornings are cooler than afternoons due to the sun's position.

Pairing 2: Investigating Plant Growth Under Different Conditions

Big Skill: Conceptual Understanding – Connecting plant growth to environmental conditions.

Small Skill: Step-by-Step Exploration – Observing and recording plant changes over time.

Teaching Technique: Guided exploration.

Actions: Have students plant seeds and document daily growth, comparing growth rates under different conditions.

Example: Students grow plants in sunlight and shade, tracking height and leaf size, then discuss why sunlight is crucial for plant growth.

Pairing 3: Tracking Weather Patterns Over Time

Big Skill: Conceptual Understanding – Recognizing how weather patterns affect daily life.

Small Skill: Step-by-Step Exploration – Recording weather conditions and identifying trends.

Teaching Technique: Problem-solving exercises.

Actions: Have students keep a weather journal for two weeks, documenting daily temperature, cloud cover, and precipitation, and then analyze trends.

Example: Students track rainy and sunny days, then discuss why rain is more common in certain months based on seasonal changes.

Pairing 4: Observing Moon Phases and Their Impact on Tides

Big Skill: Conceptual Understanding – Understanding the phases of the moon and their effect on tides.

Small Skill: Step-by-Step Exploration – Tracking moon phases and tidal changes.

Teaching Technique: Observation journals.

Actions: Guide students to observe and sketch the moon each night for a month, recording how its shape changes and correlating with tidal charts.

Example: Students notice the full moon appears every 28 days and discuss why the moon's orbit around Earth creates visible phases.

Pairing 5: Documenting Animal Adaptations Over Time

Big Skill: Conceptual Understanding – Exploring how animals adapt to different environments.

Small Skill: Step-by-Step Exploration – Observing animal characteristics and behaviors over time.

Teaching Technique: Sensory exploration.

Actions: Have students observe and document different animal adaptations using photos, videos, or live observations at a park or zoo.

Example: Students track the behavior of squirrels over a week, recording when they gather food and discussing why they store food before winter.

Pairing 6: Analyzing Food Chains and Ecosystem Relationships

Big Skill: Conceptual Understanding – Understanding food chains and ecosystems.

Small Skill: Step-by-Step Exploration – Recording predator-prey relationships in nature.

Teaching Technique: Collaborative group work.

Actions: Assign students different ecosystems and guide them to research and document how energy flows through a food chain.

Example: Students research the ocean food chain, tracking how plankton feeds small fish, which are eaten by bigger fish, and discuss why energy transfers between species.

Pairing 7: Measuring Human Impact on the Environment

Big Skill: Conceptual Understanding – Examining how human activities affect the environment.

Small Skill: Step-by-Step Exploration – Measuring and tracking pollution levels.

Teaching Technique: Real-world applications.

Actions: Guide students in a project where they measure and record pollution in different areas (schoolyard, park, street), then analyze patterns.

Example: Students collect trash in different locations, documenting which areas have more litter, and discuss why urban areas tend to have more pollution than rural areas.

Pairing 8: Observing How Forces Like Gravity Affect Motion

Big Skill: Conceptual Understanding – Understanding how forces like gravity affect motion.

Small Skill: Step-by-Step Exploration – Observing and documenting motion changes under different conditions.

Teaching Technique: Hands-on experiments.

Actions: Have students roll balls down ramps of different heights and record distance traveled, discussing how gravity and friction impact motion.

Example: Students test how a ball rolls on smooth vs. rough surfaces and discuss why friction slows down movement.

Pairing 9: Investigating Light and Shadow Changes Throughout the Day

Big Skill: Conceptual Understanding – Investigating the effect of light on shadows.

Small Skill: Step-by-Step Exploration – Observing and recording shadow changes throughout the day.

Teaching Technique: Observation journals.

Actions: Guide students in tracking their shadow length at different times of the day, noting how the position of the sun affects the shadow's size.

Example: Students record shadow lengths in the morning, noon, and evening, then discuss why shadows are longest in the morning and evening due to the sun's angle.

Pairing 10: Recognizing Patterns in Language and Storytelling

Big Skill: Conceptual Understanding – Recognizing patterns in language and communication.

Small Skill: Step-by-Step Exploration – Identifying repeating word patterns in texts.

Teaching Technique: Cultural exploration.

Actions: Guide students in analyzing different folktales or poems, tracking repeated themes or phrases, and discussing why repetition helps storytelling.

Example: Students read fables from different cultures, identifying moral patterns, and discuss why many stories teach lessons through repeated ideas.

Curiosity and Observation – (Big Skill) Conceptual Understanding vs. (Small Skill) Following Guided Observations

---

Pairing 1: Observing Temperature Patterns Throughout the Day

Big Skill: Conceptual Understanding – Seeing connections between observed patterns and larger ideas (e.g., "Why do the seasons change?").

Small Skill: Following Guided Observations – Completing teacher-led activities that highlight patterns (e.g., "Find all the circles in the classroom").

Teaching Technique: Hands-on experiments.

Actions: Guide students through a temperature tracking activity where they observe and categorize temperature changes throughout the day to understand heat patterns.

Example: Students record the temperature at different times of the day and discuss why midday is the warmest due to the sun's highest position in the sky.

Pairing 2: Connecting Plant Growth to Environmental Conditions

Big Skill: Conceptual Understanding – Connecting plant life cycles to environmental conditions.

Small Skill: Following Guided Observations – Observing and recording plant changes under teacher guidance.

Teaching Technique: Guided exploration.

Actions: Lead students through an outdoor activity where they observe and classify plants based on size, color, and shape while discussing growth patterns.

Example: Students find different types of leaves and record whether they belong to young or mature plants, discussing why older leaves turn brown and fall.

Pairing 3: Observing Shadow Changes Throughout the Day

Big Skill: Conceptual Understanding – Understanding the connection between light and shadows.

Small Skill: Following Guided Observations – Identifying shadow patterns in a controlled setting.

Teaching Technique: Problem-solving exercises.

Actions: Have students observe and trace shadows at different times of the day, noting how their length changes.

Example: Students measure their shadow in the morning, noon, and afternoon, discussing why it is shortest at noon when the sun is directly overhead.

Pairing 4: Investigating How Materials Interact with Water

Big Skill: Conceptual Understanding – Exploring how different materials interact with water.

Small Skill: Following Guided Observations – Categorizing objects based on how they absorb or repel water.

Teaching Technique: Sensory exploration.

Actions: Guide students in testing various objects to see which absorb or repel water, organizing their findings into groups.

Example: Students dip paper, fabric, plastic, and metal into water, sorting them into "absorbent" and "waterproof" categories, discussing why umbrellas are made from waterproof materials.

Pairing 5: Identifying Weather Trends and Climate Patterns

Big Skill: Conceptual Understanding – Identifying weather patterns and climate trends.

Small Skill: Following Guided Observations – Tracking daily weather conditions under teacher guidance.

Teaching Technique: Observation journals.

Actions: Have students record and compare weather conditions over a two-week period to identify patterns in temperature, precipitation, and cloud cover.

Example: Students track daily temperatures and find that colder mornings often lead to sunny afternoons, discussing why the Earth warms up as the day progresses.

Pairing 6: Recognizing Patterns in Animal Behavior

Big Skill: Conceptual Understanding – Examining patterns in animal behavior.

Small Skill: Following Guided Observations – Recognizing repeated animal behaviors with teacher-led guidance.

Teaching Technique: Role-playing.

Actions: Guide students in observing videos of animals in the wild, recording common survival behaviors like migration, hibernation, or camouflage.

Example: Students watch a documentary on bird migration and identify patterns, discussing why birds fly south in winter and return in spring.

Pairing 7: Understanding Energy and Motion Through Experiments

Big Skill: Conceptual Understanding – Understanding the relationship between energy and motion.

Small Skill: Following Guided Observations – Observing and recording movement patterns under controlled conditions.

Teaching Technique: Hands-on experiments.

Actions: Have students test how different surfaces affect the speed of a rolling ball, recording observations and identifying patterns.

Example: Students roll a toy car down ramps covered in different materials (wood, sandpaper, carpet) and discuss why friction slows the car down on rough surfaces.

Pairing 8: Recognizing Mathematical Patterns in Nature

Big Skill: Conceptual Understanding – Recognizing mathematical patterns in nature.

Small Skill: Following Guided Observations – Counting and categorizing natural objects based on shapes and arrangements.

Teaching Technique: Collaborative group work.

Actions: Lead students in an activity where they search for repeated patterns in leaves, flowers, or shells, recording their observations in groups.

Example: Students count the number of petals on flowers and find that many have 3, 5, or 8 petals, discussing how Fibonacci numbers appear in nature.

Pairing 9: Identifying Repeated Themes in Historical Events

Big Skill: Conceptual Understanding – Connecting historical events to repeated social patterns.

Small Skill: Following Guided Observations – Identifying repeated themes in historical events with teacher guidance.

Teaching Technique: Cultural exploration.

Actions: Provide students with historical case studies where they compare different events and identify repeating causes and effects.

Example: Students compare the American and French Revolutions, noting that both involved citizens rebelling against unfair taxation, discussing why economic inequality often leads to revolutions.

Pairing 10: Exploring Geometric Structures in Architecture

Big Skill: Conceptual Understanding – Exploring geometric patterns in architecture.

Small Skill: Following Guided Observations – Finding and classifying repeated structures in buildings.

Teaching Technique: Real-world applications.

Actions: Have students examine different types of bridges or buildings, categorizing them based on repeated geometric shapes.

Example: Students analyze famous bridges and observe how triangles are commonly used for support, discussing why triangular structures provide stability.

Example Teaching Strategies for Big vs. Small Thinking Skill Pairings of Basic Problem-Solving at the Grade School Level

Here are 90 unique teaching strategy examples for the big thinking and small thinking skill pairings, each described in detail to align with **Big Thinking Skills** (Concepts & Strategies) and **Small Thinking Skills** (Step-by-Step Execution) for the **Basic Problem-Solving Skill** taught at the grade school level.

Basic Problem-Solving

Big Thinking Skills (Concepts & Strategies)

- **Logical Reasoning:** Understanding cause and effect relationships (e.g., "What happens if you do not water a plant?").
- **Strategic Thinking:** Choosing the best move in simple games like **Tic-Tac-Toe**.
- **Abstract Thinking:** Understanding that different solutions can solve the same problem.

Small Thinking Skills (Step-by-Step Execution)

- **Trial and Error:** Testing different puzzle solutions to see what works.
- **If-Then Reasoning:** Practicing "If I do X, Y will happen" with simple tasks.
- **Breaking Down a Problem:** Completing a puzzle by focusing on one piece at a time.

Basic Problem-Solving – (Big Skill) Logical Reasoning vs. (Small Skill) Trial and Error

---

Pairing 1: Experimenting with Plant Growth

Big Skill: Logical Reasoning – Understanding cause and effect relationships (e.g., "What happens if you do not water a plant?").

Small Skill: Trial and Error – Testing different puzzle solutions to see what works.

Teaching Technique: Hands-on experiments.

Actions: Guide students through an experiment where they test different methods for growing a plant, adjusting variables through trial and error.

Example: Students water one plant daily, one plant every few days, and one plant not at all, observing which grows best and discussing why water is essential for survival.

Pairing 2: Discovering Stability and Balance

Big Skill: Logical Reasoning – Exploring how balance affects stability.

Small Skill: Trial and Error – Adjusting different weight placements to find balance.

Teaching Technique: Guided exploration.

Actions: Have students build structures with blocks or cups, testing different stacking methods to see which one is most stable.

Example: Students stack books in different ways to find the most stable arrangement, learning that a wider base creates better balance.

Pairing 3: Exploring the Effects of Force on Movement

Big Skill: Logical Reasoning – Understanding how force affects movement.

Small Skill: Trial and Error – Testing different pushing and pulling techniques to move objects.

Teaching Technique: Problem-solving exercises.

Actions: Provide students with toy cars and ramps of different angles, guiding them to test how slope affects speed.

Example: Students test rolling cars on steep, medium, and flat ramps, discovering that a steeper slopes make the cars go faster due to gravity.

Pairing 4: Identifying Number Patterns

Big Skill: Logical Reasoning – Identifying patterns in number sequences.

Small Skill: Trial and Error – Testing different ways to find the missing number in a pattern.

Teaching Technique: Observation journals.

Actions: Have students work through math puzzles where they predict the next number in a sequence by trying different addition and multiplication rules.

Example: Students work with sequences like 2, 4, 8, __ and test whether the missing number follows an addition or multiplication pattern, discovering that it doubles each time.

Pairing 5: Constructing Electrical Circuits

Big Skill: Logical Reasoning – Recognizing how circuits work in electricity.

Small Skill: Trial and Error – Testing different circuit connections to light a bulb.

Teaching Technique: Hands-on experiments.

Actions: Provide students with wires, batteries, and bulbs, guiding them to connect circuits in different ways to find out what works.

Example: Students test different wire placements to see which complete a circuit, discovering that a full loop is needed for the bulb to light up.

Pairing 6: Experimenting with Simple Machines

Big Skill: Logical Reasoning – Understanding how simple machines reduce effort.

Small Skill: Trial and Error – Testing different simple machines to see which makes lifting easier.

Teaching Technique: Sensory exploration.

Actions: Have students use pulleys, levers, and inclined planes to move objects, recording which tool requires the least effort.

Example: Students lift a heavy book with and without a lever, realizing that the lever reduces the force needed.

Pairing 7: Rearranging Story Events

Big Skill: Logical Reasoning – Exploring cause and effect in storytelling.

Small Skill: Trial and Error – Rearranging story events to see what makes sense.

Teaching Technique: Collaborative group work.

Actions: Provide students with mixed-up story events and have them rearrange them, testing different sequences to find the most logical order.

Example: Students work with a scrambled fairy tale and try different sequences, discovering that a cause (losing a shoe) must come before effect (the prince finds Cinderella).

Pairing 8: Investigating Light and Transparency

Big Skill: Logical Reasoning – Understanding how light interacts with objects.

Small Skill: Trial and Error – Testing different materials to see which block or pass light.

Teaching Technique: Real-world applications.

Actions: Have students shine a flashlight through various materials (plastic wrap, paper, glass) to see which are transparent, translucent, or opaque.

Example: Students test how light passes through wax paper and glass, discovering that a transparent materials allow more light through.

Pairing 9: Testing How Sound Waves Travel

Big Skill: Logical Reasoning – Exploring how sound waves travel.

Small Skill: Trial and Error – Testing different materials to see which transmit sound best.

Teaching Technique: Cultural exploration.

Actions: Guide students in making simple string telephones and testing different string types to see which transmits sound most clearly.

Example: Students use yarn, fishing line, and metal wire to make telephones, discovering that a taut fishing line carries sound better than loose yarn.

Pairing 10: Observing Changes in Cooking Ingredients

Big Skill: Logical Reasoning – Understanding how food ingredients change when cooked.

Small Skill: Trial and Error – Testing different cooking methods to observe changes.

Teaching Technique: Hands-on experiments.

Actions: Have students compare what happens when they heat butter, freeze water, and bake dough, observing cause-and-effect transformations.

Example: Students melt chocolate at different temperatures, discovering that a lower heat melts it slowly while high heat burns it.

Basic Problem-Solving – (Big Skill) Logical Reasoning vs. (Small Skill) If-Then Reasoning

---

Pairing 1: Predicting Plant Growth Outcomes

Big Skill: Logical Reasoning – Understanding cause and effect relationships (e.g., "What happens if you do not water a plant?").

Small Skill: If-Then Reasoning – Practicing "If I do X, Y will happen" with simple tasks.

Teaching Technique: Hands-on experiments.

Actions: Guide students through an experiment where they control the amount of water given to plants, predicting what will happen in each case.

Example: Students predict, "If I water the plant daily, it will grow tall, but if I do not water it, it will wilt." They track results over two weeks.

Pairing 2: Exploring Force and Motion

Big Skill: Logical Reasoning – Exploring how force affects motion.

Small Skill: If-Then Reasoning – Testing different levels of force to see their effect on movement.

Teaching Technique: Guided exploration.

Actions: Provide students with toy cars and ramps of different heights, leading them to predict what will happen if the ramp angle is increased or decreased.

Example: Students hypothesize, "If I make the ramp steeper, the car will go faster," then test and record results.

Pairing 3: Understanding Electrical Circuits

Big Skill: Logical Reasoning – Understanding how circuits work.

Small Skill: If-Then Reasoning – Predicting what will happen when a circuit is opened or closed.

Teaching Technique: Problem-solving exercises.

Actions: Provide students with batteries, wires, and light bulbs, guiding them to predict what will happen if they connect or disconnect parts of the circuit.

Example: Students test, "If I disconnect the wire, the light will turn off," confirming their hypothesis by opening and closing the circuit.

Pairing 4: Recognizing Number Patterns

Big Skill: Logical Reasoning – Recognizing patterns in number sequences.

Small Skill: If-Then Reasoning – Predicting the next number in a sequence.

Teaching Technique: Observation journals.

Actions: Have students analyze number patterns and make predictions about the next numbers based on logical rules.

Example: Students observe the sequence 2, 4, 8, 16, and predict, "If I multiply the last number by 2, the next number will be 32."

Pairing 5: Exploring Temperature and State Changes

Big Skill: Logical Reasoning – Understanding the relationship between temperature and state changes.

Small Skill: If-Then Reasoning – Predicting how substances change with temperature.

Teaching Technique: Hands-on experiments.

Actions: Guide students through heating and cooling experiments where they predict and observe how different substances change states.

Example: Students predict, "If I freeze water, it will turn into ice, and if I heat ice, it will turn back into water," then test their predictions.

Pairing 6: Investigating Sound Transmission

Big Skill: Logical Reasoning – Identifying how sound waves travel.

Small Skill: If-Then Reasoning – Predicting how different materials affect sound transmission.

Teaching Technique: Sensory exploration.

Actions: Have students use string telephones with different materials and predict how sound quality will change.

Example: Students hypothesize, "If I use a tight string, the sound will be clearer," then test with loose and tight strings to observe differences.

Pairing 7: Exploring Light and Reflection

Big Skill: Logical Reasoning – Exploring how light interacts with objects.

Small Skill: If-Then Reasoning – Predicting whether light will pass through, reflect, or be absorbed.

Teaching Technique: Collaborative group work.

Actions: Provide students with materials like glass, paper, and aluminum foil, guiding them to predict how light will behave when shined on each.

Example: Students predict, "If I shine light on a mirror, it will reflect, but if I shine it on cardboard, it will not pass through."

Pairing 8: Investigating Simple Machines and Effort

Big Skill: Logical Reasoning – Understanding how levers reduce effort.

Small Skill: If-Then Reasoning – Predicting how placement affects force.

Teaching Technique: Real-world applications.

Actions: Have students test different placements of a fulcrum on a lever to see how it affects lifting effort.

Example: Students predict, "If I move the fulcrum closer to the weight, it will be easier to lift," then test and confirm.

Pairing 9: Predicting Story Outcomes

Big Skill: Logical Reasoning – Exploring cause and effect in storytelling.

Small Skill: If-Then Reasoning – Predicting logical outcomes based on story events.

Teaching Technique: Cultural exploration.

Actions: Guide students through cause-and-effect exercises where they predict character actions based on events in a story.

Example: Students read a fable and predict, "If the fox tricks the crow, he will get the cheese," then discuss the lesson learned.

Pairing 10: Understanding Friction and Movement

Big Skill: Logical Reasoning – Understanding how friction affects movement.

Small Skill: If-Then Reasoning – Predicting how different surfaces affect speed.

Teaching Technique: Hands-on experiments.

Actions: Have students slide objects on smooth and rough surfaces, making predictions about how friction will change movement.

Example: Students predict, "If I slide the toy car on carpet, it will go slower than on tile," then test and analyze the results.

Basic Problem-Solving – (Big Skill) Logical Reasoning vs. (Small Skill) Breaking Down a Problem

---

Pairing 1: Analyzing Plant Growth Factors

Big Skill: Logical Reasoning – Understanding cause and effect relationships (e.g., "What happens if you do not water a plant?").

Small Skill: Breaking Down a Problem – Completing a puzzle by focusing on one piece at a time.

Teaching Technique: Hands-on experiments.

Actions: Guide students through a step-by-step plant growth experiment where they change one factor at a time (water, sunlight, soil) to analyze how it affects plant health.

Example: Students test, "If I remove sunlight but still water the plant, what happens?" They record that plants become weak and yellow without sunlight.

Pairing 2: Understanding Simple Machines

Big Skill: Logical Reasoning – Understanding how simple machines make work easier.

Small Skill: Breaking Down a Problem – Testing one machine part at a time to analyze its function.

Teaching Technique: Guided exploration.

Actions: Have students test different simple machines (lever, pulley, inclined plane) separately before combining them into a compound machine.

Example: Students first lift a small object with a pulley, then add an inclined plane and compare which method requires less force.

Pairing 3: Sequencing Story Events

Big Skill: Logical Reasoning – Understanding how stories follow a sequence.

Small Skill: Breaking Down a Problem – Organizing a story by identifying beginning, middle, and end.

Teaching Technique: Problem-solving exercises.

Actions: Provide students with mixed-up story events and have them sort them into logical order by identifying key cause-and-effect moments.

Example: Students arrange *The Three Little Pigs* events by first recognizing that the weak houses were blown down before the brick house stayed standing.

Pairing 4: Troubleshooting Electrical Circuits

Big Skill: Logical Reasoning – Understanding how electrical circuits work.

Small Skill: Breaking Down a Problem – Identifying one faulty connection at a time.

Teaching Technique: Observation journals.

Actions: Guide students through testing an incomplete circuit, troubleshooting step by step to determine which part prevents the bulb from lighting up.

Example: Students test, "If I replace the wire but the bulb still does not light, what should I check next?" They continue testing until they find the missing battery connection.

Pairing 5: Identifying Weather Patterns

Big Skill: Logical Reasoning – Understanding how weather changes over time.

Small Skill: Breaking Down a Problem – Tracking daily weather patterns one day at a time before identifying trends.

Teaching Technique: Hands-on experiments.

Actions: Have students record daily temperatures, cloud cover, and wind speed, then analyze how weather patterns form over a week.

Example: Students observe, "If it is cloudy today and getting colder, what might happen tomorrow?" and predict rain based on previous observations.

Pairing 6: Solving Math Word Problems

Big Skill: Logical Reasoning – Understanding how math word problems are solved step by step.

Small Skill: Breaking Down a Problem – Solving one mathematical operation at a time before completing the full equation.

Teaching Technique: Sensory exploration.

Actions: Guide students to highlight key numbers in a math problem, solving the first operation before moving on to the next.

Example: For a word problem about sharing apples, students first calculate total apples, then divide them among friends, realizing that breaking it down makes solving easier.

Pairing 7: Understanding Human Body Systems

Big Skill: Logical Reasoning – Understanding how human body systems work together.

Small Skill: Breaking Down a Problem – Studying one body system at a time before connecting them.

Teaching Technique: Collaborative group work.

Actions: Assign small groups different body systems (digestive, respiratory, circulatory) and have them explain how each works before linking them together.

Example: Students learn, "If the lungs provide oxygen to the blood, and the heart pumps blood through the body, how do they work together?"

Pairing 8: Investigating Physical Forces

Big Skill: Logical Reasoning – Understanding how physical forces affect movement.

Small Skill: Breaking Down a Problem – Testing one force at a time before analyzing combined effects.

Teaching Technique: Real-world applications.

Actions: Have students roll a toy car on different surfaces (smooth, rough, inclined), testing friction and gravity separately before analyzing them together.

Example: Students observe, "If I roll the car on sandpaper, it slows down. Why does friction make movement harder?"

Pairing 9: Studying Ecosystem Balance

Big Skill: Logical Reasoning – Understanding how ecosystems maintain balance.

Small Skill: Breaking Down a Problem – Studying one part of the ecosystem at a time before seeing the full interaction.

Teaching Technique: Cultural exploration.

Actions: Have students study food chains step by step, identifying how energy moves from plants to herbivores, then to predators.

Example: Students analyze, "If we remove all the rabbits from the ecosystem, what will happen to the fox population?"

Pairing 10: Testing Construction Materials for Strength

Big Skill: Logical Reasoning – Understanding how construction materials affect building strength.

Small Skill: Breaking Down a Problem – Testing different materials before constructing a full structure.

Teaching Technique: Hands-on experiments.

Actions: Have students build a paper bridge, testing folded, layered, and stacked paper separately before combining methods.

Example: Students realize, "If I fold the paper into a triangle, it supports more weight than a flat sheet," and apply it to their final bridge design.

Basic Problem-Solving – (Big Skill) Strategic Thinking vs. (Small Skill) Trial and Error

---

Pairing 1: Optimizing Moves in Tic-Tac-Toe

Big Skill: Strategic Thinking – Choosing the best move in simple games like Tic-Tac-Toe.

Small Skill: Trial and Error – Testing different puzzle solutions to see what works.

Teaching Technique: Hands-on experiments.

Actions: Guide students to play Tic-Tac-Toe multiple times, trying different first moves and tracking the outcomes to determine the best strategy.

Example: Students test, "If I always start in the center, do I win more often than when I start in a corner?"

Pairing 2: Navigating Mazes Efficiently

Big Skill: Strategic Thinking – Planning the best approach in mazes and logic puzzles.

Small Skill: Trial and Error – Testing different paths until finding the correct one.

Teaching Technique: Guided exploration.

Actions: Have students navigate a printed or digital maze, trying different routes and marking dead ends to learn from mistakes.

Example: Students start solving a maze, then realize, "If I hit a dead end, I should backtrack and try another path."

Pairing 3: Winning Strategies in Connect Four

Big Skill: Strategic Thinking – Understanding patterns in board games like Connect Four.

Small Skill: Trial and Error – Trying different placements to test which creates a winning pattern.

Teaching Technique: Problem-solving exercises.

Actions: Have students play Connect Four, testing various placements of pieces and adjusting strategies based on previous losses.

Example: Students realize, "If I build diagonally instead of stacking, I have more ways to win."

Pairing 4: Mastering Moves in Checkers

Big Skill: Strategic Thinking – Planning moves in checkers to gain an advantage.

Small Skill: Trial and Error – Trying different moves to test their effectiveness.

Teaching Technique: Observation journals.

Actions: Students record the outcomes of different checker moves and analyze which strategies lead to capturing the most pieces.

Example: Students note, "If I sacrifice one piece early, I can set up a double jump later."

Pairing 5: Predicting Opponent Behavior in Rock-Paper-Scissors

Big Skill: Strategic Thinking – Choosing the best strategy in rock-paper-scissors.

Small Skill: Trial and Error – Testing different patterns to predict an opponent's move.

Teaching Technique: Hands-on experiments.

Actions: Have students play multiple rounds of rock-paper-scissors and track patterns in opponents' choices to adjust their own strategy.

Example: Students observe, "If my friend picks rock three times in a row, they might switch to scissors next."

Pairing 6: Solving Sudoku Through Experimentation

Big Skill: Strategic Thinking – Planning ahead in Sudoku or number puzzles.

Small Skill: Trial and Error – Filling in possible answers and adjusting based on errors.

Teaching Technique: Sensory exploration.

Actions: Guide students through solving a Sudoku puzzle by testing possible numbers in empty spaces and adjusting based on conflicts.

Example: Students try placing a 5 in a box, then realize, "If there is already a 5 in the row, I need to try another number."

Pairing 7: Maximizing Space in Tetris

Big Skill: Strategic Thinking – Deciding where to place blocks in Tetris.

Small Skill: Trial and Error – Testing different placements to see which clears the most rows.

Teaching Technique: Collaborative group work.

Actions: Have students play Tetris and compare different block placements, discussing which strategies maximize space.

Example: Students realize, "If I leave one open column, I can drop a long piece to clear four rows at once."

Pairing 8: Strengthening Memory in Matching Games

Big Skill: Strategic Thinking – Choosing the best move in memory-matching games.

Small Skill: Trial and Error – Testing different card choices until a match is found.

Teaching Technique: Real-world applications.

Actions: Have students play a memory card game, focusing on recalling previous card locations and refining strategies based on past attempts.

Example: Students remember, "If I flipped a cat card earlier, I should try to find it instead of randomly picking a new card."

Pairing 9: Debugging in Coding Challenges

Big Skill: Strategic Thinking – Planning moves in a basic coding challenge.

Small Skill: Trial and Error – Testing different code sequences to see what works.

Teaching Technique: Cultural exploration.

Actions: Have students use block-based coding (e.g., Scratch) to program a simple movement and adjust their commands after testing.

Example: Students program a sprite to move forward and turn, realizing, "If I change the turn angle, I can make it follow a square path."

Pairing 10: Stable Engineering Structures

Big Skill: Strategic Thinking – Deciding the best way to stack objects for stability.

Small Skill: Trial and Error – Testing different stacking methods to find the most stable structure.

Teaching Technique: Hands-on experiments.

Actions: Have students build a tower using blocks or cups, testing different arrangements and adjusting based on collapses.

Example: Students observe, "If I place heavier blocks on the bottom, my tower stays up longer."

Basic Problem-Solving – (Big Skill) Strategic Thinking vs. (Small Skill) If-Then Reasoning

---

Pairing 1: Predicting Opponent Reactions in Tic-Tac-Toe

Big Skill: Strategic Thinking – Choosing the best move in simple games like Tic-Tac-Toe.

Small Skill: If-Then Reasoning – Practicing "If I do X, Y will happen" with simple tasks.

Teaching Technique: Hands-on experiments.

Actions: Have students play Tic-Tac-Toe and pause after each move to predict what will happen next based on different possible moves.

Example: Students reason, "If I place my X in this square, my opponent will block me next, so I should place it here instead."

Pairing 2: Planning Moves in Checkers

Big Skill: Strategic Thinking – Planning ahead in a simple board game like Checkers.

Small Skill: If-Then Reasoning – Predicting possible outcomes of each move.

Teaching Technique: Guided exploration.

Actions: Have students play checkers while verbalizing their reasoning before each move, predicting what their opponent will do in response.

Example: A student says, "If I move this piece here, my opponent can jump me, so I should move another piece instead."

Pairing 3: Remembering and Predicting in a Memory Game

Big Skill: Strategic Thinking – Making decisions in a memory-matching card game.

Small Skill: If-Then Reasoning – Predicting what will happen based on past choices.

Teaching Technique: Problem-solving exercises.

Actions: Guide students to track where certain cards are placed and reason through their moves.

Example: A student says, "If I flip this card and it does not match, I will remember its location for my next turn."

Pairing 4: Choosing the Best Path in a Maze

Big Skill: Strategic Thinking – Choosing the best way to complete a maze.

Small Skill: If-Then Reasoning – Testing different paths and predicting outcomes.

Teaching Technique: Observation journals.

Actions: Have students solve a printed maze while marking dead-end paths to analyze their choices.

Example: A student realizes, "If I go left, I will hit a dead end, so I should try going right instead."

Pairing 5: Recognizing Opponent Patterns in Rock-Paper-Scissors

Big Skill: Strategic Thinking – Choosing the best move in rock-paper-scissors.

Small Skill: If-Then Reasoning – Predicting the opponent's next move based on patterns.

Teaching Technique: Hands-on experiments.

Actions: Guide students to track patterns in an opponent's choices and adjust their strategy accordingly.

Example: A student observes, "If my opponent has chosen rock twice, they might switch to scissors next, so I should pick rock."

Pairing 6: Balancing a Jenga Tower

Big Skill: Strategic Thinking – Deciding where to place blocks in a stacking game like Jenga.

Small Skill: If-Then Reasoning – Predicting which blocks will keep the structure stable.

Teaching Technique: Sensory exploration.

Actions: Have students test different blocks, reasoning through which will keep the tower balanced.

Example: A student says, "If I remove this middle block, the tower will stay stable, but if I remove this side block, it might fall."

Pairing 7: Predicting Puzzle Outcomes in Sudoku

Big Skill: Strategic Thinking – Choosing the best number in Sudoku or a similar puzzle.

Small Skill: If-Then Reasoning – Predicting how placing a number will affect the puzzle's completion.

Teaching Technique: Collaborative group work.

Actions: Have students complete a Sudoku puzzle, testing different number placements to see how they affect the overall solution.

Example: A student reasons, "If I place a 4 here, it will block my chance to use 4 in the next row, so I should place a different number instead."

Pairing 8: Anticipating Outcomes in a Coding Activity

Big Skill: Strategic Thinking – Planning moves in a basic coding activity.

Small Skill: If-Then Reasoning – Predicting what will happen if a certain command is used.

Teaching Technique: Real-world applications.

Actions: Have students use a block-based coding program like Scratch to make a character move, adjusting commands based on predictions.

Example: A student predicts, "If I add a 'turn 90 degrees' block, my character will face right."

Pairing 9: Engineering a Stable Structure

Big Skill: Strategic Thinking – Deciding how to build the strongest tower using building blocks.

Small Skill: If-Then Reasoning – Predicting how different block placements will affect the tower's stability.

Teaching Technique: Cultural exploration.

Actions: Have students construct towers while testing different designs to see which is most stable.

Example: A student predicts, "If I stack smaller blocks on the bottom, my tower will be wobbly, but if I use larger blocks first, it will be stronger."

Pairing 10: Decision-Making in Role-Playing Games

Big Skill: Strategic Thinking – Making the best decision in a role-playing game.

Small Skill: If-Then Reasoning – Predicting how a character's choice will affect the outcome.

Teaching Technique: Hands-on experiments.

Actions: Have students role-play different choices in a game where they must navigate a fictional world, predicting the consequences of each action.

Example: A student predicts, "If my character takes the short path, they might run into trouble, but if I take the long path, I will stay safe."

Basic Problem-Solving – (Big Skill) Strategic Thinking vs. (Small Skill) Breaking Down a Problem

---

Pairing 1: Analyzing Moves in Tic-Tac-Toe

Big Skill: Strategic Thinking – Choosing the best move in simple games like Tic-Tac-Toe.

Small Skill: Breaking Down a Problem – Completing a puzzle by focusing on one piece at a time.

Teaching Technique: Hands-on experiments.

Actions: Guide students to play Tic-Tac-Toe by analyzing one possible move at a time before choosing the best strategy.

Example: A student thinks, "If I go here, my opponent can block me, so I should look at another move before deciding."

Pairing 2: Setting Up Future Moves in Checkers

Big Skill: Strategic Thinking – Planning ahead in Checkers to set up future moves.

Small Skill: Breaking Down a Problem – Focusing on one piece movement at a time to build a winning strategy.

Teaching Technique: Guided exploration.

Actions: Have students plan one move at a time, predicting how each move will set up their next action.

Example: A student analyzes, "If I jump this piece now, my opponent will capture me next turn. I should wait and move another piece instead."

Pairing 3: Memory Strategy in a Card Game

Big Skill: Strategic Thinking – Choosing the best strategy in a memory card game.

Small Skill: Breaking Down a Problem – Focusing on remembering one card match at a time.

Teaching Technique: Problem-solving exercises.

Actions: Guide students to focus on one section of the memory game board, remembering key cards before moving to another section.

Example: A student says, "I will focus on the top-left corner first, matching pairs there before looking at the rest of the board."

Pairing 4: Solving a Maze Step by Step

Big Skill: Strategic Thinking – Planning the best way through a maze.

Small Skill: Breaking Down a Problem – Solving one section of the maze at a time instead of looking at the entire puzzle.

Teaching Technique: Observation journals.

Actions: Have students solve a maze by marking small sections first instead of trying to solve the entire thing at once.

Example: A student realizes, "If I work from the finish backward, I can connect the correct paths one by one."

Pairing 5: Evaluating Moves in Connect Four

Big Skill: Strategic Thinking – Choosing the best move in Connect Four.

Small Skill: Breaking Down a Problem – Identifying one open space at a time and evaluating its impact.

Teaching Technique: Hands-on experiments.

Actions: Guide students to analyze potential winning moves by looking at available spaces one at a time.

Example: A student considers, "If I place my piece here, I will have two in a row, but if I place it here, I can block my opponent."

Pairing 6: Testing Stability in Jenga

Big Skill: Strategic Thinking – Planning the best stacking method in Jenga.

Small Skill: Breaking Down a Problem – Testing one block removal at a time before choosing the safest move.

Teaching Technique: Sensory exploration.

Actions: Have students gently test different blocks before deciding which one to remove.

Example: A student taps three blocks and says, "This one moves the easiest, so I should take it out first instead of guessing randomly."

Pairing 7: Step-by-Step Approach in Sudoku

Big Skill: Strategic Thinking – Selecting the best number in Sudoku.

Small Skill: Breaking Down a Problem – Filling in one number at a time instead of guessing.

Teaching Technique: Collaborative group work.

Actions: Guide students to focus on one empty space, checking all possible numbers before making a choice.

Example: A student considers, "This box can only have a 3 or 7, but since there is already a 7 in this row, it must be a 3."

Pairing 8: Debugging Code in a Block-Based Game

Big Skill: Strategic Thinking – Choosing the best coding sequence in a block-based coding game.

Small Skill: Breaking Down a Problem – Testing one block of code at a time before adding more steps.

Teaching Technique: Real-world applications.

Actions: Have students build a program by testing each block individually before connecting them.

Example: A student says, "If I make my character move forward first, I can then add a jump block to avoid the obstacle."

Pairing 9: Constructing a Stable Structure

Big Skill: Strategic Thinking – Planning the strongest structure in a building challenge.

Small Skill: Breaking Down a Problem – Testing one design concept at a time before building a complete tower.

Teaching Technique: Cultural exploration.

Actions: Guide students to build a structure by testing different support styles one at a time before committing to a final design.

Example: A student observes, "If I use triangles for the base, my structure is more stable, so I will keep building with this shape."

Pairing 10: Decision-Making in a Role-Playing Game

Big Skill: Strategic Thinking – Deciding the best sequence of moves in a role-playing adventure game.

Small Skill: Breaking Down a Problem – Evaluating one decision at a time before making a final choice.

Teaching Technique: Hands-on experiments.

Actions: Have students make small decisions in a role-playing game, testing different outcomes before committing to a full strategy.

Example: A student tests, "If I talk to the villager first instead of going into the cave, I might get more clues about my quest."

Basic Problem-Solving – (Big Skill) Abstract Thinking vs. (Small Skill) Trial and Error

---

Pairing 1: Exploring Multiple Solutions in a Bridge-Building Challenge

Big Skill: Abstract Thinking – Understanding that different solutions can solve the same problem.

Small Skill: Trial and Error – Testing different puzzle solutions to see what works.

Teaching Technique: Hands-on experiments.

Actions: Guide students through a bridge-building challenge using different materials (paper, straws, blocks) to find multiple solutions.

Example: A student tests, "If I fold the paper, it holds more weight, but if I use straws, I can build a stronger bridge in a different way."

Pairing 2: Finding Multiple Paths in a Maze

Big Skill: Abstract Thinking – Finding multiple ways to reach a goal in a maze.

Small Skill: Trial and Error – Testing different paths until finding the best one.

Teaching Technique: Guided exploration.

Actions: Have students navigate a printed maze using different approaches (starting from the finish, testing side paths) and comparing methods.

Example: A student realizes, "If I start at the end and work backward, I can solve the maze faster than guessing random paths."

Pairing 3: Testing Strategies in Tic-Tac-Toe

Big Skill: Abstract Thinking – Understanding different strategies in Tic-Tac-Toe.

Small Skill: Trial and Error – Testing different starting moves to see which works best.

Teaching Technique: Problem-solving exercises.

Actions: Guide students to play Tic-Tac-Toe multiple times, testing different first moves and tracking their success.

Example: A student observes, "If I start in the center, I have more chances to win than if I start in a corner."

Pairing 4: Exploring Multiple Solutions in Math Problems

Big Skill: Abstract Thinking – Exploring different ways to solve a math problem.

Small Skill: Trial and Error – Testing different methods (addition, subtraction, multiplication) to find a solution.

Teaching Technique: Observation journals.

Actions: Have students solve a math problem using multiple methods and discuss which is most efficient.

Example: A student solves $48 \div 6$ by testing repeated subtraction, fact families, and long division, concluding that division is fastest.

Pairing 5: Experimenting with Building Structures

Big Skill: Abstract Thinking – Understanding multiple ways to build a structure.

Small Skill: Trial and Error – Testing different building techniques to find the strongest one.

Teaching Technique: Hands-on experiments.

Actions: Guide students to build towers using different materials (LEGO, blocks, sticks) and compare which design is most stable.

Example: A student notices, "If I build a wide base, my tower is stronger than if I build straight up."

Pairing 6: Finding Balance in a Scale Experiment

Big Skill: Abstract Thinking – Finding different ways to balance weight in a scale.

Small Skill: Trial and Error – Placing different object combinations to find balance.

Teaching Technique: Sensory exploration.

Actions: Have students test different object placements on a balance scale to make both sides equal.

Example: A student tests, "If I put two small weights instead of one big weight, the scale still balances."

Pairing 7: Testing Movement Sequences in a Robot Challenge

Big Skill: Abstract Thinking – Using different ways to program a simple robot.

Small Skill: Trial and Error – Testing different movement sequences to reach a goal.

Teaching Technique: Collaborative group work.

Actions: Have students program a robot to move through a maze, trying different sequences to see what works best.

Example: A student programs, "If I make my robot turn left, it gets stuck. I will try turning right instead."

Pairing 8: Mixing Paints to Create Colors in Art

Big Skill: Abstract Thinking – Exploring different ways to make a color in art.

Small Skill: Trial and Error – Mixing different paints to achieve the same shade.

Teaching Technique: Real-world applications.

Actions: Have students mix paints, testing different amounts of red, yellow, and blue to make brown.

Example: A student finds, "If I mix blue and orange, I can also get brown, just like mixing red, yellow, and black."

Pairing 9: Comparing Strategies in Checkers

Big Skill: Abstract Thinking – Understanding different ways to win a game like Checkers.

Small Skill: Trial and Error – Trying different strategies to see which is most effective.

Teaching Technique: Cultural exploration.

Actions: Guide students to play Checkers using aggressive (quick jumps) and defensive (blocking) strategies and compare success rates.

Example: A student reflects, "If I jump too soon, I lose my piece, but if I wait and block my opponent, I can control the board."

Pairing 10: Rearranging Letters in a Word Puzzle

Big Skill: Abstract Thinking – Understanding multiple ways to solve a word puzzle.

Small Skill: Trial and Error – Testing different letter placements to form words.

Teaching Technique: Hands-on experiments.

Actions: Have students rearrange letter tiles to find multiple correct words using the same letters.

Example: A student realizes, "If I rearrange these letters, I can spell 'star' and 'rats'—both are correct answers!"

Basic Problem-Solving – (Big Skill) Abstract Thinking vs. (Small Skill) If-Then Reasoning

Pairing 1: Exploring Structural Stability in Bridge-Building

Big Skill: Abstract Thinking – Understanding that different solutions can solve the same problem.

Small Skill: If-Then Reasoning – Practicing "If I do X, Y will happen" with simple tasks.

Teaching Technique: Hands-on experiments.

Actions: Guide students through a building challenge where they test multiple ways to create a bridge using different materials and predict outcomes.

Example: A student tests, "If I use paper folded into a triangle shape, it will hold more weight, but if I use stacked blocks, it will be more stable."

Pairing 2: Strategy Planning in Tic-Tac-Toe

Big Skill: Abstract Thinking – Exploring different strategies in Tic-Tac-Toe.

Small Skill: If-Then Reasoning – Predicting what will happen based on different move choices.

Teaching Technique: Guided exploration.

Actions: Have students play Tic-Tac-Toe, pausing to analyze and predict possible outcomes before making a move.

Example: A student reasons, "If I place my X here, my opponent will block me, but if I place it there, I will win in the next turn."

Pairing 3: Predicting Outcomes in Math Problem Solving

Big Skill: Abstract Thinking – Finding different ways to solve a math problem.

Small Skill: If-Then Reasoning – Predicting the outcome when different math operations are applied.

Teaching Technique: Problem-solving exercises.

Actions: Have students solve the same math problem using multiple methods and predict if they will get the same answer.

Example: A student tests, "If I use multiplication instead of repeated addition, I will still get the same total, but it will be faster."

Pairing 4: Predicting Strategies in Puzzle Solving

Big Skill: Abstract Thinking – Understanding different ways to complete a puzzle.

Small Skill: If-Then Reasoning – Testing different puzzle-solving strategies and predicting results.

Teaching Technique: Observation journals.

Actions: Have students complete a jigsaw puzzle using different strategies (edges first vs. grouping by color) and predict which will be faster.

Example: A student records, "If I start with the corners, I can complete the frame first, but if I group colors, I can fill in sections faster."

Pairing 5: Exploring Multiple Paths in a Maze Challenge

Big Skill: Abstract Thinking – Exploring multiple ways to reach a goal in a maze.

Small Skill: If-Then Reasoning – Predicting the outcome of choosing different paths.

Teaching Technique: Hands-on experiments.

Actions: Guide students through different methods for solving a maze, predicting where each path will lead.

Example: A student thinks, "If I turn left, I might hit a dead end, but if I turn right, I can get to the exit faster."

Pairing 6: Understanding Weight Distribution in Balancing Experiments

Big Skill: Abstract Thinking – Discovering different ways to balance objects.

Small Skill: If-Then Reasoning – Predicting how weight distribution will affect stability.

Teaching Technique: Sensory exploration.

Actions: Have students test different ways to balance an object (e.g., placing blocks on different parts of a scale) and predict what will happen.

Example: A student predicts, "If I move this weight to the center, the scale will balance, but if I move it too far to one side, it will tip over."

Pairing 7: Mapping and Route Optimization

Big Skill: Abstract Thinking – Using different routes to reach the same destination.

Small Skill: If-Then Reasoning – Predicting which route will be fastest.

Teaching Technique: Collaborative group work.

Actions: Guide students through a mapping activity where they analyze different routes to a location and predict which will be shortest.

Example: A student reasons, "If I take the straight path, it will be the shortest, but if I take the curved path, I might avoid obstacles."

Pairing 8: Predicting Color Mixing in Art

Big Skill: Abstract Thinking – Exploring multiple ways to create a color.

Small Skill: If-Then Reasoning – Testing different paint mixtures to predict color results.

Teaching Technique: Real-world applications.

Actions: Have students mix different paint combinations, predicting which ones will create the same shade.

Example: A student predicts, "If I mix red and yellow, I will get orange, but if I mix yellow and a little red, I will get light orange."

Pairing 9: Predicting Strategy Outcomes in Checkers

Big Skill: Abstract Thinking – Understanding different winning strategies in Checkers.

Small Skill: If-Then Reasoning – Predicting the consequences of different move choices.

Teaching Technique: Cultural exploration.

Actions: Have students play Checkers while pausing to predict what will happen with different move options.

Example: A student reasons, "If I move this piece, I will lose it, but if I wait and move another piece, I can set up a double jump."

Pairing 10: Predicting Robot Behavior in Coding

Big Skill: Abstract Thinking – Exploring different ways to program a simple robot.

Small Skill: If-Then Reasoning – Predicting what will happen based on different coding commands.

Teaching Technique: Hands-on experiments.

Actions: Have students program a robot using block-based coding and predict what will happen when different commands are used.

Example: A student predicts, "If I make the robot move forward and turn left, it will follow a square path, but if I remove the turn, it will keep moving straight."

Basic Problem-Solving – (Big Skill) Abstract Thinking vs. (Small Skill) Breaking Down a Problem

---

Pairing 1: Strategic Approaches to Jigsaw Puzzles

Big Skill: Abstract Thinking – Understanding that different solutions can solve the same problem.

Small Skill: Breaking Down a Problem – Completing a puzzle by focusing on one piece at a time.

Teaching Technique: Hands-on experiments.

Actions: Guide students in solving a jigsaw puzzle using different strategies (edges first vs. sorting by color) and focusing on one piece at a time.

Example: A student tests, "If I start with the corner pieces, I can build the frame first, but if I sort pieces by color, I can focus on sections."

Pairing 2: Multi-Step Problem Solving in Math

Big Skill: Abstract Thinking – Finding different ways to solve a math equation.

Small Skill: Breaking Down a Problem – Solving one step of the equation at a time.

Teaching Technique: Guided exploration.

Actions: Have students solve multi-step math problems by first identifying which operation to complete before moving on to the next step.

Example: A student works through $(5 + 3) \times 2$ by first solving $5 + 3 = 8$, then multiplying $8 \times 2 = 16$ instead of trying to solve everything at once.

Pairing 3: Stepwise Navigation Through a Maze

Big Skill: Abstract Thinking – Recognizing multiple strategies in a maze challenge.

Small Skill: Breaking Down a Problem – Solving one section of the maze before moving forward.

Teaching Technique: Problem-solving exercises.

Actions: Have students work through a maze, stopping after each section to analyze the best next move rather than trying to solve it all at once.

Example: A student realizes, "If I solve the left side first, I can eliminate dead ends and find a path forward more easily."

Pairing 4: Evaluating Moves in Tic-Tac-Toe

Big Skill: Abstract Thinking – Discovering multiple ways to win in Tic-Tac-Toe.

Small Skill: Breaking Down a Problem – Analyzing each move before making a final decision.

Teaching Technique: Observation journals.

Actions: Have students track different Tic-Tac-Toe games, writing down strategies and focusing on one move at a time before committing to a winning strategy.

Example: A student observes, "If I place my X in the middle, I have the most options, but if I start in a corner, I can set up a diagonal win."

Pairing 5: Incremental Color Mixing in Art

Big Skill: Abstract Thinking – Understanding different ways to create the same color in art.

Small Skill: Breaking Down a Problem – Mixing one color at a time to observe changes before continuing.

Teaching Technique: Hands-on experiments.

Actions: Have students mix paints, testing combinations in small amounts before attempting the full color blend.

Example: A student finds, "If I mix blue and yellow, I get green, but if I add white slowly, I can make it lighter."

Pairing 6: Experimenting With Structural Stability

Big Skill: Abstract Thinking – Exploring different ways to build a strong structure.

Small Skill: Breaking Down a Problem – Testing one design element at a time before constructing the full model.

Teaching Technique: Sensory exploration.

Actions: Have students experiment with different materials and shapes before deciding on the best structure for a tower.

Example: A student tests, "If I use a triangle base, my tower is more stable, but if I use squares, I can stack higher."

Pairing 7: Sorting and Categorizing in Grouping Challenges

Big Skill: Abstract Thinking – Finding different ways to divide a set of objects.

Small Skill: Breaking Down a Problem – Sorting objects into groups step by step.

Teaching Technique: Collaborative group work.

Actions: Have students practice different grouping strategies, such as equal distribution, categorizing by type, or sorting by size.

Example: A student discovers, "If I divide by twos, I get an even split, but if I divide by color first, I can organize them better before splitting evenly."

Pairing 8: Gradual Weight Adjustments in Balance Experiments

Big Skill: Abstract Thinking – Understanding multiple ways to balance a scale.

Small Skill: Breaking Down a Problem – Adjusting weight placement one step at a time to achieve balance.

Teaching Technique: Real-world applications.

Actions: Have students test different weights on a balance scale, adding or removing small amounts to find multiple ways to balance it.

Example: A student finds, "If I put a heavy object on one side, I can balance it with two lighter objects instead of one equal weight."

Pairing 9: Stepwise Decision-Making in Map Challenges

Big Skill: Abstract Thinking – Exploring different paths to a goal in a map challenge.

Small Skill: Breaking Down a Problem – Navigating one section of the map before planning the next move.

Teaching Technique: Cultural exploration.

Actions: Have students follow a treasure map step by step, analyzing different paths before choosing the final route.

Example: A student reasons, "If I follow the river first, I avoid obstacles, but if I cut through the forest, it might be faster."

Pairing 10: Incremental Programming in Coding Challenges

Big Skill: Abstract Thinking – Testing different coding solutions for the same task.

Small Skill: Breaking Down a Problem – Writing one block of code at a time before testing the full program.

Teaching Technique: Hands-on experiments.

Actions: Have students program a character to move using block coding, testing one command at a time before running the full sequence.

Example: A student predicts, "If I add a turn block before the movement block, my character will face the right direction before moving forward."

Example Teaching Strategies for Big vs. Small Thinking Skill Pairings of Collaborative Thinking at the Grade School Level

Here are 90 unique teaching strategy examples for the big thinking and small thinking skill pairings, each described in detail to align with **Big Thinking Skills** (Concepts & Strategies) and **Small Thinking Skills** (Step-by-Step Execution) for the **Collaborative Thinking Skill** taught at the grade school level.

Collaborative Thinking

Big Thinking Skills (Concepts & Strategies)

- **Understanding Teamwork:** Learning that working together can solve problems faster.
- **Idea Sharing & Communication:** Knowing when to speak, listen, and contribute effectively.
- **Conflict Resolution:** Recognizing that different opinions exist and how to work through them.

Small Thinking Skills (Step-by-Step Execution)

- **Turn-Taking & Listening:** Practicing how to wait for one's turn in group activities.
- **Following Group Rules:** Completing a shared task using a step-by-step approach.
- **Hands-on Collaboration:** Collaborating with a partner to build something, with clear individual steps (e.g., "You place the first block, I place the second").

Collaborative Thinking – (Big) Understanding Teamwork vs. (Small) Turn-Taking & Listening

---

Pairing 1: Teamwork in STEM Challenges

Big Skill: Understanding Teamwork – Learning that working together can solve problems faster.

Small Skill: Turn-Taking & Listening – Practicing how to wait for one's turn in group activities.

Teaching Technique: Hands-on experiments.

Actions: Guide students through a STEM challenge where they must build a structure together by taking turns adding pieces.

Example: A student waits for their turn and listens to their teammate's idea: "If I place my block here, it will balance better."

Pairing 2: Teamwork in a Scavenger Hunt

Big Skill: Understanding Teamwork – Recognizing that different team members contribute different strengths.

Small Skill: Turn-Taking & Listening – Practicing active listening when teammates share ideas.

Teaching Technique: Guided exploration.

Actions: Have students work in teams to solve a scavenger hunt, taking turns reading clues aloud and listening to others' suggestions.

Example: A student listens and responds, "If you think the clue means something round, let us check near the basketballs!"

Pairing 3: Teamwork in Puzzle Solving

Big Skill: Understanding Teamwork – Learning that shared effort leads to faster success.

Small Skill: Turn-Taking & Listening – Waiting to speak while others share their thoughts.

Teaching Technique: Problem-solving exercises.

Actions: Assign students a group puzzle where each person must add one piece at a time and listen to where others think the next piece should go.

Example: A student suggests, "I think the sky pieces go together. I will wait for you to place yours before adding mine."

Pairing 4: Teamwork in Science Experiments

Big Skill: Understanding Teamwork – Recognizing that cooperation improves efficiency.

Small Skill: Turn-Taking & Listening – Waiting for a partner to finish before responding.

Teaching Technique: Observation journals.

Actions: Have students work in pairs to complete a science experiment, with one person reading instructions while the other performs each step.

Example: A student reads aloud, "Now, pour the vinegar slowly," while their partner follows the step, then they switch roles.

Pairing 5: Teamwork in Model Building

Big Skill: Understanding Teamwork – Realizing that effective communication helps teams succeed.

Small Skill: Turn-Taking & Listening – Practicing repeating what a teammate said to confirm understanding.

Teaching Technique: Hands-on experiments.

Actions: Have students build a model together, taking turns giving and following instructions.

Example: A student repeats, "So you said we should put the roof on last? I agree; that will make it stronger."

Pairing 6: Teamwork in Relay Games

Big Skill: Understanding Teamwork – Learning that supporting others strengthens the group.

Small Skill: Turn-Taking & Listening – Practicing encouraging others while waiting for a turn.

Teaching Technique: Sensory exploration.

Actions: Have students play a relay game where they must wait for their turn while cheering for their teammates.

Example: A student cheers, "You got this! I will go next and pass the ball as fast as I can!"

Pairing 7: Teamwork in Storytelling

Big Skill: Understanding Teamwork – Seeing that collaboration leads to creative solutions.

Small Skill: Turn-Taking & Listening – Practicing asking questions instead of interrupting.

Teaching Technique: Collaborative group work.

Actions: Have students write a group story, with each person adding one sentence while listening to previous contributions.

Example: A student asks, "Since the dragon found the treasure, what should happen next?"

Pairing 8: Teamwork in Debate & Discussion

Big Skill: Understanding Teamwork – Understanding that disagreements can be solved through listening.

Small Skill: Turn-Taking & Listening – Learning to compromise by listening to different viewpoints.

Teaching Technique: Real-world applications.

Actions: Guide students in a classroom debate where they must listen to their teammates before responding.

Example: A student says, "I hear your point about cats being great pets, but I think dogs are better because they can help people."

Pairing 9: Teamwork in Cooperative Board Games

Big Skill: Understanding Teamwork – Realizing that patience and cooperation improve results.

Small Skill: Turn-Taking & Listening – Taking turns while playing a cooperative game.

Teaching Technique: Cultural exploration.

Actions: Have students play a cooperative board game where they must take turns strategizing as a team.

Example: A student says, "It is your turn now! If you go left, I will go right next turn so we can win faster."

Pairing 10: Teamwork in Lego Building

Big Skill: Understanding Teamwork – Recognizing that teamwork builds friendships and trust.

Small Skill: Turn-Taking & Listening – Respecting others' ideas before contributing their own.

Teaching Technique: Hands-on experiments.

Actions: Have students work in pairs to construct a Lego structure, taking turns adding pieces while explaining their choices.

Example: A student listens and says, "That is a good idea. You build the base, and I will build the roof next."

Collaborative Thinking – (Big) Understanding Teamwork vs. (Small) Following Group Rules

---

Pairing 1: Teamwork in STEM Projects

Big Skill: Understanding Teamwork – Learning that working together can solve problems faster.

Small Skill: Following Group Rules – Completing a shared task using a step-by-step approach.

Teaching Technique: Hands-on experiments.

Actions: Guide students in a group STEM activity where they build a structure by following step-by-step instructions and taking turns.

Example: A student says, "Step 1 is to build the base. Let us follow the instructions so our tower is strong."

Pairing 2: Teamwork in Cooking Activities

Big Skill: Understanding Teamwork – Realizing that roles help a group function better.

Small Skill: Following Group Rules – Assigning roles to each student and following a process.

Teaching Technique: Guided exploration.

Actions: Have students bake a simple recipe together, assigning roles like mixing, measuring, and pouring.

Example: A student says, "I will measure the flour, and you stir. Then we switch for the next step!"

Pairing 3: Teamwork in Poster Projects

Big Skill: Understanding Teamwork – Learning that teams work best when following a plan.

Small Skill: Following Group Rules – Completing a multi-step project in order.

Teaching Technique: Problem-solving exercises.

Actions: Assign students a group poster project where they must plan and follow ordered steps (e.g., research, design, and prepare).

Example: A student reminds the group, "Let us finish writing our facts first before we start coloring the poster."

Pairing 4: Teamwork in Science Experiments

Big Skill: Understanding Teamwork – Recognizing that clear rules make teamwork smoother.

Small Skill: Following Group Rules – Taking turns to complete a group challenge in order.

Teaching Technique: Observation journals.

Actions: Have students work together in a science experiment where each step must be completed in order.

Example: A student reads the instructions, "First, we pour the vinegar, then we add baking soda. Do not skip the order!"

Pairing 5: Teamwork in Classroom Cleanup

Big Skill: Understanding Teamwork – Understanding that shared effort leads to better results.

Small Skill: Following Group Rules – Completing a classroom cleanup by dividing tasks.

Teaching Technique: Hands-on experiments.

Actions: Assign teams different cleanup jobs (sweeping, organizing, wiping desks) and have them complete them in order.

Example: A student says, "We need to put the books away before sweeping so we do not sweep dirt onto them."

Pairing 6: Teamwork in Board Games

Big Skill: Understanding Teamwork – Learning that following steps leads to success.

Small Skill: Following Group Rules – Playing a cooperative game by following turn order.

Teaching Technique: Sensory exploration.

Actions: Have students play a board game where they must follow turn-taking and game rules as a team.

Example: A student reminds, "We have to draw a card first before moving our piece. Let us not skip ahead!"

Pairing 7: Teamwork in Art Projects

Big Skill: Understanding Teamwork – Seeing that cooperation leads to efficiency.

Small Skill: Following Group Rules – Passing materials in a structured way during a shared task.

Teaching Technique: Collaborative group work.

Actions: Have students create a class mural where each team member adds their part before passing materials in order.

Example: A student says, "After I paint the tree, I will pass the brush to you so you can do the sky."

Pairing 8: Teamwork in Obstacle Course Challenges

Big Skill: Understanding Teamwork – Learning that listening and following rules keep groups organized.

Small Skill: Following Group Rules – Completing a team-building exercise with clear steps.

Teaching Technique: Real-world applications.

Actions: Have students complete a group obstacle course where each person must complete a task before the next can go.

Example: A student shouts, "You need to hop through the rings first, then I can run to the next station!"

Pairing 9: Teamwork in Group Storytelling

Big Skill: Understanding Teamwork – Realizing that group projects require patience and order.

Small Skill: Following Group Rules – Writing a group story one sentence at a time in order.

Teaching Technique: Cultural exploration.

Actions: Have students co-write a story by adding one sentence at a time in a structured order.

Example: A student adds, "Since the first sentence was about a dragon in a cave, I will write that it found a treasure."

Pairing 10: Teamwork in Group Discussions

Big Skill: Understanding Teamwork – Seeing that rules help everyone contribute fairly.

Small Skill: Following Group Rules – Rotating leadership in a group discussion.

Teaching Technique: Hands-on experiments.

Actions: Have students participate in a discussion where they must follow a speaking order, ensuring each person has a turn.

Example: A student says, "I just shared my idea, so now it is your turn to talk about your thoughts."

Collaborative Thinking – (Big) Understanding Teamwork vs. (Small) Hands-on Collaboration

---

Pairing 1: Teamwork in Building Structures

Big Skill: Understanding Teamwork – Learning that working together can solve problems faster.

Small Skill: Hands-on Collaboration – Working with a partner to build something, with clear individual steps (e.g., "You place the first block, I place the second").

Teaching Technique: Hands-on experiments.

Actions: Have students work in pairs to build a tower with blocks, taking turns adding pieces while discussing the best approach.

Example: One student says, "You place the base blocks first, then I will add the middle layer, so it does not tip over."

Pairing 2: Teamwork in Artistic Creations

Big Skill: Understanding Teamwork – Realizing that shared effort leads to better results.

Small Skill: Hands-on Collaboration – Creating a shared drawing where each partner adds specific elements.

Teaching Technique: Guided exploration.

Actions: Assign partners different roles in drawing a landscape—one draws the background, the other adds details like trees and animals.

Example: A student says, "I will draw the hills, and then you can add the sun and clouds to make the sky look complete."

Pairing 3: Teamwork in LEGO Construction

Big Skill: Understanding Teamwork – Recognizing that different skills contribute to success.

Small Skill: Hands-on Collaboration – Constructing a LEGO model where each person follows a part of the instructions.

Teaching Technique: Problem-solving exercises.

Actions: Have students follow step-by-step instructions to build a LEGO structure, with each partner responsible for a specific section.

Example: A student says, "You build the bottom part of the car, and I will work on attaching the wheels."

Pairing 4: Teamwork in Puzzle Solving

Big Skill: Understanding Teamwork – Learning that organized cooperation speeds up tasks.

Small Skill: Hands-on Collaboration – Assembling a puzzle together, alternating turns placing pieces.

Teaching Technique: Observation journals.

Actions: Have students complete a jigsaw puzzle, where each person focuses on specific sections, such as the edges or the main picture.

Example: A student says, "I will work on the border while you find pieces for the castle in the middle."

Pairing 5: Teamwork in Structural Design

Big Skill: Understanding Teamwork – Seeing that collaboration leads to creative solutions.

Small Skill: Hands-on Collaboration – Designing a paper bridge where one student folds and the other tapes.

Teaching Technique: Hands-on experiments.

Actions: Have students construct a paper bridge using different folding techniques, testing which holds the most weight.

Example: One student folds, saying, "I will fold the paper into triangles, and you tape them together to keep it strong."

Pairing 6: Teamwork in Gardening

Big Skill: Understanding Teamwork – Learning that working together improves efficiency.

Small Skill: Hands-on Collaboration – Planting a small garden where partners take turns digging and planting.

Teaching Technique: Sensory exploration.

Actions: Have students work in pairs to plant seeds, with one digging holes and the other placing the seeds.

Example: A student says, "You dig the hole, and I will drop the seed in and cover it with soil."

Pairing 7: Teamwork in Paper Airplane Engineering

Big Skill: Understanding Teamwork – Recognizing that cooperation makes complex tasks easier.

Small Skill: Hands-on Collaboration – Constructing a paper airplane together, each completing different steps.

Teaching Technique: Collaborative group work.

Actions: Have students take turns folding and creasing a paper airplane, testing different designs.

Example: A student suggests, "You fold the wings while I make the nose pointed for better flight."

Pairing 8: Teamwork in Cooking

Big Skill: Understanding Teamwork – Seeing that structured teamwork prevents mistakes.

Small Skill: Hands-on Collaboration – Cooking a simple recipe together, dividing tasks.

Teaching Technique: Real-world applications.

Actions: Have students work in pairs to make a no-bake snack, with one person measuring and the other mixing.

Example: A student says, "You measure the peanut butter, and I will stir it with the oats, so it blends well."

Pairing 9: Teamwork in Clay Sculpting

Big Skill: Understanding Teamwork – Realizing that shared responsibilities make work more enjoyable.

Small Skill: Hands-on Collaboration – Creating a clay sculpture together, one shaping and the other detailing.

Teaching Technique: Cultural exploration.

Actions: Have students mold a simple clay figure, with one making the base and the other adding details like eyes and patterns.

Example: A student says, "I will roll the clay into a ball for the head, and you can shape the arms and legs."

Pairing 10: Teamwork in Poster Creation

Big Skill: Understanding Teamwork – Learning that following a plan makes teamwork smoother.

Small Skill: Hands-on Collaboration – Decorating a classroom poster where each partner is assigned a specific section.

Teaching Technique: Hands-on experiments.

Actions: Have students create a group poster, with one partner drawing and the other coloring and writing text.

Example: A student says, "I will write the title in big letters while you color in the background."

Collaborative Thinking – (Big) Idea Sharing & Communication vs. (Small) Turn-Taking & Listening

---

Pairing 1: Speaking and Listening in Science Discussions

Big Skill: Idea Sharing & Communication – Knowing when to speak, listen, and contribute effectively.

Small Skill: Turn-Taking & Listening – Practicing how to wait for one's turn in group activities.

Teaching Technique: Hands-on experiments.

Actions: Have students take turns presenting their observations during a science experiment while others listen and ask questions.

Example: A student shares, "I noticed that the plant without sunlight is turning yellow," while another waits and then adds, "I think that is because plants need light for photosynthesis."

Pairing 2: Taking Turns in Class Discussions

Big Skill: Idea Sharing & Communication – Learning to express ideas clearly in a group.

Small Skill: Turn-Taking & Listening – Waiting to contribute until another speaker has finished.

Teaching Technique: Guided exploration.

Actions: Have students participate in a classroom discussion where they must raise their hands and wait before sharing their thoughts.

Example: A student raises their hand and listens as a classmate shares an idea before adding, "I agree with Sarah because I also think teamwork makes problem-solving easier."

Pairing 3: Communication in Group Problem-Solving

Big Skill: Idea Sharing & Communication – Understanding that good communication helps teamwork.

Small Skill: Turn-Taking & Listening – Practicing listening before responding.

Teaching Technique: Problem-solving exercises.

Actions: Have students work in small groups to solve a logic puzzle, taking turns explaining their thought process.

Example: A student says, "I think the missing number is 12 because…" before another adds, "Let us check if that pattern works before moving to the next step."

Pairing 4: Encouraging Thoughtful Discussions

Big Skill: Idea Sharing & Communication – Learning that effective speaking includes giving others time to talk.

Small Skill: Turn-Taking & Listening – Taking turns contributing to a shared discussion.

Teaching Technique: Observation journals.

Actions: Guide students in a literature discussion where they must summarize their thoughts and wait for the next speaker to contribute.

Example: A student states, "I think the main character was brave because…" and then asks, "What do you think, Alex?"

Pairing 5: Turn-Taking in Science Demonstrations

Big Skill: Idea Sharing & Communication – Knowing when to speak and when to listen.

Small Skill: Turn-Taking & Listening – Pausing before responding to allow reflection.

Teaching Technique: Hands-on experiments.

Actions: Have students conduct a group science demonstration where they take turns explaining observations while others listen.

Example: A student explains, "When we mixed vinegar and baking soda, bubbles formed because of a reaction," and the next student builds on it: "That must be because of carbon dioxide gas."

Pairing 6: Building Stories Together

Big Skill: Idea Sharing & Communication – Learning to build on others' ideas.

Small Skill: Turn-Taking & Listening – Waiting for a peer to finish speaking before responding.

Teaching Technique: Sensory exploration.

Actions: Have students participate in a storytelling circle where they take turns adding to a story.

Example: One student begins, "Once upon a time, a boy found a magic key," and another waits before adding, "He did not know that the key would unlock a secret door."

Pairing 7: Structured Speaking in Group Work

Big Skill: Idea Sharing & Communication – Understanding that effective communication involves respecting others' speaking time.

Small Skill: Turn-Taking & Listening – Following a structured speaking order in group work.

Teaching Technique: Collaborative group work.

Actions: Have students brainstorm ideas for a group project, ensuring that each person speaks before repeating ideas.

Example: A student listens and then responds, "I like your idea of a recycling poster. Maybe we could add pictures to show different materials we can recycle."

Pairing 8: Strengthening Listening in Debates

Big Skill: Idea Sharing & Communication – Learning that good listening improves discussions.

Small Skill: Turn-Taking & Listening – Practicing repeating what a teammate said before adding their own idea.

Teaching Technique: Real-world applications.

Actions: Have students engage in a structured debate, where they summarize what the previous speaker said before adding their point.

Example: A student states, "I understand that you think solar energy is expensive, but I believe it is worth the investment because it helps the environment."

Pairing 9: Encouraging Active Listening

Big Skill: Idea Sharing & Communication – Understanding that teamwork involves shared conversation.

Small Skill: Turn-Taking & Listening – Making eye contact and nodding while others speak before responding.

Teaching Technique: Cultural exploration.

Actions: Have students practice active listening during a group discussion by nodding and using phrases like "I see your point" before sharing their own thoughts.

Example: A student nods and says, "That is a good point. I had not thought about how water pollution affects animals before."

Pairing 10: Clear Communication in Group Activities

Big Skill: Idea Sharing & Communication – Learning that clear communication leads to better teamwork.

Small Skill: Turn-Taking & Listening – Taking turns explaining steps in a group activity.

Teaching Technique: Hands-on experiments.

Actions: Have students take turns explaining how to complete a multi-step task, such as assembling a craft or conducting an experiment.

Example: One student says, "First, we need to cut out the shapes," and another follows with, "Then we glue them together to make the final project."

Collaborative Thinking – (Big) Idea Sharing & Communication vs. (Small) Following Group Rules

Pairing 1: Step-by-Step Science Experiments

Big Skill: Idea Sharing & Communication – Knowing when to speak, listen, and contribute effectively.

Small Skill: Following Group Rules – Completing a shared task using a step-by-step approach.

Teaching Technique: Hands-on experiments.

Actions: Have students work in teams to conduct a science experiment, taking turns explaining each step before moving to the next.

Example: A student says, "First, we measure the water. Then, we add food coloring before stirring."

Pairing 2: Assigning Roles in Group Projects

Big Skill: Idea Sharing & Communication – Understanding that clear roles help teams function better.

Small Skill: Following Group Rules – Assigning group members specific steps in a process.

Teaching Technique: Guided exploration.

Actions: Have students complete a class project by assigning different tasks, ensuring everyone follows the process in order.

Example: A student says, "I will research the topic, you write the introduction, and Alex will add the pictures."

Pairing 3: Structured Problem-Solving with Puzzles

Big Skill: Idea Sharing & Communication – Learning that structured communication improves collaboration.

Small Skill: Following Group Rules – Completing a group puzzle by following a structured approach.

Teaching Technique: Problem-solving exercises.

Actions: Have students complete a jigsaw puzzle, where they must first organize the pieces, then build the edges, and finally fill in the center.

Example: A student suggests, "Let us find all the edge pieces first, then sort by color before putting them together."

Pairing 4: Collaborative Art Projects

Big Skill: Idea Sharing & Communication – Recognizing that taking turns improves teamwork.

Small Skill: Following Group Rules – Completing a group drawing step by step.

Teaching Technique: Observation journals.

Actions: Have students work together on a shared drawing where one student sketches, another outlines, and another colors.

Example: A student says, "First, I will draw the tree. Then, you outline it before we both color it in."

Pairing 5: Step-by-Step Crafting

Big Skill: Idea Sharing & Communication – Understanding that following instructions leads to a successful outcome.

Small Skill: Following Group Rules – Assembling a craft project step by step.

Teaching Technique: Hands-on experiments.

Actions: Have students create an origami project by following a specific order of steps, with each student taking turns leading.

Example: A student says, "First, we fold the paper into a triangle. Then, we fold the corners inward."

Pairing 6: Structured Speaking in Group Discussions

Big Skill: Idea Sharing & Communication – Learning that speaking in order keeps discussions productive.

Small Skill: Following Group Rules – Using a structured speaking order in group conversations.

Teaching Technique: Sensory exploration.

Actions: Have students follow a structured discussion format where they take turns sharing ideas based on a speaking list.

Example: A student says, "After Amy shares her idea, it will be my turn, then Carlos's turn."

Pairing 7: Organizing Materials for Group Tasks

Big Skill: Idea Sharing & Communication – Seeing that structured collaboration leads to efficiency.

Small Skill: Following Group Rules – Organizing materials before beginning a group task.

Teaching Technique: Collaborative group work.

Actions: Have students gather and set up materials before starting a shared project, ensuring all steps are followed in order.

Example: A student says, "Before we start, let us get all our supplies. We need scissors, glue, and paper."

Pairing 8: Turn-Taking in Cooperative Games

Big Skill: Idea Sharing & Communication – Recognizing that group projects require step-by-step coordination.

Small Skill: Following Group Rules – Completing a cooperative game by following turn order.

Teaching Technique: Real-world applications.

Actions: Have students play a cooperative board game where they must follow turn-taking and teamwork strategies.

Example: A student reminds, "We have to roll the dice before moving, and then we can decide what action to take as a team."

Pairing 9: Practicing Clear Instructions

Big Skill: Idea Sharing & Communication – Understanding that teamwork involves giving and receiving instructions.

Small Skill: Following Group Rules – Practicing giving clear, step-by-step instructions.

Teaching Technique: Cultural exploration.

Actions: Have students participate in a blindfolded challenge where one student gives step-by-step directions to help their partner complete a task.

Example: A student says, "Take two steps forward, then turn right. Now reach out to grab the object."

Pairing 10: Structured Story Writing

Big Skill: Idea Sharing & Communication – Learning that planning leads to more effective teamwork.

Small Skill: Following Group Rules – Writing a shared story with a structured format.

Teaching Technique: Hands-on experiments.

Actions: Have students co-write a short story where each person follows a specific format (one writes the beginning, another writes the middle, and the last writes the ending).

Example: A student says, "I will write the introduction, and then you can add what happens next before Sam writes the ending."

Collaborative Thinking – (Big) Idea Sharing & Communication vs. (Small) Hands-on Collaboration

---

Pairing 1: Structured LEGO Building

Big Skill: Idea Sharing & Communication – Knowing when to speak, listen, and contribute effectively.

Small Skill: Hands-on Collaboration – Working with a partner to build something, with clear individual steps (e.g., "You place the first block, I place the second").

Teaching Technique: Hands-on experiments.

Actions: Have students work in pairs to build a LEGO structure, taking turns placing blocks while discussing the best design.

Example: A student says, "I will build the base to keep it strong, then you add the windows, so it looks realistic."

Pairing 2: Collaborative Tower Construction

Big Skill: Idea Sharing & Communication – Understanding that good communication makes teamwork more efficient.

Small Skill: Hands-on Collaboration – Constructing a tower using cups or blocks by following a structured turn-taking process.

Teaching Technique: Guided exploration.

Actions: Have students alternate placing a block while discussing which placement will make the tower the tallest and most stable.

Example: A student suggests, "If we put this bigger block at the bottom, it will hold the top layers better."

Pairing 3: Step-by-Step Crafting

Big Skill: Idea Sharing & Communication – Learning that clear instructions improve collaboration.

Small Skill: Hands-on Collaboration – Assembling a simple craft together, where each partner has specific steps.

Teaching Technique: Problem-solving exercises.

Actions: Have students create a paper chain where one partner cuts the strips and the other loops and glues them together.

Example: A student says, "You cut the strips while I start gluing them into loops, and then we will switch."

Pairing 4: Teamwork in Poster Design

Big Skill: Idea Sharing & Communication – Recognizing that listening is as important as speaking in a team.

Small Skill: Hands-on Collaboration – Designing a class poster where each partner contributes different sections.

Teaching Technique: Observation journals.

Actions: Have students divide poster-making tasks so that one draws images while the other writes the text.

Example: A student says, "I will draw the title banner at the top, and then you can write the facts below it."

Pairing 5: Cooperative Puzzle Assembly

Big Skill: Idea Sharing & Communication – Understanding that turn-taking helps avoid mistakes.

Small Skill: Hands-on Collaboration – Following a step-by-step process to assemble a puzzle together.

Teaching Technique: Hands-on experiments.

Actions: Have students work on a puzzle, taking turns placing pieces while discussing which ones fit.

Example: A student says, "You find all the corner pieces first, and I will look for the edge pieces so we can build the frame."

Pairing 6: Paper Bridge Engineering

Big Skill: Idea Sharing & Communication – Learning to share ideas while following a structured process.

Small Skill: Hands-on Collaboration – Building a simple paper bridge where one partner folds and the other tapes.

Teaching Technique: Sensory exploration.

Actions: Have students take turns folding and taping parts of a paper bridge, discussing how to make it stronger.

Example: A student suggests, "If I fold the paper into triangles, the bridge will hold more weight before you tape the sides."

Pairing 7: Teamwork in Baking

Big Skill: Idea Sharing & Communication – Recognizing that taking turns keeps a project organized.

Small Skill: Hands-on Collaboration – Baking a simple recipe where one person measures and the other mixes.

Teaching Technique: Collaborative group work.

Actions: Have students take turns measuring ingredients, pouring them in, and stirring while discussing the best technique.

Example: A student says, "You measure the flour, I will mix it into the batter, then we will check if it is the right consistency."

Pairing 8: Clay Sculpting as a Team

Big Skill: Idea Sharing & Communication – Understanding that structured teamwork prevents errors.

Small Skill: Hands-on Collaboration – Creating a clay model where one partner shapes and the other adds details.

Teaching Technique: Real-world applications.

Actions: Have students take turns forming the basic shape and then sculpting finer details like eyes and decorations.

Example: A student says, "I will roll the clay into a ball for the head, and you can add the ears and nose after."

Pairing 9: Diorama Planning and Execution

Big Skill: Idea Sharing & Communication – Seeing that careful planning leads to a better final product.

Small Skill: Hands-on Collaboration – Making a class diorama, with one partner setting up the background while the other arranges the figures.

Teaching Technique: Cultural exploration.

Actions: Have students design a diorama scene by deciding who works on each part and ensuring their work fits together.

Example: A student says, "I will paint the sky, and then you place the trees, so it looks realistic."

Pairing 10: Writing and Illustrating a Story

Big Skill: Idea Sharing & Communication – Recognizing that teams succeed when they follow a logical order.

Small Skill: Hands-on Collaboration – Writing a group story where one student writes, and the other illustrates.

Teaching Technique: Hands-on experiments.

Actions: Have students co-create a short book, taking turns writing and drawing while making sure both parts match.

Example: A student says, "I will write the first sentence, and then you draw a picture to go with it before we move to the next part."

Collaborative Thinking – (Big) Conflict Resolution vs. (Small) Turn-Taking & Listening

---

Pairing 1: Role-Playing Different Opinions

Big Skill: Conflict Resolution – Recognizing that different opinions exist and how to work through them.

Small Skill: Turn-Taking & Listening – Practicing how to wait for one's turn in group activities.

Teaching Technique: Hands-on experiments.

Actions: Have students role-play different opinions in a group discussion and take turns presenting their viewpoints before working toward a compromise.

Example: A student listens and then responds, "I understand you think we should build a rocket for our project, but I think a volcano would be fun too. Maybe we can vote on it."

Pairing 2: Understanding Different Perspectives

Big Skill: Conflict Resolution – Learning to understand different perspectives.

Small Skill: Turn-Taking & Listening – Practicing active listening before responding.

Teaching Technique: Guided exploration.

Actions: Guide students through a structured debate where they must listen to the opposing side before making their own points.

Example: A student listens carefully before responding, "I hear that you believe dogs are the best pets because they are loyal. I think cats are great too because they are independent."

Pairing 3: Compromise in Group Decisions

Big Skill: Conflict Resolution – Understanding that compromise helps resolve conflicts.

Small Skill: Turn-Taking & Listening – Allowing each group member to share their idea before making a decision.

Teaching Technique: Problem-solving exercises.

Actions: Have students work together to decide on a class activity by letting each person share their preference before finding a fair solution.

Example: A student says, "I know some of us want to play soccer and others want to play tag. What if we split our time between both games?"

Pairing 4: Recognizing and Fixing Misunderstandings

Big Skill: Conflict Resolution – Recognizing that misunderstandings can happen and how to fix them.

Small Skill: Turn-Taking & Listening – Waiting for a peer to explain before responding.

Teaching Technique: Observation journals.

Actions: Have students participate in a storytelling activity where they must clarify misheard details before reacting.

Example: A student asks, "Did you say the story was about a dragon or a dinosaur? I want to make sure I understood before I add to it."

Pairing 5: Valuing Every Opinion in a Team Challenge

Big Skill: Conflict Resolution – Learning that everyone's opinion matters.

Small Skill: Turn-Taking & Listening – Waiting for others to finish speaking before responding.

Teaching Technique: Hands-on experiments.

Actions: Have students participate in a team challenge where they must discuss and agree on the best approach before starting.

Example: A student listens before saying, "I like your idea to stack the cups like a pyramid, but maybe we should try a wider base first."

Pairing 6: Managing Emotions in Conflict

Big Skill: Conflict Resolution – Understanding that emotions play a role in conflict.

Small Skill: Turn-Taking & Listening – Pausing to acknowledge another person's feelings before responding.

Teaching Technique: Sensory exploration.

Actions: Guide students in a reflection activity where they discuss how they feel when someone interrupts versus when they are listened to.

Example: A student says, "I feel frustrated when I do not get to finish my sentence. Next time, can we make sure everyone gets to speak?"

Pairing 7: Keeping Disagreements Respectful

Big Skill: Conflict Resolution – Recognizing that disagreements do not have to turn into arguments.

Small Skill: Turn-Taking & Listening – Taking turns to express viewpoints without interrupting.

Teaching Technique: Collaborative group work.

Actions: Have students work in pairs to decide on a classroom rule, ensuring they take turns sharing their ideas before reaching an agreement.

Example: A student waits before saying, "I hear that you think no homework would be great, but I think a little homework helps us remember lessons. Maybe we can agree on only one page of homework per night."

Pairing 8: Respectful Discussions for Solutions

Big Skill: Conflict Resolution – Learning that respectful discussion leads to solutions.

Small Skill: Turn-Taking & Listening – Summarizing what another person said before responding.

Teaching Technique: Real-world applications.

Actions: Have students practice summarizing another person's argument before adding their own thoughts in a class debate.

Example: A student says, "I understand that you think recycling is important for the planet. I also believe we should use less plastic to help reduce waste."

Pairing 9: Finding Fair Solutions Through Listening

Big Skill: Conflict Resolution – Understanding that fair solutions involve listening to all sides.

Small Skill: Turn-Taking & Listening – Ensuring each group member has a chance to speak before making a decision.

Teaching Technique: Cultural exploration.

Actions: Have students role-play a town hall meeting where they must listen to different community concerns before producing a group decision.

Example: A student says, "I hear that building a playground would be fun, but I also understand that we need space for a garden. Maybe we can design a park with both."

Pairing 10: Practicing Patience in Conflict Resolution

Big Skill: Conflict Resolution – Learning that patience and listening help resolve conflicts.

Small Skill: Turn-Taking & Listening – Encouraging others to speak before responding.

Teaching Technique: Hands-on experiments.

Actions: Have students pair up and practice conflict resolution scenarios, ensuring that each person listens fully before offering their solution.

Example: A student says, "I understand you are upset because I borrowed your marker without asking. Next time, I will make sure to ask first."

Collaborative Thinking – (Big) Conflict Resolution vs. (Small) Following Group Rules

---

Pairing 1: Resolving Disagreements While Building Together

Big Skill: Conflict Resolution – Recognizing that different opinions exist and how to work through them.

Small Skill: Following Group Rules – Completing a shared task using a step-by-step approach.

Teaching Technique: Hands-on experiments.

Actions: Have students work in pairs to build a structure using LEGO or blocks, requiring them to take turns and follow a planned sequence even if they disagree.

Example: A student says, "I want to build the tower with a wide base, and you want to make it tall first. Let us follow the instructions step by step and see which works better."

Pairing 2: Expressing Disagreements Respectfully

Big Skill: Conflict Resolution – Learning how to express disagreements respectfully.

Small Skill: Following Group Rules – Completing a group activity where each member follows assigned steps.

Teaching Technique: Guided exploration.

Actions: Have students work together to solve a classroom puzzle, following assigned steps even if they disagree on the approach.

Example: A student says, "I think we should finish the edge pieces first, but since the rule is to work on the corners first, let us follow that and then decide together."

Pairing 3: Compromising While Completing a Task

Big Skill: Conflict Resolution – Understanding that working together requires compromise.

Small Skill: Following Group Rules – Taking turns to complete steps in an ordered process.

Teaching Technique: Problem-solving exercises.

Actions: Have students take turns setting up an art project, following each step in order while discussing any disagreements.

Example: A student says, "I want to paint the background first, but you think we should draw the details first. Since our instructions say to outline before coloring, let us do that."

Pairing 4: Following Fair Processes to Avoid Conflict

Big Skill: Conflict Resolution – Recognizing that fair solutions involve listening to all sides.

Small Skill: Following Group Rules – Participating in a classroom cleanup where tasks are completed in a set order.

Teaching Technique: Observation journals.

Actions: Assign students different cleaning roles and ensure they follow the order of tasks while resolving conflicts fairly.

Example: A student says, "I want to sweep first, but the rule says we should put the books away first, so we do not sweep dust onto them. Let us follow that order."

Pairing 5: Preventing Disagreements by Following Rules

Big Skill: Conflict Resolution – Learning that rules help prevent disagreements.

Small Skill: Following Group Rules – Playing a cooperative game where players must take turns and follow the rules.

Teaching Technique: Hands-on experiments.

Actions: Have students play a board game where they must follow turn-taking rules even if they disagree on the best strategy.

Example: A student says, "I think we should move this piece, but the rules say we have to roll the dice first. Let us follow the steps and decide after that."

Pairing 6: Preventing Arguments by Sticking to Instructions

Big Skill: Conflict Resolution – Understanding that taking turns and following steps can prevent arguments.

Small Skill: Following Group Rules – Completing a class science experiment in a step-by-step order.

Teaching Technique: Sensory exploration.

Actions: Have students conduct an experiment where they must follow steps in order and resolve any disagreements respectfully.

Example: A student says, "I want to mix the ingredients all at once, but the instructions say we should do it slowly. Let us follow the steps and see if it works."

Pairing 7: Staying Cooperative While Assembling a Project

Big Skill: Conflict Resolution – Learning to stay calm and cooperative when solving disagreements.

Small Skill: Following Group Rules – Working in pairs to assemble a class project by following written instructions.

Teaching Technique: Collaborative group work.

Actions: Have students construct a classroom model by following a set of instructions and negotiating when they disagree.

Example: A student says, "I think we should add the doors before the windows, but the instructions say to do windows first. Let us stick to the plan and discuss it afterwards.

Pairing 8: Maintaining Order While Co-Creating a Story

Big Skill: Conflict Resolution – Seeing that respecting rules helps groups function smoothly.

Small Skill: Following Group Rules – Taking turns to write and illustrate a group story by following a sequence.

Teaching Technique: Real-world applications.

Actions: Have students co-write a story where each person follows a set order for writing and illustrating, ensuring disagreements are handled respectfully.

Example: A student says, "I want to write the ending, but the rule says we have to finish the middle first. Let us follow the order so the story makes sense."

Pairing 9: Using a Fair System to Make Decisions

Big Skill: Conflict Resolution – Understanding that following a fair process makes decisions easier.

Small Skill: Following Group Rules – Using a structured voting system to decide on a group activity.

Teaching Technique: Cultural exploration.

Actions: Have students follow a step-by-step voting process to choose a class reward, ensuring all voices are heard.

Example: A student says, "I want to choose a movie, but the rule says we vote first. Let us follow the steps so everyone gets a fair chance."

Pairing 10: Resolving Differences While Building as a Team

Big Skill: Conflict Resolution – Recognizing that teamwork is more effective when rules are followed.

Small Skill: Following Group Rules – Completing a structured building challenge with a team.

Teaching Technique: Hands-on experiments.

Actions: Have students work in small teams to build a tower using provided materials, ensuring they follow step-by-step instructions despite different opinions.

Example: A student says, "I think we should add decorations now, but the rule says we finish building first. Let us follow the plan and decorate at the end."

Collaborative Thinking – (Big) Conflict Resolution vs. (Small) Hands-on Collaboration

---

Pairing 1: Resolving Disagreements While Building Together

Big Skill: Conflict Resolution – Recognizing that different opinions exist and how to work through them.

Small Skill: Hands-on Collaboration – Working with a partner to build something, with clear individual steps (e.g., "You place the first block, I place the second").

Teaching Technique: Hands-on experiments.

Actions: Have students work in pairs to build a block tower, ensuring they take turns and negotiate disagreements about structure and design.

Example: A student says, "I want to make the tower taller, but you want it wider. Let us try your idea first, and then we can evaluate mine next."

Pairing 2: Expressing Disagreements While Designing a Poster

Big Skill: Conflict Resolution – Understanding that disagreements can be solved through discussion.

Small Skill: Hands-on Collaboration – Creating a shared drawing where each partner contributes specific elements.

Teaching Technique: Guided exploration.

Actions: Have students co-create a poster, with one drawing and the other coloring, taking turns making decisions about design.

Example: A student says, "I think the title should be at the top, and you think it should go in the middle. Let us sketch both and see which looks better."

Pairing 3: Compromising While Crafting a Paper Chain

Big Skill: Conflict Resolution – Learning how to express different opinions respectfully.

Small Skill: Hands-on Collaboration – Constructing a simple craft together, following a step-by-step plan.

Teaching Technique: Problem-solving exercises.

Actions: Have students make a paper chain, with one cutting and the other looping and gluing, resolving any disagreements on color or length.

Example: A student says, "You want shorter loops, and I want longer ones. Let us alternate so we use both ideas."

Pairing 4: Taking Turns to Complete a Puzzle

Big Skill: Conflict Resolution – Recognizing that patience helps in solving conflicts.

Small Skill: Hands-on Collaboration – Building a simple puzzle together by alternating turns.

Teaching Technique: Observation journals.

Actions: Have students complete a jigsaw puzzle, ensuring each takes turns and communicates about where pieces should go.

Example: A student says, "I think this piece fits here, but you think it goes somewhere else. Let us evaluate both ideas and see."

Pairing 5: Problem-Solving While Building a Paper Bridge

Big Skill: Conflict Resolution – Understanding that compromise helps teams succeed.

Small Skill: Hands-on Collaboration – Constructing a bridge out of paper and tape, with one folding and the other taping.

Teaching Technique: Hands-on experiments.

Actions: Have students build a paper bridge, resolving differences in folding techniques through testing.

Example: A student says, "Let us try your folding method first and test if it holds more weight before we change anything."

Pairing 6: Negotiating Design Choices in a Clay Model

Big Skill: Conflict Resolution – Learning that taking turns helps avoid conflict.

Small Skill: Hands-on Collaboration – Creating a clay model together, with one partner shaping and the other adding details.

Teaching Technique: Sensory exploration.

Actions: Have students work together to mold an animal figure, ensuring they alternate shaping and detailing.

Example: A student says, "I want to make big ears, and you want small ones. Let us make one of each and see which looks better."

Pairing 7: Balancing Different Baking Techniques

Big Skill: Conflict Resolution – Seeing that cooperation leads to better teamwork.

Small Skill: Hands-on Collaboration – Baking a simple recipe together, dividing steps.

Teaching Technique: Collaborative group work.

Actions: Have students take turns measuring, pouring, and mixing while discussing any disagreements about technique.

Example: A student says, "You want to mix slowly, and I want to mix fast. Let us start slowly and speed up if needed."

Pairing 8: Adjusting Design Choices in a Diorama

Big Skill: Conflict Resolution – Recognizing that listening to others' ideas can improve outcomes.

Small Skill: Hands-on Collaboration – Assembling a class diorama with each partner responsible for different sections.

Teaching Technique: Real-world applications.

Actions: Have students work on different parts of a diorama, discussing and adjusting details to fit the overall design.

Example: A student says, "I think the river should be in the middle, and you think it should be on the side. Let us sketch both and decide together."

Pairing 9: Improving a Paper Airplane Design Together

Big Skill: Conflict Resolution – Understanding that working through conflicts makes projects stronger.

Small Skill: Hands-on Collaboration – Designing a paper airplane where one partner folds and the other adjusts the wings.

Teaching Technique: Cultural exploration.

Actions: Have students test different folding techniques, compromising on adjustments to improve flight.

Example: A student says, "You think the wings should be flat, and I think they should curve. Let us evaluate both designs and see which flies farther."

Pairing 10: Resolving Storyline Conflicts While Writing Together

Big Skill: Conflict Resolution – Learning that discussing differences calmly leads to better solutions.

Small Skill: Hands-on Collaboration – Writing a group story where one student writes and the other illustrates.

Teaching Technique: Hands-on experiments.

Actions: Have students create a short book, taking turns writing and drawing while resolving any disagreements about the story.

Example: A student says, "You want the main character to be a dog, and I want it to be a cat. What if we make it a story about both?"

Example Teaching Strategies for Big vs. Small Thinking Skill Pairings of Logical Thinking at the Middle School Level

Here are 90 unique teaching strategy examples for the big thinking and small thinking skill pairings, each described in detail to align with **Big Thinking Skills** (Concepts & Strategies) and **Small Thinking Skills** (Step-by-Step Execution) for the **Logical Thinking Skill** taught at the middle school level.

Logical Thinking

Big Thinking Skills (Concepts & Strategies)

- **If-Then Reasoning:** Understanding how logical conditions work in different contexts (e.g., "If I do not study, then I will not perform well on the test").
- **Evaluating Evidence:** Learning to differentiate facts from opinions and identify reliable sources.
- **Drawing Conclusions:** Using logical steps to make well-reasoned decisions rather than relying on assumptions.

Small Thinking Skills (Step-by-Step Execution)

- **Breaking Down Arguments:** Identifying premises, evidence, and conclusions in a structured way.
- **Applying Logical Rules:** Solving logic puzzles, practicing syllogisms, and simple coding exercises.
- **Organizing Information:** Using graphic organizers to categorize evidence in debates or essays.

Logical Thinking – (Big Skill) If-Then Reasoning vs. (Small Skill) Breaking Down Arguments

---

Pairing 1: Identifying Logical Arguments in Persuasive Writing

Big Skill: If-Then Reasoning – Understanding how logical conditions work in different contexts (e.g., "If I do not study, then I will not perform well on the test").

Small Skill: Breaking Down Arguments – Identifying premises, evidence, and conclusions in a structured way.

Teaching Technique: Hands-on experiments.

Actions: Have students analyze a short persuasive essay, identifying the main argument and how the "if-then" statements support the reasoning.

Example: A student highlights, "If students get more sleep, then they perform better in school," and identifies the evidence supporting this claim.

Pairing 2: Evaluating If-Then Claims in Advertising

Big Skill: If-Then Reasoning – Recognizing cause-and-effect relationships in real-world scenarios.

Small Skill: Breaking Down Arguments – Separating claims from supporting evidence.

Teaching Technique: Guided exploration.

Actions: Provide students with an advertisement and ask them to determine the claim, the supporting evidence, and whether the "if-then" reasoning is logical.

Example: A student analyzes, "If you use this shampoo, then your hair will grow faster," and assesses the scientific basis behind the claim.

Pairing 3: Using If-Then Reasoning in Scientific Hypotheses

Big Skill: If-Then Reasoning – Understanding how conditional logic applies in scientific hypotheses.

Small Skill: Breaking Down Arguments – Structuring an argument to support a hypothesis.

Teaching Technique: Problem-solving exercises.

Actions: Have students create a scientific hypothesis using if-then reasoning and outline the evidence needed to prove or disprove it.

Example: A student writes, "If plants are given more sunlight, then they will grow taller," and identifies the variables and expected results.

Pairing 4: Comparing Arguments Using If-Then Logic

Big Skill: If-Then Reasoning – Applying conditional logic in decision-making.

Small Skill: Breaking Down Arguments – Evaluating the strength of reasoning in different arguments.

Teaching Technique: Observation journals.

Actions: Present students with two arguments about a school rule change and have them determine which has stronger if-then reasoning and better supporting evidence.

Example: A student compares, "If students are allowed to listen to music, then they will concentrate better" versus "If students listen to music, then they will be distracted," and decides which argument is more valid based on evidence.

Pairing 5: Detecting Logical Fallacies in If-Then Arguments

Big Skill: If-Then Reasoning – Identifying logical fallacies in conditional reasoning.

Small Skill: Breaking Down Arguments – Spotting flaws in reasoning and unsupported claims.

Teaching Technique: Hands-on experiments.

Actions: Have students review examples of flawed if-then arguments and determine where the reasoning fails.

Example: A student critiques, "If you eat chocolate, then you will become a better athlete," identifying that there is no causal relationship.

Pairing 6: Constructing If-Then Arguments in Ethical Debates

Big Skill: If-Then Reasoning – Using logical conditions in ethical debates.

Small Skill: Breaking Down Arguments – Structuring arguments for and against a position.

Teaching Technique: Sensory exploration.

Actions: Have students debate an ethical issue, ensuring each side presents an if-then argument and identifies supporting evidence.

Example: A student argues, "If we reduce plastic waste, then ocean life will thrive," and presents statistics on pollution.

Pairing 7: Analyzing If-Then Reasoning in Legal Cases

Big Skill: If-Then Reasoning – Understanding how logic is used in legal arguments.

Small Skill: Breaking Down Arguments – Identifying premises, counterarguments, and conclusions.

Teaching Technique: Collaborative group work.

Actions: Have students analyze a mock court case, identifying how lawyers use if-then statements and breaking down their reasoning.

Example: A student states, "If the defendant was at school during the crime, then they could not have committed it," and examines the evidence presented.

Pairing 8: Examining If-Then Reasoning in History

Big Skill: If-Then Reasoning – Understanding how logic applies to historical cause and effect.

Small Skill: Breaking Down Arguments – Analyzing how historians use evidence to support conclusions.

Teaching Technique: Real-world applications.

Actions: Have students analyze a historical event and use if-then reasoning to explain its causes and consequences.

Example: A student writes, "If the Boston Tea Party had not happened, then the American Revolution might have been delayed," and finds evidence to support or refute the claim.

Pairing 9: Applying If-Then Logic to Engineering and Technology

Big Skill: If-Then Reasoning – Using logic in technological and engineering problem-solving.

Small Skill: Breaking Down Arguments – Evaluating the effectiveness of different engineering solutions.

Teaching Technique: Cultural exploration.

Actions: Have students analyze different bridge designs and determine which one logically provides the most stability based on if-then conditions.

Example: A student reasons, "If the bridge has a triangular truss design, then it will support more weight," and uses physics principles to support the argument.

Pairing 10: Understanding If-Then Statements in Computer Science

Big Skill: If-Then Reasoning – Applying logic in computer science and coding.

Small Skill: Breaking Down Arguments – Understanding how algorithms follow logical conditions.

Teaching Technique: Hands-on experiments.

Actions: Have students write a simple algorithm using if-then statements and analyze how each condition affects the program's outcome.

Example: A student writes, "If the player clicks the green flag, then the robot moves forward," and explains how the code follows logical conditions.

Logical Thinking – (Big Skill) If-Then Reasoning vs. (Small Skill) Applying Logical Rules

---

Pairing 1: Solving Logic Puzzles with If-Then Statements

Big Skill: If-Then Reasoning – Understanding how logical conditions work in different contexts (e.g., "If I do not study, then I will not perform well on the test").

Small Skill: Applying Logical Rules – Solving logic puzzles, practicing syllogisms, and simple coding exercises.

Teaching Technique: Hands-on experiments.

Actions: Provide students with a logic puzzle that requires them to follow a series of if-then statements to solve a problem.

Example: A student solves, "If the red block is above the blue block, and the green block is below the blue block, then where is the green block?"

Pairing 2: Practicing Syllogisms to Determine Logical Validity

Big Skill: If-Then Reasoning – Recognizing how conditional statements apply in real-world scenarios.

Small Skill: Applying Logical Rules – Practicing syllogisms to determine the logical validity of statements.

Teaching Technique: Guided exploration.

Actions: Have students evaluate syllogisms and determine whether the conclusions logically follow the premises.

Example: A student analyzes, "If all birds have feathers, and robins are birds, then robins have feathers," confirming the logical conclusion.

Pairing 3: Using If-Then Logic in Basic Coding

Big Skill: If-Then Reasoning – Understanding how logic is applied in computer science.

Small Skill: Applying Logical Rules – Writing simple if-then conditional statements in block-based coding.

Teaching Technique: Problem-solving exercises.

Actions: Have students create a simple program in Scratch where an if-then statement controls an animation or character movement.

Example: A student codes, "If the player presses the spacebar, then the character jumps," and tests the condition in Scratch.

Pairing 4: Predicting Outcomes Using Logic Grid Puzzles

Big Skill: If-Then Reasoning – Using logical thinking to predict outcomes.

Small Skill: Applying Logical Rules – Practicing grid-based logic puzzles to determine missing information.

Teaching Technique: Observation journals.

Actions: Have students complete a logic grid puzzle, ensuring they apply if-then conditions to deduce missing facts.

Example: A student solves, "If John is older than Sarah and Sarah is older than Mia, then who is the oldest?"

Pairing 5: Recognizing If-Then Patterns in Number Sequences

Big Skill: If-Then Reasoning – Recognizing conditional logic in problem-solving.

Small Skill: Applying Logical Rules – Identifying patterns in number sequences using if-then relationships.

Teaching Technique: Hands-on experiments.

Actions: Provide students with a number pattern and ask them to determine the rule that governs it using if-then logic.

Example: A student determines, "If each number is double the previous one, then the next number in the sequence (2, 4, 8, 16) must be 32."

Pairing 6: Sorting Information Using If-Then Venn Diagrams

Big Skill: If-Then Reasoning – Understanding the role of conditions in decision-making.

Small Skill: Applying Logical Rules – Using Venn diagrams to sort information based on logical conditions.

Teaching Technique: Sensory exploration.

Actions: Have students use a Venn diagram to categorize animals based on traits using if-then logic.

Example: A student reasons, "If an animal has feathers, then it must be in the 'birds' category. If it also swims, then it fits in both 'birds' and 'water animals' categories."

Pairing 7: Applying If-Then Reasoning in Strategy-Based Games

Big Skill: If-Then Reasoning – Applying logical thinking to strategy-based games.

Small Skill: Applying Logical Rules – Using conditional logic in simple board games or puzzles.

Teaching Technique: Collaborative group work.

Actions: Have students play a strategy-based board game where they must predict and analyze possible moves using if-then reasoning.

Example: A student considers, "If my opponent moves their piece here, then I can block them and win in the next move."

Pairing 8: Using If-Then Logic in Probability and Statistics

Big Skill: If-Then Reasoning – Recognizing how logic applies to probability and statistics.

Small Skill: Applying Logical Rules – Using probability experiments to determine likely outcomes.

Teaching Technique: Real-world applications.

Actions: Have students conduct probability experiments (rolling dice, drawing cards) and use if-then statements to predict results.

Example: A student predicts, "If I roll a six-sided die, then the probability of rolling a 4 is 1 out of 6."

Pairing 9: Mapping Out Decision Trees with If-Then Logic

Big Skill: If-Then Reasoning – Understanding logic in real-world decision-making.

Small Skill: Applying Logical Rules – Creating simple flowcharts to map out decision trees.

Teaching Technique: Cultural exploration.

Actions: Have students create a flowchart for a real-world decision, such as choosing what to wear based on the weather.

Example: A student maps out, "If it is raining, then wear a raincoat. If it is sunny, then wear sunglasses."

Pairing 10: Identifying Logical Fallacies in If-Then Statements

Big Skill: If-Then Reasoning – Using logic to evaluate arguments.

Small Skill: Applying Logical Rules – Determining logical fallacies in flawed reasoning.

Teaching Technique: Hands-on experiments.

Actions: Provide students with flawed if-then arguments and have them identify the logical fallacies.

Example: A student critiques, "If you drink milk, then you will become a professional athlete," identifying the lack of a direct causal relationship.

Logical Thinking – (Big Skill) If-Then Reasoning vs. (Small Skill) Organizing Information

---

Pairing 1: Mapping If-Then Scenarios in History

Big Skill: If-Then Reasoning – Understanding how logical conditions work in different contexts (e.g., "If I do not study, then I will not perform well on the test").

Small Skill: Organizing Information – Using graphic organizers to categorize evidence in debates or essays.

Teaching Technique: Hands-on experiments.

Actions: Have students use a cause-and-effect graphic organizer to map out if-then scenarios related to a historical event.

Example: A student maps, "If the American colonies had not protested British taxes, then the Revolutionary War might not have started."

Pairing 2: Using T-Charts to Analyze Decision-Making

Big Skill: If-Then Reasoning – Applying logical conditions to decision-making.

Small Skill: Organizing Information – Categorizing pros and cons in a decision-making chart.

Teaching Technique: Guided exploration.

Actions: Have students use a T-chart to list pros and cons of a school policy and determine the logical outcome based on if-then reasoning.

Example: A student writes, "If we extend lunch by 10 minutes, then students will have more time to eat, but classes may be shorter."

Pairing 3: Structuring Arguments Using Graphic Organizers

Big Skill: If-Then Reasoning – Using logical reasoning to structure a persuasive argument.

Small Skill: Organizing Information – Structuring an argumentative essay using a graphic outline.

Teaching Technique: Problem-solving exercises.

Actions: Have students complete a graphic organizer that categorizes their claim, supporting evidence, and counterarguments before writing an essay.

Example: A student outlines, "If schools require uniforms, then students will feel less peer pressure. Evidence: Studies show that dress codes reduce bullying."

Pairing 4: Analyzing Cause-and-Effect in Literature

Big Skill: If-Then Reasoning – Understanding logical consequences in literature.

Small Skill: Organizing Information – Creating a cause-and-effect diagram to analyze a novel.

Teaching Technique: Observation journals.

Actions: Have students track major events in a novel, using if-then logic to connect decisions to consequences.

Example: A student records, "If Romeo had waited before acting, then he and Juliet might not have died."

Pairing 5: Using If-Then Logic in Scientific Hypotheses

Big Skill: If-Then Reasoning – Using logic to evaluate scientific hypotheses.

Small Skill: Organizing Information – Structuring experiment data using a scientific method chart.

Teaching Technique: Hands-on experiments.

Actions: Have students use a graphic organizer to categorize hypotheses, experiments, observations, and conclusions in a science lab.

Example: A student writes, "If salt is added to water, then the freezing point will lower. Observations: Water with salt took longer to freeze."

Pairing 6: Creating Timelines to Show If-Then Historical Connections

Big Skill: If-Then Reasoning – Recognizing logical structures in historical events.

Small Skill: Organizing Information – Creating a timeline to show how one event led to another.

Teaching Technique: Sensory exploration.

Actions: Have students construct a timeline of major historical events, using if-then reasoning to explain their connections.

Example: A student writes, "If the Great Depression had not happened, then World War II might have been delayed due to stronger economies."

Pairing 7: Preparing Arguments for a Debate Using If-Then Reasoning

Big Skill: If-Then Reasoning – Structuring arguments in a debate.

Small Skill: Organizing Information – Using a debate outline to categorize claims, evidence, and counterarguments.

Teaching Technique: Collaborative group work.

Actions: Have students prepare for a debate by filling in a graphic organizer that categorizes their main argument, supporting details, and rebuttals.

Example: A student writes, "Claim: If schools switch to a four-day week, then students will have more time to rest. Counterargument: Some students may fall behind without five days of instruction."

Pairing 8: Using Concept Maps to Explore Ethical Perspectives

Big Skill: If-Then Reasoning – Understanding how logic applies in social and ethical issues.

Small Skill: Organizing Information – Sorting ethical perspectives in a concept map.

Teaching Technique: Real-world applications.

Actions: Have students categorize different ethical viewpoints on a controversial issue and predict logical outcomes.

Example: A student categorizes, "If we ban plastic bags, then pollution will decrease. Counterpoint: Some businesses rely on plastic bag production."

Pairing 9: Designing Flowcharts to Solve Global Issues

Big Skill: If-Then Reasoning – Using logic to solve real-world problems.

Small Skill: Organizing Information – Creating a flowchart to outline possible solutions to a problem.

Teaching Technique: Cultural exploration.

Actions: Have students design a flowchart mapping out the possible solutions and consequences of addressing a global issue.

Example: A student maps, "If deforestation continues, then habitats will be destroyed. If reforestation efforts increase, then ecosystems will recover."

Pairing 10: Structuring Step-by-Step Reasoning in Math Proofs

Big Skill: If-Then Reasoning – Understanding logic in mathematical proofs.

Small Skill: Organizing Information – Structuring a step-by-step reasoning process in a math proof organizer.

Teaching Technique: Hands-on experiments.

Actions: Have students use a graphic organizer to structure their thought process in solving a geometry proof.

Example: A student writes, "If two angles add up to 180°, then they are supplementary. Given: $\angle A = 110°$ and $\angle B = 70°$. Conclusion: $\angle A + \angle B = 180°$, so they are supplementary."

Logical Thinking – (Big Skill) Evaluating Evidence vs. (Small Skill) Breaking Down Arguments

---

Pairing 1: Identifying Facts vs. Opinions in an Article

Big Skill: Evaluating Evidence – Learning to differentiate facts from opinions and identify reliable sources.

Small Skill: Breaking Down Arguments – Identifying premises, evidence, and conclusions in a structured way.

Teaching Technique: Hands-on experiments.

Actions: Provide students with an article containing both factual statements and opinions. Have them highlight facts, identify supporting evidence, and locate conclusions.

Example: A student reads, "Homework is essential for success," and determines whether it is a fact or an opinion, then looks for supporting data in the article.

Pairing 2: Comparing Sources for Credibility

Big Skill: Evaluating Evidence – Understanding how to verify the credibility of a source.

Small Skill: Breaking Down Arguments – Determining whether an argument is well-supported by reliable evidence.

Teaching Technique: Guided exploration.

Actions: Have students compare two articles on the same topic, one from a peer-reviewed journal and one from a blog, and analyze the strength of the evidence in each.

Example: A student identifies, "The journal article provides scientific studies as evidence, while the blog uses personal opinions, making the journal more reliable."

Pairing 3: Recognizing Bias in Arguments

Big Skill: Evaluating Evidence – Recognizing bias in arguments.

Small Skill: Breaking Down Arguments – Identifying how word choice and selective evidence can shape conclusions.

Teaching Technique: Problem-solving exercises.

Actions: Provide students with two opposing opinion pieces and ask them to analyze how each selects evidence to support its claim.

Example: A student notes, "One article only presents negative effects of video games, while the other only highlights benefits. Neither gives a balanced view."

Pairing 4: Differentiating Strong and Weak Evidence

Big Skill: Evaluating Evidence – Learning to distinguish between strong and weak evidence.

Small Skill: Breaking Down Arguments – Structuring arguments based on the strength of evidence provided.

Teaching Technique: Observation journals.

Actions: Have students categorize pieces of evidence from an article as strong, weak, or irrelevant and determine how they affect the argument's validity.

Example: A student writes, "A government health study is strong evidence, while a personal story from one person is weak evidence."

Pairing 5: Fact-Checking Claims Using Credible Sources

Big Skill: Evaluating Evidence – Understanding how to fact-check information.

Small Skill: Breaking Down Arguments – Verifying whether a conclusion logically follows from the evidence.

Teaching Technique: Hands-on experiments.

Actions: Give students a list of claims and ask them to research and fact-check each one using credible sources.

Example: A student fact-checks, "The moon is made of cheese," and finds NASA articles confirming it is composed of rock, proving the claim is false.

Pairing 6: Identifying Logical Fallacies in Arguments

Big Skill: Evaluating Evidence – Recognizing logical fallacies in arguments.

Small Skill: Breaking Down Arguments – Identifying faulty reasoning that undermines an argument.

Teaching Technique: Sensory exploration.

Actions: Have students analyze advertisements or political speeches to identify logical fallacies such as hasty generalizations or false cause.

Example: A student identifies, "This ad says, 'If you do not buy this product, you will be unhappy,' which is an emotional appeal instead of factual evidence."

Pairing 7: Understanding the Importance of Citing Sources

Big Skill: Evaluating Evidence – Understanding the importance of citing sources.

Small Skill: Breaking Down Arguments – Identifying arguments that lack proper evidence or citations.

Teaching Technique: Collaborative group work.

Actions: Have students review an essay and highlight statements that require citations, determining whether the argument is weakened without sources.

Example: A student notes, "The author says, 'most scientists agree' but does not provide a source, so we do not know if this claim is true."

Pairing 8: Analyzing Statistics in Arguments

Big Skill: Evaluating Evidence – Recognizing the role of statistics in supporting claims.

Small Skill: Breaking Down Arguments – Differentiating between misleading and meaningful statistics.

Teaching Technique: Real-world applications.

Actions: Provide students with two articles, one using reliable statistics and one using misleading percentages, and have them evaluate the difference.

Example: A student identifies, "Saying '80% of people surveyed agree' is misleading if only five people were surveyed."

Pairing 9: Comparing Conflicting Reports on the Same Topic

Big Skill: Evaluating Evidence – Learning how to analyze conflicting reports on the same topic.

Small Skill: Breaking Down Arguments – Comparing the evidence used in opposing viewpoints to determine which is more convincing.

Teaching Technique: Cultural exploration.

Actions: Have students compare arguments from different news sources about a controversial issue, evaluating which provides the strongest evidence.

Example: A student writes, "One article provides five studies on climate change, while the other only uses personal opinion, making the first argument stronger."

Pairing 10: Identifying Fake News and Misinformation

Big Skill: Evaluating Evidence – Understanding the impact of fake news and misinformation.

Small Skill: Breaking Down Arguments – Identifying misleading arguments and checking sources for reliability.

Teaching Technique: Hands-on experiments.

Actions: Have students analyze viral social media claims and use fact-checking websites to determine their accuracy.

Example: A student checks, "A viral post claims drinking lemon water cures cancer, but medical websites show no scientific evidence supporting this claim."

Logical Thinking – (Big Skill) Evaluating Evidence vs. (Small Skill) Applying Logical Rules

---

Pairing 1: Logic in Action: Solving Problems with Facts and Rules

Big Skill: Evaluating Evidence – Learning to differentiate facts from opinions and identify reliable sources.

Small Skill: Applying Logical Rules – Solving logic puzzles, practicing syllogisms, and simple coding exercises.

Teaching Technique: Hands-on experiments.

Actions: Have students complete a logic puzzle where they must use fact-based clues to determine the correct solution.

Example: A student analyzes, "If John is older than Sarah, and Sarah is older than Mia, then John must be the oldest," verifying the reasoning with logical rules.

Pairing 2: Trust or Bust: Testing Arguments with Syllogistic Logic

Big Skill: Evaluating Evidence – Learning to distinguish between reliable and unreliable sources.

Small Skill: Applying Logical Rules – Practicing syllogisms to understand logical conclusions.

Teaching Technique: Guided exploration.

Actions: Provide students with syllogisms where they must verify whether the conclusion is supported by true premises.

Example: A student evaluates, "All trustworthy sources cite evidence. This article provides no evidence. Therefore, this article is not a trustworthy source."

Pairing 3: If-Then Thinking: Fact-Checking in a Clickbait World

Big Skill: Evaluating Evidence – Understanding how to fact-check information.

Small Skill: Applying Logical Rules – Using conditional statements to verify claims.

Teaching Technique: Problem-solving exercises.

Actions: Have students fact-check a social media claim and use if-then logic to determine whether the evidence supports the statement.

Example: A student checks, "If this statistic is true, then it should be reported in multiple sources. Since only one blog reports it, it might be unreliable."

Pairing 4: Spot the Flaw: Bias, Fallacies, and Logical Clarity

Big Skill: Evaluating Evidence – Recognizing bias in sources and arguments.

Small Skill: Applying Logical Rules – Identifying fallacies in arguments using logic.

Teaching Technique: Observation journals.

Actions: Have students analyze advertisements and political statements to find logical fallacies, such as appeals to emotion or false causation.

Example: A student notes, "This ad says, 'If you do not buy this product, you will not be happy,' which is an emotional appeal, not factual evidence."

Pairing 5: Coincidence or Cause? Using Logic to Unravel the Truth

Big Skill: Evaluating Evidence – Learning to differentiate correlation from causation.

Small Skill: Applying Logical Rules – Using logic puzzles to understand conditional relationships.

Teaching Technique: Hands-on experiments.

Actions: Have students solve a puzzle where they must determine whether one event truly caused another or if it was just correlated.

Example: A student evaluates, "More ice cream is sold in summer, and more shark attacks happen in summer, but ice cream does not cause shark attacks."

Pairing 6: Source Code: Sorting Information with Logical Flow

Big Skill: Evaluating Evidence – Understanding the importance of primary vs. secondary sources.

Small Skill: Applying Logical Rules – Using coding logic to differentiate between types of sources.

Teaching Technique: Sensory exploration.

Actions: Have students create an if-then flowchart that categorizes sources as primary or secondary based on their characteristics.

Example: A student programs, "If the source is an eyewitness account, then it is primary. If it is an analysis of another source, then it is secondary."

Pairing 7: Statistical Smarts: Reading Numbers with Logical Eyes

Big Skill: Evaluating Evidence – Recognizing misleading statistics in arguments.

Small Skill: Applying Logical Rules – Using mathematical logic to assess statistical claims.

Teaching Technique: Collaborative group work.

Actions: Provide students with a misleading statistic (e.g., "Crime increased by 100%" without context) and have them use logic to interpret the data correctly.

Example: A student explains, "If the number of crimes went from 1 to 2, that is a 100% increase, but it does not mean crime is a big problem."

Pairing 8: Scientific Logic: Turning Evidence into Understanding

Big Skill: Evaluating Evidence – Understanding how scientists use evidence in experiments.

Small Skill: Applying Logical Rules – Practicing hypothesis testing through logical statements.

Teaching Technique: Real-world applications.

Actions: Have students evaluate a scientific experiment by breaking it into a logical if-then format.

Example: A student writes, "If plants receive more sunlight, then they will grow taller. The data shows the plants in sunlight grew 5 inches more than those in shade."

Pairing 9: ruth Tracker: Detecting Misinformation with Logic Trees

Big Skill: Evaluating Evidence – Learning to detect fake news and misinformation.

Small Skill: Applying Logical Rules – Creating logical flowcharts to verify claims.

Teaching Technique: Cultural exploration.

Actions: Have students create a decision tree to verify if an article is trustworthy, following a logical sequence of questions.

Example: A student maps, "If the source is a scientific journal, then it is reliable. If it is a random social media post, then check for bias."

Pairing 10: Building Arguments: From Evidence to If-Then Insight

Big Skill: Evaluating Evidence – Understanding how to construct a strong argument based on facts.

Small Skill: Applying Logical Rules – Using logic statements to form well-structured arguments.

Teaching Technique: Hands-on experiments.

Actions: Have students write an argument and then convert it into a logical if-then structure to ensure it follows valid reasoning.

Example: A student writes, "If students exercise daily, then they will have better concentration. Studies show that physical activity improves focus."

Logical Thinking – (Big Skill) Evaluating Evidence vs. (Small Skill) Organizing Information

---

Pairing 1: Identifying Facts vs. Opinions

Big Skill: Evaluating Evidence – Learning to differentiate facts from opinions and identify reliable sources.

Small Skill: Organizing Information – Using graphic organizers to categorize evidence in debates or essays.

Teaching Technique: Hands-on experiments.

Actions: Have students read an article and use a T-chart to separate facts from opinions, ensuring they analyze sources for reliability.

Example: A student places "The Earth revolves around the Sun" under facts and "Space exploration is too expensive" under opinions after checking a scientific source.

Pairing 2: Verifying Credibility of Sources

Big Skill: Evaluating Evidence – Understanding how to verify the credibility of a source.

Small Skill: Organizing Information – Categorizing sources as primary or secondary in a graphic organizer.

Teaching Technique: Guided exploration.

Actions: Have students use a Venn diagram to compare characteristics of primary and secondary sources based on reliability and directness of evidence.

Example: A student places "diary entry from a Civil War soldier" under Primary Source and "history textbook about the Civil War" under Secondary Source.

Pairing 3: Recognizing Bias in Sources

Big Skill: Evaluating Evidence – Recognizing bias in different types of arguments.

Small Skill: Organizing Information – Creating a bias evaluation chart to categorize sources based on neutrality.

Teaching Technique: Problem-solving exercises.

Actions: Have students evaluate two news articles covering the same topic and use a bias scale to rate them.

Example: A student writes, "Source A only presents positive data about electric cars, while Source B includes both pros and cons, so Source B is more balanced."

Pairing 4: Ranking Evidence by Strength

Big Skill: Evaluating Evidence – Distinguishing between strong and weak evidence.

Small Skill: Organizing Information – Using an evidence evaluation chart to rank sources based on credibility.

Teaching Technique: Observation journals.

Actions: Have students sort different types of evidence (scientific study, expert opinion, personal story) into a credibility pyramid from most to least reliable.

Example: A student places peer-reviewed medical research at the top and celebrity opinion on health at the bottom, explaining why expert-reviewed studies are more reliable.

Pairing 5: Identifying Misleading Statistics

Big Skill: Evaluating Evidence – Understanding how statistics can be manipulated.

Small Skill: Organizing Information – Creating a "misleading statistics" graphic organizer to categorize different types of statistical misrepresentation.

Teaching Technique: Hands-on experiments.

Actions: Have students analyze misleading statistics from advertisements and categorize them into types (cherry-picking data, missing context, exaggerated percentages).

Example: A student identifies, "Saying 90% of customers prefer a product is misleading if only 10 people were surveyed."

Pairing 6: Fact-Checking Media Claims

Big Skill: Evaluating Evidence – Fact-checking claims in media and social platforms.

Small Skill: Organizing Information – Using a fact-checking checklist to categorize sources and verify claims.

Teaching Technique: Sensory exploration.

Actions: Have students research viral social media claims and use a checklist to evaluate credibility (author, date, citations, multiple sources).

Example: A student checks, "A post claims a celebrity donated $1M, but no credible news sources confirm it, so the claim is unreliable."

Pairing 7: Using Evidence in Argumentative Writing

Big Skill: Evaluating Evidence – Understanding the role of evidence in constructing persuasive arguments.

Small Skill: Organizing Information – Structuring an argumentative essay using an evidence-based outline.

Teaching Technique: Collaborative group work.

Actions: Have students outline an essay using a claim-evidence-reasoning (CER) chart, ensuring all arguments are backed by strong sources.

Example: A student writes, "Claim: School should start later. Evidence: Research shows teenagers need more sleep for brain function. Reasoning: More sleep improves test scores and health."

Pairing 8: Identifying Logical Fallacies

Big Skill: Evaluating Evidence – Identifying logical fallacies in arguments.

Small Skill: Organizing Information – Categorizing fallacies using a logical fallacy organizer.

Teaching Technique: Real-world applications.

Actions: Have students analyze political speeches and advertisements, then categorize different fallacies (straw man, slippery slope, false cause).

Example: A student recognizes, "This ad says, 'If you do not use this shampoo, you will never have friends,' which is a false cause fallacy."

Pairing 9: Comparing Opposing Viewpoints

Big Skill: Evaluating Evidence – Comparing different viewpoints on a controversial issue.

Small Skill: Organizing Information – Using a two-column chart to categorize supporting and opposing viewpoints.

Teaching Technique: Cultural exploration.

Actions: Have students research and categorize arguments for and against a topic like climate change policies, ensuring they cite sources.

Example: A student writes, "Supporting: Scientists show rising CO2 causes warming. Opposing: Some industries argue changes are natural."

Pairing 10: Organizing Scientific Evidence

Big Skill: Evaluating Evidence – Learning how scientific evidence supports theories.

Small Skill: Organizing Information – Using a scientific evidence matrix to categorize observations, hypotheses, and conclusions.

Teaching Technique: Hands-on experiments.

Actions: Have students conduct a simple experiment and categorize their observations, data, and conclusions in a graphic organizer.

Example: A student records, "Observation: Ice melts faster in warm air. Data: 10° C melts ice in 3 minutes. Conclusion: Higher temperature increases melting rate."

Logical Thinking – (Big Skill) Drawing Conclusions vs. (Small Skill) Breaking Down Arguments

---

Pairing 1: Identifying Missing Premises in Logical Reasoning

Big Skill: Drawing Conclusions – Using logical steps to make well-reasoned decisions rather than relying on assumptions.

Small Skill: Breaking Down Arguments – Identifying premises, evidence, and conclusions in a structured way.

Teaching Technique: Hands-on experiments.

Actions: Provide students with an incomplete argument and have them identify missing premises before drawing a logical conclusion.

Example: A student fills in, "Premise 1: The library is closed on Sundays. Premise 2: Today is Sunday. Conclusion: The library is closed today."

Pairing 2: Distinguishing Cause from Unrelated Factors

Big Skill: Drawing Conclusions – Analyzing cause-and-effect relationships to determine logical outcomes.

Small Skill: Breaking Down Arguments – Separating causes from unrelated factors in an argument.

Teaching Technique: Guided exploration.

Actions: Have students analyze a historical event and determine which causes directly led to the outcome versus those that were unrelated.

Example: A student writes, "The Great Depression led to bank failures, but it did not directly cause World War II. Other factors contributed to that conflict."

Pairing 3: Applying Logic to Real-World Problem-Solving

Big Skill: Drawing Conclusions – Understanding how logic applies in real-world problem-solving.

Small Skill: Breaking Down Arguments – Structuring a problem-solving argument with clear evidence.

Teaching Technique: Problem-solving exercises.

Actions: Have students analyze a real-world issue (e.g., should schools require uniforms?) and construct a logical argument with supporting premises.

Example: A student argues, "Premise: Schools with uniforms report less bullying. Conclusion: If our school adopts uniforms, bullying may decrease."

Pairing 4: Evaluating Viewpoints Based on Evidence

Big Skill: Drawing Conclusions – Evaluating different viewpoints before deciding on the most logical conclusion.

Small Skill: Breaking Down Arguments – Identifying when arguments rely on opinion rather than fact.

Teaching Technique: Observation journals.

Actions: Provide students with two opposing opinion pieces and ask them to determine which is supported by stronger evidence.

Example: A student writes, "This article says video games cause violence, but it has no research citations. The other article uses FBI crime data to show no connection."

Pairing 5: Recognizing Patterns to Support Logical Reasoning

Big Skill: Drawing Conclusions – Identifying patterns in evidence to form a reasonable conclusion.

Small Skill: Breaking Down Arguments – Recognizing patterns in arguments that indicate logical reasoning.

Teaching Technique: Hands-on experiments.

Actions: Have students observe a scientific experiment and use a claim-evidence-reasoning (CER) chart to draw conclusions from data.

Example: A student concludes, "If salt is added to water, then it boils faster. Data shows water with salt boiled 30 seconds faster in three trials."

Pairing 6: Validating Conclusions Based on Logical Statements

Big Skill: Drawing Conclusions – Using logic to predict outcomes based on given information.

Small Skill: Breaking Down Arguments – Evaluating if conclusions logically follow from premises.

Teaching Technique: Sensory exploration.

Actions: Provide students with a set of if-then statements and have them determine whether the conclusions are valid.

Example: A student analyzes, "If students who practice math daily score higher on tests, then students who do not practice will score lower. This is not necessarily true, as other factors affect scores."

Pairing 7: Determining Reliability of Conflicting Information

Big Skill: Drawing Conclusions – Determining the most reasonable decision when presented with conflicting information.

Small Skill: Breaking Down Arguments – Assessing the credibility of different sources before concluding.

Teaching Technique: Collaborative group work.

Actions: Have students compare news reports from different sources on the same event, identifying which presents the strongest argument.

Example: A student notes, "Source A interviews experts and presents statistics, while Source B only uses social media opinions. Source A is more reliable."

Pairing 8: Identifying and Challenging Hidden Assumptions

Big Skill: Drawing Conclusions – Recognizing when assumptions interfere with logical thinking.

Small Skill: Breaking Down Arguments – Identifying assumptions and distinguishing them from facts.

Teaching Technique: Real-world applications.

Actions: Provide students with an argument that contains hidden assumptions and ask them to identify and challenge those assumptions.

Example: A student finds, "This argument says, 'If students do not go to college, they will not be successful.' But it assumes college is the only path to success, which is not true."

Pairing 9: Evaluating Deductive Reasoning in Arguments

Big Skill: Drawing Conclusions – Using deductive reasoning to evaluate a claim.

Small Skill: Breaking Down Arguments – Identifying when an argument uses deductive reasoning correctly.

Teaching Technique: Cultural exploration.

Actions: Have students analyze famous philosophical or legal arguments and determine whether their conclusions are logically sound.

Example: A student evaluates, "All humans are mortal. Socrates is human. Therefore, Socrates is mortal," recognizing this as valid deductive reasoning.

Pairing 10: Differentiating Correlation from Causation

Big Skill: Drawing Conclusions – Understanding when correlation does not imply causation.

Small Skill: Breaking Down Arguments – Distinguishing between causal and correlational relationships in arguments.

Teaching Technique: Hands-on experiments.

Actions: Have students examine examples of correlations and determine whether they indicate real causation or just coincidence.

Example: A student writes, "Ice cream sales and drowning incidents increase in summer, but eating ice cream does not cause drowning. Warm weather leads to both."

Logical Thinking – (Big Skill) Drawing Conclusions vs. (Small Skill) Applying Logical Rules

---

Pairing 1: Identifying Logical Steps in Reasoning

Big Skill: Drawing Conclusions – Using logical steps to make well-reasoned decisions rather than relying on assumptions.

Small Skill: Applying Logical Rules – Solving logic puzzles, practicing syllogisms, and simple coding exercises.

Teaching Technique: Hands-on experiments.

Actions: Have students complete a logic puzzle where they must use a set of given clues to determine the correct solution, ensuring they apply deductive reasoning.

Example: A student solves, "If Tom is older than Jane, and Jane is older than Mike, then who is the oldest?" **Answer:** Tom is the oldest.

Pairing 2: Determining Logical Validity in Syllogisms

Big Skill: Drawing Conclusions – Recognizing patterns in data and making logical decisions.

Small Skill: Applying Logical Rules – Practicing syllogisms to determine logical validity.

Teaching Technique: Guided exploration.

Actions: Provide students with a list of syllogisms and ask them to determine whether the conclusions are logically valid based on the premises.

Example: A student evaluates,

1. **Premise 1:** All mammals have fur.
2. **Premise 2:** A whale is a mammal.
3. **Conclusion:** A whale has fur.
4. **Answer:** The conclusion is false because the premise is incorrect—not all mammals have fur.

Pairing 3: Using If-Then Logic in Coding

Big Skill: Drawing Conclusions – Analyzing cause-and-effect relationships to determine logical outcomes.

Small Skill: Applying Logical Rules – Using conditional statements in coding exercises.

Teaching Technique: Problem-solving exercises.

Actions: Have students write a simple coding program using if-then statements to create an interactive sequence.

Example: A student writes,

```
if button_is_clicked:
    print("Hello, world!")
```

and tests whether the program follows the correct logic.

Pairing 4: Organizing Information with Logic Grids

Big Skill: Drawing Conclusions – Evaluating multiple sources of evidence before making a decision.

Small Skill: Applying Logical Rules – Using logic grids to organize information.

Teaching Technique: Observation journals.

Actions: Provide students with a logic grid puzzle, where they must determine which person owns which pet based on clues.

Example: A student solves, "If Lisa has a dog, and the person with the cat is not Mike, then who owns the bird?" **Answer:** Mike owns the bird.

Pairing 5: Distinguishing Correlation from Causation

Big Skill: Drawing Conclusions – Understanding the difference between correlation and causation.

Small Skill: Applying Logical Rules – Identifying fallacies in logical reasoning.

Teaching Technique: Hands-on experiments.

Actions: Have students evaluate correlations vs. causations by analyzing misleading claims from advertisements or media.

Example: A student evaluates,

1. **Claim:** People who eat more chocolate score higher on math tests.
2. **Conclusion:** Eating chocolate improves math skills.
3. **Answer:** This is a correlation, not causation—other factors may contribute to the math scores.

Pairing 6: Evaluating Logical Statements with Truth Tables

Big Skill: Drawing Conclusions – Using logical structures to evaluate an argument.

Small Skill: Applying Logical Rules – Using truth tables to analyze statements.

Teaching Technique: Sensory exploration.

Actions: Have students construct truth tables for given statements and determine whether they are logically valid.

Example: A student creates a truth table for:

1. **Statement:** If it rains, then the ground is wet.

2. **Answer:** The statement is only false if it rains and the ground is not wet, which means the claim holds true under logical conditions.

Pairing 7: Mapping Decision Processes with Flowcharts

Big Skill: Drawing Conclusions – Applying step-by-step reasoning in decision-making.

Small Skill: Applying Logical Rules – Using flowcharts to map logical processes.

Teaching Technique: Collaborative group work.

Actions: Have students create flowcharts to outline how a decision-making process works, using if-then logic.

Example: A student draws,

- **Question:** Should I bring an umbrella?
    - If it is raining, then bring an umbrella.
    - If it is not raining, then do not bring an umbrella.

Pairing 8: Applying Logic in Scientific Hypothesis Testing

Big Skill: Drawing Conclusions – Understanding how logic applies in scientific reasoning.

Small Skill: Applying Logical Rules – Practicing hypothesis testing through logical statements.

Teaching Technique: Real-world applications.

Actions: Have students analyze a scientific experiment and break down the logic behind hypothesis testing.

Example: A student writes,

1. **Hypothesis:** If plants receive sunlight, then they will grow.
2. **Experiment:** Two plants, one with sunlight, one without.
3. **Conclusion:** The plant in sunlight grew, supporting the hypothesis.

Pairing 9: Using Decision Trees for Logical Comparisons

Big Skill: Drawing Conclusions – Evaluating choices using step-by-step logical comparisons.

Small Skill: Applying Logical Rules – Using decision trees to visualize reasoning.

Teaching Technique: Cultural exploration.

Actions: Have students create a decision tree for a real-world problem, such as selecting a school project topic.

Example:

- Decision tree:
    - Do you like science? → Yes → Choose a science fair project.
    - Do you prefer writing? → Yes → Choose a research paper.

Pairing 10: Debugging Logic Errors in Coding

Big Skill: Drawing Conclusions – Applying deductive reasoning to solve problems.

Small Skill: Applying Logical Rules – Practicing coding logic through debugging exercises.

Teaching Technique: Hands-on experiments.

Actions: Have students analyze a flawed coding script and debug it by following logical rules.

Example: A student fixes,

```
if temperature > 100:
    print("It is hot outside!")
else:
    print("It is cold outside!")
```

Fix: Add a condition for moderate temperature to prevent faulty conclusions, such as:

```
if temperature > 100:
    print("It is hot outside!")
else-if temperature < 50:
    print("It is cold outside!")
else
    print("It is a moderate day.")
```

Logical Thinking – (Big Skill) Drawing Conclusions vs. (Small Skill) Organizing Information

---

Pairing 1: Mapping Cause-and-Effect Relationships

Big Skill: Drawing Conclusions – Using logical steps to make well-reasoned decisions rather than relying on assumptions.

Small Skill: Organizing Information – Using graphic organizers to categorize evidence in debates or essays.

Teaching Technique: Hands-on experiments.

Actions: Have students use a cause-and-effect diagram to map out the logical steps leading to a conclusion based on scientific data.

Example: A student writes,

1. **Observation:** A plant in direct sunlight grows faster.
2. **Evidence:** Plants need sunlight for photosynthesis.
3. **Conclusion:** More sunlight leads to increased plant growth.

Pairing 2: Weighing Pros and Cons for Decision-Making

Big Skill: Drawing Conclusions – Analyzing multiple perspectives before making a logical decision.

Small Skill: Organizing Information – Using a pros-and-cons chart to compare arguments.

Teaching Technique: Guided exploration.

Actions: Have students create a pros-and-cons chart to evaluate different viewpoints on a school policy change.

Example: A student compares,

1. **Pros:** Longer recess improves focus in class.
2. **Cons:** Extending recess may reduce learning time.
3. **Conclusion:** A balanced schedule with slightly longer recess could be beneficial.

Pairing 3: Identifying Patterns in Historical Events

Big Skill: Drawing Conclusions – Identifying patterns in information to make logical inferences.

Small Skill: Organizing Information – Using a Venn diagram to compare similarities and differences.

Teaching Technique: Problem-solving exercises.

Actions: Have students compare two historical events using a Venn diagram to identify common causes and unique factors.

Example: A student writes,

1. **Similarities:** Both the American and French Revolutions fought for freedom.
2. **Differences:** The American Revolution led to democracy, while the French Revolution resulted in political instability.

Pairing 4: Ranking Evidence for Argument Strength

Big Skill: Drawing Conclusions – Evaluating evidence to determine the strength of an argument.

Small Skill: Organizing Information – Using an evidence-ranking chart to assess the reliability of sources.

Teaching Technique: Observation journals.

Actions: Have students rate different types of evidence (scientific studies, expert opinions, anecdotes) in a graphic organizer.

Example: A student categorizes,

1. **Strong Evidence:** Peer-reviewed research study.
2. **Weak Evidence:** Personal blog post.
3. **Conclusion:** Arguments should rely on strong, well-supported evidence.

Pairing 5: Visualizing Decision Processes with Flowcharts

Big Skill: Drawing Conclusions – Understanding the logical connections between causes and effects.

Small Skill: Organizing Information – Using a flowchart to visualize the decision-making process.

Teaching Technique: Hands-on experiments.

Actions: Have students construct a flowchart to analyze the logical sequence of decisions in a real-world problem.

Example: A student creates,

- Decision Tree:
  - **Question:** Should students have homework every day?
  - **If Yes:** Helps reinforce learning but may cause stress.
  - **If No:** Reduces stress but may limit practice.

Pairing 6: Structuring Logical Arguments for Debate

Big Skill: Drawing Conclusions – Using deductive reasoning to analyze a claim.

Small Skill: Organizing Information – Structuring a debate outline with claims, evidence, and counterarguments.

Teaching Technique: Sensory exploration.

Actions: Have students organize a debate outline where each argument is supported by logical evidence.

Example: A student writes,

1. **Claim:** School uniforms improve discipline.
2. **Evidence:** Studies show fewer behavior issues in schools with uniforms.
3. **Counterargument:** Uniforms limit self-expression.
4. **Conclusion:** Uniforms have benefits but should allow for some individuality.

Pairing 7: Distinguishing Between Correlation and Causation

Big Skill: Drawing Conclusions – Distinguishing between correlation and causation.

Small Skill: Organizing Information – Using a comparison chart to separate linked events from direct causes.

Teaching Technique: Collaborative group work.

Actions: Have students analyze misleading claims in advertisements using a comparison chart.

Example: A student categorizes,

1. **Claim:** "Drinking energy drinks improves grades."
2. **Correlation:** People who study more may also drink energy drinks.
3. **Causation:** Studying improves grades, not the drink itself.

Pairing 8: Predicting Future Outcomes Based on Evidence

Big Skill: Drawing Conclusions – Making evidence-based predictions.

Small Skill: Organizing Information – Using a hypothesis-evidence-outcome chart.

Teaching Technique: Real-world applications.

Actions: Have students predict future trends based on current data using a structured chart.

Example: A student writes,

1. **Hypothesis:** If social media use increases, then attention spans will decrease.
2. **Evidence:** Studies show students who spend more time on social media have lower focus in school.
3. **Outcome:** Schools may need to adjust learning methods.

Pairing 9: Evaluating Conflicting Evidence from Multiple Sources

Big Skill: Drawing Conclusions – Evaluating conflicting evidence before making a decision.

Small Skill: Organizing Information – Using a multiple-perspectives organizer.

Teaching Technique: Cultural exploration.

Actions: Have students compare news articles from different sources to determine the most reliable conclusions.

Example: A student writes,

1. **Source A:** Focuses only on benefits of technology.
2. **Source B:** Shows both advantages and drawbacks.
3. **Conclusion:** Source B is more balanced and trustworthy.

Pairing 10: Using Logical Reasoning to Solve Complex Problems

Big Skill: Drawing Conclusions – Applying logical reasoning to solve complex problems.

Small Skill: Organizing Information – Using a structured problem-solving chart to analyze solutions.

Teaching Technique: Hands-on experiments.

Actions: Have students create a problem-solution-evidence chart to evaluate potential solutions to a real-world issue.

Example: A student writes,

1. **Problem:** Traffic congestion in cities.
2. **Solution 1:** Build more roads.
3. **Solution 2:** Improve public transportation.
4. **Evidence:** Studies show that better public transit reduces congestion.
5. **Conclusion:** Investing in public transit is a more effective solution.

Example Teaching Strategies for Big vs. Small Thinking Skill Pairings of Goal Setting at the Middle School Level

Here are 90 unique teaching strategy examples for the big thinking and small thinking skill pairings, each described in detail to align with **Big Thinking Skills** (Concepts & Strategies) and **Small Thinking Skills** (Step-by-Step Execution) for the **Goal Setting Skill** taught at the middle school level.

Goal Setting

Big Thinking Skills (Concepts & Strategies)

- **Long-Term Vision:** Understanding how setting smaller goals contributes to larger life achievements.
- **Strategic Thinking:** Developing a roadmap for achieving personal or academic goals.
- **Self-Monitoring & Adaptation:** Evaluating progress and making necessary adjustments.

Small Thinking Skills (Step-by-Step Execution)

- **Breaking Goals into Actionable Steps:** Learning how to deconstruct a big goal into smaller, achievable milestones.
- **Tracking Progress:** Using checklists, habit trackers, or goal charts to measure success.
- **Time Management:** Practicing prioritization and scheduling tasks effectively.

Goal Setting – (Big Skill) Long-Term Vision vs. (Small Skill) Breaking Goals into Actionable Steps

---

Pairing 1: Creating a Milestone Timeline for Career Goals

Big Skill: Long-Term Vision – Understanding how setting smaller goals contributes to larger life achievements.

Small Skill: Breaking Goals into Actionable Steps – Learning how to deconstruct a big goal into smaller, achievable milestones.

Teaching Technique: Hands-on experiments.

Actions: Have students choose a career goal and break it into education, skill-building, and experience milestones on a timeline.

Example: A student chooses "Becoming a doctor" and sets milestones:

1. **Get good grades** in science classes.
2. **Volunteer** at a hospital.
3. **Research** medical school requirements.

Pairing 2: Mapping Out an Academic Success Plan

Big Skill: Long-Term Vision – Recognizing the importance of planning for future success.

Small Skill: Breaking Goals into Actionable Steps – Identifying short-term tasks that lead to long-term success.

Teaching Technique: Guided exploration.

Actions: Have students create a personal academic success plan with specific yearly goals leading to their desired high school or college path.

Example: A student who wants to attend a top university sets milestones:

1. **Maintain** a 3.8+ GP.
2. **Join** a leadership club.
3. **Participate** in science competitions.

Pairing 3: Designing a Weekly Routine for Skill Development

Big Skill: Long-Term Vision – Understanding that consistency leads to progress over time.

Small Skill: Breaking Goals into Actionable Steps – Developing a daily or weekly routine for progress.

Teaching Technique: Problem-solving exercises.

Actions: Have students design a weekly schedule that includes time for homework, hobbies, and self-improvement activities.

Example: A student working on learning to code indicates:

1. **30 minutes** of coding practice each day.
2. **Completing** one coding project per month.
3. **Applying** for a summer coding camp.

Pairing 4: Using Journals to Track Progress and Stay Motivated

Big Skill: Long-Term Vision – Learning how to stay motivated when working toward big goals.

Small Skill: Breaking Goals into Actionable Steps – Setting milestones with rewards to maintain motivation.

Teaching Technique: Observation journals.

Actions: Have students track their progress toward a long-term goal in a journal and set small rewards for reaching key milestones.

Example: A student aiming to write a book indicates:

1. **Finish one chapter per month** → Reward: Favorite snack.
2. **Complete first draft** → Reward: New notebook.
3. **Publish final version** → Reward: Celebration dinner.

Pairing 5: Visualizing a Goal with a Step-by-Step Ladder

Big Skill: Long-Term Vision – Understanding that personal growth happens in stages.

Small Skill: Breaking Goals into Actionable Steps – Using a goal ladder to break a large ambition into step-by-step progress.

Teaching Technique: Hands-on experiments.

Actions: Have students visualize a career or personal goal as a ladder, writing each step needed to climb toward success.

Example: A student who wants to start a YouTube channel indicates:

1. **Learn** basic video editing.
2. **Plan** 10 video topics.
3. **Post** once a week for 3 months.

Pairing 6: Identifying Obstacles and Solutions for Goal Achievement

Big Skill: Long-Term Vision – Understanding that achieving a dream requires discipline and persistence.

Small Skill: Breaking Goals into Actionable Steps – Identifying potential obstacles and ways to overcome them.

Teaching Technique: Sensory exploration.

Actions: Have students list possible setbacks for a goal and create solutions to stay on track.

Example: A student aiming to run a marathon identifies:

1. **Lack of stamina** → Solution: Start with 1-mile runs.
2. **Losing motivation** → Solution: Find a running partner.
3. **Time management** → Solution: Schedule runs in advance.

Pairing 7: Building Habits Through a 30-Day Challenge

Big Skill: Long-Term Vision – Seeing how small, repeated actions lead to expertise.

Small Skill: Breaking Goals into Actionable Steps – Practicing deliberate improvement through daily habit tracking.

Teaching Technique: Collaborative group work.

Actions: Have students set a 30-day challenge to build a habit related to a long-term goal.

Example: A student wanting to improve public speaking indicates:

1. **Practice** speaking in front of the mirror.
2. **Record** a speech and evaluate it.
3. **Present** in front of friends once a week.

Pairing 8: Creating a Flexible Roadmap for Goal Adaptability

Big Skill: Long-Term Vision – Recognizing that goals evolve over time.

Small Skill: Breaking Goals into Actionable Steps – Creating a flexible roadmap to adjust goals as needed.

Teaching Technique: Real-world applications.

Actions: Have students make a roadmap for a future goal, adding alternative routes if plans change.

Example: A student who wants to become an architect indicates:

1. **Plan A:** Study architecture in college.
2. **Plan B:** Take online courses in design.
3. **Plan C:** Shadow an architect for hands-on experience.

Pairing 9: Prioritizing Tasks Using a Priority Matrix

Big Skill: Long-Term Vision – Understanding that achieving a goal requires prioritizing tasks.

Small Skill: Breaking Goals into Actionable Steps – Using a priority matrix to organize tasks effectively.

Teaching Technique: Cultural exploration.

Actions: Have students categorize their tasks into important vs. urgent quadrants to decide what to tackle first.

Example: A student planning to start a school club categorizes:

1. **Important & Urgent:** Get teacher approval.
2. **Important & Not Urgent:** Create a club website.
3. **Not Important & Not Urgent:** Design club T-shirts.

Pairing 10: Reviewing and Adjusting Goals for Long-Term Success

Big Skill: Long-Term Vision – Understanding that success is built on consistency and adaptability.

Small Skill: Breaking Goals into Actionable Steps – Reviewing and adjusting progress monthly to stay on track.

Teaching Technique: Hands-on experiments.

Actions: Have students set a monthly goal review system to reflect on what is working and what needs improvement.

Example: A student learning a new language indicates:

1. **Month 1:** Learn 50 words → **Adjust:** Focus more on conversation skills.
2. **Month 2:** Hold a short conversation → **Adjust:** Watch movies in the language.
3. **Month 3:** Write a short essay → **Adjust:** Get feedback from a native speaker.

Goal Setting – (Big Skill) Long-Term Vision vs. (Small Skill) Tracking Progress

---

Pairing 1: Creating a Personal Goal Chart

Big Skill: Long-Term Vision – Understanding how setting smaller goals contributes to larger life achievements.

Small Skill: Tracking Progress – Using checklists, habit trackers, or goal charts to measure success.

Teaching Technique: Hands-on experiments.

Actions: Have students create a personal goal chart with milestones and check off tasks as they complete them.

Example: A student tracking "learning to play the guitar" indicates:

1. **Learn three basic chords** → Completed.
2. **Play a full song** → Completed.
3. **Perform in front of family** → In progress.

Pairing 2: Maintaining a 30-Day Habit Tracker

Big Skill: Long-Term Vision – Understanding how progress builds over time.

Small Skill: Tracking Progress – Using a habit tracker to maintain consistency.

Teaching Technique: Guided exploration.

Actions: Have students use a 30-day habit tracker to track their daily efforts toward a goal.

Example: A student aiming to exercise regularly tracks:

1. **Monday:** Complete.
2. **Tuesday:** Complete.
3. **Wednesday:** Not complete (adjust for next week).

Pairing 3: Using a Step-by-Step Goal Checklist

Big Skill: Long-Term Vision – Learning that breaking goals into small tasks makes them more achievable.

Small Skill: Tracking Progress – Creating a step-by-step goal checklist.

Teaching Technique: Problem-solving exercises.

Actions: Have students outline a step-by-step checklist for a long-term goal and track progress weekly.

Example: A student working toward writing a short story indicates:

1. **Brainstorm story ideas** → Complete.
2. **Write the first draft** → Complete.
3. **Edit and get feedback** → In Progress.

Pairing 4: Tracking Growth with a Goal Journal

Big Skill: Long-Term Vision – Recognizing that consistent tracking leads to long-term success.

Small Skill: Tracking Progress – Using a goal journal to document reflections and improvements.

Teaching Technique: Observation journals.

Actions: Have students write weekly reflections on their progress toward a personal or academic goal.

Example: A student practicing public speaking writes:

1. **Week 1:** "Nervous, but spoke clearly."
2. **Week 2:** "Improved eye contact."
3. **Week 3:** "Need to work on voice projection."

Pairing 5: Using a Vision Board to Track Milestones

Big Skill: Long-Term Vision – Understanding how small efforts lead to big achievements.

Small Skill: Tracking Progress – Using a vision board to visually track milestones.

Teaching Technique: Hands-on experiments.

Actions: Have students create a goal vision board and add pictures or quotes as they complete milestones.

Example: A student aiming to save money for a new bike adds:

1. **Picture of a piggy bank** → First $50 saved!
2. **Picture of the bike** → Halfway to goal.

Pairing 6: Evaluating Progress with a Monthly Review

Big Skill: Long-Term Vision – Understanding that goals require ongoing assessment.

Small Skill: Tracking Progress – Using a monthly progress review to adjust strategies.

Teaching Technique: Sensory exploration.

Actions: Have students evaluate their progress monthly and adjust their plan if needed.

Example: A student learning Spanish writes:

1. **Month 1:** Learned 50 words → Complete.
2. **Month 2:** Held a 2-minute conversation → Complete.
3. **Month 3:** Struggling with grammar → Adjust strategy.

Pairing 7: Staying Motivated with a Milestone Reward System

Big Skill: Long-Term Vision – Seeing how goal tracking leads to motivation.

Small Skill: Tracking Progress – Creating a milestone reward system for motivation.

Teaching Technique: Collaborative group work.

Actions: Have students set small rewards for completing different goal levels.

Example: A student preparing to run a 5K indicates:

1. **Run 1 mile without stopping** → Complete. Get a new water bottle.
2. **Run 3 miles** → Complete. Get new running shoes.
3. **Run 5K race** → Complete. Celebrate with friends.

Pairing 8: Using a Progress Thermometer for Motivation

Big Skill: Long-Term Vision – Understanding how long-term success depends on short-term tracking.

Small Skill: Tracking Progress – Using a progress thermometer to visually track improvement.

Teaching Technique: Real-world applications.

Actions: Have students color in a progress thermometer as they reach different milestones.

Example: A student working toward reading 10 books in a year fills in:

1. **2 books read** → Complete.
2. **5 books read** → Halfway there!

Pairing 9: Using Self-Assessment Surveys for Reflection

Big Skill: Long-Term Vision – Learning that tracking goals helps identify areas for improvement.

Small Skill: Tracking Progress – Using self-assessment surveys to measure progress.

Teaching Technique: Cultural exploration.

Actions: Have students complete self-reflection surveys to assess how they are progressing toward their goals.

Example: A student preparing for a math competition evaluates:

1. Confidence level in problem-solving: 6/10 → Need more practice.
2. Speed in solving equations: 8/10 → Strong area.

Pairing 10: Building Confidence with a Success Jar

Big Skill: Long-Term Vision – Recognizing that tracking small successes builds confidence.

Small Skill: Tracking Progress – Creating a "Success Jar" to record achievements.

Teaching Technique: Hands-on experiments.

Actions: Have students write small successes on slips of paper and place them in a jar to review at the end of the year.

Example: A student working on overcoming shyness writes:

1. "**Gave** a presentation in class."
2. "**Introduced** myself to a new student."
3. "**Asked** a question in a group discussion."

Goal Setting – (Big Skill) Long-Term Vision vs. (Small Skill) Time Management

---

Pairing 1: Creating a Long-Term Study Plan

Big Skill: Long-Term Vision – Understanding how setting smaller goals contributes to larger life achievements.

Small Skill: Time Management – Practicing prioritization and scheduling tasks effectively.

Teaching Technique: Hands-on experiments.

Actions: Have students create a long-term study plan for a subject they want to master by breaking it into weekly study sessions.

Example: A student preparing for a math competition schedules:

1. **Week 1:** Review algebra basics.
2. **Week 2:** Solve 10 practice problems per day.
3. **Week 3:** Take a timed practice test.

Pairing 2: Using a Weekly Planner for Time Allocation

Big Skill: Long-Term Vision – Learning that breaking a goal into tasks makes it more manageable.

Small Skill: Time Management – Using a weekly planner to allocate time for different tasks.

Teaching Technique: Guided exploration.

Actions: Have students design a weekly planner to balance schoolwork, extracurriculars, and personal development.

Example: A student balancing school and soccer practice plans:

1. **Monday:** Study for science quiz (1 hour), Soccer practice (2 hours).
2. **Tuesday:** Read for English (30 min), Homework (1 hour).

Pairing 3: Prioritizing Tasks Using a Priority Matrix

Big Skill: Long-Term Vision – Understanding that time management improves long-term success.

Small Skill: Time Management – Learning to prioritize urgent vs. important tasks using a priority matrix.

Teaching Technique: Problem-solving exercises.

Actions: Have students categorize their daily tasks into a 4-quadrant priority matrix (e.g., Urgent/Important, Important/Not Urgent).

Example: A student planning for a big test and a soccer game categorizes:

1. **Urgent & Important:** Study for the test.
2. **Important but Not Urgent**: Soccer practice.
3. **Not Important & Not Urgent:** Watching TV.

Pairing 4: Creating a Daily Routine for Productivity

Big Skill: Long-Term Vision – Understanding that efficient scheduling maximizes productivity.

Small Skill: Time Management – Creating a daily routine to build productive habits.

Teaching Technique: Observation journals.

Actions: Have students track their daily habits for a week and adjust their schedules for better productivity.

Example: A student preparing to read more books adjusts their routine:

1. **Read** for 15 minutes before bed.
2. **Reduce** social media time by 30 minutes.

Pairing 5: Using a Countdown Calendar for Big Projects

Big Skill: Long-Term Vision – Recognizing that managing time wisely prevents last-minute stress.

Small Skill: Time Management – Using a countdown calendar for big projects.

Teaching Technique: Hands-on experiments.

Actions: Have students create a countdown calendar for an upcoming project and plan tasks for each phase.

Example: A student preparing a science fair project schedules:

1. **Week 1:** Choose a topic.
2. **Week 2:** Conduct research.
3. **Week 3:** Start the experiment.

Pairing 6: Reviewing Weekly Progress to Adjust Schedules

Big Skill: Long-Term Vision – Understanding that success requires planning and adjusting schedules.

Small Skill: Time Management – Reviewing weekly and monthly progress to adjust time allocation.

Teaching Technique: Sensory exploration.

Actions: Have students evaluate their weekly time management and adjust based on what worked or did not.

Example: A student tracking fitness goals finds:

1. **Week 1:** Exercised 5 days → Felt great!
2. **Week 2:** Only exercised 2 days → Adjust plan to schedule workouts earlier.

Pairing 7: Managing Daily Responsibilities with Time-Blocking

Big Skill: Long-Term Vision – Understanding that time spent wisely builds long-term habits.

Small Skill: Time Management – Using time-blocking techniques to manage daily responsibilities.

Teaching Technique: Collaborative group work.

Actions: Have students break their day into time blocks for different types of tasks and reflect on their effectiveness.

Example: A student juggling homework and hobbies creates:

1. **4:00-5:00 PM:** Homework.
2. **5:00-6:00 PM:** Break.
3. **6:00-7:00 PM:** Music practice.

Pairing 8: Setting Personal Deadlines to Avoid Procrastination

Big Skill: Long-Term Vision – Recognizing that setting realistic deadlines leads to better outcomes.

Small Skill: Time Management – Creating a personal deadline system to stay on track.

Teaching Technique: Real-world applications.

Actions: Have students set personal deadlines ahead of actual deadlines to avoid procrastination.

Example: A student preparing for a history report indicates:

1. **Research due** one week before the actual deadline.
2. **First draft due** four days before submission.

Pairing 9: Analyzing Time Usage with a Time Log

Big Skill: Long-Term Vision – Understanding that balancing different priorities improves efficiency.

Small Skill: Time Management – Using a time log to analyze time spent on various activities.

Teaching Technique: Cultural exploration.

Actions: Have students log their activities for a week and identify areas to reallocate time for better balance.

Example: A student logs their time and realizes:

1. **Screen time: 3 hours/day** → Reduce to 1 hour.
2. **Study time: 1 hour/day** → Increase to 2 hours.

Pairing 10: Using Weekly Reflections to Improve Time Management

Big Skill: Long-Term Vision – Learning that consistency in time management leads to long-term success.

Small Skill: Time Management – Using a weekly reflection sheet to assess progress.

Teaching Technique: Hands-on experiments.

Actions: Have students reflect on their weekly productivity and make adjustments for the next week.

Example: A student reviewing piano practice goals writes:

1. **This week:** Practiced only 3 days instead of 5.
2. **Next week:** Set practice alarms to stay on track.

Goal Setting – (Big Skill) Strategic Thinking vs. (Small Skill) Breaking Goals into Actionable Steps

---

Pairing 1: Creating a Roadmap for Achieving Goals

Big Skill: Strategic Thinking – Developing a roadmap for achieving personal or academic goals.

Small Skill: Breaking Goals into Actionable Steps – Learning how to deconstruct a big goal into smaller, achievable milestones.

Teaching Technique: Hands-on experiments.

Actions: Have students create a step-by-step roadmap for a personal or academic goal using a graphic organizer.

Example: A student planning to join the school basketball team indicates:

1. **Practice** dribbling daily.
2. **Improve** shooting accuracy.
3. **Attend** tryouts.

Pairing 2: Planning for Long-Term Growth

Big Skill: Strategic Thinking – Understanding that long-term goals require detailed planning.

Small Skill: Breaking Goals into Actionable Steps – Setting weekly and monthly targets to stay on track.

Teaching Technique: Guided exploration.

Actions: Have students map out a three-month plan to improve in a specific subject.

Example: A student struggling with math indicates:

1. **Month 1:** Review basic algebra.
2. **Month 2:** Solve 10 problems daily.
3. **Month 3:** Take a full practice test.

Pairing 3: Anticipating Challenges and Backup Plans

Big Skill: Strategic Thinking – Learning how to anticipate obstacles and adjust goals.

Small Skill: Breaking Goals into Actionable Steps – Identifying backup plans for each milestone.

Teaching Technique: Problem-solving exercises.

Actions: Have students list possible challenges and create solutions for staying on track.

Example: A student aiming to get better at public speaking indicates:

1. **If nervous,** practice with a friend.
2. **If stumbling on words,** slow down speech.
3. **If forgetting key points,** use note cards.

Pairing 4: Revising Goals Based on Progress

Big Skill: Strategic Thinking – Recognizing that success requires long-term vision and flexibility.

Small Skill: Breaking Goals into Actionable Steps – Revising goals based on progress.

Teaching Technique: Observation journals.

Actions: Have students track progress monthly and adjust goals if necessary.

Example: A student learning a new language reflects:

1. **Month 1:** Learned 100 words → Increase goal to 150 words.
2. **Month 2:** Struggled with pronunciation → Focus on listening practice.

Pairing 5: Using the Ladder Approach to Reach Goals

Big Skill: Strategic Thinking – Understanding that breaking down goals improves motivation.

Small Skill: Breaking Goals into Actionable Steps – Using a "ladder approach" to reach goals step by step.

Teaching Technique: Hands-on experiments.

Actions: Have students create a goal ladder, placing small steps on each rung.

Example: A student preparing for a science fair project indicates:

1. **Choose** a topic.
2. **Research** and gather materials.
3. **Conduct** an experiment.

Pairing 6: Prioritizing High-Impact Steps for Progress

Big Skill: Strategic Thinking – Seeing the connection between short-term and long-term success.

Small Skill: Breaking Goals into Actionable Steps – Prioritizing high-impact steps for quicker progress.

Teaching Technique: Sensory exploration.

Actions: Have students list the top three actions that will make the biggest difference in achieving a goal.

Example: A student training for a 5K race indicates:

1. **Run** three times a week.
2. **Improve** endurance with strength training.
3. **Track** running times weekly.

Pairing 7: Ordering Steps for Maximum Efficiency

Big Skill: Strategic Thinking – Recognizing that achieving a goal requires a logical sequence.

Small Skill: Breaking Goals into Actionable Steps – Ordering steps in the most efficient way.

Teaching Technique: Collaborative group work.

Actions: Have students rearrange scrambled goal steps into the correct order.

Example: A student planning to write a book correctly orders:

1. **Brainstorm** ideas.
2. **Outline** chapters.
3. **Write** the first draft.

Pairing 8: Using a Checklist for Goal Tracking

Big Skill: Strategic Thinking – Understanding that success depends on tracking progress effectively.

Small Skill: Breaking Goals into Actionable Steps – Using a checklist to stay on schedule.

Teaching Technique: Real-world applications.

Actions: Have students create a goal-tracking checklist and update it weekly.

Example: A student working on building a robot for a competition indicates:

1. **Research** design options.
2. **Gather** materials.
3. **Assemble** components.

Pairing 9: Creating a Reflection Process for Continuous Improvement

Big Skill: Strategic Thinking – Understanding that setbacks require re-evaluation of goals.

Small Skill: Breaking Goals into Actionable Steps – Creating a reflection process for continuous improvement.

Teaching Technique: Cultural exploration.

Actions: Have students write weekly reflections on their goal journey.

Example: A student preparing for a debate competition writes:

1. **This week:** Improved delivery but need better rebuttals.
2. **Next step:** Research counterarguments more deeply.

Pairing 10: Structuring Goals Using a Pyramid Approach

Big Skill: Strategic Thinking – Learning that detailed plans increase the likelihood of success.

Small Skill: Breaking Goals into Actionable Steps – Using a goal pyramid to structure steps from easy to complex.

Teaching Technique: Hands-on experiments.

Actions: Have students draw a pyramid of milestones, starting with the easiest steps at the bottom.

Example: A student aiming to start a school club indicates:

1. **Find** a faculty advisor.
2. **Recruit** members.
3. **Plan** the first event.

Goal Setting – (Big Skill) Strategic Thinking vs. (Small Skill) Tracking Progress

---

Pairing 1: Using Checklists for Goal Tracking

Big Skill: Strategic Thinking – Developing a roadmap for achieving personal or academic goals.

Small Skill: Tracking Progress – Using checklists, habit trackers, or goal charts to measure success.

Teaching Technique: Hands-on experiments.

Actions: Have students create a goal roadmap with checkpoints, using a checklist to track progress.

Example: A student planning to get an "A" in science indicates:

1. **Complete** all homework assignments.
2. **Study** for quizzes weekly.
3. **Join** a study group for extra help.

Pairing 2: Developing Habits for Long-Term Success

Big Skill: Strategic Thinking – Understanding the importance of progress tracking in goal achievement.

Small Skill: Tracking Progress – Using a habit tracker to monitor daily activities.

Teaching Technique: Guided exploration.

Actions: Have students design a habit tracker to develop consistent study or personal growth habits.

Example: A student wanting to read more books tracks:

1. **Monday:** Complete.
2. **Tuesday:** Complete.
3. **Wednesday:** (adjust for next week).

Pairing 3: Breaking Goals into Milestones

Big Skill: Strategic Thinking – Learning how to divide a long-term goal into trackable short-term objectives.

Small Skill: Tracking Progress – Using a milestone chart to break a large goal into steps.

Teaching Technique: Problem-solving exercises.

Actions: Have students create a progress chart with mini-milestones.

Example: A student working toward running a 5K race indicates:

1. **Run 1 mile** without stopping.
2. **Run 2 miles** comfortably.
3. **Complete** a full 5K practice run.

Pairing 4: Using Goal Journals for Reflection

Big Skill: Strategic Thinking – Understanding that tracking progress helps adjust strategies.

Small Skill: Tracking Progress – Using a goal journal to document reflections and changes.

Teaching Technique: Observation journals.

Actions: Have students write weekly reflections on their progress and adjust their strategies if needed.

Example: A student learning to play the piano writes:

1. **Week 1:** Practiced 15 minutes a day but struggled with hand coordination.
2. **Week 2:** Increased practice to 20 minutes and improved.

Pairing 5: Reward-Based Progress Tracking

Big Skill: Strategic Thinking – Recognizing that tracking helps maintain motivation.

Small Skill: Tracking Progress – Using a reward-based progress system to stay motivated.

Teaching Technique: Hands-on experiments.

Actions: Have students set small rewards for reaching different milestones.

Example: A student working toward writing a short story indicates:

1. **Complete** character outlines → **Reward:** Favorite snack.
2. **Write** the first chapter → **Reward:** Watch a movie.
3. **Finish** the full draft.

Pairing 6: Reviewing and Adjusting Strategies

Big Skill: Strategic Thinking – Understanding that consistent tracking ensures continuous improvement.

Small Skill: Tracking Progress – Using monthly goal reviews to analyze progress and make changes.

Teaching Technique: Sensory exploration.

Actions: Have students review their goal progress monthly and adjust strategies if needed.

Example: A student preparing for a debate competition reflects:

1. **Month 1:** Focused on speech delivery.
2. **Month 2:** Needed better argument structuring, so shifted to research practice.

Pairing 7: Visualizing Progress Through Graphs

Big Skill: Strategic Thinking – Recognizing that tracking helps identify strengths and weaknesses.

Small Skill: Tracking Progress – Using a progress graph to visualize achievements.

Teaching Technique: Collaborative group work.

Actions: Have students **create a progress graph** to monitor improvement over time.

Example: A student tracking **typing speed** charts:

1. **Week 1:** 25 words per minute (wpm).
2. **Week 2:** 30 wpm.
3. **Week 3:** 35 wpm.

Pairing 8: Color-Coded Progress Trackers

Big Skill: Strategic Thinking – Understanding that breaking a goal into trackable segments prevents burnout.

Small Skill: Tracking Progress – Using a color-coded progress tracker to organize different aspects of a goal.

Teaching Technique: Real-world applications.

Actions: Have students categorize progress by color to track multiple aspects of a goal.

Example: A student managing a school project uses:

1. **Green:** Research completed.
2. **Yellow:** Outline in progress.
3. **Red:** Need to start writing.

Pairing 9: Peer Accountability for Goal Success

Big Skill: Strategic Thinking – Learning that tracking creates accountability for goal success.

Small Skill: Tracking Progress – Using peer accountability charts to stay on track.

Teaching Technique: Cultural exploration.

Actions: Have students partner with a classmate to track progress and check in weekly.

Example: A student practicing Spanish vocabulary checks in with a peer:

1. **Week 1:** Learned 20 new words.
2. **Week 2:** Practiced speaking but struggled with pronunciation.

Pairing 10: Documenting Small Wins for Motivation

Big Skill: Strategic Thinking – Understanding that reviewing past progress builds future success.

Small Skill: Tracking Progress – Using a "Success Jar" to document small wins.

Teaching Technique: Hands-on experiments.

Actions: Have students write each small success on a slip of paper and add it to a jar.

Example: A student improving public speaking writes:

1. "**Gave** a speech in class."
2. "**Used** better eye contact."
3. "**Spoke** confidently in a group."

Goal Setting – (Big Skill) Strategic Thinking vs. (Small Skill) Time Management

---

Pairing 1: Creating a Timeline Roadmap for Goal Achievement

Big Skill: Strategic Thinking – Developing a roadmap for achieving personal or academic goals.

Small Skill: Time Management – Practicing prioritization and scheduling tasks effectively.

Teaching Technique: Hands-on experiments.

Actions: Have students create a timeline roadmap for a long-term academic or personal goal, breaking it into weekly priorities.

Example: A student preparing for a school play audition indicates:

1. **Week 1:** Memorize lines.
2. **Week 2:** Practice voice projection.
3. **Week 3:** Work on body language and stage presence.

Pairing 2: Prioritizing Tasks Using a Matrix

Big Skill: Strategic Thinking – Understanding that achieving goals requires structured planning.

Small Skill: Time Management – Using a priority matrix to determine which tasks to complete first.

Teaching Technique: Guided exploration.

Actions: Have students categorize their daily tasks into quadrants (e.g., urgent/important, important/not urgent).

Example: A student balancing homework and sports establishes these categories:

1. **Urgent & Important:** Finish math assignment.
2. **Important but Not Urgent:** Practice for soccer tryouts.
3. **Not Important & Not Urgent:** Watch YouTube videos.

Pairing 3: Structuring Daily and Weekly Planning

Big Skill: Strategic Thinking – Learning how to break a big goal into smaller, time-based actions.

Small Skill: Time Management – Using a structured planner to map out daily and weekly tasks.

Teaching Technique: Problem-solving exercises.

Actions: Have students fill in a weekly planner that includes school, personal projects, and relaxation time.

Example: A student working toward writing a novel schedules:

1. **Monday:** Outline character backgrounds.
2. **Wednesday:** Write 500 words.
3. **Friday:** Revise first draft.

Pairing 4: Using Time Blocking for Efficient Scheduling

Big Skill: Strategic Thinking – Recognizing that time management reduces stress and improves focus.

Small Skill: Time Management – Using time blocking to allocate specific periods for different tasks.

Teaching Technique: Observation journals.

Actions: Have students divide their daily schedule into time blocks and stick to a structured routine.

Example: A student preparing for final exams indicates:

1. **4:00-5:00 PM:** Study math.
2. **5:00-5:30 PM:** Break.
3. **5:30-6:30 PM:** Review science notes.

Pairing 5: Tracking Daily Tasks with Checklist

Big Skill: Strategic Thinking – Understanding that long-term success requires daily discipline

Small Skill: Time Management – Using checklists to ensure daily progress toward goals

Teaching Technique: Hands-on experiment.

Actions: Have students create a daily checklist for habits and goal-related tasks

Example: A student training for a 5K race marks:

1. Warm-up exercises.
2. Run for 20 minutes.
3. Stretching routine (**not completed today**).

Pairing 6: Adjusting Schedules Based on Weekly Reflections

Big Skill: Strategic Thinking – Learning that effective time management leads to increased productivity.

Small Skill: Time Management – Reviewing weekly progress and adjusting schedules if necessary.

Teaching Technique: Sensory exploration.

Actions: Have students reflect on how well they managed their time each week and make adjustments.

Example: A student tracking study habits writes:

1. **Week 1:** Did not finish history reading → **Need to start earlier.**
2. **Week 2:** Studied science in the morning → **Worked better.**

Pairing 7: Using Countdown Trackers for Deadlines

Big Skill: Strategic Thinking – Recognizing that setting priorities helps maintain focus.

Small Skill: Time Management – Using a countdown tracker for upcoming deadlines.

Teaching Technique: Collaborative group work.

Actions: Have students create a countdown tracker for a big assignment or project.

Example: A student completing a research paper indicates:

1. **10 days left:** Finish research.
2. **5 days left:** Write the draft.
3. **2 days left:** Edit and proofread.

Pairing 8: Building Flexibility into Schedules

Big Skill: Strategic Thinking – Understanding that flexibility is key in time management.

Small Skill: Time Management – Creating backup time slots in case unexpected issues arise.

Teaching Technique: Real-world applications.

Actions: Have students build "catch-up time" into their schedules for missed tasks.

Example: A student learning a new instrument schedules:

1. **Monday:** Guitar lesson.
2. **Wednesday:** Practice session (if missed Monday, reschedule here).

Pairing 9: Tracking Progress on Long-Term Assignments

Big Skill: Strategic Thinking – Recognizing that well-planned schedules prevent last-minute stress.

Small Skill: Time Management – Using a project timeline to track progress on long-term assignments.

Teaching Technique: Cultural exploration.

Actions: Have students create a visual project timeline and check off completed stages.

Example: A student working on a history presentation tracks the following:

1. **Research phase** completed.
2. **PowerPoint slides** created.
3. **Final speech** practice.

Pairing 10: Conducting Weekly Time Audits for Productivity

Big Skill: Strategic Thinking – Understanding that structured time management leads to long-term success.

Small Skill: Time Management – Setting daily and weekly reflections to analyze time use.

Teaching Technique: Hands-on experiments.

Actions: Have students complete a weekly time audit and adjust their schedules.

Example: A student balancing extracurriculars and schoolwork realizes:

1. **Spent 2 hours on social media daily** → Reduce to 1 hour.
2. **Not enough time for math practice** → Add 30 minutes before bedtime.

Goal Setting – (Big Skill) Self-Monitoring & Adaptation vs. (Small Skill) Breaking Goals into Actionable Steps

---

Pairing 1: Evaluating and Adjusting Milestones

Big Skill: Self-Monitoring & Adaptation – Evaluating progress and making necessary adjustments.

Small Skill: Breaking Goals into Actionable Steps – Learning how to deconstruct a big goal into smaller, achievable milestones.

Teaching Technique: Hands-on experiments.

Actions: Have students set a long-term goal and create a self-evaluation system to adjust milestones along the way.

Example: A student working toward running a mile in under 8 minutes indicates:

1. **Week 1:** Run ¼ mile without stopping.
2. **Week 2:** Increase to ½ mile.
3. **Week 3:** Run full mile, but adjust if needed.

Pairing 2: Reflecting on What Works and What Does not

Big Skill: Self-Monitoring & Adaptation – Understanding when a strategy is not working and adjusting it.

Small Skill: Breaking Goals into Actionable Steps – Identifying key challenges and revising steps.

Teaching Technique: Guided exploration.

Actions: Have students reflect weekly on what is working and what needs to change.

Example: A student aiming to improve test scores evaluates:

1. **Week 1:** Flashcards did not help → Switch to practice tests.
2. **Week 2:** Study groups improved understanding → **Keep attending.**

Pairing 3: Developing Alternative Strategies for Success

Big Skill: Self-Monitoring & Adaptation – Learning that flexibility leads to success.

Small Skill: Breaking Goals into Actionable Steps – Creating alternative strategies for reaching a goal.

Teaching Technique: Problem-solving exercises.

Actions: Have students list alternative paths to reach their goal if their first attempt fails.

Example: A student struggling to memorize vocabulary indicates:

1. **Try writing** words repeatedly.
2. **If that does not work**, create flashcards.
3. **If still struggling**, use a vocabulary app.

Pairing 4: Adjusting Goal Timelines Based on Progress

Big Skill: Self-Monitoring & Adaptation – Recognizing when to slow down or speed up based on progress.

Small Skill: Breaking Goals into Actionable Steps – Adjusting goal timelines based on results.

Teaching Technique: Observation journals.

Actions: Have students track progress over time and shift deadlines if needed.

Example: A student learning to play the piano realizes:

1. **Week 1:** Learned two songs quickly → **Add a third.**
2. **Week 2:** Struggled with one song → **Extend practice time.**

Pairing 5: Using Peer Feedback to Improve Strategies

Big Skill: Self-Monitoring & Adaptation – Learning to use feedback to improve.

Small Skill: Breaking Goals into Actionable Steps – Using peer review to refine goal-setting.

Teaching Technique: Hands-on experiments.

Actions: Have students exchange progress reports with a classmate for suggestions.

Example: A student practicing public speaking gets feedback:

1. "Good posture but need to project voice more."
2. "Adjust practice to include voice exercises."

Pairing 6: Conducting Weekly Check-Ins to Reassess Goals

Big Skill: Self-Monitoring & Adaptation – Understanding that reviewing progress helps refine strategies.

Small Skill: Breaking Goals into Actionable Steps – Weekly check-ins to reassess the effectiveness of steps.

Teaching Technique: Sensory exploration.

Actions: Have students set weekly progress goals and revise if necessary.

Example: A student working on writing a short story tracks:

1. **Week 1:** Wrote 3 pages → **On track.**
2. **Week 2:** Only wrote 1 page → Adjust writing time.

Pairing 7: Planning for Setbacks and Recovery Strategies

Big Skill: Self-Monitoring & Adaptation – Recognizing that mistakes are part of growth.

Small Skill: Breaking Goals into Actionable Steps – Creating a plan for overcoming setbacks.

Teaching Technique: Collaborative group work.

Actions: Have students identify potential setbacks and create backup plans.

Example: A student struggling with math homework indicates:

1. **Try solving** alone.
2. **If stuck**, ask a classmate.
3. **If still unclear**, seek help from the teacher.

Pairing 8: Testing Different Strategies for Personal Success

Big Skill: Self-Monitoring & Adaptation – Understanding that different strategies work for different people.

Small Skill: Breaking Goals into Actionable Steps – Trying different approaches to see what works best.

Teaching Technique: Real-world applications.

Actions: Have students experiment with multiple strategies before choosing the best one.

Example: A student preparing for a history exam tests:

1. **Study notes** → Did not retain much.
2. **Watch history videos** → Understood better.
3. **Join a study group** → Best method!

Pairing 9: Rewriting Goals Based on New Information

Big Skill: Self-Monitoring & Adaptation – Recognizing when to make a major change in approach.

Small Skill: Breaking Goals into Actionable Steps – Rewriting goal steps based on new information.

Teaching Technique: Cultural exploration.

Actions: Have students rewrite goals after reflecting on progress.

Example: A student originally planning to write a research paper on space travel realizes:

1. **Too broad** → Narrow topic to Mars exploration.

Pairing 10: Setting New Goals After Completing a Milestone

Big Skill: Self-Monitoring & Adaptation – Understanding that goal-setting is an ongoing process.

Small Skill: Breaking Goals into Actionable Steps – Reviewing success and setting new goals.

Teaching Technique: Hands-on experiments.

Actions: Have students reflect on a completed goal and set their next challenge.

Example: A student who learned 50 new words in Spanish decides:

1. **Next goal**: Hold a five-minute conversation in Spanish.

Goal Setting – (Big Skill) Self-Monitoring & Adaptation vs. (Small Skill) Tracking Progress

---

Pairing 1: Using Goal Trackers for Measurable Progress

Big Skill: Self-Monitoring & Adaptation – Evaluating progress and making necessary adjustments.

Small Skill: Tracking Progress – Using checklists, habit trackers, or goal charts to measure success.

Teaching Technique: Hands-on experiments.

Actions: Have students create a personal goal tracker with milestones and check off tasks as they complete them.

Example: A student working toward learning 50 new vocabulary words in a month tracks:

1. **Week 1:** Learned 10 words.
2. **Week 2:** Learned 15 words.
3. **Week 3:** Only learned 5 words → Adjust study plan.

Pairing 2: Daily Habit Tracking for Consistency

Big Skill: Self-Monitoring & Adaptation – Understanding when a strategy is ineffective and changing it.

Small Skill: Tracking Progress – Using a habit tracker to monitor daily activities.

Teaching Technique: Guided exploration.

Actions: Have students keep a 30-day habit tracker to track their daily efforts toward a goal and revise strategies if necessary.

Example: A student aiming to practice piano daily logs:

1. **Monday:** Completed.
2. **Tuesday:** Completed.
3. **Wednesday:** (adjust schedule for next week).

Pairing 3: Self-Assessments for Performance Improvements

Big Skill: Self-Monitoring & Adaptation – Learning to identify strengths and weaknesses in goal progress.

Small Skill: Tracking Progress – Using weekly self-assessments to measure improvements.

Teaching Technique: Problem-solving exercises.

Actions: Have students complete a weekly self-assessment form and make adjustments based on results.

Example: A student preparing for a spelling competition evaluates:

1. **Week 1:** Scored 70% on practice test → **Need more review time.**
2. **Week 2:** Scored 85% → Improvement, but still work to do.

Pairing 4: Visual Goal Charts for Motivation

Big Skill: Self-Monitoring & Adaptation – Recognizing that tracking progress helps set realistic goals.

Small Skill: Tracking Progress – Using a goal chart to visually measure success.

Teaching Technique: Observation journals.

Actions: Have students design a progress chart with milestones for a long-term project.

Example: A student tracking book-reading progress charts:

1. **2 books read** → Completed.
2. **4 books read** → Completed.
3. **6 books read** → Adjusted goal to 8 books instead of 10.

Pairing 5: Success vs. Challenge Logs for Identifying Patterns

Big Skill: Self-Monitoring & Adaptation – Learning that progress is not always linear and adjustments are needed.

Small Skill: Tracking Progress – Using a success vs. challenge log to document what works and what does not.

Teaching Technique: Hands-on experiments.

Actions: Have students keep a log of successes and setbacks to identify patterns in performance.

Example: A student working on time management logs:

1. "**Morning study sessions** were productive."
2. **"Evening sessions did not work—too tired."** → Shifted study time to mornings.

Pairing 6: Monthly Progress Reviews for Adjustments

Big Skill: Self-Monitoring & Adaptation – Understanding that regular reflection leads to improved strategies.

Small Skill: Tracking Progress – Using monthly progress reviews to assess goal achievement.

Teaching Technique: Sensory exploration.

Actions: Have students complete a monthly review sheet and adjust their strategies for the next month.

Example: A student tracking fitness goals reviews:

1. **Month 1:** Exercised 4 times a week → **On track.**
2. **Month 2:** Exercised only twice a week → Need to plan workouts better.

Pairing 7: Reward-Based Tracking for Motivation

Big Skill: Self-Monitoring & Adaptation – Recognizing that tracking provides motivation and accountability.

Small Skill: Tracking Progress – Using a reward-based tracking system for motivation.

Teaching Technique: Collaborative group work.

Actions: Have students set small rewards for achieving different goal levels.

Example: A student working toward writing a short story indicates:

1. **Complete** character profiles → **Reward**: Favorite snack.
2. **Write** three chapters → **Reward**: Watch a movie.
3. **Finish** final draft → **Celebration** with friends.

Pairing 8: Color-Coded Trackers for Multi-Step Goals

Big Skill: Self-Monitoring & Adaptation – Understanding that consistent tracking helps maintain focus.

Small Skill: Tracking Progress – Using a color-coded tracker to separate different aspects of a goal.

Teaching Technique: Real-world applications.

Actions: Have students categorize progress into color-coded sections to track different parts of a goal.

Example: A student working on a science project uses:

1. **Green:** Research phase completed.
2. **Yellow:** Experiment in progress.
3. **Red:** Need to finish presentation slides.

Pairing 9: Progress Thermometers for Incremental Success

Big Skill: Self-Monitoring & Adaptation – Recognizing when goals need to be revised based on progress tracking.

Small Skill: Tracking Progress – Using a progress thermometer to measure incremental success.

Teaching Technique: Cultural exploration.

Actions: Have students create a progress thermometer to visually track improvement.

Example: A student learning a new language indicates:

1. **20 words learned** → Completed.
2. **50 words learned** → Completed.
3. **100 words learned** → Adjusted goal to 75 words instead.

Pairing 10: Success Jars for Documenting Small Wins

Big Skill: Self-Monitoring & Adaptation – Understanding that tracking small successes leads to long-term improvement.

Small Skill: Tracking Progress – Using a "Success Jar" to record progress.

Teaching Technique: Hands-on experiments.

Actions: Have students write each success on a slip of paper and review them at the end of the term.

Example: A student working on overcoming shyness in class writes:

1. "**Asked** a question during class."
2. "**Spoke** in a group discussion."
3. "**Gave** a class presentation."

Goal Setting – (Big Skill) Self-Monitoring & Adaptation vs. (Small Skill) Time Management

---

Pairing 1: Weekly Schedule Reviews for Efficiency

Big Skill: Self-Monitoring & Adaptation – Evaluating progress and making necessary adjustments.

Small Skill: Time Management – Practicing prioritization and scheduling tasks effectively.

Teaching Technique: Hands-on experiments.

Actions: Have students track their weekly schedules and identify areas where they need better time management.

Example: A student reviewing their homework routine finds:

1. **Monday:** Spent too much time on math, rushed science work.
2. **Adjustment:** Set a timer for each subject to stay balanced.

Pairing 2: Prioritizing Tasks for Maximum Productivity

Big Skill: Self-Monitoring & Adaptation – Recognizing when to adjust schedules based on productivity.

Small Skill: Time Management – Using a priority matrix to identify the most important tasks.

Teaching Technique: Guided exploration.

Actions: Have students categorize tasks into four quadrants (e.g., urgent/important, important/not urgent) and adjust their schedules accordingly.

Example: A student planning for a test and an art project prioritizes:

1. **Urgent & Important:** Study for test.
2. **Important but Not Urgent:** Work on art project.
3. **Adjustment:** Focus on studying first, then schedule time for art.

Pairing 3: Revising Deadlines for Better Efficiency

Big Skill: Self-Monitoring & Adaptation – Learning when to revise deadlines for better efficiency.

Small Skill: Time Management – Using a structured planner to map out tasks.

Teaching Technique: Problem-solving exercises.

Actions: Have students create a daily planner with deadlines and evaluate if they need adjustments.

Example: A student preparing for a debate competition realizes:

1. **Planned:** Research Monday, write speech Tuesday, practice Wednesday.
2. **Reality:** Research took longer than expected → Adjust schedule to allow more research time.

Pairing 4: Using Time-Blocking for Focused Productivity

Big Skill: Self-Monitoring & Adaptation – Understanding how flexibility improves time management.

Small Skill: Time Management – Implementing time-blocking techniques to improve productivity.

Teaching Technique: Observation journals.

Actions: Have students divide their study time into focused work blocks with breaks and adjust as needed.

Example: A student working on math and history homework indicates:

1. **4:00-4:45 PM:** Math.
2. **4:45-5:00 PM:** Break.
3. **5:00-5:45 PM:** History.
4. **Adjustment:** Extend math block if struggling, shorten history time if work is done.

Pairing 5: Balancing Work and Rest to Prevent Burnout

Big Skill: Self-Monitoring & Adaptation – Recognizing that small adjustments prevent burnout.

Small Skill: Time Management – Creating a balanced schedule with work and rest.

Teaching Technique: Hands-on experiments.

Actions: Have students review their daily schedule and adjust if too much time is spent on one activity.

Example: A student practicing for a music recital realizes:

1. **Practicing** 2 hours straight is exhausting.
2. **Adjustment:** Break practice into 45-minute sessions with short rests.

Pairing 6: Weekly Check-Ins to Adjust Schedules

Big Skill: Self-Monitoring & Adaptation – Understanding that reviewing progress helps refine time use.

Small Skill: Time Management – Setting weekly check-ins to adjust schedules.

Teaching Technique: Sensory exploration.

Actions: Have students reflect on how well they followed their schedules each week and adjust for better efficiency.

Example: A student tracking exercise goals writes:

1. **Week 1:** Worked out 3 times but skipped one session due to homework.
2. **Adjustment:** Move workouts to earlier in the day to prevent skipping.

Pairing 7: Using Color-Coded Calendars for Balance

Big Skill: Self-Monitoring & Adaptation – Learning how to balance multiple responsibilities effectively.

Small Skill: Time Management – Using a color-coded calendar to organize different tasks.

Teaching Technique: Collaborative group work.

Actions: Have students assign colors to different types of tasks (school, hobbies, chores) and adjust if certain areas are neglected.

Example: A student balancing school, soccer, and chores finds:

1. **Too much time on soccer,** little time for studying.
2. **Adjustment:** Reduce practice by 30 minutes and reassign time to studying.

Pairing 8: Weekly Review Charts for Progress Tracking

Big Skill: Self-Monitoring & Adaptation – Recognizing that tracking progress keeps schedules on track.

Small Skill: Time Management – Creating a weekly review chart to assess task completion.

Teaching Technique: Real-world applications.

Actions: Have students list their weekly goals and mark completed, partially completed, or unfinished tasks.

Example: A student preparing for a science fair tracks:

1. Research completed.
2. **Experiment started**, but not finished.
3. **Presentation not started** → Adjust schedule to prioritize presentation.

Pairing 9: Replacing Time-Wasting Habits with Productivity

Big Skill: Self-Monitoring & Adaptation – Understanding that making small daily changes leads to better time efficiency.

Small Skill: Time Management – Tracking time-wasting activities and replacing them with productive habits.

Teaching Technique: Cultural exploration.

Actions: Have students log their daily screen time, distractions, and wasted time to identify areas to improve.

Example: A student logs:

1. **Social media:** 2 hours/day → Reduce to 1 hour.
2. **Studying:** 1 hour/day → Increase to 2 hours.

Pairing 10: Adjusting Time Allocations for Maximum Productivity

Big Skill: Self-Monitoring & Adaptation – Learning that goal-setting is a continuous process of improvement.

Small Skill: Time Management – Adjusting time allocations to maximize productivity.

Teaching Technique: Hands-on experiments.

Actions: Have students reflect on what time of day they are most productive and adjust their schedules.

Example: A student working on a coding project realizes:

1. Nighttime sessions lead to distractions.
2. **Adjustment**: Move project work to morning when more focused.

Example Teaching Strategies for Big vs. Small Thinking Skill Pairings of Systems Thinking Introduction at the Middle School Level

Here are 90 unique teaching strategy examples for the big thinking and small thinking skill pairings, each described in detail to align with **Big Thinking Skills** (Concepts & Strategies) and **Small Thinking Skills** (Step-by-Step Execution) for the **Systems Thinking Introduction Skill** taught at the middle school level.

Systems Thinking Introduction

Big Thinking Skills (Concepts & Strategies)

- **Understanding Interconnections:** Recognizing how different parts of a system (e.g., ecosystems, economies, communities) influence each other.
- **Cause-and-Effect on a Large Scale:** Analyzing how small changes in a system impact the whole (e.g., pollution affecting marine life).
- **Feedback Loops & Complexity:** Understanding how actions create cycles (e.g., supply and demand in economics).

Small Thinking Skills (Step-by-Step Execution)

- **Mapping Systems Visually:** Drawing diagrams, mind maps, or flowcharts to represent interconnections.
- **Step-by-Step System Analysis:** Breaking down how a system works in small, sequential steps.
- **Testing Small Changes:** Running simulations or role-playing different scenarios to see how variables affect a system.

Systems Thinking Introduction – (Big Skill) Understanding Interconnections vs. (Small Skill) Mapping Systems Visually

---

Pairing 1: Visualizing Ecosystem Interactions

Big Skill: Understanding Interconnections – Recognizing how different parts of a system (e.g., ecosystems, economies, communities) influence each other.

Small Skill: Mapping Systems Visually – Drawing diagrams, mind maps, or flowcharts to represent interconnections.

Teaching Technique: Hands-on experiments.

Actions: Have students create a food web diagram to show how species in an ecosystem depend on one another.

Example: A student drawing a forest ecosystem maps:

Plants → Rabbits → Foxes → Eagles, showing how energy flows through the system.

Pairing 2: Understanding Economic Systems Through Flowcharts

Big Skill: Understanding Interconnections – Recognizing the relationships between different components in an economic system.

Small Skill: Mapping Systems Visually – Using a flowchart to represent economic transactions.

Teaching Technique: Guided exploration.

Actions: Have students map the journey of a product, from raw materials to consumer purchase.

Example: A student traces:

Cotton farm → Clothing factory → Retail store → Consumer to see how each stage connects.

Pairing 3: Exploring School Community Roles

Big Skill: Understanding Interconnections – Learning how different parts of a school community interact.

Small Skill: Mapping Systems Visually – Creating a school community diagram.

Teaching Technique: Problem-solving exercises.

Actions: Have students map out how different roles (students, teachers, principals, cafeteria staff) work together.

Example: A student charts:

Teachers provide lessons → Principals manage operations → Cafeteria staff provide food → Students learn and participate.

Pairing 4: Cause-and-Effect in Climate Change

Big Skill: Understanding Interconnections – Recognizing how climate and weather patterns are connected.

Small Skill: Mapping Systems Visually – Using a cause-and-effect diagram for climate change.

Teaching Technique: Observation journals.

Actions: Have students create a climate change flowchart that connects human actions with environmental impacts.

Example: A student maps the following:

More cars on the road → More CO_2 emissions → Increased global temperature → Melting ice caps.

Pairing 5: Tracing Global Trade Networks

Big Skill: Understanding Interconnections – Recognizing how supply chains work in global trade.

Small Skill: Mapping Systems Visually – Using a global trade mind map.

Teaching Technique: Hands-on experiments.

Actions: Have students map where different products come from and how they travel globally.

Example: A student explores:

Coffee beans → Grown in Brazil → Shipped to U.S. → Brewed at cafés.

Pairing 6: Understanding Government Systems

Big Skill: Understanding Interconnections – Learning how government, laws, and citizens interact.

Small Skill: Mapping Systems Visually – Creating a government system diagram.

Teaching Technique: Sensory exploration.

Actions: Have students illustrate the branches of government and how they balance each other.

Example: A student maps the following:

1. **Legislative Branch** → Makes laws.
2. **Judicial Branch** → Interprets laws.
3. **Executive Branch** → Enforces laws.

Pairing 7: Mapping City Infrastructure and Public Services

Big Skill: Understanding Interconnections – Recognizing how different parts of a city's infrastructure are connected.

Small Skill: Mapping Systems Visually – Drawing a city system flowchart.

Teaching Technique: Collaborative group work.

Actions: Have students map how roads, electricity, water, and public transport interact.

Example: A student maps the following:

Subways and buses → Reduce traffic → Lower pollution → Improved city life.

Pairing 8: Visualizing Internet Networks and Data Flow

Big Skill: Understanding Interconnections – Learning how technology networks function.

Small Skill: Mapping Systems Visually – Using a diagram to explain the Internet's infrastructure.

Teaching Technique: Real-world applications.

Actions: Have students draw a simplified diagram of how the Internet works (servers, networks, and users).

Example: A student maps the following:

User clicks a website → Data request sent to a server → Server processes data → Page appears on screen.

Pairing 9: Tracing the Impact of Pollution on Ecosystems

Big Skill: Understanding Interconnections – Recognizing how pollution affects different ecosystems.

Small Skill: Mapping Systems Visually – Creating a pollution impact chart.

Teaching Technique: Cultural exploration.

Actions: Have students illustrate how different types of pollution (air, water, soil) spread and impact ecosystems.

Example: A student maps the following:

Factory waste enters river → Fish absorb toxins → Humans eat fish → Health problems increase.

Pairing 10: Connecting Historical Events Through Timelines

Big Skill: Understanding Interconnections – Recognizing how historical events are linked.

Small Skill: Mapping Systems Visually – Using a cause-and-effect history timeline.

Teaching Technique: Hands-on experiments.

Actions: Have students map out how one event led to another, showing key turning points.

Example: A student creates a World War II timeline:

Germany invades Poland → Allies declare war → Major battles occur → War ends with atomic bombings.

Systems Thinking Introduction – (Big Skill) Understanding Interconnections vs. (Small Skill) Step-by-Step System Analysis

---

Pairing 1: Energy Flow in Ecosystems

Big Skill: Understanding Interconnections – Recognizing how different parts of a system (e.g., ecosystems, economies, communities) influence each other.

Small Skill: Step-by-Step System Analysis – Breaking down how a system works in small, sequential steps.

Teaching Technique: Hands-on experiments.

Actions: Have students analyze a food chain by identifying how energy moves through different organisms in a step-by-step manner.

Example: A student analyzing a desert food chain maps:

Cactus stores water and provides food → Insects eat cactus → Lizard eats insects → Hawk eats lizards.

Pairing 2: The Supply and Demand Process

Big Skill: Understanding Interconnections – Recognizing how economic systems depend on supply and demand.

Small Skill: Step-by-Step System Analysis – Breaking down how goods and services move through an economy.

Teaching Technique: Guided exploration.

Actions: Have students trace the path of a product from raw material to final purchase.

Example: A student follows the process of making a loaf of bread:

Wheat is grown on a farm → Processed into flour at a mill → Baked into bread at a factory → Sold in stores → Bought and eaten by consumers.

Pairing 3: The Water Cycle in Action

Big Skill: Understanding Interconnections – Learning how water cycles through the environment.

Small Skill: Step-by-Step System Analysis – Identifying the sequential stages of the water cycle.

Teaching Technique: Problem-solving exercises.

Actions: Have students break down the water cycle into clear steps to understand how water moves through ecosystems.

Example: A student traces:

1. **Evaporation** – Sun heats water, turning it into vapor.
2. **Condensation** – Vapor cools and forms clouds.
3. **Precipitation** – Water falls as rain, snow, or hail.
4. **Collection** – Water returns to rivers, lakes, and oceans.

Pairing 4: Public Transportation Systems

Big Skill: Understanding Interconnections – Recognizing how transportation systems connect people and goods.

Small Skill: Step-by-Step System Analysis – Mapping how a public transportation system functions.

Teaching Technique: Observation journals.

Actions: Have students analyze a subway or bus system and list each step in a passenger's journey.

Example: A student traces:

1. **Person waits** at the bus stop.
2. **Bus arrives** and picks up passengers.
3. **Bus follows** a set route, stopping at locations.
4. **Passengers exit** at their destinations.

Pairing 5: Electricity from Power Plants to Homes

Big Skill: Understanding Interconnections – Understanding how electricity moves through a power grid.

Small Skill: Step-by-Step System Analysis – Breaking down the process of electricity generation and distribution.

Teaching Technique: Hands-on experiments.

Actions: Have students trace the journey of electricity from its source to their home.

Example: A student explains:

1. Power plant generates electricity.
2. **Electricity moves** through transmission lines.
3. **Electricity reaches homes** through local power lines.
4. **Used** to power appliances and devices.

Pairing 6: How Ecosystems Recover from Disasters

Big Skill: Understanding Interconnections – Learning how ecosystems recover after disasters.

Small Skill: Step-by-Step System Analysis – Breaking down the process of ecological succession.

Teaching Technique: Sensory exploration.

Actions: Have students analyze how a forest regenerates after a wildfire.

Example: A student traces:

1. **Wildfire burns** through a forest.
2. **Small plants** like moss and grasses grow first.
3. Shrubs and small trees start growing.
4. **Mature forest** returns after decades.

Pairing 7: The Impact of Farming on the Environment

Big Skill: Understanding Interconnections – Recognizing how food production impacts the environment.

Small Skill: Step-by-Step System Analysis – Analyzing the effects of farming practices.

Teaching Technique: Collaborative group work.

Actions: Have students examine how agriculture affects soil, water, and biodiversity.

Example: A student traces:

1. **Farmers plow land** and plant crops.
2. **Crops require** water and nutrients.
3. **Harvested crops** are sold for food.
4. **Waste from farms** impacts water systems.

Pairing 8: How Diseases Spread in Communities

Big Skill: Understanding Interconnections – Understanding how diseases spread in communities.

Small Skill: Step-by-Step System Analysis – Analyzing the path of disease transmission.

Teaching Technique: Real-world applications.

Actions: Have students model the spread of disease using a flowchart.

Example: A student tracks:

1. **Person gets infected** with a virus.

2. **They go to school** and interact with others.
3. **Germs spread** through contact.
4. **More people get sick** and visit hospitals.

Pairing 9: The Spread of Information Through Social Media

Big Skill: Understanding Interconnections – Recognizing how social media networks spread information.

Small Skill: Step-by-Step System Analysis – Breaking down the flow of information online.

Teaching Technique: Cultural exploration.

Actions: Have students analyze how a social media post spreads from one person to thousands.

Example: A student tracks:

1. **User posts** a message online.
2. **Friends and followers** see it.
3. **People share the post**, increasing reach.
4. The message spreads globally.

Pairing 10: How Recycling Reduces Waste

Big Skill: Understanding Interconnections – Recognizing how recycling systems reduce waste.

Small Skill: Step-by-Step System Analysis – Identifying how materials move through a recycling center.

Teaching Technique: Hands-on experiments.

Actions: Have students analyze the recycling process from start to finish.

Example: A student explains:

1. **Person places plastic** in a recycling bin.
2. **Truck collects** recyclables and transports them.
3. **Recycling center sorts** and processes materials.
4. **New products** are made from recycled materials.

Systems Thinking Introduction – (Big Skill) Understanding Interconnections vs. (Small Skill) Testing Small Changes

---

Pairing 1: How Species Impact an Ecosystem

Big Skill: Understanding Interconnections – Recognizing how different parts of a system (e.g., ecosystems, economies, communities) influence each other.

Small Skill: Testing Small Changes – Running simulations or role-playing different scenarios to see how variables affect a system.

Teaching Technique: Hands-on experiments.

Actions: Have students change one factor in a food web simulation and observe the effects.

Example: A student removes all bees from an ecosystem and records the impact:

1. **Fewer plants grow** (no pollination).
2. Less food for herbivores.
3. Predator populations decline.

Pairing 2: How Price Changes Affect Demand

Big Skill: Understanding Interconnections – Recognizing how supply and demand shape economies.

Small Skill: Testing Small Changes – Simulating economic conditions to observe the effects of price changes.

Teaching Technique: Guided exploration.

Actions: Have students simulate the price of a product and see how demand changes.

Example: A student raises the price of a toy in a class economy game:

1. **Higher price** → Fewer students buy.
2. **Lower price** → More demand.

Pairing 3: Climate Effects on Ecosystems

Big Skill: Understanding Interconnections – Recognizing how weather changes impact ecosystems.

Small Skill: Testing Small Changes – Simulating how climate changes affect plant and animal populations.

Teaching Technique: Problem-solving exercises.

Actions: Have students change temperature or rainfall in a classroom plant-growing experiment.

Example: A student increases heat and reduces water for a plant:

1. **Too much heat** → Leaves dry out.
2. **Less water** → Plant growth slows.

Pairing 4: Managing Traffic Flow in Cities

Big Skill: Understanding Interconnections – Recognizing how traffic flow affects cities.

Small Skill: Testing Small Changes – Simulating how adjusting road layouts impacts congestion.

Teaching Technique: Observation journals.

Actions: Have students use a traffic simulation tool to test how changes in speed limits or road closures affect congestion.

Example: A student closes one major road in a simulation:

1. **More traffic** in surrounding areas.
2. **Longer wait times** at intersections.

Pairing 5: The Spread of Pollution in Water & Air

Big Skill: Understanding Interconnections – Recognizing how pollution spreads in environments.

Small Skill: Testing Small Changes – Conducting experiments to see how pollutants travel through water and air.

Teaching Technique: Hands-on experiments.

Actions: Have students drop food coloring in water and observe how pollution spreads with movement.

Example: A student adds oil to water and stirs:

1. **Pollution spreads** faster with motion.
2. **Harder for marine life** to escape affected areas.

Pairing 6: Population Growth & Limited Resources

Big Skill: Understanding Interconnections – Recognizing how population growth affects resources.

Small Skill: Testing Small Changes – Simulating how limited resources impact growing populations.

Teaching Technique: Sensory exploration.

Actions: Have students play a survival game where resources (food, water) are reduced while the population increases.

Example: A student reduces available food in a class simulation:

1. Fewer resources → More competition.
2. Some groups get more; others struggle.

Pairing 7: The Impact of Technology on Society

Big Skill: Understanding Interconnections – Recognizing how technological changes impact society.

Small Skill: Testing Small Changes – Simulating how introducing new technology affects daily life.

Teaching Technique: Collaborative group work.

Actions: Have students role-play life before and after a technological invention (e.g., the Internet).

Example: A student removes access to the Internet in a school setting:

1. Students rely more on books.
2. Increased face-to-face discussions.
3. Slower research process.

Pairing 8: Government Policies & Economic Impact

Big Skill: Understanding Interconnections – Recognizing how government policies affect economies.

Small Skill: Testing Small Changes – Running an economic policy simulation to observe effects.

Teaching Technique: Real-world applications.

Actions: Have students simulate raising or lowering taxes and measure economic outcomes.

Example: A student lowers taxes in a class economy game:

1. People have more spending money.
2. **Less government revenue** for public services.

Pairing 9: How Social Media Spreads Information

Big Skill: Understanding Interconnections – Recognizing how social media spreads information.

Small Skill: Testing Small Changes – Simulating how different types of posts affect engagement.

Teaching Technique: Cultural exploration.

Actions: Have students evaluate how different types of messages spread in a class "news" simulation.

Example: A student shares positive vs. negative news in a class activity:

1. Positive news spreads slowly.
2. Controversial news spreads faster.

Pairing 10: Disaster Response Strategies & Effectiveness

Big Skill: Understanding Interconnections – Recognizing how natural disasters impact human communities.

Small Skill: Testing Small Changes – Simulating disaster response strategies and their effectiveness.

Teaching Technique: Hands-on experiments.

Actions: Have students simulate an emergency response plan for a natural disaster.

Example: A student delays emergency response time in a flood simulation:

1. More damage occurs.
2. Harder to evacuate people.
3. Recovery takes longer.

Systems Thinking Introduction – (Big Skill) Cause-and-Effect on a Large Scale vs. (Small Skill) Mapping Systems Visually

---

Pairing 1: The Chain Reaction of Ocean Pollution

Big Skill: Cause-and-Effect on a Large Scale – Analyzing how small changes in a system impact the whole (e.g., pollution affecting marine life).

Small Skill: Mapping Systems Visually – Drawing diagrams, mind maps, or flowcharts to represent interconnections.

Teaching Technique: Hands-on experiments.

Actions: Have students illustrate how ocean pollution spreads and affects marine ecosystems.

Example: A student creates a flowchart showing:

Factories release plastic waste → Waste enters rivers and oceans → Fish consume microplastics → Humans eat contaminated fish → Health problems increase.

Pairing 2: How Deforestation Affects the Planet

Big Skill: Cause-and-Effect on a Large Scale – Understanding how deforestation affects climate, animals, and people.

Small Skill: Mapping Systems Visually – Using a cause-and-effect diagram to track environmental impacts.

Teaching Technique: Guided exploration.

Actions: Have students map out the chain reaction of deforestation using arrows to show interconnected effects.

Example: A student draws a diagram showing:

Trees cut down → Animal habitats lost → Carbon dioxide levels rise → Climate change intensifies.

Pairing 3: The Impact of Overfishing

Big Skill: Cause-and-Effect on a Large Scale – Recognizing how overfishing disrupts marine ecosystems.

Small Skill: Mapping Systems Visually – Using a web diagram to show food chain disruptions.

Teaching Technique: Problem-solving exercises.

Actions: Have students trace the impact of removing a species from the food chain.

Example: A student maps out what happens when sharks are overfished:

Fewer sharks → More small fish survive → Small fish overeat plankton → Ocean ecosystem collapses.

Pairing 4: The Environmental Effects of Urbanization

Big Skill: Cause-and-Effect on a Large Scale – Understanding how urbanization affects natural environments.

Small Skill: Mapping Systems Visually – Using a flowchart to show city expansion and its consequences.

Teaching Technique: Observation journals.

Actions: Have students map out how cities grow and affect ecosystems over time.

Example: A student creates a city expansion map showing:

More buildings → Less green space → More flooding → Higher temperatures (urban heat islands).

Pairing 5: Global Warming & Extreme Weather Patterns

Big Skill: Cause-and-Effect on a Large Scale – Recognizing how global warming affects weather patterns.

Small Skill: Mapping Systems Visually – Using a climate impact mind map to show changes in temperature and storms.

Teaching Technique: Hands-on experiments.

Actions: Have students track how rising global temperatures affect weather events.

Example: A student draws a climate change map showing:

> **Warmer oceans → More hurricanes → Increased flooding → More infrastructure damage.**

Pairing 6: How Invasive Species Disrupt Ecosystems

Big Skill: Cause-and-Effect on a Large Scale – Understanding how invasive species affect ecosystems.

Small Skill: Mapping Systems Visually – Using a species invasion diagram to track ecosystem changes.

Teaching Technique: Sensory exploration.

Actions: Have students trace how an invasive species disrupts native species.

Example: A student maps the spread of zebra mussels in freshwater lakes:

> **Mussels introduced accidentally → Native fish populations decline → Water plants overgrow → Ecosystem balance shifts.**

Pairing 7: The Environmental Cost of Food Waste

Big Skill: Cause-and-Effect on a Large Scale – Recognizing how food waste contributes to environmental issues.

Small Skill: Mapping Systems Visually – Using a decomposition timeline to illustrate landfill pollution.

Teaching Technique: Collaborative group work.

Actions: Have students map out how food waste affects landfills, methane emissions, and global warming.

Example: A student creates a food waste timeline showing:

> **Food thrown away → Decomposes in landfill → Methane gas released → Warms the planet.**

Pairing 8: The Devastating Impact of Oil Spills

Big Skill: Cause-and-Effect on a Large Scale – Recognizing how oil spills impact marine ecosystems.

Small Skill: Mapping Systems Visually – Using a contamination spread diagram to track oil movement.

Teaching Technique: Real-world applications.

Actions: Have students map how oil spreads and affects marine life.

Example: A student illustrates oil spread after an oil spillage:

Oil spills into the ocean → Spreads across the water surface → Coats marine animals → Enters the food chain.

Pairing 9: Industrial Waste & Water Pollution

Big Skill: Cause-and-Effect on a Large Scale – Understanding how industrial waste affects freshwater sources.

Small Skill: Mapping Systems Visually – Using a river pollution map to show contamination pathways.

Teaching Technique: Cultural exploration.

Actions: Have students track the journey of pollution from a factory to a community water source.

Example: A student creates a flowchart showing:

Factory releases chemicals → Chemicals enter rivers → Polluted water reaches homes → Increases health problems.

Pairing 10: The Consequences of Biodiversity Loss

Big Skill: Cause-and-Effect on a Large Scale – Recognizing how loss of biodiversity affects ecosystems.

Small Skill: Mapping Systems Visually – Using a biodiversity web to show ecosystem balance.

Teaching Technique: Hands-on experiments.

Actions: Have students map how species interact and what happens when one disappears.

Example: A student maps the effects of bee population decline:

Fewer bees → Less pollination → Fewer crops → Higher food prices.

Systems Thinking Introduction – (Big Skill) Cause-and-Effect on a Large Scale vs. (Small Skill) Step-by-Step System Analysis

---

Pairing 1: The Journey of Plastic Waste to the Ocean

Big Skill: Cause-and-Effect on a Large Scale – Analyzing how small changes in a system impact the whole (e.g., pollution affecting marine life).

Small Skill: Step-by-Step System Analysis – Breaking down how a system works in small, sequential steps.

Teaching Technique: Hands-on experiments.

Actions: Have students trace how plastic waste moves from land to the ocean and impacts marine life.

Example: A student breaks down the plastic waste cycle:

Plastic thrown away → Rainwater carries plastic to drains → Waste enters rivers and oceans → Marine animals mistake plastic for food → Disrupts the food chain.

Pairing 2: How Deforestation Impacts the Climate

Big Skill: Cause-and-Effect on a Large Scale – Understanding how deforestation affects the environment.

Small Skill: Step-by-Step System Analysis – Tracking the impact of tree loss on climate, animals, and communities.

Teaching Technique: Guided exploration.

Actions: Have students analyze how cutting down a forest leads to climate change effects.

Example: A student maps the deforestation process:

Trees cut down → Less CO_2 absorbed → More greenhouse gases remain → Temperatures rise → More droughts and extreme weather.

Pairing 3: The Spread of Air Pollution & Acid Rain

Big Skill: Cause-and-Effect on a Large Scale – Recognizing how air pollution spreads and affects ecosystems.

Small Skill: Step-by-Step System Analysis – Breaking down how pollutants travel through the atmosphere.

Teaching Technique: Problem-solving exercises.

Actions: Have students simulate how vehicle emissions lead to acid rain formation.

Example: A student tracks the acid rain cycle:

Car emissions release sulfur dioxide → Pollutants mix with water in clouds → Acid rain falls to Earth → Soil and water become acidic → Plants and animals struggle to survive.

Pairing 4: The Domino Effect of Overfishing

Big Skill: Cause-and-Effect on a Large Scale – Understanding how overfishing affects marine ecosystems.

Small Skill: Step-by-Step System Analysis – Analyzing how removing a species changes the food chain.

Teaching Technique: Observation journals.

Actions: Have students trace the effects of overfishing on ocean life.

Example: A student follows the overfishing chain reaction:

Too many fish caught → Predators lose food source → Prey fish overpopulate → Algae and plankton levels drop → Ocean ecosystem collapses.

Pairing 5: The Connection Between Climate Change & Rising Sea Levels

Big Skill: Cause-and-Effect on a Large Scale – Recognizing how climate change leads to rising sea levels.

Small Skill: Step-by-Step System Analysis – Breaking down the melting of polar ice caps.

Teaching Technique: Hands-on experiments.

Actions: Have students observe ice melting and track its effects on water levels.

Example: A student maps how ice melting affects the planet:

Rising temperatures melt glaciers → More freshwater enters oceans → Coastal areas flood → People relocate inland → Changing ecosystems force species to adapt or perish.

Pairing 6: The Spread and Consequences of Wildfires

Big Skill: Cause-and-Effect on a Large Scale – Understanding how wildfires impact the environment.

Small Skill: Step-by-Step System Analysis – Analyzing how fires spread and their long-term consequences.

Teaching Technique: Sensory exploration.

Actions: Have students trace how a small fire spreads into a large-scale disaster.

Example: A student follows the wildfire impact sequence:

Dry conditions ignite a small fire → Wind spreads flames rapidly → Homes and habitats destroyed → Soil erosion increases → New vegetation takes decades to grow.

Pairing 7: How Urbanization Triggers Local Climate Changes

Big Skill: Cause-and-Effect on a Large Scale – Recognizing how urbanization changes local climates.

Small Skill: Step-by-Step System Analysis – Analyzing the urban heat island effect.

Teaching Technique: Collaborative group work.

Actions: Have students track how city expansion affects temperature and air quality.

Example: A student maps the urban heat island cycle:

Buildings replace trees → Dark surfaces absorb heat → City temperatures rise → More air conditioning use → Increased energy demand worsens air pollution.

Pairing 8: How Farming Contributes to Soil Erosion

Big Skill: Cause-and-Effect on a Large Scale – Recognizing how agricultural practices affect soil quality.

Small Skill: Step-by-Step System Analysis – Breaking down how farming techniques lead to soil erosion.

Teaching Technique: Real-world applications.

Actions: Have students analyze how overuse of farmland reduces soil fertility.

Example: A student follows the erosion process:

Heavy farming removes nutrients → Soil becomes loose and dry → Wind and rain carry soil away → Crops struggle to grow → Food production declines.

Pairing 9: The Lasting Effects of Oil Spills on Marine Life

Big Skill: Cause-and-Effect on a Large Scale – Understanding how oil spills impact ecosystems.

Small Skill: Step-by-Step System Analysis – Breaking down how oil spreads and affects marine life.

Teaching Technique: Cultural exploration.

Actions: Have students trace how oil moves through the environment and its consequences.

Example: A student maps an oil spill's effects:

Oil tanker accident releases oil → Spills spread with ocean currents → Coats marine animals and birds → Rescue efforts struggle to clean affected areas → Declining marine populations disrupt food chains.

Pairing 10: The Human Impact on Freshwater Resources

Big Skill: Cause-and-Effect on a Large Scale – Recognizing how human water consumption affects freshwater supplies.

Small Skill: Step-by-Step System Analysis – Analyzing how water moves through supply systems.

Teaching Technique: Hands-on experiments.

Actions: Have students simulate a city's water supply system and how overuse leads to shortages.

Example: A student tracks water usage:

> **Water is pumped from lakes and rivers → Treated for human use → Used in homes, factories, and farms → Wastewater is treated and returned → Overuse lowers reservoir levels, affecting supply.**

Systems Thinking Introduction – (Big Skill) Cause-and-Effect on a Large Scale vs. (Small Skill) Testing Small Changes

---

Pairing 1: The Impact of Pollution on Aquatic Life

Big Skill: Cause-and-Effect on a Large Scale – Analyzing how small changes in a system impact the whole (e.g., pollution affecting marine life).

Small Skill: Testing Small Changes – Running simulations or role-playing different scenarios to see how variables affect a system.

Teaching Technique: Hands-on experiments.

Actions: Have students simulate different levels of pollution in water and observe the effects on plant and animal life.

Example: A student adds small amounts of **oil to water** and records how it affects **floating plants and small aquatic creatures** over time.

Pairing 2: How Deforestation Alters Ecosystems

Big Skill: Cause-and-Effect on a Large Scale – Understanding how deforestation changes ecosystems.

Small Skill: Testing Small Changes – Simulating the loss of trees in an area to see how it affects soil, animals, and climate.

Teaching Technique: Guided exploration.

Actions: Have students remove plants from a soil box and test how erosion changes compared to a box with plants.

Example: A student pours water over two soil samples (one with plants, one without) and observes that the sample without plants erodes faster.

Pairing 3: Rising Temperatures and Their Effect on Weather

Big Skill: Cause-and-Effect on a Large Scale – Recognizing how rising temperatures impact climate patterns.

Small Skill: Testing Small Changes – Simulating temperature increases and their effects on weather and water cycles.

Teaching Technique: Problem-solving exercises.

Actions: Have students use a heat lamp to warm a model ocean and observe how evaporation and precipitation change.

Example: A student raises the temperature of a water dish and records that more water evaporates, simulating increased rainfall and storms.

Pairing 4: The Impact of Plastic Waste on Oceans

Big Skill: Cause-and-Effect on a Large Scale – Understanding how plastic waste affects ocean ecosystems.

Small Skill: Testing Small Changes – Simulating how different waste management solutions impact pollution levels.

Teaching Technique: Observation journals.

Actions: Have students test different disposal methods (recycling, composting, trash) to see how much plastic waste remains after a week.

Example: A student records how quickly different materials break down in soil and finds that plastic does not decompose at all.

Pairing 5: How Urbanization Affects Wildlife

Big Skill: Cause-and-Effect on a Large Scale – Recognizing how urbanization impacts wildlife.

Small Skill: Testing Small Changes – Simulating how adding or removing green spaces affects animals in a city.

Teaching Technique: Hands-on experiments.

Actions: Have students design a city model with different levels of green space and test where toy animals "survive" best.

Example: A student removes parks from a model city and observes that animals have fewer places to live, and food sources disappear.

Pairing 6: The Effects of Oil Spills on Marine Ecosystems

Big Skill: Cause-and-Effect on a Large Scale – Understanding how oil spills damage ecosystems.

Small Skill: Testing Small Changes – Running a simulation to see how cleanup methods affect an oil spill.

Teaching Technique: Sensory exploration.

Actions: Have students add oil to water and test different cleanup methods (sponges, dish soap, cotton).

Example: A student uses dish soap to clean oil from feathers and observes that some oil remains, showing long-term damage to animals.

Pairing 7: Carbon Emissions and Air Quality

Big Skill: Cause-and-Effect on a Large Scale – Recognizing how carbon emissions affect air quality.

Small Skill: Testing Small Changes – Simulating how reducing car usage lowers air pollution.

Teaching Technique: Collaborative group work.

Actions: Have students compare air quality measurements in areas with high vs. low traffic.

Example: A student places air filters near a highway and a park and finds that the highway filter collects more dirt, showing higher pollution levels.

Pairing 8: How Overfishing Affects Marine Life

Big Skill: Cause-and-Effect on a Large Scale – Understanding how overfishing affects marine life.

Small Skill: Testing Small Changes – Running a fishing economy simulation to see how fishing limits change sustainability.

Teaching Technique: Real-world applications.

Actions: Have students act as fishers in a simulated ocean and adjust fishing quotas to see how it affects fish populations.

Example: A student allows unlimited fishing and observes that the fish population crashes, while setting limits helps the population recover.

Pairing 9: Global Warming and Melting Ice Caps

Big Skill: Cause-and-Effect on a Large Scale – Recognizing how global warming affects glaciers and sea levels.

Small Skill: Testing Small Changes – Simulating how melting ice affects ocean levels.

Teaching Technique: Cultural exploration.

Actions: Have students place ice in a model ocean and measure the water rise as it melts.

Example: A student compares floating ice (Arctic) vs. land ice (Antarctic) and observes that only land ice melting raises sea levels.

Pairing 10: Water Shortages and Farming Efficiency

Big Skill: Cause-and-Effect on a Large Scale – Understanding how water shortages impact farming.

Small Skill: Testing Small Changes – Simulating how different irrigation methods affect water use.

Teaching Technique: Hands-on experiments.

Actions: Have students test different watering techniques (drip irrigation, flooding, sprinklers) on soil and measure water loss.

Example: A student finds that drip irrigation conserves more water compared to flooding, showing why some farmers use this technique.

Systems Thinking Introduction – (Big Skill) Feedback Loops & Complexity vs. (Small Skill) Mapping Systems Visually

---

Pairing 1: The Supply and Demand Cycle

Big Skill: Feedback Loops & Complexity – Understanding how actions create cycles (e.g., supply and demand in economics).

Small Skill: Mapping Systems Visually – Drawing diagrams, mind maps, or flowcharts to represent interconnections.

Teaching Technique: Hands-on experiments.

Actions: Have students map a simple supply-and-demand cycle to show how product prices change based on availability.

Example: A student creates a supply-demand flowchart:

More video game consoles produced → Stores have a large supply → Prices drop → More people buy → Supply decreases, prices rise again.

Pairing 2: The Deforestation Cycle and Climate Impact

Big Skill: Feedback Loops & Complexity – Recognizing how deforestation impacts climate and ecosystems in a cycle.

Small Skill: Mapping Systems Visually – Using a cause-and-effect diagram to show deforestation feedback loops.

Teaching Technique: Guided exploration.

Actions: Have students illustrate how cutting down trees leads to a cycle of negative effects.

Example: A student creates a deforestation loop:

Fewer trees → More CO_2 in the air → Climate warms → More droughts → Even fewer trees survive.

Pairing 3: Predator-Prey Population Cycles

Big Skill: Feedback Loops & Complexity – Understanding how predator-prey relationships create balancing loops in ecosystems.

Small Skill: Mapping Systems Visually – Using a food chain diagram to show population cycles.

Teaching Technique: Problem-solving exercises.

Actions: Have students map out how changes in predator populations affect prey numbers.

Example: A student creates a wolf-rabbit feedback loop:

More wolves → Fewer rabbits → Less food for wolves → Fewer wolves → Rabbit population grows again.

Pairing 4: Urbanization and Traffic Congestion

Big Skill: Feedback Loops & Complexity – Recognizing how urbanization creates cycles of traffic congestion.

Small Skill: Mapping Systems Visually – Using a city transportation loop diagram.

Teaching Technique: Observation journals.

Actions: Have students illustrate how building more roads does not always fix traffic congestion.

Example: A student maps out a traffic congestion loop:

> **More people drive → More roads built → More urban sprawl → More cars on the road → Traffic congestion increases again.**

Pairing 5: Debt Cycles and Financial Burden

Big Skill: Feedback Loops & Complexity – Recognizing how debt cycles impact economies.

Small Skill: Mapping Systems Visually – Using a personal finance loop diagram.

Teaching Technique: Hands-on experiments.

Actions: Have students map a cycle showing how borrowing and interest create debt cycles.

Example: A student creates a debt cycle diagram:

> **Borrow money → Pay interest → Less money available → Borrow more → Debt increases.**

Pairing 6: Ocean Temperatures and Hurricane Strength

Big Skill: Feedback Loops & Complexity – Understanding how ocean temperature affects hurricane strength.

Small Skill: Mapping Systems Visually – Using a climate feedback loop diagram.

Teaching Technique: Sensory exploration.

Actions: Have students illustrate how warmer ocean water leads to stronger hurricanes.

Example: A student creates a hurricane intensity loop:

> **Warmer ocean water → Stronger hurricanes → More coastal flooding → Infrastructure damage → More heat-trapping gases released → Ocean warms further.**

Pairing 7: The Recycling Feedback Loop

Big Skill: Feedback Loops & Complexity – Recognizing how recycling affects resource use and pollution.

Small Skill: Mapping Systems Visually – Using a waste management loop diagram.

Teaching Technique: Collaborative group work.

Actions: Have students trace how recycling reduces landfill waste and keeps resources in use.

Example: A student maps a recycling loop:

Plastic bottles recycled → Turned into new products → Used again → Less landfill waste → Less pollution in the environment.

Pairing 8: Misinformation and Social Media Cycles

Big Skill: Feedback Loops & Complexity – Understanding how misinformation spreads on social media.

Small Skill: Mapping Systems Visually – Using a social media feedback loop diagram.

Teaching Technique: Real-world applications.

Actions: Have students illustrate how false news spreads and grows through sharing.

Example: A student maps a misinformation loop:

False information posted → People share without fact-checking → More people believe it → More engagement encourages new false stories → The cycle repeats.

Pairing 9: The Unhealthy Eating and Obesity Cycle

Big Skill: Feedback Loops & Complexity – Recognizing how obesity rates increase due to unhealthy food consumption.

Small Skill: Mapping Systems Visually – Using a nutrition and health cycle diagram.

Teaching Technique: Cultural exploration.

Actions: Have students map out how poor food choices lead to long-term health problems.

Example: A student creates an unhealthy eating loop:

More fast food consumption → Higher obesity rates → More medical costs → More stress → Increased reliance on fast food.

Pairing 10: Melting Ice and Climate Change

Big Skill: Feedback Loops & Complexity – Understanding how melting polar ice affects global warming.

Small Skill: Mapping Systems Visually – Using a climate change loop diagram.

Teaching Technique: Hands-on experiments.

Actions: Have students trace how melting ice leads to further warming.

Example: A student maps an ice melt loop:

Ice melts → Dark ocean absorbs more heat → Temperatures rise → More ice melts → Ice cover continues decreasing.

Systems Thinking Introduction – (Big Skill) Feedback Loops & Complexity vs. (Small Skill) Step-by-Step System Analysis

---

Pairing 1: The Supply and Demand Cycle

Big Skill: Feedback Loops & Complexity – Understanding how actions create cycles (e.g., supply and demand in economics).

Small Skill: Step-by-Step System Analysis – Breaking down how a system works in small, sequential steps.

Teaching Technique: Hands-on experiments.

Actions: Have students analyze how prices fluctuate based on supply and demand by simulating a market.

Example: A student creates a step-by-step breakdown:

Factory produces a popular toy. Stores sell out quickly → Demand increases. Prices rise due to scarcity. More factories start producing the toy. Supply increases → prices drop.

Pairing 2: The Greenhouse Effect and Climate Change

Big Skill: Feedback Loops & Complexity – Recognizing how climate change cycles intensify over time.

Small Skill: Step-by-Step System Analysis – Breaking down how greenhouse gases trap heat in the atmosphere.

Teaching Technique: Guided exploration.

Actions: Have students trace the impact of increased CO_2 levels step by step.

Example: A student maps the greenhouse effect:

Cars release CO_2. CO_2 traps heat in the atmosphere. Temperatures rise, causing ice caps to melt. More heat absorbed by oceans. Cycle repeats with more extreme effects.

Pairing 3: The Debt Cycle

Big Skill: Feedback Loops & Complexity – Understanding how debt cycles impact individuals and economies.

Small Skill: Step-by-Step System Analysis – Analyzing how borrowing leads to financial instability.

Teaching Technique: Problem-solving exercises.

Actions: Have students break down the cycle of credit card debt.

Example: A student follows the debt loop:

> **Person borrows money on a credit card. Monthly payments include high interest. Less money left for essentials. More borrowing to cover costs. Debt increases over time.**

Pairing 4: Overfishing and Ocean Ecosystems

Big Skill: Feedback Loops & Complexity – Recognizing how overfishing disrupts marine ecosystems.

Small Skill: Step-by-Step System Analysis – Breaking down how removing one species affects the food chain.

Teaching Technique: Observation journals.

Actions: Have students analyze how removing a predator affects ocean balance.

Example: A student creates a food chain breakdown:

> **Large fish are overfished. Small fish populations increase. Small fish overconsume plankton. Ocean ecosystems shift. Some fish species disappear, worsening the cycle.**

Pairing 5: Traffic Congestion and Urban Growth

Big Skill: Feedback Loops & Complexity – Understanding how traffic congestion worsens in urban areas.

Small Skill: Step-by-Step System Analysis – Breaking down how road expansions impact congestion.

Teaching Technique: Hands-on experiments.

Actions: Have students simulate different traffic patterns in a class model.

Example: A student traces the congestion cycle:

> **More people move to cities. More roads built to reduce traffic. More people relocate due to easier commuting. More cars on the road. Traffic jams return.**

Pairing 6: The Spread of Misinformation

Big Skill: Feedback Loops & Complexity – Recognizing how misinformation spreads rapidly.

Small Skill: Step-by-Step System Analysis – Breaking down how false news circulates.

Teaching Technique: Sensory exploration.

Actions: Have students track how a false rumor spreads on social media.

Example: A student creates a misinformation cycle:

> **False news is posted. People share without fact-checking. More people believe it is true. It spreads further, reaching media outlets. Even after being debunked, some people still believe it.**

Pairing 7: The Recycling Process

Big Skill: Feedback Loops & Complexity – Understanding how recycling reduces landfill waste.

Small Skill: Step-by-Step System Analysis – Breaking down how recycling keeps materials in use.

Teaching Technique: Collaborative group work.

Actions: Have students follow the journey of a plastic bottle through the recycling process.

Example: A student tracks recycling in five steps:

> **Plastic bottle is placed in a recycling bin. Collected and transported to a facility. Melted down and reshaped into a new product. Sold and used again. Fewer bottles end up in landfills.**

Pairing 8: Antibiotic Resistance in Bacteria

Big Skill: Feedback Loops & Complexity – Recognizing how antibiotic overuse leads to resistant bacteria.

Small Skill: Step-by-Step System Analysis – Breaking down how bacteria evolve resistance over time.

Teaching Technique: Real-world applications.

Actions: Have students model bacterial resistance through a classroom experiment.

Example: A student creates an antibiotic resistance cycle:

> **Antibiotic kills weak bacteria. Some bacteria survive and multiply. Stronger bacteria spread. New, stronger antibiotics are needed. Cycle repeats, making infections harder to treat.**

Pairing 9: Soil Erosion and Food Shortages

Big Skill: Feedback Loops & Complexity – Understanding how soil erosion worsens farming conditions.

Small Skill: Step-by-Step System Analysis – Analyzing how soil loss leads to food shortages.

Teaching Technique: Cultural exploration.

Actions: Have students trace how deforestation leads to poor soil conditions.

Example: A student outlines the erosion cycle:

> **Trees cut down for farming. Rain washes away topsoil. Crops struggle to grow. Less food produced. More deforestation for farmland.**

Pairing 10: Melting Ice and Global Warming

Big Skill: Feedback Loops & Complexity – Recognizing how melting polar ice accelerates climate change.

Small Skill: Step-by-Step System Analysis – Breaking down the heat absorption cycle.

Teaching Technique: Hands-on experiments.

Actions: Have students compare heat absorption between ice and water.

Example: A student illustrates how melting ice speeds up warming:

> **Ice reflects sunlight. Warmer temperatures melt ice. Darker water absorbs more heat. Ocean temperatures rise. More ice melts, continuing the cycle.**

Systems Thinking Introduction – (Big Skill) Feedback Loops & Complexity vs. (Small Skill) Testing Small Changes

---

Pairing 1: Supply and Demand in an Economy

Big Skill: Feedback Loops & Complexity – Understanding how actions create cycles (e.g., supply and demand in economics).

Small Skill: Testing Small Changes – Running simulations or role-playing different scenarios to see how variables affect a system.

Teaching Technique: Hands-on experiments.

Actions: Have students simulate an economy where product prices fluctuate based on supply and demand.

Example: A student increases the price of a popular product in a simulation:

> **Demand decreases → Stores lower prices → Demand rises again → The cycle continues.**

Pairing 2: CO_2 Emissions and Climate Change

Big Skill: Feedback Loops & Complexity – Understanding how climate change cycles intensify over time.

Small Skill: Testing Small Changes – Simulating how reducing carbon emissions affects temperature rise.

Teaching Technique: Guided exploration.

Actions: Have students run a climate change simulation where CO_2 levels are adjusted to see how it affects global temperatures.

Example: A student reduces CO_2 emissions in the simulation:

Temperatures stabilize. Ocean levels rise more slowly. Ice caps remain intact longer. Less extreme weather events occur.

Pairing 3: Deforestation vs. Reforestation

Big Skill: Feedback Loops & Complexity – Recognizing how deforestation leads to increased carbon levels.

Small Skill: Testing Small Changes – Simulating the effects of tree planting versus deforestation.

Teaching Technique: Problem-solving exercises.

Actions: Have students create two small terrariums—one with plants and one without—and measure CO_2 levels over time.

Example: A student removes plants from one terrarium and finds:

CO_2 levels rise faster. Temperature increases. Oxygen levels drop.

Pairing 4: Public Transportation and Traffic Congestion

Big Skill: Feedback Loops & Complexity – Understanding how urbanization increases traffic congestion.

Small Skill: Testing Small Changes – Simulating how different transportation policies affect congestion.

Teaching Technique: Observation journals.

Actions: Have students run a simulation where public transportation is improved and car use is reduced.

Example: A student adds a subway system in the model city and observes:

Fewer cars on the road. Traffic flow improves. Businesses see increased customers. Fewer emissions lead to better air quality.

Pairing 5: Antibiotic Resistance and Bacteria Evolution

Big Skill: Feedback Loops & Complexity – Recognizing how antibiotic overuse leads to resistant bacteria.

Small Skill: Testing Small Changes – Simulating bacterial resistance through a class experiment.

Teaching Technique: Hands-on experiments.

Actions: Have students model bacterial evolution by exposing bacteria samples to increasing doses of antibiotics.

Example: A student applies different antibiotic levels to bacteria cultures:

Weak bacteria die first. Some bacteria survive and multiply. Higher doses of antibiotics are needed over time.

Pairing 6: Recycling and Waste Reduction

Big Skill: Feedback Loops & Complexity – Understanding how recycling reduces landfill waste.

Small Skill: Testing Small Changes – Simulating how different waste management techniques impact trash levels.

Teaching Technique: Sensory exploration.

Actions: Have students sort classroom waste into categories and measure what percentage is recyclable.

Example: A student removes recyclables from classroom waste and tracks:

Less landfill waste. More materials reused. Less environmental pollution.

Pairing 7: The Spread of Misinformation

Big Skill: Feedback Loops & Complexity – Recognizing how misinformation spreads online.

Small Skill: Testing Small Changes – Simulating how different types of news spread at different rates.

Teaching Technique: Collaborative group work.

Actions: Have students role-play as journalists, social media users, and fact-checkers to track how false news spreads.

Example: A student shares an exaggerated story in the simulation:

False story spreads faster than real news. People believe and share without checking. Cycle repeats with increased misinformation.

Pairing 8: Supply Chains and Global Trade

Big Skill: Feedback Loops & Complexity – Understanding how supply chains work in a global economy.

Small Skill: Testing Small Changes – Simulating how supply disruptions affect product availability.

Teaching Technique: Real-world applications.

Actions: Have students simulate a toy production company and adjust variables like factory shutdowns or shipping delays.

Example: A student shuts down a major factory in the simulation:

Fewer toys reach stores. Prices increase. Factories reopen, supply rebounds.

Pairing 9: Fishing Quotas and Marine Ecosystems

Big Skill: Feedback Loops & Complexity – Recognizing how overfishing disrupts marine ecosystems.

Small Skill: Testing Small Changes – Simulating fishing limits and their effects on fish populations.

Teaching Technique: Cultural exploration.

Actions: Have students use a fishing simulation where quotas can be adjusted.

Example: A student sets strict fishing limits:

Fish populations recover. Marine ecosystems stabilize. Fishing industries stay sustainable.

Pairing 10: Melting Ice Caps and Rising Sea Levels

Big Skill: Feedback Loops & Complexity – Understanding how melting ice caps contribute to rising sea levels.

Small Skill: Testing Small Changes – Simulating how changes in temperature affect ice melt rates.

Teaching Technique: Hands-on experiments.

Actions: Have students place ice in a model ocean and measure water levels as it melts under different heat conditions.

Example: A student raises the temperature in one test:

Ice melts faster. Water levels rise. Coastal areas begin flooding.

Example Teaching Strategies for Big vs. Small Thinking Skill Pairings of Critical Thinking at the Middle School Level

Here are 90 unique teaching strategy examples for the big thinking and small thinking skill pairings, each described in detail to align with **Big Thinking Skills** (Concepts & Strategies) and **Small Thinking Skills** (Step-by-Step Execution) for the **Critical Thinking Skill** taught at the middle school level.

Critical Thinking

Big Thinking Skills (Concepts & Strategies)

- **Questioning Assumptions:** Encouraging students to ask, "How do we know this is true?" and seek deeper answers.
- **Analyzing Arguments:** Learning how to spot bias, fallacies, and weak reasoning.
- **Weighing Different Perspectives:** Understanding that different viewpoints exist and considering multiple sides of an issue.

Small Thinking Skills (Step-by-Step Execution)

- **Identifying Key Claims & Evidence:** Learning how to break down an argument into main points and supporting details.
- **Comparing Sources:** Practicing how to evaluate different pieces of evidence for reliability.
- **Structured Argumentation:** Using frameworks like the Claim-Evidence-Reasoning (CER) method to organize thoughts logically.

Critical Thinking – (Big Skill) Questioning Assumptions vs. (Small Skill) Identifying Key Claims & Evidence

---

Pairing 1: Investigating Common Myths

Big Skill: Questioning Assumptions – Encouraging students to ask, "How do we know this is true?" and seek deeper answers.

Small Skill: Identifying Key Claims & Evidence – Learning how to break down an argument into main points and supporting details.

Teaching Technique: Hands-on experiments.

Actions: Have students investigate a common myth by identifying the claim and researching evidence.

Example: A student examines the claim, "Goldfish only have a three-second memory."

Identifies the claim. Researches scientific studies on goldfish memory. Discovers goldfish can remember tasks for months.

Pairing 2: Analyzing News Articles

Big Skill: Questioning Assumptions – Encouraging skepticism about everyday information.

Small Skill: Identifying Key Claims & Evidence – Analyzing media articles for claims and supporting facts.

Teaching Technique: Guided exploration.

Actions: Have students choose an online news article and highlight its key claims and evidence.

Example: A student examines an article on climate change:

1. **Identifies claim:** "The Earth is warming at an unprecedented rate."
2. **Finds supporting evidence:** NASA temperature data, expert quotes, historical comparisons.

Pairing 3: Challenging Historical Narratives

Big Skill: Questioning Assumptions – Challenging widely accepted beliefs by looking at different perspectives.

Small Skill: Identifying Key Claims & Evidence – Analyzing arguments from multiple sources.

Teaching Technique: Problem-solving exercises.

Actions: Have students debate a historical event by identifying key claims from different viewpoints.

Example: A student explores multiple perspectives on the American Revolution:

1. **British argument:** "Colonists were rebelling against fair taxation."
2. **Colonial argument:** "Taxation without representation was unfair."
3. Examines historical documents to find evidence for both.

Pairing 4: Deconstructing Advertising Claims

Big Skill: Questioning Assumptions – Encouraging deeper analysis of advertising claims.

Small Skill: Identifying Key Claims & Evidence – Breaking down persuasive techniques used in marketing.

Teaching Technique: Observation journals.

Actions: Have students analyze a commercial, identifying its key claims and the evidence used to support them.

Example: A student watches a toothpaste ad:

1. **Claim:** "9 out of 10 dentists recommend this brand."
2. **Examines the fine print:** "Based on a survey of 50 dentists."
3. Questions whether the sample size is reliable.

Pairing 5: Investigating Stereotypes

Big Skill: Questioning Assumptions – Encouraging students to challenge stereotypes.

Small Skill: Identifying Key Claims & Evidence – Investigating whether common stereotypes are based on facts.

Teaching Technique: Hands-on experiments.

Actions: Have students research stereotypes about intelligence, sports, or careers and analyze the evidence behind them.

Example: A student investigates the stereotype that *boys are better at math than girls*:

1. **Reviews** international test scores.
2. **Reads** about cultural influences on education.
3. **Discovers** that gender does not determine math ability.

Pairing 6: Verifying Sources for Reliability

Big Skill: Questioning Assumptions – Encouraging students to verify sources of information.

Small Skill: Identifying Key Claims & Evidence – Differentiating between reliable and unreliable sources.

Teaching Technique: Sensory exploration.

Actions: Have students compare an academic article and a social media post on the same topic.

Example: A student analyzes two sources on vaccines:

1. **CDC website provides** research-based claims with sources.
2. **Social media post provides** anecdotal evidence without studies.
3. **Questions** which source is more credible.

Pairing 7: Analyzing Historical Bias

Big Skill: Questioning Assumptions – Encouraging critical thinking in historical narratives.

Small Skill: Identifying Key Claims & Evidence – Analyzing how history is written and whose perspective is highlighted.

Teaching Technique: Collaborative group work.

Actions: Have students examine how different history textbooks present the same event.

Example: A student compares U.S. and British textbooks on the Boston Tea Party:

1. **U.S. textbook** calls it *a heroic protest.*
2. **British textbook** calls it *destruction of property.*
3. **Examines documents** from the time to assess the truth.

Pairing 8: Bias in Storytelling

Big Skill: Questioning Assumptions – Encouraging students to think about bias in storytelling.

Small Skill: Identifying Key Claims & Evidence – Examining how characters in literature present different perspectives.

Teaching Technique: Real-world applications.

Actions: Have students analyze how different characters in a novel describe the same event.

Example: A student reads *"To Kill a Mockingbird:"*

1. **Identifies claims** from Atticus, Tom Robinson, and Mayella.
2. **Analyzes** how personal experiences shape their perspectives.
3. **Discusses** how bias affects storytelling.

Pairing 9: Fact-Checking Conspiracy Theories

Big Skill: Questioning Assumptions – Encouraging skepticism toward conspiracy theories.

Small Skill: Identifying Key Claims & Evidence – Analyzing whether theories have credible evidence.

Teaching Technique: Cultural exploration.

Actions: Have students examine a popular conspiracy theory and fact-check its claims.

Example: A student investigates the *moon landing hoax* theory:

1. **Claim:** "The moon landing was faked."
2. **Evidence**: Shadows in photos, flag movement, NASA rebuttals.
3. **Analyzes** real physics to debunk the theory.

Pairing 10: Testing Everyday Science Claims

Big Skill: Questioning Assumptions – Encouraging curiosity about everyday science.

Small Skill: Identifying Key Claims & Evidence – Investigating scientific claims and evaluating them.

Teaching Technique: Hands-on experiments.

Actions: Have students evaluate a claim like *"sugar makes kids hyper"* by observing real effects.

Example: A student compares behavior in two groups—one given sugar, one not:

1. **Identifies claim:** "Sugar causes hyperactivity."
2. **Observes behaviors**.
3. **Researches studies** that show no link.

Critical Thinking – (Big Skill) Questioning Assumptions vs. (Small Skill) Comparing Sources

---

Pairing 1: Investigating Health Claims

Big Skill: Questioning Assumptions – Encouraging students to ask, "How do we know this is true?" and seek deeper answers.

Small Skill: Comparing Sources – Practicing how to evaluate different pieces of evidence for reliability.

Teaching Technique: Hands-on experiments.

Actions: Have students compare scientific articles and social media posts on a health-related claim.

Example: A student investigates the claim that *"drinking lemon water detoxifies the body:"*

1. **Source 1:** A viral social media post → No sources cited.
2. **Source 2:** A medical journal → Explains liver function and detox processes.
3. **Conclusion:** Detoxing is a bodily function, not caused by lemon water.

Pairing 2: Analyzing Historical Narratives

Big Skill: Questioning Assumptions – Encouraging students to analyze common historical narratives.

Small Skill: Comparing Sources – Examining different historical accounts of the same event.

Teaching Technique: Guided exploration.

Actions: Have students compare U.S. and international textbooks on a major historical event.

Example: A student examines how **World War II** is portrayed:

1. **U.S. textbook:** Emphasizes military heroism.
2. **German textbook:** Discusses economic struggles.
3. **Conclusion:** History is shaped by different perspectives.

Pairing 3: Evaluating News Bias

Big Skill: Questioning Assumptions – Encouraging students to verify claims in news media.

Small Skill: Comparing Sources – Checking how different news outlets report the same story.

Teaching Technique: Problem-solving exercises.

Actions: Have students analyze a current news event across multiple sources.

Example: A student researches an environmental policy change:

1. **Green energy news site:** Supports the policy.
2. **Business news site:** Criticizes economic impact.
3. **Major newspaper:** Presents neutral analysis.
4. **Conclusion:** Bias exists in news reporting.

Pairing 4: Questioning Advertising Claims

Big Skill: Questioning Assumptions – Encouraging skepticism toward advertising claims.

Small Skill: Comparing Sources – Evaluating corporate vs. independent product reviews.

Teaching Technique: Observation journals.

Actions: Have students compare company advertisements with third-party product reviews.

Example: A student examines claims about a "miracle" skin cream:

1. **Company's ad:** Promises instant results.
2. **Dermatologist's review:** States it takes weeks to see changes.
3. **Customer reviews:** Show mixed results.
4. **Conclusion:** The ad exaggerated results.

Pairing 5: Investigating Scientific Studies

Big Skill: Questioning Assumptions – Encouraging students to think critically about scientific studies.

Small Skill: Comparing Sources – Checking if claims are backed by peer-reviewed research.

Teaching Technique: Hands-on experiments.

Actions: Have students evaluate a "scientific" claim by comparing real studies with unsupported claims.

Example: A student researches the claim *"eating chocolate improves memory:"*

1. **Blog claim:** Chocolate boosts brain function.
2. **Peer-reviewed study:** Shows mixed results.

3. **Conclusion:** The blog misrepresents the science.

Pairing 6: Fact-Checking Social Media Claims

Big Skill: Questioning Assumptions – Encouraging students to evaluate credibility in social media.

Small Skill: Comparing Sources – Identifying fact-based vs. opinion-based reporting.

Teaching Technique: Sensory exploration.

Actions: Have students analyze viral social media posts and compare them with professional journalism.

Example: A student investigates the viral claim *"schools are banning books:"*

1. **Social media post:** Emotional but lacks sources.
2. **Local news:** Verifies a school removed a single book.
3. **Conclusion:** The social media post exaggerated the truth.

Pairing 7: Analyzing Cultural Myths

Big Skill: Questioning Assumptions – Encouraging students to analyze stereotypes and cultural myths.

Small Skill: Comparing Sources – Researching whether commonly held beliefs are fact-based.

Teaching Technique: Collaborative group work.

Actions: Have students examine cultural myths (e.g., *"People only use 10% of their brains"*) and check academic sources.

Example: A student investigates the "10% brain myth:"

1. **TV shows:** Claim untapped brain power exists.
2. **Neuroscience textbooks:** Explain full-brain activity.
3. **Conclusion:** The myth is not backed by science.

Pairing 8: Evaluating Financial Advice

Big Skill: Questioning Assumptions – Encouraging students to evaluate financial advice and economic claims.

Small Skill: Comparing Sources – Checking expert opinions vs. unverified claims.

Teaching Technique: Real-world applications.

Actions: Have students compare financial advice from influencers and economic experts.

Example: A student analyzes the viral claim *"crypto is the best investment:"*

1. **Influencer:** Says "invest now, get rich fast."
2. **Economist:** Warns about volatility and risks.
3. **Conclusion:** Investing requires careful research.

Pairing 9: Exploring Historical "What If" Scenarios

Big Skill: Questioning Assumptions – Encouraging students to analyze historical "what if" scenarios.

Small Skill: Comparing Sources – Evaluating historical speculation vs. actual evidence.

Teaching Technique: Cultural exploration.

Actions: Have students examine theories about alternative historical outcomes.

Example: A student explores "What if the South won the Civil War?"

1. **Fictional book:** Imagines a divided America.
2. **Historians:** Cite economic and military realities.
3. **Conclusion:** Speculation is interesting but must be grounded in facts.

Pairing 10: Fact-Checking Medical Misinformation

Big Skill: Questioning Assumptions – Encouraging curiosity about medical and health claims.

Small Skill: Comparing Sources – Checking medical sources against misinformation.

Teaching Technique: Hands-on experiments.

Actions: Have students research a controversial health claim, comparing medical websites and misinformation sources.

Example: A student analyzes the claim *"natural remedies cure the flu:"*

1. **Holistic websites:** Promote herbal teas as cures.
2. **CDC:** States only rest and hydration help.
3. **Conclusion:** Natural remedies help symptoms but do not cure the flu.

Critical Thinking – (Big Skill) Questioning Assumptions vs. (Small Skill) Structured Argumentation

---

Pairing 1: Evaluating Scientific Claims

Big Skill: Questioning Assumptions – Encouraging students to ask, "How do we know this is true?" and seek deeper answers.

Small Skill: Structured Argumentation – Using frameworks like the Claim-Evidence-Reasoning (CER) method to organize thoughts logically.

Teaching Technique: Hands-on experiments.

Actions: Have students test the claim that plants grow better with music using the CER method.

Example: A student conducts an experiment on plant growth and music exposure:

1. **Claim:** "Music helps plants grow faster."
2. **Evidence:** Plants exposed to music grew 2 cm taller than those without music.
3. **Reasoning:** Sound vibrations may stimulate plant cells, but further research is needed.

Pairing 2: Questioning Advertising Claims

Big Skill: Questioning Assumptions – Encouraging skepticism in media claims.

Small Skill: Structured Argumentation – Using the CER method to analyze advertising statements.

Teaching Technique: Guided exploration.

Actions: Have students analyze a commercial claim, breaking it down into claim, evidence, and reasoning.

Example: A student examines a sports drink ad claiming to "boost energy instantly:"

1. **Claim:** "This drink provides instant energy."
2. **Evidence:** Ad features athletes performing after drinking it, but no scientific data.
3. **Reasoning:** Sugar may cause a short-term spike, but real energy comes from nutrients.

Pairing 3: Debating Historical Narratives

Big Skill: Questioning Assumptions – Challenging widely accepted historical narratives.

Small Skill: Structured Argumentation – Organizing historical debates using the CER method.

Teaching Technique: Problem-solving exercises.

Actions: Have students debate the real causes of the American Revolution using claim-evidence-reasoning.

Example: A student analyzes whether taxation or colonial self-rule was the main cause:

1. **Claim:** "The American Revolution was primarily about taxation."
2. **Evidence:** British tax policies like the Stamp Act fueled protests.
3. **Reasoning:** Economic hardship led to resistance, but political independence was also a factor.

Pairing 4: Evaluating Conspiracy Theories

Big Skill: Questioning Assumptions – Encouraging deeper analysis of conspiracy theories.

Small Skill: Structured Argumentation – Using the CER method to evaluate claims for logical validity.

Teaching Technique: Observation journals.

Actions: Have students analyze a popular conspiracy theory and construct an argument based on evidence.

Example: A student investigates the "moon landing was faked" theory:

1. **Claim:** "The moon landing was real."
2. **Evidence:** NASA's physical samples, independent satellite photos, astronaut testimonies.
3. **Reasoning:** Conspiracy theories lack verifiable evidence, while multiple sources confirm the landing.

Pairing 5: Debating Economic Policies

Big Skill: Questioning Assumptions – Encouraging skepticism toward political or social claims.

Small Skill: Structured Argumentation – Constructing logical arguments in classroom debates.

Teaching Technique: Hands-on experiments.

Actions: Have students analyze and debate real-world policies, using CER to support their positions.

Example: A student debates the effectiveness of raising the minimum wage:

1. **Claim:** "Raising the minimum wage reduces poverty."
2. **Evidence:** Studies show increased wages improve worker well-being.
3. **Reasoning:** Higher wages help but may cause inflation—so policy design matters.

Pairing 6: Investigating Scientific "Facts"

Big Skill: Questioning Assumptions – Encouraging students to evaluate scientific "facts."

Small Skill: Structured Argumentation – Using the CER method to evaluate scientific claims.

Teaching Technique: Sensory exploration.

Actions: Have students research a claim like "organic food is healthier" and organize their findings.

Example: A student examines the health benefits of organic vs. non-organic food:

1. **Claim:** "Organic food is healthier than conventional food."
2. **Evidence:** Some studies show fewer pesticides, but no significant nutritional differences.

3. **Reasoning:** Organic food may reduce chemical exposure, but overall nutrition is similar.

Pairing 7: Analyzing Literary Themes

Big Skill: Questioning Assumptions – Encouraging critical thinking in literary analysis.

Small Skill: Structured Argumentation – Using CER to analyze themes in literature.

Teaching Technique: Collaborative group work.

Actions: Have students construct arguments about a novel's central theme using claim-evidence-reasoning.

Example: A student analyzes the theme of racism in "To Kill a Mockingbird:"

1. **Claim:** "Racism is the central theme of the novel."
2. **Evidence:** Tom Robinson's unfair trial, Scout's observations about race.
3. **Reasoning:** The novel critiques racial injustice through multiple character experiences.

Pairing 8: Evaluating Health Myths

Big Skill: Questioning Assumptions – Encouraging students to analyze health myths.

Small Skill: Structured Argumentation – Using the CER method to evaluate health-related claims.

Teaching Technique: Real-world applications.

Actions: Have students evaluate a popular health myth using scientific research.

Example: A student investigates the claim "drinking 8 glasses of water a day is necessary:"

1. **Claim:** "Drinking exactly 8 glasses of water is not required for good health."
2. **Evidence:** Studies show hydration needs vary by person.
3. **Reasoning:** Water intake depends on diet, climate, and activity level.

Pairing 9: Researching Environmental Policies

Big Skill: Questioning Assumptions – Encouraging students to evaluate environmental claims.

Small Skill: Structured Argumentation – Constructing logical arguments based on climate science.

Teaching Technique: Cultural exploration.

Actions: Have students research and argue different positions on renewable energy using CER.

Example: A student examines whether solar energy is the best renewable source:

1. **Claim:** "Solar energy is the most sustainable renewable resource."

2. **Evidence:** It reduces carbon emissions and works in most regions.
3. **Reasoning:** While effective, energy storage and cost remain challenges.

Pairing 10: Debating Artificial Intelligence Ethics

Big Skill: Questioning Assumptions – Encouraging curiosity about technology and AI ethics.

Small Skill: Structured Argumentation – Using CER to analyze the pros and cons of artificial intelligence.

Teaching Technique: Hands-on experiments.

Actions: Have students debate whether AI is beneficial or harmful using structured arguments.

Example: A student explores AI in job automation:

1. **Claim:** "AI will replace many jobs but also create new ones."
2. **Evidence:** AI automates repetitive tasks but increases demand for tech-based jobs.
3. **Reasoning:** The economy must adapt to AI-driven job shifts rather than resist change.

Critical Thinking – (Big Skill) Analyzing Arguments vs. (Small Skill) Identifying Key Claims & Evidence

---

Pairing 1: Spotting Bias in News Articles

Big Skill: Analyzing Arguments – Learning how to spot bias, fallacies, and weak reasoning.

Small Skill: Identifying Key Claims & Evidence – Learning how to break down an argument into main points and supporting details.

Teaching Technique: Hands-on experiments.

Actions: Have students compare two news articles from different sources covering the same event, identifying key claims and possible bias.

Example:

1. **Claim (News Source 1):** "The protest was peaceful and well-organized."
2. **Claim (News Source 2):** "The protest turned into chaos and disruption."
3. **Evidence Comparison:** One article includes interviews with peaceful demonstrators; the other focuses on a single act of vandalism.
4. **Conclusion:** Students recognize how selective reporting affects public perception.

Pairing 2: Identifying Logical Fallacies in Political Speeches

Big Skill: Analyzing Arguments – Recognizing flawed reasoning in arguments.

Small Skill: Identifying Key Claims & Evidence – Breaking down political speeches into main claims and supporting arguments.

Teaching Technique: Guided exploration.

Actions: Have students listen to or read a political speech and highlight fallacies such as ad hominem attacks or false cause reasoning.

Example:

1. **Claim:** "If we do not increase military funding, our country will be overrun by enemies." (Slippery slope fallacy).
2. **Evidence Analysis:** No direct proof is given that reducing military funding leads to invasion.
3. **Conclusion:** Students see how fear-based reasoning manipulates audiences.

Pairing 3: Analyzing Advertisements for Persuasive Techniques

Big Skill: Analyzing Arguments – Understanding how arguments are used to persuade consumers.

Small Skill: Identifying Key Claims & Evidence – Breaking down commercial claims and their evidence.

Teaching Technique: Problem-solving exercises.

Actions: Have students evaluate the claims made in TV or online advertisements.

Example:

1. **Claim:** "This toothpaste will make your teeth 5 shades whiter in one week!"
2. **Evidence Analysis:** The ad shows before-and-after photos but does not cite clinical studies.
3. **Conclusion:** Students recognize how advertising often lacks reliable evidence.

Pairing 4: Evaluating Social Media Misinformation

Big Skill: Analyzing Arguments – Spotting weak reasoning and false information.

Small Skill: Identifying Key Claims & Evidence – Fact-checking viral posts.

Teaching Technique: Observation journals.

Actions: Have students analyze a viral claim and determine whether it is supported by reliable sources.

Example:

1. **Claim:** "Drinking celery juice cures cancer!"
2. **Evidence Analysis:** A medical journal states there is no scientific proof of this claim.

3. **Conclusion:** Students learn to question sources and verify claims with credible evidence.

Pairing 5: Examining Historical Speeches for Rhetorical Strategies

Big Skill: Analyzing Arguments – Understanding how historical figures crafted persuasive arguments.

Small Skill: Identifying Key Claims & Evidence – Distilling key points from historical speeches.

Teaching Technique: Hands-on experiments.

Actions: Have students analyze famous speeches (e.g., MLK's "I Have a Dream") and identify claims and supporting evidence.

Example:

1. **Claim:** "All people should be judged by the content of their character, not by the color of their skin."
2. **Evidence:** Examples of racial injustice and calls for equality.
3. **Conclusion:** Students understand how strong arguments are backed by ethical and emotional appeals.

Pairing 6: Comparing Opinion vs. Fact-Based Writing

Big Skill: Analyzing Arguments – Differentiating between subjective opinions and fact-based arguments.

Small Skill: Identifying Key Claims & Evidence – Distinguishing emotional appeals from logical arguments.

Teaching Technique: Sensory exploration.

Actions: Have students read an editorial and a research article on the same topic and compare how claims are supported.

Example:

1. **Editorial Claim:** "School should start later because waking up early is annoying." (Opinion-based).
2. **Research Claim:** "Students who start school later get better grades and sleep more." (Fact-based, with evidence from sleep studies).
3. **Conclusion:** Facts are supported by research, while opinions rely on personal experience.

Pairing 7: Identifying Cherry-Picked Evidence in Debates

Big Skill: Analyzing Arguments – Recognizing how arguments can be manipulated by selecting only favorable evidence.

Small Skill: Identifying Key Claims & Evidence – Evaluating whether arguments consider all perspectives.

Teaching Technique: Collaborative group work.

Actions: Have students analyze a debate and see if both sides present full evidence or selectively highlight data.

Example:

1. **Claim:** "Video games cause violence because crime rates increased when gaming became popular."
2. **Evidence Analysis:** FBI data shows crime rates actually decreased while gaming increased.
3. **Conclusion:** Cherry-picking data can lead to misleading arguments.

Pairing 8: Fact-Checking Political Campaign Promises

Big Skill: Analyzing Arguments – Evaluating political statements for accuracy.

Small Skill: Identifying Key Claims & Evidence – Researching whether claims are backed by real data.

Teaching Technique: Real-world applications.

Actions: Have students research a politician's campaign promise and compare it to factual data.

Example:

1. **Claim:** "Lowering taxes will create millions of jobs."
2. **Evidence Analysis:** Economic studies show mixed results on whether tax cuts directly lead to job creation.
3. **Conclusion:** Political claims often oversimplify complex economic issues.

Pairing 9: Examining Conspiracy Theories for Logical Gaps

Big Skill: Analyzing Arguments – Recognizing weak reasoning in conspiracy theories.

Small Skill: Identifying Key Claims & Evidence – Checking whether conspiracy claims are supported by credible sources.

Teaching Technique: Cultural exploration.

Actions: Have students analyze a common conspiracy theory and determine if it has valid evidence.

Example:

1. **Claim:** "The moon landing was faked."
2. **Evidence Analysis:** NASA's evidence includes thousands of photos, moon rock samples, and testimonies from astronauts.
3. **Conclusion:** The moon landing-hoax theory lacks credible evidence and relies on speculation.

Pairing 10: Investigating Health Myths and Science-Based Claims

Big Skill: Analyzing Arguments – Distinguishing between pseudoscience and real scientific research.

Small Skill: Identifying Key Claims & Evidence – Checking sources for medical claims.

Teaching Technique: Hands-on experiments.

Actions: Have students analyze a health claim (e.g., "organic food is healthier") and compare scientific studies to general beliefs.

Example:

1. **Claim:** "Organic food is healthier than conventional food."
2. **Evidence Analysis:** Studies show organic food has fewer pesticides but similar nutrition levels.
3. **Conclusion:** Organic food may have environmental benefits but is not always nutritionally superior.

Critical Thinking – (Big Skill) Analyzing Arguments vs. (Small Skill) Comparing Sources

---

Pairing 1: Spotting Bias in News Articles

Big Skill: Analyzing Arguments – Learning how to spot bias, fallacies, and weak reasoning.

Small Skill: Comparing Sources – Practicing how to evaluate different pieces of evidence for reliability.

Teaching Technique: Hands-on experiments.

Actions: Have students compare two news articles on the same topic, identifying biased language and logical fallacies.

Example: A student analyzes two articles on climate change:

1. **Article 1** uses emotional language ("climate catastrophe is unstoppable").
2. **Article 2** presents scientific data and expert opinions.
3. **Concludes** that the first article is biased, while the second is more balanced.

Pairing 2: Evaluating Viral Internet Claims

Big Skill: Analyzing Arguments – Identifying logical fallacies in everyday discussions.

Small Skill: Comparing Sources – Evaluating the credibility of sources supporting different arguments.

Teaching Technique: Guided exploration.

Actions: Have students examine a viral internet claim and compare sources that confirm or debunk it.

Example: A student investigates the claim that "5G causes health problems:"

1. **Social media post** cites no scientific studies.
2. **WHO website** explains no proven health risks from 5G.
3. **Identifies** fear-based reasoning and lack of credible evidence in the original claim.

Pairing 3: Recognizing Bias in Historical Narratives

Big Skill: Analyzing Arguments – Recognizing bias in historical narratives.

Small Skill: Comparing Sources – Looking at multiple perspectives on historical events.

Teaching Technique: Problem-solving exercises.

Actions: Have students compare U.S. and international perspectives on the American Revolution.

Example: A student examines textbooks from the U.S. and Britain:

1. **U.S. textbook** calls the revolution a fight for freedom.
2. **British textbook** highlights economic consequences for Britain.
3. **Concludes** that history is shaped by the perspective of the writer.

Pairing 4: Evaluating Advertising Exaggerations

Big Skill: Analyzing Arguments – Learning to detect exaggerations in advertisements.

Small Skill: Comparing Sources – Verifying marketing claims using independent reviews.

Teaching Technique: Observation journals.

Actions: Have students analyze an ad's claims and compare them to third-party reviews.

Example: A student examines a shampoo commercial promising "healthier hair in 3 days:"

1. **The ad** features celebrity endorsements but no scientific studies.
2. **Dermatologists** say hair health takes weeks to improve.
3. **Concludes** that the claim is exaggerated.

Pairing 5: Identifying Emotional vs. Factual Reasoning

Big Skill: Analyzing Arguments – Identifying appeal to emotion vs. factual reasoning.

Small Skill: Comparing Sources – Checking how different sources support or challenge emotional arguments.

Teaching Technique: Hands-on experiments

Actions: Have students examine political speeches and compare them with fact-checking sources.

Example: A student analyzes a speech claiming "crime is skyrocketing:"

1. **Politician** uses fear-based language.
2. **FBI crime data** shows a decline in crime rates.
3. **Identifies** emotional appeal without supporting evidence.

Pairing 6: Recognizing Cherry-Picked Data

Big Skill: Analyzing Arguments – Recognizing cherry-picking of data.

Small Skill: Comparing Sources – Evaluating whether data is taken out of context.

Teaching Technique: Sensory exploration.

Actions: Have students compare how different sources present the same statistics.

Example: A student investigates whether "video games cause violence:"

1. **News article** cites one study linking games to aggression.
2. **Scientific meta-analysis** finds no consistent evidence of causation.
3. **Concludes** that the first article cherry-picked its data.

Pairing 7: Spotting Personal Attacks in Arguments

Big Skill: Analyzing Arguments – Spotting ad hominem attacks vs. logical critiques.

Small Skill: Comparing Sources – Looking for personal attacks vs. issue-based criticism.

Teaching Technique: Collaborative group work.

Actions: Have students compare two opposing editorials and evaluate their argument styles.

Example: A student analyzes opinion pieces on school uniforms:

1. **One article** argues uniforms improve focus (fact-based).
2. **Another attacks** proponents of uniforms as "control freaks" (ad hominem).
3. **Concludes** that personal attacks weaken an argument.

Pairing 8: Identifying Sensationalism in Media

Big Skill: Analyzing Arguments – Identifying sensationalism in media.

Small Skill: Comparing Sources – Looking at neutral vs. sensationalized reporting.

Teaching Technique: Real-world applications.

Actions: Have students compare a clickbait headline with a factual news report.

Example: A student analyzes headlines about a celebrity:

1. **Clickbait:** "Celebrity A's career is OVER after shocking event!"
2. **Reputable news:** "Celebrity A addresses controversy, continues career."
3. **Identifies** exaggeration used to gain clicks.

Pairing 9: Recognizing False Equivalence in Debates

Big Skill: Analyzing Arguments – Recognizing false equivalence in debates.

Small Skill: Comparing Sources – Checking whether opposing views are equally valid.

Teaching Technique: Cultural exploration.

Actions: Have students examine opposing arguments on scientific consensus topics.

Example: A student investigates climate change denial vs. scientific consensus:

1. **99% of climate scientists** agree on human-caused warming.
2. **One blog claims** "scientists are divided" but cites no data.
3. **Identifies** false equivalence—one side has overwhelming evidence.

Pairing 10: Detecting Logical Inconsistencies in Politics

Big Skill: Analyzing Arguments – Detecting logical inconsistencies in debates.

Small Skill: Comparing Sources – Evaluating whether sources contradict themselves.

Teaching Technique: Hands-on experiments.

Actions: Have students find contradictions in political arguments over time.

Example: A student examines past statements by a politician:

1. A leader supports Policy A in one speech.
2. Later, they oppose Policy A for political reasons.
3. **Recognizes** the shift and evaluates possible motivations.

Critical Thinking – (Big Skill) Analyzing Arguments vs. (Small Skill) Structured Argumentation

---

Pairing 1: Evaluating Political Arguments Using CER

Big Skill: Analyzing Arguments – Learning how to spot bias, fallacies, and weak reasoning.

Small Skill: Structured Argumentation – Using frameworks like the Claim-Evidence-Reasoning (CER) method to organize thoughts logically.

Teaching Technique: Hands-on experiments.

Actions: Have students analyze a political speech for logical fallacies and construct a response using the CER framework.

Example: A student evaluates the claim that "banning junk food will make all students healthier:"

1. **Claim:** "Junk food bans improve student health."
2. **Evidence:** Some schools saw lower obesity rates, but others did not.
3. **Reasoning:** Health depends on overall diet and exercise, not just school policies.

Pairing 2: Analyzing Emotional Manipulation in Ads

Big Skill: Analyzing Arguments – Identifying emotional manipulation in advertising.

Small Skill: Structured Argumentation – Breaking down ad claims using CER to determine validity.

Teaching Technique: Guided exploration.

Actions: Have students evaluate a commercial's claim using logical reasoning instead of emotional response.

Example: A student analyzes a skincare ad claiming to "erase wrinkles in one use:"

1. **Claim:** "This cream erases wrinkles instantly."
2. **Evidence:** The ad shows before-and-after images but lacks scientific studies.
3. **Reasoning:** Short-term effects could be from lighting, not actual skin improvement.

Pairing 3: Recognizing Fallacies in Online Debates

Big Skill: Analyzing Arguments – Recognizing fallacies in online debates.

Small Skill: Structured Argumentation – Using CER to construct a logical counterargument.

Teaching Technique: Problem-solving exercises.

Actions: Have students engage in a structured online debate using logical argumentation.

Example: A student rebuts a claim that "video games cause violence:"

1. **Claim:** "Video games do not cause real-world violence."
2. **Evidence:** Multiple studies show no consistent link.
3. **Reasoning:** Correlation does not equal causation—other factors like home environment play a role.

Pairing 4: Clarifying Correlation vs. Causation

Big Skill: Analyzing Arguments – Detecting false cause-and-effect reasoning.

Small Skill: Structured Argumentation – Using CER to clarify correlation vs. causation.

Teaching Technique: Observation journals.

Actions: Have students analyze a claim and determine whether it confuses correlation with causation.

Example: A student examines the argument that "ice cream causes crime:"

1. **Claim:** "Eating ice cream leads to more crime."
2. **Evidence:** Crime rates rise in the summer when ice cream sales also rise.
3. **Reasoning:** The real cause is hot weather, which increases outdoor activity—not ice cream.

Pairing 5: Evaluating Biased News Articles

Big Skill: Analyzing Arguments – Evaluating biased news articles.

Small Skill: Structured Argumentation – Constructing a balanced argument using verified sources.

Teaching Technique: Hands-on experiments.

Actions: Have students compare news coverage from different sources and write a CER-based summary.

Example: A student analyzes two news articles about a protest:

1. **Claim:** "News coverage of protests is often biased."
2. **Evidence:** One article describes "violent rioters," another describes "peaceful demonstrators."
3. **Reasoning:** Word choice influences perception, showing how bias affects reporting.

Pairing 6: Identifying Personal Attacks in Arguments

Big Skill: Analyzing Arguments – Identifying ad hominem attacks in discussions.

Small Skill: Structured Argumentation – Using CER to structure logical arguments without personal attacks.

Teaching Technique: Sensory exploration.

Actions: Have students rewrite an argument that contains personal attacks into a logical, evidence-based statement.

Example: A student rewrites an argument about school dress codes:

1. **Claim:** "Dress codes should focus on professionalism, not punishment."
2. **Evidence:** Schools with flexible dress codes report fewer discipline issues.
3. **Reasoning:** The focus should be on learning, not strict clothing enforcement.

Pairing 7: Recognizing Slippery Slope Fallacies

Big Skill: Analyzing Arguments – Recognizing slippery slope fallacies.

Small Skill: Structured Argumentation – Rebuilding arguments using CER to avoid exaggeration.

Teaching Technique: Collaborative group work.

Actions: Have students find examples of slippery slope arguments and correct them using CER.

Example: A student challenges the claim that "if we allow cell phones in class, no one will ever pay attention:"

1. **Claim:** "Allowing phones in class can be managed responsibly."
2. **Evidence:** Some schools use phone time as an educational tool.
3. **Reasoning:** Policies can encourage balance rather than outright bans.

Pairing 8: Evaluating Scientific Claims in the Media

Big Skill: Analyzing Arguments – Evaluating scientific claims in the media.

Small Skill: Structured Argumentation – Using CER to determine if claims are backed by peer-reviewed research.

Teaching Technique: Real-world applications.

Actions: Have students analyze a health claim from a popular website and cross-check with scientific studies.

Example: A student investigates the claim that "organic food is healthier:"

1. **Claim:** "Organic food is healthier than conventional food."

2. **Evidence:** Some studies show fewer pesticides, but nutrition levels are similar.
3. **Reasoning:** Organic food may reduce exposure to chemicals but is not always more nutritious.

Pairing 9: Recognizing Hasty Generalizations

Big Skill: Analyzing Arguments – Recognizing hasty generalizations.

Small Skill: Structured Argumentation – Using CER to revise overly broad claims into accurate ones.

Teaching Technique: Cultural exploration.

Actions: Have students find and correct generalizations in media and personal arguments.

Example: A student modifies the claim "teenagers are addicted to social media:"

1. **Claim:** "Some teenagers struggle with social media overuse."
2. **Evidence:** Studies show screen time varies widely by individual.
3. **Reasoning:** Not all teens are affected the same way, so blanket statements are inaccurate.

Pairing 10: Spotting Contradictions in Public Statements

Big Skill: Analyzing Arguments – Spotting contradictions in public statements.

Small Skill: Structured Argumentation Using CER to hold arguments to logical consistency.

Teaching Technique: Hands-on experiments.

Actions: Have students track and compare politicians' statements over time to find inconsistencies.

Example: A student examines changing positions on environmental policies:

1. **Claim:** "Some politicians change their stance based on public opinion."
2. **Evidence:** A senator once opposed environmental regulations but now supports them.
3. **Reasoning:** Position shifts may reflect changing public attitudes, rather than personal conviction.

Critical Thinking – (Big Skill) Weighing Different Perspectives vs. (Small Skill) Identifying Key Claims & Evidence

---

Pairing 1: School Uniform Debate

Big Skill: Weighing Different Perspectives – Understanding that different viewpoints exist and considering multiple sides of an issue.

Small Skill: Identifying Key Claims & Evidence – Learning how to break down an argument into main points and supporting details.

Teaching Technique: Hands-on experiments.

Actions: Have students analyze two contrasting news articles on school uniforms, identifying key claims and supporting evidence.

Example:

1. **Pro:** "Uniforms promote equality and reduce bullying."
    - **Claim:** Creates a sense of unity.
    - **Evidence:** Schools with uniform policies report fewer bullying incidents.
2. **Con:** "Uniforms restrict personal expression."
    - **Claim:** Limits student individuality.
    - **Evidence:** Student surveys show a preference for self-expression in clothing.

Pairing 2: Historical Conflict Analysis

Big Skill: Weighing Different Perspectives – Examining historical conflicts from multiple viewpoints.

Small Skill: Identifying Key Claims & Evidence – Breaking down arguments from primary sources.

Teaching Technique: Guided exploration.

Actions: Have students compare American and British perspectives on the American Revolution using primary documents.

Example:

1. **Colonists' View:** "No taxation without representation!"
    - **Claim:** British taxes were unfair.
    - **Evidence:** The Stamp Act imposed taxes without colonial input.
2. **British View:** "Taxes were necessary to fund protection."
    - **Claim:** Colonists benefited from British defense.
    - **Evidence:** British troops protected the colonies from attacks.

Pairing 3: Extending the School Day Debate

Big Skill: Weighing Different Perspectives – Understanding the pros and cons of a policy change.

Small Skill: Identifying Key Claims & Evidence – Structuring arguments with supporting details.

Teaching Technique: Problem-solving exercises.

Actions: Have students debate whether schools should extend the school day, identifying key claims and evidence for both sides.

Example:

1. **Pro:** "A longer school day improves academic performance."
   - **Claim:** More learning time equals better test scores.
   - **Evidence:** Research shows extended hours boost student achievement.
2. **Con:** "Students need time for extracurriculars and rest."
   - **Claim:** Longer days can lead to burnout.
   - **Evidence:** Pediatric studies link school stress to mental health issues.

Pairing 4: Climate Change Policies

Big Skill: Weighing Different Perspectives – Evaluating different views on environmental policies.

Small Skill: Identifying Key Claims & Evidence – Analyzing opposing claims in climate policy debates.

Teaching Technique: Observation journals.

Actions: Have students compare perspectives on banning plastic bags.

Example:

1. **Environment indicates:** "Plastic bags pollute ecosystems."
   - **Claim:** Harmful to wildlife.
   - **Evidence:** Studies show marine animals ingest plastic waste.
2. **Business Owners:** "Plastic bans hurt small businesses."
   - **Claim:** Higher costs for alternative packaging.
   - **Evidence:** Case studies of stores with increased expenses.

Pairing 5: Death Penalty Debate

Big Skill: Weighing Different Perspectives – Analyzing ethical dilemmas from multiple cultural viewpoints.

Small Skill: Identifying Key Claims & Evidence – Comparing moral arguments.

Teaching Technique: Hands-on experiments.

Actions: Have students evaluate arguments for and against the death penalty.

Example:

1. **Pro:** "The death penalty provides justice for victims."
   - **Claim:** Deterrence against crime.
   - **Evidence:** Some studies suggest lower crime rates in states with capital punishment.
2. **Con:** "The death penalty risks executing innocent people."
   - **Claim:** Judicial errors occur.
   - **Evidence:** Cases of wrongful convictions being overturned.

Pairing 6: Animal Testing Debate

Big Skill: Weighing Different Perspectives – Understanding ethical debates in science.

Small Skill: Identifying Key Claims & Evidence – Examining evidence for both sides.

Teaching Technique: Sensory exploration.

Actions: Have students analyze arguments on animal testing in medical research.

Example:

1. **Pro:** "Animal testing saves human lives."
   - **Claim:** Necessary for drug safety.
   - **Evidence:** Vaccines and medications have been developed using animal testing.
2. **Con:** "Animal testing is cruel and outdated."
   - **Claim:** Alternatives exist.
   - **Evidence:** Companies now use lab-grown tissues and computer simulations.

Pairing 7: Novel Analysis (To Kill a Mockingbird)

Big Skill: Weighing Different Perspectives – Understanding different character viewpoints in literature.

Small Skill: Identifying Key Claims & Evidence – Breaking down character arguments.

Teaching Technique: Collaborative group work.

Actions: Have students analyze different perspectives in a novel.

Example:

1. **Atticus Finch:** "Justice should be fair for all."
   - **Claim:** Everyone deserves a fair trial.
   - **Evidence:** His defense of Tom Robinson despite racial bias.
2. **Maycomb Society:** "Social norms must be maintained."

- **Claim:** Some people are considered “less than.”
- **Evidence:** Townspeople’s biased views against Tom Robinson.

Pairing 8: Artificial Intelligence in the Workplace

Big Skill: Weighing Different Perspectives – Understanding the impact of AI on jobs.

Small Skill: Identifying Key Claims & Evidence – Evaluating different sides in a tech debate.

Teaching Technique: Real-world applications.

Actions: Have students debate whether AI benefits or harms employment.

Example:

1. **Pro-AI:** "AI increases productivity."
   - **Claim:** Automates repetitive tasks.
   - **Evidence:** AI-driven companies report higher efficiency.
2. **Anti-AI:** "AI replaces human jobs."
   - **Claim:** Many workers are at risk of losing employment.
   - **Evidence:** Reports predict millions of jobs will be automated.

Pairing 9: Banning Controversial Books

Big Skill: Weighing Different Perspectives – Examining censorship in education.

Small Skill: Identifying Key Claims & Evidence – Analyzing arguments in book banning.

Teaching Technique: Cultural exploration.

Actions: Have students research perspectives on banning books in schools.

Example:

1. **Pro-Ban:** "Some books contain inappropriate content."
   - **Claim:** Protects children from harmful material.
   - **Evidence:** Schools restrict violent or explicit books.
2. **Anti-Ban:** "Censorship limits free thought."
   - **Claim:** Students should learn critical thinking.
   - **Evidence:** Historically, banned books include important literature.

Pairing 10: Social Media and Free Speech

Big Skill: Weighing Different Perspectives – Evaluating limits on free expression.

Small Skill: Identifying Key Claims & Evidence – Breaking down censorship debates.

Teaching Technique: Hands-on experiments.

Actions: Have students examine perspectives on regulating hate speech online.

Example:

1. **Pro-Regulation:** "Hate speech should be limited."
   - **Claim:** Prevents real-world harm.
   - **Evidence:** Studies link online hate speech to violence.
2. **Anti-Regulation:** "Free speech should be absolute."
   - **Claim:** Censorship is a slippery slope.
   - **Evidence:** Historical cases of government overreach in speech restrictions.

Critical Thinking – (Big Skill) Weighing Different Perspectives vs. (Small Skill) Comparing Sources

---

Pairing 1: Climate Change Debate

Big Skill: Weighing Different Perspectives – Understanding that different viewpoints exist and considering multiple sides of an issue.

Small Skill: Comparing Sources – Practicing how to evaluate different pieces of evidence for reliability.

Teaching Technique: Hands-on experiments.

Actions: Have students compare sources that present different views on climate change, evaluating credibility based on scientific research.

Example:

1. **Source 1:** A peer-reviewed journal article from NASA citing temperature data.
2. **Source 2:** A blog post claiming climate change is a hoax, with no citations.
3. **Conclusion:** The NASA article provides measurable, scientific evidence, while the blog is opinion-based.

Pairing 2: Historical Events in Textbooks

Big Skill: Weighing Different Perspectives – Recognizing how historical events are presented differently in various countries.

Small Skill: Comparing Sources – Evaluating different historical accounts for reliability and bias.

Teaching Technique: Guided exploration.

Actions: Have students compare how U.S. history textbooks and British history textbooks describe the American Revolution.

Example:

1. **U.S. Textbook:** Frames the revolution as a heroic fight for independence.
2. **British Textbook:** Describes it as a colonial rebellion against lawful governance.
3. **Conclusion:** Different perspectives highlight national biases in historical writing.

Pairing 3: News Reporting on Protests

Big Skill: Weighing Different Perspectives – Understanding that media bias exists.

Small Skill: Comparing Sources – Evaluating news outlets' coverage of the same event.

Teaching Technique: Problem-solving exercises.

Actions: Have students analyze how different news organizations report on the same protest.

Example:

1. **News Outlet 1:** Describes protestors as "peaceful activists."
2. **News Outlet 2:** Calls them "violent rioters."
3. **Conclusion:** Language choice shapes public perception of events.

Pairing 4: Advertising vs. Scientific Studies

Big Skill: Weighing Different Perspectives – Understanding that marketing materials present biased perspectives.

Small Skill: Comparing Sources – Evaluating company claims vs. independent studies.

Teaching Technique: Observation journals.

Actions: Have students compare a skincare ad's claims with dermatologists' reviews.

Example:

1. **Ad Claim:** "Eliminates wrinkles instantly!"
2. **Scientific Study:** States that results take 4+ weeks and depend on skin type.
3. **Conclusion:** Ads often exaggerate effectiveness for sales purposes.

Pairing 5: Social Media Misinformation

Big Skill: Weighing Different Perspectives – Recognizing that not all sources are trustworthy.

Small Skill: Comparing Sources – Identifying fact-based vs. opinion-based information.

Teaching Technique: Hands-on experiments.

Actions: Have students fact-check a viral social media post on a trending issue.

Example:

1. **Tweet:** "Drinking celery juice cures cancer!" (No sources cited).
2. **Medical Journal:** No evidence supports the claim.
3. **Conclusion:** Viral posts spread misinformation when they lack credible sources.

Pairing 6: Minimum Wage Debate

Big Skill: Weighing Different Perspectives – Understanding economic policies' different impacts.

Small Skill: Comparing Sources – Evaluating government reports vs. personal opinions.

Teaching Technique: Sensory exploration.

Actions: Have students compare arguments for and against raising the minimum wage.

Example:

1. **Economist Study:** Raising wages increases consumer spending but may reduce jobs.
2. **Business Owner Interview:** Claims higher wages would force layoffs but provides no statistics.
3. **Conclusion:** Data-driven sources are more reliable than individual opinions.

Pairing 7: AI and Job Loss Debate

Big Skill: Weighing Different Perspectives – Understanding technological advancements' pros and cons.

Small Skill: Comparing Sources – Evaluating tech companies' claims vs. labor organizations' concerns.

Teaching Technique: Collaborative group work.

Actions: Have students analyze whether AI is good or bad for the job market.

Example:

1. **Tech Company Report:** AI increases productivity and creates new jobs.
2. **Worker Union Report:** AI automates jobs and leads to unemployment.
3. **Conclusion:** AI has both benefits and drawbacks, depending on perspective.

Pairing 8: Censorship and Free Speech

Big Skill: Weighing Different Perspectives – Recognizing different views on free expression.

Small Skill: Comparing Sources – Evaluating arguments from government policies vs. civil rights groups.

Teaching Technique: Real-world applications.

Actions: Have students compare perspectives on banning controversial books.

Example:

1. **Government Report:** Argues bans protect children from inappropriate material.
2. **Civil Rights Organization:** Argues bans limit free thought and discussion.
3. **Conclusion:** Censorship debates involve balancing safety and freedom.

Pairing 9: Health Claims in the Media

Big Skill: Weighing Different Perspectives – Recognizing how health advice varies across sources.

Small Skill: Comparing Sources – Identifying credible vs. misleading health claims.

Teaching Technique: Cultural exploration.

Actions: Have students analyze conflicting advice on diet trends.

Example:

1. **Fitness Influencer:** Promotes a diet without citing research.
2. **Mayo Clinic Study:** Provides evidence-based dietary recommendations.
3. **Conclusion:** Research-backed medical sources are more reliable than influencers' claims.

Pairing 10: Space Exploration Funding Debate

Big Skill: Weighing Different Perspectives – Evaluating whether governments should invest in space programs.

Small Skill: Comparing Sources – Examining arguments from scientific institutions vs. taxpayers.

Teaching Technique: Hands-on experiments.

Actions: Have students compare arguments for and against funding NASA.

Example:

1. **NASA Report:** Argues space exploration leads to technological advancements.
2. **Taxpayer Group Statement:** Claims funding should go to pressing issues like education and healthcare.
3. **Conclusion:** Funding priorities depend on long-term vs. immediate societal needs.

Critical Thinking – (Big Skill) Weighing Different Perspectives vs. (Small Skill) Structured Argumentation

---

Pairing 1: School Start Times Debate

Big Skill: Weighing Different Perspectives – Understanding that different viewpoints exist and considering multiple sides of an issue.

Small Skill: Structured Argumentation – Using frameworks like the Claim-Evidence-Reasoning (CER) method to organize thoughts logically.

Teaching Technique: Hands-on experiments.

Actions: Have students research and construct CER arguments for and against later school start times.

Example:

1. **Claim:** "Later school start times improve student performance."
2. **Evidence:** Studies show students who start school later get better sleep and higher grades.
3. **Reasoning:** More sleep enhances memory and concentration, leading to better academic outcomes.

Pairing 2: Should Junk Food Be Banned in Schools?

Big Skill: Weighing Different Perspectives – Understanding health, personal choice, and school policies.

Small Skill: Structured Argumentation – Using the CER method to examine policy decisions.

Teaching Technique: Guided exploration.

Actions: Have students construct arguments for and against banning junk food in schools.

Example:

1. **Claim:** "Junk food should be banned in schools to promote health."
2. **Evidence:** High sugar intake leads to obesity and diabetes.
3. **Reasoning:** Schools should encourage healthy habits, as student health affects learning.

Pairing 3: The Benefits and Risks of Social Media

Big Skill: Weighing Different Perspectives – Understanding the role of technology in social life and mental health.

Small Skill: Structured Argumentation – Using CER to analyze benefits and risks.

Teaching Technique: Problem-solving exercises.

Actions: Have students develop structured arguments about whether social media is beneficial or harmful.

Example:

1. **Claim:** "Social media negatively impacts mental health."
2. **Evidence:** Research links excessive social media use to anxiety and depression.
3. **Reasoning:** Social media can create unrealistic expectations and encourage unhealthy comparisons.

Pairing 4: Animal Testing in Medical Research

Big Skill: Weighing Different Perspectives – Considering ethical and scientific viewpoints.

Small Skill: Structured Argumentation – Constructing logical arguments using CER.

Teaching Technique: Observation journals.

Actions: Have students argue both for and against animal testing using structured reasoning.

Example:

1. **Claim:** "Animal testing is necessary for medical advancements."
2. **Evidence:** Many life-saving drugs and vaccines were developed using animal testing.
3. **Reasoning:** Testing on animals helps ensure safety before testing on humans.

Pairing 5: Should Students Have Homework?

Big Skill: Weighing Different Perspectives – Evaluating educational policies from multiple angles.

Small Skill: Structured Argumentation – Developing logical reasoning frameworks.

Teaching Technique: Hands-on experiments.

Actions: Have students construct structured arguments about the necessity of homework.

Example:

1. **Claim:** "Homework reinforces learning and responsibility."
2. **Evidence:** Studies show students who do homework perform better on tests.
3. **Reasoning:** Homework provides practice that helps solidify classroom concepts.

Pairing 6: Universal Basic Income (UBI) Debate

Big Skill: Weighing Different Perspectives – Understanding economic and social implications of UBI.

Small Skill: Structured Argumentation – Using CER to analyze potential benefits and drawbacks.

Teaching Technique: Sensory exploration.

Actions: Have students create structured arguments about whether UBI should be implemented.

Example:

1. **Claim:** "UBI would reduce poverty and boost economic stability."
2. **Evidence:** Pilot programs show people spend UBI on essential needs like food and housing.
3. **Reasoning:** A stable income reduces financial stress, leading to healthier communities.

Pairing 7: Space Exploration Funding

Big Skill: Weighing Different Perspectives – Evaluating government spending on space programs.

Small Skill: Structured Argumentation – Constructing logical reasoning using CER.

Teaching Technique: Collaborative group work.

Actions: Have students argue for or against increasing funding for NASA.

Example:

1. **Claim:** "Space exploration benefits humanity and should be funded."
2. **Evidence:** NASA research has led to technological advancements like GPS and medical devices.
3. **Reasoning:** Space research improves life on Earth and prepares us for future challenges.

Pairing 8: The Ethics of Self-Driving Cars

Big Skill: Weighing Different Perspectives – Understanding the risks and benefits of AI in transportation.

Small Skill: Structured Argumentation – Using CER to analyze safety, economic, and ethical concerns.

Teaching Technique: Real-world applications.

Actions: Have students construct arguments if self-driving cars should replace human drivers.

Example:

1. **Claim:** "Self-driving cars are safer than human drivers."
2. **Evidence:** Studies show autonomous vehicles reduce accidents caused by human error.
3. **Reasoning:** Unlike humans, AI does not get distracted or impaired while driving.

Pairing 9: The Right to Free Speech vs. Hate Speech Laws

Big Skill: Weighing Different Perspectives – Evaluating the balance between free expression and protecting individuals.

Small Skill: Structured Argumentation – Constructing CER arguments about policy decisions.

Teaching Technique: Cultural exploration.

Actions: Have students analyze the pros and cons of limiting free speech.

Example:

1. **Claim:** "Some speech should be limited to prevent harm."
2. **Evidence:** Hate speech has been linked to increased violence and discrimination.
3. **Reasoning:** While free speech is essential, it should not enable harm to others.

Pairing 10: Should College Be Free?

Big Skill: Weighing Different Perspectives – Analyzing the impact of government-funded higher education.

Small Skill: Structured Argumentation – Developing logical arguments using the CER framework.

Teaching Technique: Hands-on experiments.

Actions: Have students argue for or against free college tuition.

Example:

1. **Claim:** "College should be free for all students."
2. **Evidence:** Countries with free colleges have higher graduation rates.
3. **Reasoning:** Free college increases accessibility and economic mobility, benefiting society.

Example Teaching Strategies for Big vs. Small Thinking Skill Pairings of Strategic Planning at the High School Level

Here are 90 unique teaching strategy examples for the big thinking and small thinking skill pairings, each described in detail to align with **Big Thinking Skills** (Concepts & Strategies) and **Small Thinking Skills** (Step-by-Step Execution) for the **Strategic Planning Skill** taught at the high school level.

Strategic Planning

Big Thinking Skills (Concepts & Strategies)

- **Long-Term Vision:** Understanding how small steps contribute to long-term success (e.g., college/career planning).
- **Prioritization & Resource Management:** Identifying the most critical steps in a plan and allocating time/resources effectively.
- **Contingency Thinking:** Developing backup plans in case the original plan fails.

Small Thinking Skills (Step-by-Step Execution)

- **Breaking Down Plans:** Creating detailed action steps to achieve a goal.
- **Tracking & Adjusting Progress:** Using planners, Gantt charts, or checklists to stay on course.
- **Time Management:** Scheduling tasks efficiently to maximize productivity.

Strategic Planning – (Big Skill) Long-Term Vision vs. (Small Skill) Breaking Down Plans

---

Pairing 1: College Application Planning

Big Skill: Long-Term Vision – Understanding how small steps contribute to long-term success (e.g., college/career planning).

Small Skill: Breaking Down Plans – Creating detailed action steps to achieve a goal.

Teaching Technique: Hands-on experiments.

Actions: Have students create a step-by-step plan for applying to their top-choice colleges.

Example:

1. Research colleges and admission requirements.
2. Schedule SAT/ACT exams.
3. Write and revise personal statements.
4. Submit applications before deadlines.

Pairing 2: Starting a Small Business

Big Skill: Long-Term Vision – Understanding the steps required to build a successful business.

Small Skill: Breaking Down Plans – Mapping out necessary actions.

Teaching Technique: Guided exploration.

Actions: Have students develop a business plan with short-term and long-term goals.

Example:

1. Identify a product/service.
2. Research target customers.
3. Create a budget.
4. Launch and market the business.

Pairing 3: Preparing for a Competitive Scholarship

Big Skill: Long-Term Vision – Understanding the importance of early preparation for scholarships.

Small Skill: Breaking Down Plans – Creating a structured timeline for application requirements.

Teaching Technique: Problem-solving exercises.

Actions: Have students identify a scholarship and outline the necessary steps to apply successfully.

Example:

1. Maintain a strong GPA.
2. Collect recommendation letters.
3. Draft essays and edit them.
4. Submit applications ahead of deadlines.

Pairing 4: Planning for Financial Independence

Big Skill: Long-Term Vision – Understanding financial security in adulthood.

Small Skill: Breaking Down Plans – Setting detailed financial goals.

Teaching Technique: Observation journals.

Actions: Have students create a budget and savings plan for living independently after high school.

Example:

1. Research costs of rent, utilities, and food.
2. Determine necessary savings before moving out.
3. Set up a bank account and start saving.
4. Plan for additional income sources.

Pairing 5: Learning a New Language for Future Career Opportunities

Big Skill: Long-Term Vision – Understanding how language skills enhance career options.

Small Skill: Breaking Down Plans – Structuring a learning schedule.

Teaching Technique: Hands-on experiments.

Actions: Have students create a plan to become fluent in a second language.

Example:

1. Choose a language relevant to career goals.
2. Set up a study schedule (e.g., daily practice).
3. Engage in speaking and writing exercises.
4. Take proficiency exams.

Pairing 6: Training for a Sports Scholarship

Big Skill: Long-Term Vision – Understanding the commitment required for athletic scholarships.

Small Skill: Breaking Down Plans – Developing a structured training regimen.

Teaching Technique: Sensory exploration.

Actions: Have students develop a training and academic schedule to qualify for a sports scholarship.

Example:

1. Attend practice consistently.
2. Follow a strength and conditioning program.
3. Maintain academic performance for eligibility.
4. Reach out to college recruiters.

Pairing 7: Preparing for a Medical Career

Big Skill: Long-Term Vision – Understanding the long-term path to becoming a doctor.

Small Skill: Breaking Down Plans – Outlining educational and professional milestones.

Teaching Technique: Collaborative group work.

Actions: Have students map out the steps needed to enter medical school.

Example:

1. Take advanced science courses.
2. Volunteer in healthcare settings.
3. Prepare for the MCAT.
4. Apply to medical school.

Pairing 8: Writing and Publishing a Book

Big Skill: Long-Term Vision – Understanding the process of writing and publishing.

Small Skill: Breaking Down Plans – Setting writing and revision goals.

Teaching Technique: Real-world applications.

Actions: Have students develop a writing schedule and publishing strategy.

Example:

1. Outline book chapters.
2. Write a draft and revise.
3. Seek feedback and edit.
4. Publish independently or submit to publishers.

Pairing 9: Organizing a Community Service Project

Big Skill: Long-Term Vision – Understanding the impact of community service on personal and professional growth.

Small Skill: Breaking Down Plans – Structuring the project into actionable steps.

Teaching Technique: Cultural exploration.

Actions: Have students plan and execute a meaningful community service project.

Example:

1. Identify a community need.
2. Gather volunteers.
3. Set a timeline for execution.
4. Assess the project's impact.

Pairing 10: Preparing for a Career in Technology

Big Skill: Long-Term Vision – Understanding career paths in the tech industry.

Small Skill: Breaking Down Plans – Creating a roadmap for skill development.

Teaching Technique: Hands-on experiments.

Actions: Have students create a career roadmap for entering a tech-related field.

Example:

1. Learn coding basics through online courses.
2. Build a personal project portfolio.
3. Apply for internships.
4. Obtain industry certifications.

Strategic Planning – (Big Skill) Long-Term Vision vs. (Small Skill) Tracking & Adjusting Progress

---

Pairing 1: College Admission Timeline

Big Skill: Long-Term Vision – Understanding how small steps contribute to long-term success (e.g., college/career planning).

Small Skill: Tracking & Adjusting Progress – Using planners, Gantt charts, or checklists to stay on course.

Teaching Technique: Hands-on experiments.

Actions: Have students develop a college application timeline using a planner or Gantt chart.

Example:

1. List application deadlines.
2. Schedule test prep for SAT/ACT.
3. Set dates for essay drafts and reviews.
4. Adjust timeline based on school workload.

Pairing 2: Earning a Certification

Big Skill: Long-Term Vision – Planning for career certifications needed for a specific job.

Small Skill: Tracking & Adjusting Progress – Using checklists to complete coursework and practice exams.

Teaching Technique: Guided exploration.

Actions: Have students create a checklist for earning an industry certification (e.g., IT, medical, finance).

Example:

1. Register for an online course.
2. Set weekly study goals.
3. Take practice tests and adjust study time accordingly.
4. Schedule and complete certification exam.

Pairing 3: Saving for a Major Purchase (i.e., Car, College)

Big Skill: Long-Term Vision – Understanding how financial planning contributes to long-term success.

Small Skill: Tracking & Adjusting Progress – Using budgeting tools to monitor savings.

Teaching Technique: Problem-solving exercises.

Actions: Have students set a financial goal and track their savings using budgeting apps or spreadsheets.

Example:

1. Determine total cost of purchase.
2. Create a monthly savings plan.
3. Track spending and adjust savings habits if needed.
4. Reassess and adjust based on unexpected expenses.

Pairing 4: Preparing for a Competitive Exam

Big Skill: Long-Term Vision – Understanding the importance of early preparation for standardized tests.

Small Skill: Tracking & Adjusting Progress – Using study planners and mock tests to adjust strategies.

Teaching Technique: Observation journals.

Actions: Have students use a planner to track study hours, mock test scores, and areas needing improvement.

Example:

1. Create a study schedule for SAT/ACT/AP exams.
2. Take practice tests and log scores.
3. Adjust study topics based on weak areas.
4. Review and refine strategy before the exam.

Pairing 5: Organizing a Personal Fitness Plan

Big Skill: Long-Term Vision – Understanding how consistent exercise leads to long-term health.

Small Skill: Tracking & Adjusting Progress – Using fitness trackers or journals to measure improvement.

Teaching Technique: Hands-on experiments.

Actions: Have students set fitness goals and track progress through journals or apps.

Example:

1. Set goals (e.g., run a mile in 8 minutes).
2. Record workout routines.
3. Track improvement and adjust intensity.
4. Reflect on changes and modify goals.

Pairing 6: Learning a New Skill (e.g., Coding, Public Speaking)

Big Skill: Long-Term Vision – Understanding how skill-building contributes to future career options.

Small Skill: Tracking & Adjusting Progress – Using planners to structure learning milestones.

Teaching Technique: Sensory exploration.

Actions: Have students set a skill mastery goal and track progress through a checklist or Gantt chart.

Example:

1. Choose a programming language (e.g., Python).
2. Complete beginner-level tutorials.
3. Start a small project and track milestones.
4. Adjust learning focus based on difficulty.

Pairing 7: Planning a Study-Abroad Experience

Big Skill: Long-Term Vision – Understanding the steps required for international education.

Small Skill: Tracking & Adjusting Progress – Using spreadsheets to manage application deadlines and visa requirements.

Teaching Technique: Collaborative group work.

Actions: Have students develop a roadmap for studying abroad, adjusting based on application requirements.

Example:

1. Research study-abroad programs.
2. Track application deadlines.
3. Complete visa and travel paperwork.
4. Adjust plans based on funding and admission status.

Pairing 8: Launching a Personal Project (Podcast, Blog, YouTube Channel)

Big Skill: Long-Term Vision – Understanding how personal projects contribute to skill-building and branding.

Small Skill: Tracking & Adjusting Progress – Using project management tools to refine content schedules.

Teaching Technique: Real-world applications.

Actions: Have students plan and track progress for a creative project, making adjustments based on engagement.

Example:

1. Create a content calendar for videos/blogs.
2. Track views, engagement, and feedback.
3. Adjust content strategy based on audience response.
4. Refine posting schedule and format.

Pairing 9: Developing a Leadership Role in School or Community

Big Skill: Long-Term Vision – Understanding leadership development for future opportunities.

Small Skill: Tracking & Adjusting Progress – Using reflection journals to refine leadership skills.

Teaching Technique: Cultural exploration.

Actions: Have students track leadership activities and reflect on personal growth.

Example:

1. Identify leadership opportunities (e.g., club president, volunteer coordinator).
2. Record tasks, challenges, and successes.
3. Adjust leadership approach based on feedback.
4. Evaluate progress and set future leadership goals.

Pairing 10: Planning a Gap Year for Skill Development

Big Skill: Long-Term Vision – Understanding how a gap year can provide career-related experience.

Small Skill: Tracking & Adjusting Progress – Using planning tools to structure activities.

Teaching Technique: Hands-on experiments.

Actions: Have students create a structured plan for a gap year with learning and work experiences.

Example:

1. Research gap year programs and travel options.
2. Track application deadlines for internships or volunteer programs.
3. Adjust plans based on availability and funding.
4. Reflect on experiences to refine career goals.

Strategic Planning – (Big Skill) Long-Term Vision vs. (Small Skill) Time Management

---

Pairing 1: College Application Preparation

Big Skill: Long-Term Vision – Understanding how small steps contribute to long-term success (e.g., college/career planning).

Small Skill: Time Management – Scheduling tasks efficiently to maximize productivity.

Teaching Technique: Hands-on experiments.

Actions: Have students create a college application timeline, breaking tasks into deadlines.

Example:

1. Research colleges and application deadlines.
2. Schedule SAT/ACT prep sessions.
3. Set deadlines for personal statement drafts.
4. Allocate time for letters of recommendation.

Pairing 2: Balancing School and Part-Time Work

Big Skill: Long-Term Vision – Understanding how managing work and studies contributes to financial stability.

Small Skill: Time Management – Creating a work-study schedule to maintain academic performance.

Teaching Technique: Guided exploration.

Actions: Have students develop a weekly planner balancing school, work, and personal time.

Example:

1. Identify available study hours.
2. Allocate work shifts without affecting school responsibilities.

3. Set time for assignments and test prep.
4. Adjust schedule during peak exam periods.

Pairing 3: Planning for an Internship or Apprenticeship

Big Skill: Long-Term Vision – Understanding how internships enhance career readiness.

Small Skill: Time Management – Scheduling applications, networking, and preparation.

Teaching Technique: Problem-solving exercises.

Actions: Have students plan a timeline for securing and managing an internship.

Example:

1. Research companies offering internships.
2. Schedule time for resume writing and applications.
3. Plan transportation and working hours.
4. Allocate time for networking and skill-building.

Pairing 4: Preparing for Standardized Tests

Big Skill: Long-Term Vision – Understanding how preparation leads to better college admission outcomes.

Small Skill: Time Management – Structuring a study schedule to optimize retention.

Teaching Technique: Observation journals.

Actions: Have students develop a test prep schedule with designated study blocks.

Example:

1. Identify weaker subject areas and schedule focused study sessions.
2. Incorporate practice tests into the schedule.
3. Set review sessions before test day.
4. Plan for breaks to avoid burnout.

Pairing 5: Developing a Fitness Routine for Long-Term Health

Big Skill: Long-Term Vision – Understanding how consistent habits contribute to lifelong fitness.

Small Skill: Time Management – Allocating time for workouts alongside school and extracurriculars.

Teaching Technique: Hands-on experiments.

Actions: Have students design a weekly workout schedule that fits with their lifestyle.

Example:

1. Set specific goals (e.g., improve endurance, strength).
2. Schedule workout sessions on school-free hours.
3. Balance rest days and active days.
4. Adjust workouts based on progress and school workload.

Pairing 6: Learning a New Skill or Hobby

Big Skill: Long-Term Vision – Understanding how developing skills leads to personal and professional growth.

Small Skill: Time Management – Creating a structured learning schedule.

Teaching Technique: Sensory exploration.

Actions: Have students schedule time for skill development, balancing practice, and learning.

Example:

1. Identify a skill (e.g., playing guitar, coding, photography).
2. Allocate time for daily or weekly practice.
3. Set learning milestones (e.g., complete a project).
4. Adjust schedule based on progress.

Pairing 7: Managing a Community Service Project

Big Skill: Long-Term Vision – Understanding how volunteering enhances leadership and impact.

Small Skill: Time Management – Creating a project timeline with milestones.

Teaching Technique: Collaborative group work.

Actions: Have students plan and execute a community service project with a structured timeline.

Example:

1. Define the project goal (e.g., food drive, tutoring program).
2. Schedule time for planning, outreach, and execution.
3. Allocate volunteer tasks efficiently.
4. Reflect and adjust based on project progress.

Pairing 8: Writing and Publishing a Research Paper or Blog

Big Skill: Long-Term Vision – Understanding how consistent research and writing lead to publication.

Small Skill: Time Management – Organizing a writing and revision schedule.

Teaching Technique: Real-world applications.

Actions: Have students break down the writing process into manageable deadlines.

Example:

1. Set deadlines for topic selection and research.
2. Allocate time for drafting and revisions.
3. Schedule peer reviews or mentor feedback.
4. Adjust time allocation for unexpected challenges.

Pairing 9: Organizing a Club or Extracurricular Event

Big Skill: Long-Term Vision – Understanding leadership and organizational development.

Small Skill: Time Management – Creating a structured timeline for club activities.

Teaching Technique: Cultural exploration.

Actions: Have students develop a semester-long plan for club activities and meetings.

Example:

1. Schedule recurring meetings and major events.
2. Assign deadlines for planning fundraising or outreach.
3. Ensure balance between academic work and club responsibilities.
4. Adjust activity timeline based on participation and workload.

Pairing 10: Developing a Long-Term Financial Plan

Big Skill: Long-Term Vision – Understanding the importance of budgeting for future goals.

Small Skill: Time Management – Allocating time for financial literacy and planning.

Teaching Technique: Hands-on experiments.

Actions: Have students set up a budgeting schedule for tracking income and expenses.

Example:

1. Identify financial goals (e.g., saving for college or a car).
2. Set a budget and track expenses weekly.
3. Allocate time monthly for financial planning.
4. Adjust spending habits based on progress.

Strategic Planning – (Big Skill) Prioritization & Resource Management vs. (Small Skill) Breaking Down Plans

---

Pairing 1: Planning a College Application Strategy

Big Skill: Prioritization & Resource Management – Identifying the most critical steps in a plan and allocating time/resources effectively.

Small Skill: Breaking Down Plans – Creating detailed action steps to achieve a goal.

Teaching Technique: Hands-on experiments.

Actions: Have students prioritize application components (test scores, essays, recommendations) and break them into manageable tasks.

Example:

1. Identify priority deadlines for early decision or regular decision.
2. Allocate time for standardized test prep.
3. Set dates for drafting and revising personal statements.
4. Submit applications ahead of deadlines.

Pairing 2: Developing a Study Plan for Final Exams

Big Skill: Prioritization & Resource Management – Understanding which subjects need more focus based on difficulty and importance.

Small Skill: Breaking Down Plans – Structuring study sessions efficiently.

Teaching Technique: Guided exploration.

Actions: Have students assess subjects where they need the most improvement and create a structured study timeline.

Example:

1. Identify weak subjects and allocate more study time to them.
2. Break down study sessions into focused 30-45 minute blocks.
3. Prioritize review of past exams and teacher feedback.
4. Adjust study schedule based on practice test scores.

Pairing 3: Managing a Personal Budget for a Major Purchase

Big Skill: Prioritization & Resource Management – Allocating financial resources effectively for long-term goals.

Small Skill: Breaking Down Plans – Structuring financial steps toward saving.

Teaching Technique: Problem-solving exercises.

Actions: Have students develop a savings plan for purchasing a car, a laptop, or another major expense.

Example:

1. Determine total cost and set a savings goal.
2. Identify necessary expenses vs. optional spending.
3. Set weekly/monthly savings targets.
4. Track progress and adjust spending habits as needed.

Pairing 4: Preparing for a Leadership Role in a Club or Organization

Big Skill: Prioritization & Resource Management – Learning to focus on key responsibilities and delegate tasks.

Small Skill: Breaking Down Plans – Structuring leadership goals and activities.

Teaching Technique: Observation journals.

Actions: Have students plan a leadership development strategy by focusing on key skills and organizing team activities.

Example:

1. Identify key responsibilities and challenges of the leadership role.
2. Create a step-by-step action plan for implementing changes or new initiatives.
3. Delegate minor tasks to team members.
4. Evaluate leadership effectiveness and adjust strategies.

Pairing 5: Creating a Balanced Daily Schedule

Big Skill: Prioritization & Resource Management – Managing time effectively to balance academics, extracurriculars, and personal life.

Small Skill: Breaking Down Plans – Structuring daily routines for efficiency.

Teaching Technique: Hands-on experiments.

Actions: Have students develop a structured schedule that prioritizes essential activities while avoiding burnout.

Example:

1. List all daily tasks (school, work, social time, self-care).
2. Rank them by importance and urgency.
3. Allocate time slots based on productivity levels.
4. Make adjustments based on actual experience.

Pairing 6: Organizing a Fundraising Event

Big Skill: Prioritization & Resource Management – Identifying key tasks and effectively distributing financial and human resources.

Small Skill: Breaking Down Plans – Creating a structured step-by-step event plan.

Teaching Technique: Sensory exploration.

Actions: Have students design a fundraiser for a school or community cause, breaking it down into key phases.

Example:

1. Choose an event type (e.g., bake sale, charity run).
2. Identify necessary resources (budget, volunteers, venue).
3. Create a timeline for promotion and execution.
4. Evaluate fundraising results and adjust for future events.

Pairing 7: Mapping Out a Fitness or Sports Training Plan

Big Skill: Prioritization & Resource Management – Structuring workouts based on fitness goals and available time/resources.

Small Skill: Breaking Down Plans – Setting progressive training milestones.

Teaching Technique: Collaborative group work.

Actions: Have students develop a customized fitness plan, prioritizing key workouts and tracking improvements.

Example:

1. Set a goal (e.g., increase endurance, strength, flexibility).
2. Identify necessary workouts and frequency.
3. Allocate training time around school and responsibilities.
4. Monitor progress and adjust workouts as needed.

Pairing 8: Planning a Gap Year or Career Exploration Year

Big Skill: Prioritization & Resource Management – Choosing experiences that align with career aspirations.

Small Skill: Breaking Down Plans – Structuring the year with actionable steps.

Teaching Technique: Real-world applications.

Actions: Have students outline a structured gap-year plan that balances work, learning, and travel.

Example:

1. Research opportunities (internships, travel programs, online courses).
2. Set a timeline for applications and enrollment.
3. Identify funding sources and budget accordingly.
4. Adjust plans based on acceptance and feasibility.

Pairing 9: Launching a Small Business or Side Hustle

Big Skill: Prioritization & Resource Management – Allocating time and financial resources efficiently to start a business.

Small Skill: Breaking Down Plans – Creating a business plan with short-term and long-term goals.

Teaching Technique: Cultural exploration.

Actions: Have students develop a step-by-step strategy for starting a small business, balancing school, and entrepreneurship.

Example:

1. Identify a product/service and target audience.
2. Research costs and budget accordingly.
3. Develop a production and marketing timeline.
4. Adjust strategy based on initial sales and feedback.

Pairing 10: Writing and Publishing a Research Paper or Creative Work

Big Skill: Prioritization & Resource Management – Managing time and research materials effectively to complete a publication.

Small Skill: Breaking Down Plans – Structuring writing and revision phases.

Teaching Technique: Hands-on experiments.

Actions: Have students outline a research or creative writing project and allocate time efficiently for each stage.

Example:

1. Choose a topic and conduct preliminary research.
2. Create an outline and timeline for writing drafts.
3. Schedule revisions and peer feedback sessions.
4. Submit for publication or presentation.

Strategic Planning – (Big Skill) Prioritization & Resource Management vs. (Small Skill) Tracking & Adjusting Progress

---

Pairing 1: College Application Management

Big Skill: Prioritization & Resource Management – Identifying the most critical steps in a plan and allocating time/resources effectively.

Small Skill: Tracking & Adjusting Progress – Using planners, Gantt charts, or checklists to stay on course.

Teaching Technique: Hands-on experiments.

Actions: Have students use a checklist to track each step of their college applications.

Example:

1. Research top-choice schools.
2. Track SAT/ACT score submission deadlines.
3. Set deadlines for essays, recommendations, and applications.
4. Adjust timelines based on application requirements.

Pairing 2: Budgeting for a Major Purchase

Big Skill: Prioritization & Resource Management – Allocating financial resources effectively toward a long-term purchase.

Small Skill: Tracking & Adjusting Progress – Using budgeting tools to monitor savings.

Teaching Technique: Guided exploration.

Actions: Have students track monthly savings toward a major purchase (e.g., car, laptop, travel).

Example:

1. Identify total cost and set a savings goal.
2. Use a budget spreadsheet to track income vs. expenses.
3. Adjust spending based on progress and unexpected costs.
4. Modify savings plan, as necessary.

Pairing 3: Managing a Study Plan for Final Exams

Big Skill: Prioritization & Resource Management – Focusing on subjects that need the most improvement.

Small Skill: Tracking & Adjusting Progress – Using planners to allocate study time efficiently.

Teaching Technique: Problem-solving exercises.

Actions: Have students develop a study plan and adjust it based on mock exam results.

Example:

1. Identify subjects needing extra review.
2. Schedule weekly study sessions and practice tests.
3. Track progress using a Gantt chart.
4. Adjust study time if weaker subjects need more focus.

Pairing 4: Organizing an Extracurricular Project or Event

Big Skill: Prioritization & Resource Management – Managing team roles, materials, and deadlines efficiently.

Small Skill: Tracking & Adjusting Progress – Using Gantt charts or project management tools to monitor deadlines.

Teaching Technique: Observation journals.

Actions: Have students plan a school event and track key milestones.

Example:

1. Define event goals (e.g., fundraising, awareness).
2. Assign tasks using a shared project planner.
3. Track completion of tasks like booking venues and marketing.
4. Adjust plans based on volunteer availability and event needs.

Pairing 5: Developing a Fitness & Health Plan

Big Skill: Prioritization & Resource Management – Allocating time and energy toward a structured health routine.

Small Skill: Tracking & Adjusting Progress – Using workout logs or fitness trackers to monitor improvements.

Teaching Technique: Hands-on experiments.

Actions: Have students track their fitness goals and adjust based on progress.

Example:

1. Set a long-term fitness goal (e.g., running a 5K).
2. Log weekly workouts and diet choices.
3. Use a fitness tracker to measure improvement.
4. Adjust training intensity based on results.

Pairing 6: Learning a New Skill (e.g., Coding, Public Speaking, Music)

Big Skill: Prioritization & Resource Management – Structuring time and resources to master a new skill.

Small Skill: Tracking & Adjusting Progress – Using progress logs to refine learning techniques.

Teaching Technique: Sensory exploration.

Actions: Have students track their skill development and modify learning strategies.

Example:

1. Break skill into weekly learning milestones.
2. Record progress in a journal or tracking app.
3. Adjust practice routines based on difficulty level.
4. Set new goals based on progress and feedback.

Pairing 7: Preparing for a Summer Job or Internship

Big Skill: Prioritization & Resource Management – Managing time for applications, resume building, and interview prep.

Small Skill: Tracking & Adjusting Progress – Using a checklist to complete necessary steps.

Teaching Technique: Collaborative group work.

Actions: Have students develop a plan to apply for and secure an internship.

Example:

1. Identify companies or organizations offering internships.
2. Track resume submission deadlines.
3. Prepare for interviews and schedule practice sessions.
4. Adjust strategy based on application feedback.

Pairing 8: Writing and Publishing a Research Paper

Big Skill: Prioritization & Resource Management – Organizing research, writing, and editing for a successful publication.

Small Skill: Tracking & Adjusting Progress – Using writing schedules and progress charts.

Teaching Technique: Real-world applications.

Actions: Have students track research, drafting, and submission stages.

Example:

1. Choose a research topic and gather sources.
2. Track progress in a research log.

3. Schedule time for drafting, revising, and peer review.
4. Adjust research depth based on available data.

Pairing 9: Planning a Long-Term Travel Experience (e.g., Study Abroad, Gap Year)

Big Skill: Prioritization & Resource Management – Managing budget, paperwork, and preparation for travel.

Small Skill: Tracking & Adjusting Progress – Using planners to track deadlines for visas, scholarships, and accommodations.

Teaching Technique: Cultural exploration.

Actions: Have students create a structured plan for a study-abroad or travel experience.

Example:

1. Research destinations and required paperwork.
2. Track application deadlines for visas and programs.
3. Adjust travel budget based on real-time expenses.
4. Modify itinerary based on availability and personal goals.

Pairing 10: Launching a Personal Brand or Business

Big Skill: Prioritization & Resource Management – Managing time and financial resources to build a business.

Small Skill: Tracking & Adjusting Progress – Using business plans and progress charts.

Teaching Technique: Hands-on experiments.

Actions: Have students develop a structured business launch plan, tracking milestones.

Example:

1. Identify target market and product/service offerings.
2. Track business expenses and sales.
3. Schedule marketing and promotional efforts.
4. Adjust pricing and strategies based on customer feedback.

Strategic Planning – (Big Skill) Prioritization & Resource Management vs. (Small Skill) Time Management

---

Pairing 1: College Application Planning

Big Skill: Prioritization & Resource Management – Identifying the most critical steps in a plan and allocating time/resources effectively.

Small Skill: Time Management – Scheduling tasks efficiently to maximize productivity.

Teaching Technique: Hands-on experiments.

Actions: Have students create a college application timeline, setting priority deadlines and allocating study/writing time.

Example:

1. Research top-choice schools.
2. Allocate time for SAT/ACT prep.
3. Set deadlines for drafting and revising essays.
4. Balance application work with school assignments.

Pairing 2: Managing Academics and Extracurriculars

Big Skill: Prioritization & Resource Management – Balancing schoolwork and after-school activities.

Small Skill: Time Management – Allocating time for both academics and extracurriculars.

Teaching Technique: Guided exploration.

Actions: Have students plan a structured weekly schedule for school, sports, clubs, and free time.

Example:

1. Identify the most important weekly responsibilities.
2. Use a planner to divide time effectively.
3. Schedule breaks to prevent burnout.
4. Adjust schedule based on workload.

Pairing 3: Preparing for Final Exams Efficiently

Big Skill: Prioritization & Resource Management – Identifying key study areas and allocating focus accordingly.

Small Skill: Time Management – Structuring study sessions for maximum retention.

Teaching Technique: Problem-solving exercises.

Actions: Have students develop a study plan based on subject difficulty and exam schedule.

Example:

1. Rank subjects based on difficulty.
2. Create a daily study schedule.
3. Allocate time for practice exams.
4. Adjust based on mock test performance.

Pairing 4: Managing a Personal Budget Alongside Work & School

Big Skill: Prioritization & Resource Management – Allocating financial resources efficiently.

Small Skill: Time Management – Scheduling work shifts while ensuring time for school.

Teaching Technique: Observation journals.

Actions: Have students track their work schedule and align it with study time and personal finances.

Example:

1. Identify available work hours without affecting school.
2. Allocate time for financial planning and savings.
3. Schedule bill payments and track expenses.
4. Adjust work-study balance if grades are impacted.

Pairing 5: Launching and Managing a Passion Project

Big Skill: Prioritization & Resource Management – Allocating time and materials for a successful project.

Small Skill: Time Management – Structuring project work to meet deadlines.

Teaching Technique: Hands-on experiments.

Actions: Have students plan and execute a creative or entrepreneurial project.

Example:

1. Identify project goals and deadlines.
2. Allocate work hours in a planner.
3. Monitor progress and ensure tasks stay on track.
4. Adjust deadlines if unexpected challenges arise.

Pairing 6: Training for an Athletic Competition

Big Skill: Prioritization & Resource Management – Allocating practice time for peak performance.

Small Skill: Time Management – Structuring workouts while balancing recovery.

Teaching Technique: Sensory exploration.

Actions: Have students create a structured training plan that balances practice, school, and rest.

Example:

1. Identify peak competition dates.

2. Schedule workouts and strength training sessions.
3. Allocate rest and recovery days.
4. Adjust training intensity based on performance feedback.

Pairing 7: Organizing a Community Service Initiative

Big Skill: Prioritization & Resource Management – Identifying key tasks to run a successful project.

Small Skill: Time Management – Creating a structured plan to complete the project efficiently.

Teaching Technique: Collaborative group work.

Actions: Have students plan and execute a community service event with specific roles and deadlines.

Example:

1. Choose a service project and set a timeline.
2. Assign volunteers tasks with clear deadlines.
3. Balance planning meetings with execution.
4. Adjust tasks as necessary based on participation.

Pairing 8: Preparing for an Internship or Job Application Process

Big Skill: Prioritization & Resource Management – Organizing application materials efficiently.

Small Skill: Time Management – Scheduling tasks like resume writing and interview prep.

Teaching Technique: Real-world applications.

Actions: Have students develop a structured application timeline, balancing research, and preparation.

Example:

1. Research internships and job opportunities.
2. Allocate time for resume and cover letter writing.
3. Schedule interview practice sessions.
4. Adjust applications based on feedback.

Pairing 9: Learning a New Language for Future Career Opportunities

Big Skill: Prioritization & Resource Management – Identifying effective language-learning strategies.

Small Skill: Time Management – Scheduling consistent practice sessions.

Teaching Technique: Cultural exploration.

Actions: Have students develop a structured language-learning schedule with immersion techniques.

Example:

1. Identify useful vocabulary and grammar areas.
2. Set daily practice time for reading, speaking, and listening.
3. Adjust learning schedule based on proficiency growth.
4. Incorporate real-world language experiences (e.g., conversation practice).

Pairing 10: Writing and Publishing a Research Paper or Blog

Big Skill: Prioritization & Resource Management – Managing research, writing, and editing efficiently.

Small Skill: Time Management – Structuring time to ensure steady progress toward deadlines.

Teaching Technique: Hands-on experiments.

Actions: Have students develop a writing plan, ensuring structured time for research, drafting, and revision.

Example:

1. Choosc a research topic and set milestones.
2. Allocate time for research, outlining, and writing.
3. Set deadlines for revisions and peer feedback.
4. Adjust timeline based on research complexity.

Strategic Planning – (Big Skill) Contingency Thinking vs. (Small Skill) Breaking Down Plans

---

Pairing 1: College Application Backup Plan

Big Skill: Contingency Thinking – Developing backup plans in case the original plan fails.

Small Skill: Breaking Down Plans – Creating detailed action steps to achieve a goal.

Teaching Technique: Hands-on experiments.

Actions: Have students create a Plan B and Plan C for their college applications in case they do not get into their first-choice school.

Example:

1. Identify top-choice, second-choice, and safety schools.

2. Set application deadlines and requirements for each.
3. Research gap-year options if no acceptances come through.
4. Adjust decisions based on early acceptance outcomes.

Pairing 2: Alternative Career Planning

Big Skill: Contingency Thinking – Preparing alternative career paths in case a preferred option is unavailable.

Small Skill: Breaking Down Plans – Outlining steps for each career path.

Teaching Technique: Guided exploration.

Actions: Have students develop three different career plans based on different success scenarios.

Example:

1. Plan A: Direct college-to-career pathway.
2. Plan B: Vocational or trade school alternative.
3. Plan C: Internship or apprenticeship approach.
4. Adjust focus depending on scholarship, admission, or financial constraints.

Pairing 3: Emergency Backup Plan for a School Project

Big Skill: Contingency Thinking – Preparing an alternative solution if a key element of the project fails.

Small Skill: Breaking Down Plans – Structuring tasks to ensure completion.

Teaching Technique: Problem-solving exercises.

Actions: Have students design a backup plan for a group project in case team members miss deadlines.

Example:

1. Assign primary and secondary responsibilities to each member.
2. Set milestone deadlines and progress checks.
3. Identify alternative ways to present the project if a major component fails.
4. Adjust roles if someone is absent or does not complete their tasks.

Pairing 4: Managing Unexpected Exam Changes

Big Skill: Contingency Thinking – Preparing for unexpected academic challenges.

Small Skill: Breaking Down Plans – Structuring a flexible study plan.

Teaching Technique: Observation journals.

Actions: Have students prepare for possible changes in an exam format or content focus.

Example:

1. Identify essential topics likely to be covered.
2. Allocate study time across different possible formats (essay, multiple-choice, oral).
3. Prepare a backup study guide with key concepts.
4. Adjust focus based on class hints or professor guidance.

Pairing 5: Financial Planning for Unexpected Expenses

Big Skill: Contingency Thinking – Creating financial backup plans for unexpected expenses.

Small Skill: Breaking Down Plans – Structuring a budget with alternative options.

Teaching Technique: Hands-on experiments.

Actions: Have students set up a financial plan with backup strategies for unexpected costs.

Example:

1. Set a primary savings goal for a specific purchase.
2. Create a separate emergency fund.
3. Identify ways to cut spending if necessary.
4. Adjust financial goals based on unexpected income changes.

Pairing 6: Preparing for a Sports Competition with Injury Risks

Big Skill: Contingency Thinking – Developing alternative training or participation strategies.

Small Skill: Breaking Down Plans – Structuring training steps and recovery options.

Teaching Technique: Sensory exploration.

Actions: Have students develop a training plan that includes injury prevention and backup participation roles.

Example:

1. Create a primary training routine.
2. Develop modified workouts in case of minor injuries.
3. Identify an alternative role (assistant coach, team support) if unable to compete.
4. Adjust based on physical condition close to the competition date.

Pairing 7: Organizing a School Event with Venue & Weather Risks

Big Skill: Contingency Thinking – Planning alternative logistics in case of unexpected disruptions.

Small Skill: Breaking Down Plans – Structuring event execution steps.

Teaching Technique: Collaborative group work.

Actions: Have students create a Plan B for a major event in case of cancellations, weather issues, or technical failures.

Example:

1. Secure a primary venue and an alternative location.
2. Develop an indoor and outdoor activity plan.
3. Identify virtual alternatives if in-person participation is not possible.
4. Adjust final event structure based on last-minute conditions.

Pairing 8: Preparing for Job or Internship Interview Uncertainty

Big Skill: Contingency Thinking – Planning for different types of interview scenarios.

Small Skill: Breaking Down Plans – Structuring preparation for various interview styles.

Teaching Technique: Real-world applications.

Actions: Have students prepare for different possible job interview formats and questions.

Example:

1. Practice standard in-person interview questions.
2. Prepare responses for behavioral and technical questions.
3. Develop strategies for virtual and phone interviews.
4. Adjust preparation focus based on employer communication.

Pairing 9: Travel Planning with Flight Delays or Cancellations

Big Skill: Contingency Thinking – Preparing alternative transportation and lodging plans.

Small Skill: Breaking Down Plans – Structuring a travel itinerary with flexibility.

Teaching Technique: Cultural exploration.

Actions: Have students plan a trip while considering possible transportation or lodging disruptions.

Example:

1. Identify primary and backup travel routes.
2. Book refundable accommodation options.
3. Create a list of alternative nearby destinations.
4. Adjust itinerary based on real-time travel updates.

Pairing 10: Adapting to Changing College Major or Course Availability

Big Skill: Contingency Thinking – Preparing for shifts in academic pathways.

Small Skill: Breaking Down Plans – Structuring alternative course plans.

Teaching Technique: Hands-on experiments.

Actions: Have students develop alternative course pathways in case their desired major is unavailable.

Example:

1. Identify three major potential options.
2. Research course requirements and transfer policies.
3. Develop an alternative academic plan for each scenario.
4. Adjust based on college policies and personal interest changes.

Strategic Planning – (Big Skill) Contingency Thinking vs. (Small Skill) Tracking & Adjusting Progress

---

Pairing 1: College Application Backup Tracking

Big Skill: Contingency Thinking – Developing backup plans in case the original plan fails.

Small Skill: Tracking & Adjusting Progress – Monitoring progress and adjusting plans when needed.

Teaching Technique: Hands-on experiments.

Actions: Have students track their college applications, with alternative options if a school rejects them.

Example:

1. Apply to a mix of reach, match, and safety schools.
2. Track application progress using a planner.
3. Adjust based on early decision responses.
4. Modify plans to include gap year or community college options.

Pairing 2: Career Planning with Multiple Pathways

Big Skill: Contingency Thinking – Preparing for different career options.

Small Skill: Tracking & Adjusting Progress – Regularly assessing steps toward career goals.

Teaching Technique: Guided exploration.

Actions: Have students track their career progress and adjust their focus based on opportunities.

Example:

1. Identify primary and secondary career interests.

2. Track progress in gaining necessary skills.
3. Adjust career path based on internship/job availability.
4. Modify plan if original industry choice is no longer feasible.

Pairing 3: Managing Academic Challenges with Alternative Study Plans

Big Skill: Contingency Thinking – Preparing alternative approaches if the initial study method fails.

Small Skill: Tracking & Adjusting Progress – Monitoring grades and adapting strategies accordingly.

Teaching Technique: Problem-solving exercises.

Actions: Have students track their academic performance and adjust study methods if they struggle.

Example:

1. Set grade goals for each subject.
2. Use a planner to track assignments and study hours.
3. Adjust study techniques if test scores do not improve.
4. Seek tutoring or change focus if performance continues to decline.

Pairing 4: Managing a Sports Training Schedule with Injury Risks

Big Skill: Contingency Thinking – Developing alternative training plans in case of injury or schedule conflicts.

Small Skill: Tracking & Adjusting Progress – Monitoring performance and making necessary changes.

Teaching Technique: Observation journals.

Actions: Have students track their training progress and adjust for injuries or unexpected obstacles.

Example:

1. Log weekly workouts and progress.
2. Adjust intensity based on performance.
3. Develop an alternative plan for low-impact training if injured.
4. Modify schedule based on school workload.

Pairing 5: Planning a Budget with Emergency Adjustments

Big Skill: Contingency Thinking – Preparing financial alternatives for unexpected expenses.

Small Skill: Tracking & Adjusting Progress – Monitoring spending and adjusting financial goals.

Teaching Technique: Hands-on experiments.

Actions: Have students track their budget and adjust expenses if unexpected costs arise.

Example:

1. Set a primary savings goal.
2. Use a budget planner to track spending.
3. Adjust savings if unexpected expenses occur.
4. Develop alternative income strategies if needed.

Pairing 6: Event Planning with Backup Logistics

Big Skill: Contingency Thinking – Preparing alternative logistics for unexpected issues.

Small Skill: Tracking & Adjusting Progress – Monitoring event planning milestones.

Teaching Technique: Collaborative group work.

Actions: Have students track an event planning process and adjust for unforeseen issues.

Example:

1. Develop an event timeline.
2. Track progress on venue, vendors, and promotion.
3. Adjust plans if a vendor cancels or venue is unavailable.
4. Implement an alternative event format if necessary.

Pairing 7: Preparing for Multiple Job or Internship Opportunities

Big Skill: Contingency Thinking – Developing alternative plans in case a preferred job/internship falls through.

Small Skill: Tracking & Adjusting Progress – Monitoring job applications and adjusting efforts.

Teaching Technique: Real-world applications.

Actions: Have students track their job/internship applications and adjust strategies based on responses.

Example:

1. Apply to multiple job opportunities.
2. Track responses and follow-up interviews.
3. Adjust application strategy if rejection rates are high.
4. Consider alternative job sectors if needed.

Pairing 8: Alternative Travel Planning

Big Skill: Contingency Thinking – Preparing alternative travel options in case of disruptions.

Small Skill: Tracking & Adjusting Progress – Monitoring itinerary and making adjustments.

Teaching Technique: Cultural exploration.

Actions: Have students plan a trip while tracking potential changes and making backup plans.

Example:

1. Research transportation options.
2. Track weather conditions and potential delays.
3. Adjust lodging and transportation if disruptions occur.
4. Modify itinerary based on real-time conditions.

Pairing 9: Adapting to Changing College Course Availability

Big Skill: Contingency Thinking – Preparing alternative academic plans if preferred courses are unavailable.

Small Skill: Tracking & Adjusting Progress – Monitoring course registration and adapting schedules.

Teaching Technique: Hands-on experiments.

Actions: Have students track their course selections and make backup plans in case a class is full.

Example:

1. Identify primary and secondary course choices.
2. Track registration dates and deadlines.
3. Adjust class schedule if a course is unavailable.
4. Select an alternative course if necessary.

Pairing 10: Adjusting a Personal Project Timeline

Big Skill: Contingency Thinking – Developing alternative deadlines in case of project setbacks.

Small Skill: Tracking & Adjusting Progress – Monitoring progress and making changes when needed.

Teaching Technique: Guided exploration.

Actions: Have students track the progress of a personal project and adjust based on challenges.

Example:

1. Set initial project milestones.
2. Track completion of tasks.
3. Adjust deadlines if delays occur.
4. Modify project scope if needed to meet deadlines.

Strategic Planning – (Big Skill) Contingency Thinking vs. (Small Skill) Time Management

---

Pairing 1: College Application Scheduling with Backup Plans

Big Skill: Contingency Thinking – Developing backup plans in case the original plan fails.

Small Skill: Time Management – Scheduling tasks efficiently to maximize productivity.

Teaching Technique: Hands-on experiments.

Actions: Have students create a timeline for college applications while scheduling alternative options if plans change.

Example:

1. Set deadlines for primary and backup college applications.
2. Schedule essay writing and review sessions.
3. Allocate time for scholarship applications in case financial aid falls through.
4. Adjust schedule based on acceptance/rejection letters.

Pairing 2: Balancing School & Extracurriculars with Plan B Options

Big Skill: Contingency Thinking – Having a backup strategy if schedule conflicts arise.

Small Skill: Time Management – Efficiently allocating time between academics and extracurriculars.

Teaching Technique: Guided exploration.

Actions: Have students develop alternative schedules in case they need to shift priorities between school and activities.

Example:

1. Identify key academic deadlines.
2. Allocate time blocks for clubs, sports, and other commitments.
3. Develop a backup schedule in case extracurriculars become overwhelming.
4. Adjust time allocation based on academic performance.

Pairing 3: Study Plan with Flexible Backup Sessions

Big Skill: Contingency Thinking – Adjusting study plans if original schedules are disrupted.

Small Skill: Time Management – Structuring study sessions efficiently.

Teaching Technique: Problem-solving exercises.

Actions: Have students set a primary study schedule with built-in flexibility.

Example:

1. Create a subject-specific study timetable.
2. Schedule extra sessions in case of last-minute difficulties.
3. Set buffer days before exams to catch up if needed.
4. Adjust schedule if unplanned events affect study time.

Pairing 4: Managing a Job While in High School

Big Skill: Contingency Thinking – Developing a backup plan in case job hours conflict with schoolwork.

Small Skill: Time Management – Efficiently scheduling work shifts alongside academics.

Teaching Technique: Observation journals.

Actions: Have students plan work shifts with alternate study schedules.

Example:

1. Set work hours that allow time for homework.
2. Create an emergency backup study schedule in case of extra work shifts.
3. Plan days off before major exams.
4. Adjust weekly schedules based on work demands.

Pairing 5: Fitness Training with Alternative Workouts

Big Skill: Contingency Thinking – Creating a backup training plan in case of injuries or conflicts.

Small Skill: Time Management – Scheduling workouts around academic responsibilities.

Teaching Technique: Hands-on experiments.

Actions: Have students develop a workout schedule with alternative plans for interruptions.

Example:

1. Plan regular training sessions for peak performance.
2. Include lighter workouts in case of fatigue or injury.
3. Schedule additional time for recovery if needed.
4. Adjust intensity based on academic workload.

Pairing 6: Planning a Fundraising Event with Backup Timelines

Big Skill: Contingency Thinking – Preparing an alternative event schedule if issues arise.

Small Skill: Time Management – Allocating time efficiently for event planning.

Teaching Technique: Sensory exploration.

Actions: Have students plan a school fundraising event with a contingency timeline.

Example:

1. Set a primary event date with alternative dates if weather or scheduling conflicts arise.
2. Allocate time for promotion, ticket sales, and logistics.
3. Develop an emergency response plan for unexpected issues.
4. Adjust event schedule based on participation and interest.

Pairing 7: Internship or Job Application Plan B

Big Skill: Contingency Thinking – Preparing for rejection or scheduling conflicts.

Small Skill: Time Management – Structuring application and interview preparation efficiently.

Teaching Technique: Collaborative group work.

Actions: Have students schedule internship/job applications with alternatives in case they do not get their first choice.

Example:

1. Apply to multiple internships/jobs with staggered deadlines.
2. Schedule interview preparation and resume workshops.
3. Develop an alternative summer plan (volunteering, certifications).
4. Adjust job application strategy based on responses.

Pairing 8: Preparing for a Debate or Competition with Time Buffers

Big Skill: Contingency Thinking – Developing alternative preparation strategies in case of team or topic changes.

Small Skill: Time Management – Scheduling practice efficiently to maximize readiness.

Teaching Technique: Real-world applications.

Actions: Have students schedule practice rounds with backup prep in case of sudden debate topic changes.

Example:

1. Set a regular debate practice schedule.
2. Create an emergency study session plan for unexpected rule changes.
3. Allocate additional time in case of last-minute topic switches.

4. Adjust strategy based on competition format updates.

Pairing 9: Managing a Travel Plan with Backup Itineraries

Big Skill: Contingency Thinking – Preparing alternative routes and schedules for travel disruptions.

Small Skill: Time Management – Scheduling travel efficiently to avoid last-minute problems.

Teaching Technique: Cultural exploration.

Actions: Have students develop an adaptable travel plan that accounts for potential delays.

Example:

1. Plan a primary and alternative travel route.
2. Schedule key travel activities with flexibility.
3. Set buffer times for flight delays or missed connections.
4. Adjust itinerary based on real-time travel conditions.

Pairing 10: Academic Course Planning with Alternative Schedules

Big Skill: Contingency Thinking – Planning for class availability issues.

Small Skill: Time Management – Scheduling courses to optimize workload balance.

Teaching Technique: Hands-on experiments.

Actions: Have students plan their academic schedule with backup course options.

Example:

1. Select required classes with multiple time slot options.
2. Identify alternative electives in case of scheduling conflicts.
3. Allocate study time for different class formats (online vs. in-person).
4. Adjust course plan based on availability and academic needs.

Example Teaching Strategies for Big vs. Small Thinking Skill Pairings of Decision Making at the High School Level

Here are 90 unique teaching strategy examples for the big thinking and small thinking skill pairings, each described in detail to align with **Big Thinking Skills** (Concepts & Strategies) and **Small Thinking Skills** (Step-by-Step Execution) for the **Decision-Making Skill** taught at the high school level.

Decision Making

Big Thinking Skills (Concepts & Strategies)

- **Weighing Pros & Cons:** Analyzing benefits and drawbacks before making a choice.
- **Risk Assessment:** Evaluating potential consequences and uncertainties in decisions.
- **Logical and Ethical Considerations:** Considering logical, emotional, and ethical factors in choices.

Small Thinking Skills (Step-by-Step Execution)

- **Structured Decision Frameworks:** Applying decision-making models like SWOT analysis or the Decision Matrix.
- **Data Gathering & Comparison:** Collecting rclcvant information to compare different options.
- **Justifying Decisions:** Writing or presenting rational explanations for choices.

Decision Making – (Big Skill) Weighing Pros & Cons vs. (Small Skill) Structured Decision Frameworks

---

Pairing 1: Choosing a College

Big Skill: Weighing Pros & Cons – Analyzing benefits and drawbacks before making a choice.

Small Skill: Structured Decision Frameworks – Applying decision-making models like SWOT analysis or the Decision Matrix.

Teaching Technique: Hands-on experiments.

Actions: Have students use a Decision Matrix to compare multiple college options based on cost, location, major availability, and campus life.

Example:

1. List top three college choices.

2. Assign a score (1-10) for factors such as tuition, scholarships, and job prospects.
3. Calculate totals and compare results.
4. Adjust choices based on financial aid offers or personal preferences.

Pairing 2: Selecting an Extracurricular Activity

Big Skill: Weighing Pros & Cons – Evaluating how an activity aligns with personal goals.

Small Skill: Structured Decision Frameworks – Using a Decision Matrix to weigh different factors.

Teaching Technique: Guided exploration.

Actions: Have students evaluate extracurricular options (sports, clubs, volunteering) based on time commitment, skill-building, and personal interest.

Example:

1. List available extracurricular activities.
2. Score each activity based on schedule fit, enjoyment, and future benefits.
3. Use the Decision Matrix to rank top options.
4. Adjust choice based on personal workload.

Pairing 3: Deciding Whether to Get a Part-Time Job

Big Skill: Weighing Pros & Cons – Balancing income benefits with potential time constraints.

Small Skill: Structured Decision Frameworks – Applying SWOT analysis to assess the impact of working while in school.

Teaching Technique: Problem-solving exercises.

Actions: Have students use SWOT analysis to determine whether taking a job is beneficial.

Example:

1. Identify Strengths (extra money, job experience).
2. Identify Weaknesses (less study time, fatigue).
3. Identify Opportunities (networking, skill-building).
4. Identify Threats (falling grades, missing social events).

Pairing 4: Picking an AP or Dual Enrollment Course

Big Skill: Weighing Pros & Cons – Understanding course difficulty versus long-term benefits.

Small Skill: Structured Decision Frameworks – Using the Decision Matrix to assess the best course option.

Teaching Technique: Observation journals.

Actions: Have students compare potential AP and dual enrollment classes based on workload, subject interest, and college credit value.

Example:

1. List course options (e.g., AP Chemistry, AP U.S. History).
2. Score each based on difficulty, interest level, and future relevance.
3. Select the best fit based on calculated totals.
4. Adjust based on workload tolerance and career aspirations.

Pairing 5: Choosing a College Major

Big Skill: Weighing Pros & Cons – Comparing different academic paths before committing.

Small Skill: Structured Decision Frameworks – Using SWOT analysis to assess personal strengths, career prospects, and market demand.

Teaching Technique: Hands-on experiments.

Actions: Have students analyze different majors using SWOT analysis to determine the best fit.

Example:

1. Strengths: Interest in STEM vs. Humanities.
2. Weaknesses: Academic struggles in specific subjects.
3. Opportunities: Job market growth in related fields.
4. Threats: High competition, potential career shifts.

Pairing 6: Deciding Whether to Take a Leadership Role

Big Skill: Weighing Pros & Cons – Analyzing time commitment and leadership growth.

Small Skill: Structured Decision Frameworks – Using a Decision Matrix to evaluate responsibilities.

Teaching Technique: Sensory exploration.

Actions: Have students compare pros and cons of running for student government or club president.

Example:

1. List leadership opportunities (Student Council, Yearbook Editor).
2. Score based on time commitment, leadership growth, and enjoyment.
3. Compare totals and select the best fit.
4. Adjust based on personal workload.

Pairing 7: Selecting a Summer Program or Internship

Big Skill: Weighing Pros & Cons – Understanding experience benefits versus logistical challenges.

Small Skill: Structured Decision Frameworks – Using SWOT analysis to assess opportunities.

Teaching Technique: Collaborative group work.

Actions: Have students evaluate different summer programs based on cost, location, and career benefits.

Example:

1. Strengths: Hands-on experience, resume boost.
2. Weaknesses: High cost, limited summer break.
3. Opportunities: Networking, college recommendations.
4. Threats: Missing social events, travel restrictions.

Pairing 8: Deciding Whether to Buy or Lease a Car

Big Skill: Weighing Pros & Cons – Comparing short-term and long-term costs.

Small Skill: Structured Decision Frameworks – Applying the Decision Matrix to financial decisions.

Teaching Technique: Real-world applications.

Actions: Have students compare buying versus leasing a car using a financial decision framework.

Example:

1. List financial factors (monthly cost, insurance, depreciation).
2. Score each option based on affordability and flexibility.
3. Choose based on calculated scores and financial situation.
4. Adjust decision based on income changes.

Pairing 9: Evaluating a Social Media Presence for College Admissions

Big Skill: Weighing Pros & Cons – Understanding how online behavior affects opportunities.

Small Skill: Structured Decision Frameworks – Using SWOT analysis to assess digital reputation.

Teaching Technique: Cultural exploration.

Actions: Have students analyze their social media profiles and how colleges/employers may perceive them.

Example:

1. Strengths: Positive posts about achievements.
2. Weaknesses: Inappropriate content or controversial topics.
3. Opportunities: Networking, scholarship visibility.
4. Threats: Misinterpretation, cyber reputation risks.

Pairing 10: Choosing Between Community College and a Four-Year University

Big Skill: Weighing Pros & Cons – Analyzing financial and academic trade-offs.

Small Skill: Structured Decision Frameworks – Using a Decision Matrix for long-term educational planning.

Teaching Technique: Hands-on experiments.

Actions: Have students compare attending community college first versus going straight to a four-year university.

Example:

1. List factors (cost, flexibility, career opportunities).
2. Assign scores to each based on personal and financial goals.
3. Calculate totals and compare choices.
4. Adjust decision based on financial aid and career plans.

Decision Making – (Big Skill) Weighing Pros & Cons vs. (Small Skill) Data Gathering & Comparison

---

Pairing 1: Choosing a College Based on Financial Aid & Programs

Big Skill: Weighing Pros & Cons – Analyzing benefits and drawbacks before making a choice.

Small Skill: Data Gathering & Comparison – Collecting relevant information to compare different options.

Teaching Technique: Hands-on experiments.

Actions: Have students research and compare multiple colleges based on cost, scholarships, and academic programs.

Example:

1. Gather tuition and scholarship data from multiple schools.
2. Compare program offerings related to intended major.
3. Analyze campus amenities and student life.
4. Weigh overall cost versus potential job opportunities.

Pairing 2: Selecting the Best Smartphone or Laptop for School

Big Skill: Weighing Pros & Cons – Understanding trade-offs between features and cost.

Small Skill: Data Gathering & Comparison – Researching specifications, customer reviews, and prices.

Teaching Technique: Guided exploration.

Actions: Have students compare different devices based on performance, durability, and affordability.

Example:

1. Collect reviews and technical specifications from different brands.
2. Compare battery life, storage, and speed for schoolwork.
3. Weigh upfront cost versus long-term durability.
4. Make a final selection based on needs and budget.

Pairing 3: Deciding Between AP, IB, or Dual Enrollment Courses

Big Skill: Weighing Pros & Cons – Comparing different academic paths.

Small Skill: Data Gathering & Comparison – Researching credit transferability, workload, and difficulty.

Teaching Technique: Problem-solving exercises.

Actions: Have students compare the benefits of AP, IB, and dual enrollment courses based on academic and career goals.

Example:

1. Research how different programs are transferred to colleges.
2. Compare difficulty levels and coursework.
3. Assess flexibility of each program with extracurricular activities.
4. Make an informed decision based on academic aspirations.

Pairing 4: Deciding Whether to Get a Part-Time Job in High School

Big Skill: Weighing Pros & Cons – Balancing financial benefits with time commitment.

Small Skill: Data Gathering & Comparison – Comparing job options, wages, and work schedules.

Teaching Technique: Observation journals.

Actions: Have students analyze different part-time job opportunities and weigh their impact on school performance.

Example:

1. Research wages and work hours for local jobs.
2. Compare job flexibility and potential skill-building.
3. Evaluate how a job would affect school and extracurriculars.
4. Decide if and when to start working based on priorities.

Pairing 5: Choosing Between Buying a Used vs. New Car

Big Skill: Weighing Pros & Cons – Evaluating cost, maintenance, and reliability.

Small Skill: Data Gathering & Comparison – Researching prices, insurance costs, and depreciation rates.

Teaching Technique: Hands-on experiments.

Actions: Have students compare car prices, insurance costs, and reliability ratings to make an informed decision.

Example:

1. Research vehicle prices from dealerships and private sellers.
2. Compare insurance costs for new models vs. used.
3. Look at long-term maintenance expenses.
4. Decide based on budget and expected usage.

Pairing 6: Evaluating College Majors Based on Job Prospects

Big Skill: Weighing Pros & Cons – Comparing future salary potential and job satisfaction.

Small Skill: Data Gathering & Comparison – Researching job market trends and salary expectations.

Teaching Technique: Sensory exploration.

Actions: Have students compare different college majors based on employment rates, income, and industry demand.

Example:

1. Gather job placement statistics for different fields.
2. Compare starting salaries and career growth opportunities.
3. Weigh job stability and personal interest.
4. Select a major based on long-term career outlook.

Pairing 7: Choosing a Summer Program or Internship

Big Skill: Weighing Pros & Cons – Understanding time commitment versus skill-building.

Small Skill: Data Gathering & Comparison – Researching program costs, locations, and career relevance.

Teaching Technique: Collaborative group work.

Actions: Have students evaluate summer programs or internships by comparing learning experiences, networking opportunities, and costs.

Example:

1. Research internships in different industries.
2. Compare stipend offers versus unpaid opportunities.
3. Analyze networking potential and mentorship access.
4. Choose the best option based on career aspirations.

Pairing 8: Choosing a Social Media Platform for Personal Branding

Big Skill: Weighing Pros & Cons – Understanding engagement differences across platforms.

Small Skill: Data Gathering & Comparison – Analyzing audience demographics, reach, and content format.

Teaching Technique: Real-world applications.

Actions: Have students compare different social media platforms based on content type and personal branding goals.

Example:

1. Research audience demographics for Instagram, TikTok, LinkedIn, and YouTube.
2. Compare engagement rates for different types of content.
3. Analyze monetization potential on each platform.
4. Choose the platform best suited for personal or business branding.

Pairing 9: Choosing Between Renting or Buying a Home in the Future

Big Skill: Weighing Pros & Cons – Comparing financial implications of renting vs. buying.

Small Skill: Data Gathering & Comparison – Researching home prices, mortgage rates, and rental costs.

Teaching Technique: Cultural exploration.

Actions: Have students analyze housing costs and financial commitment required for renting vs. homeownership.

Example:

1. Compare monthly costs of renting vs. buying.
2. Research mortgage interest rates and home appreciation.
3. Analyze flexibility and long-term financial benefits.

4. Decide which option fits future lifestyle goals.

Pairing 10: Evaluating Online vs. In-Person College Learning

Big Skill: Weighing Pros & Cons – Assessing flexibility, cost, and social experience.

Small Skill: Data Gathering & Comparison – Researching tuition differences, program effectiveness, and student satisfaction.

Teaching Technique: Hands-on experiments.

Actions: Have students compare in-person and online degree programs based on tuition, job outcomes, and learning experiences.

Example:

1. Research tuition costs and scholarships for both options.
2. Compare flexibility of online learning with in-person experiences.
3. Evaluate job placement rates for graduates of online vs. in-person programs.
4. Make a decision based on financial and academic goals.

Decision Making – (Big Skill) Weighing Pros & Cons vs. (Small Skill) Justifying Decisions

---

Pairing 1: Choosing Between Two College Acceptance Offers

Big Skill: Weighing Pros & Cons – Analyzing benefits and drawbacks before making a choice.

Small Skill: Justifying Decisions – Writing or presenting rational explanations for choices.

Teaching Technique: Hands-on experiments.

Actions: Have students compare their college acceptance offers and write a justification for their final decision.

Example:

1. List tuition costs, program quality, and location for both colleges.
2. Identify personal priorities (career opportunities, student life, financial aid).
3. Write an essay or speech justifying their final choice.
4. Present the decision and reasoning in a class discussion.

Pairing 2: Selecting an Extracurricular Activity for Personal Growth

Big Skill: Weighing Pros & Cons – Evaluating time commitment and personal benefits.

Small Skill: Justifying Decisions – Defending why a specific activity is the best choice.

Teaching Technique: Guided exploration.

Actions: Have students analyze different extracurricular options and present their choice in a persuasive format.

Example:

1. Compare benefits of joining a sports team vs. debate club vs. volunteering.
2. Consider long-term skill development and networking opportunities.
3. Write a persuasive essay or deliver a presentation defending the choice.
4. Adjust decision if feedback reveals overlooked factors.

Pairing 3: Deciding Whether to Get a Part-Time Job

Big Skill: Weighing Pros & Cons – Comparing financial benefits with time constraints.

Small Skill: Justifying Decisions – Writing a structured explanation for the choice.

Teaching Technique: Problem-solving exercises.

Actions: Have students explain why they would or would not take on a part-time job.

Example:

1. Analyze potential work hours, pay, and school impact.
2. Weigh career-building opportunities against social/life balance.
3. Write a justification statement based on personal priorities.
4. Present and discuss the decision in small groups.

Pairing 4: Choosing an AP or Dual Enrollment Course

Big Skill: Weighing Pros & Cons – Evaluating difficulty level vs. college credit benefits.

Small Skill: Justifying Decisions – Defending the choice in an essay or debate.

Teaching Technique: Observation journals.

Actions: Have students compare course options and justify their selection in writing.

Example:

1. Research benefits of AP vs. dual enrollment.
2. Consider long-term advantages for college applications.
3. Write an argumentative essay explaining the choice.
4. Adjust the argument based on peer feedback.

Pairing 5: Selecting a Summer Program or Internship

Big Skill: Weighing Pros & Cons – Comparing cost, experience, and career benefits.

Small Skill: Justifying Decisions – Presenting a logical explanation for the choice.

Teaching Technique: Hands-on experiments.

Actions: Have students prepare a written justification for choosing one program over another.

Example:

1. Compare multiple summer programs based on cost, location, and relevance.
2. Rank the most important decision factors.
3. Write an essay explaining why one program is the best fit.
4. Present in a mock scholarship interview format.

Pairing 6: Deciding Between Buying or Leasing a Car

Big Skill: Weighing Pros & Cons – Understanding long-term financial implications.

Small Skill: Justifying Decisions – Defending the financial choice in a debate.

Teaching Technique: Sensory exploration.

Actions: Have students evaluate leasing vs. buying a car and justify their decision in a persuasive format.

Example:

1. Research costs, maintenance, and depreciation for each option.
2. Consider personal financial situation and future plans.
3. Write a persuasive speech defending the decision.
4. Debate with classmates who chose the opposite option.

Pairing 7: Choosing a College Major Based on Career Goals

Big Skill: Weighing Pros & Cons – Considering job market trends and personal interests.

Small Skill: Justifying Decisions – Writing a research-backed justification.

Teaching Technique: Collaborative group work.

Actions: Have students compare majors and defend their choice using data and logic.

Example:

1. Research job market growth for different fields.
2. Analyze potential salary and career stability.
3. Write a research paper justifying the major selection.
4. Adjust argument based on class discussion.

Pairing 8: Choosing Between Online vs. In-Person Learning

Big Skill: Weighing Pros & Cons – Comparing flexibility, cost, and effectiveness.

Small Skill: Justifying Decisions – Defending the best option in a structured argument.

Teaching Technique: Real-world applications.

Actions: Have students examine online vs. in-person learning and write a persuasive essay.

Example:

1. Research learning outcomes and student satisfaction rates.
2. Compare time management challenges and benefits.
3. Write a structured argument supporting one format.
4. Present findings in a class debate.

Pairing 9: Deciding Whether to Take a Leadership Role

Big Skill: Weighing Pros & Cons – Analyzing responsibility versus personal growth.

Small Skill: Justifying Decisions – Writing a rationale for pursuing or declining leadership.

Teaching Technique: Cultural exploration.

Actions: Have students explain why they would or would not take on a leadership role in a school club.

Example:

1. Compare time commitment, stress, and leadership experience.
2. Consider long-term college and career benefits.
3. Write an application letter justifying the decision.
4. Peer-review and refine the justification.

Pairing 10: Choosing Between Two Job Offers

Big Skill: Weighing Pros & Cons – Comparing salary, benefits, and growth opportunities.

Small Skill: Justifying Decisions – Defending the final decision in a professional format.

Teaching Technique: Hands-on experiments.

Actions: Have students simulate choosing between two job offers and write a formal justification.

Example:

1. Compare salaries, benefits, and company culture.
2. Rank most important decision factors.
3. Write a formal email justifying the final decision.
4. Present the decision in a mock interview setting.

Decision Making – (Big Skill) Risk Assessment vs. (Small Skill) Structured Decision Frameworks

---

Pairing 1: Choosing a College with Financial & Career Risks

Big Skill: Risk Assessment – Evaluating potential consequences and uncertainties in decisions.

Small Skill: Structured Decision Frameworks – Applying decision-making models like SWOT analysis or the Decision Matrix.

Teaching Technique: Hands-on experiments.

Actions: Have students compare colleges using a SWOT analysis, factoring in tuition costs, job placement rates, and long-term financial risks.

Example:

1. Strengths: Strong academic program, career connections.
2. Weaknesses: High tuition, possible student loan debt.
3. Opportunities: Internship programs, alumni network.
4. Threats: Limited financial aid, unstable job market in the field.

Pairing 2: Evaluating the Risks of a Part-Time Job During School

Big Skill: Risk Assessment – Considering potential downsides such as lower grades, stress, or time management issues.

Small Skill: Structured Decision Frameworks – Using a Decision Matrix to weigh pros and cons.

Teaching Technique: Guided exploration.

Actions: Have students compare part-time job options and risks using a Decision Matrix.

Example:

1. Assign scores for factors such as pay, scheduling flexibility, and impact on grades.
2. Weigh benefits against potential academic struggles.
3. Analyze how the job supports long-term goals.
4. Adjust decision based on school workload.

Pairing 3: Deciding Between AP vs. Honors Courses

Big Skill: Risk Assessment – Analyzing difficulty level, GPA impact, and long-term college benefits.

Small Skill: Structured Decision Frameworks – Using SWOT analysis to determine the best course selection.

Teaching Technique: Problem-solving exercises.

Actions: Have students assess the risks and benefits of taking an AP course versus an Honors course.

Example:

1. Strengths: AP courses provide college credit.
2. Weaknesses: More demanding coursework.
3. Opportunities: Weighted GPA boost, better college applications.
4. Threats: Risk of a lower grade if the course is too difficult.

Pairing 4: Assessing the Risks of Buying vs. Leasing a Car

Big Skill: Risk Assessment – Weighing financial and long-term commitments.

Small Skill: Structured Decision Frameworks – Using a Decision Matrix to compare car ownership options.

Teaching Technique: Observation journals.

Actions: Have students compare financial risks using a structured decision-making tool.

Example:

1. Compare costs of leasing vs. buying a car.
2. Consider long-term value and resale potential.
3. Weigh risks of maintenance vs. mileage restrictions.
4. Adjust decision based on personal finances and transportation needs.

Pairing 5: Choosing a College Major with Career Uncertainty

Big Skill: Risk Assessment – Understanding the job market demand and stability of different fields.

Small Skill: Structured Decision Frameworks – Using SWOT analysis to evaluate options.

Teaching Technique: Hands-on experiments.

Actions: Have students research career fields and apply SWOT analysis to assess risks.

Example:

1. Strengths: Passion for the field, strong academic performance.
2. Weaknesses: Limited job market for certain majors.
3. Opportunities: Growth in related industries.
4. Threats: Automation, outsourcing of jobs.

Pairing 6: Deciding Whether to Invest in Stocks vs. Savings Accounts

Big Skill: Risk Assessment – Weighing short-term vs. long-term financial stability.

Small Skill: Structured Decision Frameworks – Applying a Decision Matrix to compare investment risks.

Teaching Technique: Sensory exploration.

Actions: Have students compare risks of different investment strategies using a structured framework.

Example:

1. List options (stocks, savings, cryptocurrency).
2. Assign risk scores based on volatility and potential returns.
3. Compare liquidity and long-term financial impact.
4. Select the best financial plan based on goals.

Pairing 7: Assessing the Risks of Studying Abroad

Big Skill: Risk Assessment – Evaluating financial, academic, and cultural challenges.

Small Skill: Structured Decision Frameworks – Using SWOT analysis to assess the feasibility.

Teaching Technique: Collaborative group work.

Actions: Have students apply SWOT analysis to determine whether studying abroad is a good decision.

Example:

1. Strengths: Exposure to new cultures, academic prestige.
2. Weaknesses: High costs, language barriers.
3. Opportunities: Networking, career growth.
4. Threats: Homesickness, adapting to new academic systems.

Pairing 8: Choosing Between In-Person vs. Online College Courses

Big Skill: Risk Assessment – Weighing flexibility against learning effectiveness.

Small Skill: Structured Decision Frameworks – Applying a Decision Matrix to compare learning formats.

Teaching Technique: Real-world applications.

Actions: Have students analyze online vs. in-person courses using structured comparison.

Example:

1. Compare costs and accessibility of online vs. in-person programs.
2. Weigh risks of reduced engagement or lack of in-person networking.

3. Assign scores based on effectiveness in learning style.
4. Make a final decision based on career needs.

Pairing 9: Deciding Whether to Start a Small Business in High School

Big Skill: Risk Assessment – Understanding financial risks, time commitment, and failure rates.

Small Skill: Structured Decision Frameworks – Using SWOT analysis to identify potential challenges.

Teaching Technique: Cultural exploration.

Actions: Have students analyze entrepreneurship risks and benefits before launching a small business.

Example:

1. Strengths: Independence, potential for profit.
2. Weaknesses: Requires time management skills.
3. Opportunities: Growth potential, unique market niche.
4. Threats: Competition, financial risk.

Pairing 10: Choosing a Safe but Challenging Travel Destination

Big Skill: Risk Assessment – Evaluating safety risks, cultural differences, and budget constraints.

Small Skill: Structured Decision Frameworks – Using a Decision Matrix to compare destinations.

Teaching Technique: Hands-on experiments.

Actions: Have students compare travel options based on safety, cost, and cultural experience.

Example:

1. List destinations with safety ratings and travel restrictions.
2. Compare potential costs and language barriers.
3. Assign scores based on travel goals and risk level.
4. Select the most suitable travel location.

Decision Making – (Big Skill) Risk Assessment vs. (Small Skill) Data Gathering & Comparison

---

Pairing 1: Choosing a College Based on Financial & Career Risks

Big Skill: Risk Assessment – Evaluating potential consequences and uncertainties in decisions.

Small Skill: Data Gathering & Comparison – Collecting relevant information to compare different options.

Teaching Technique: Hands-on experiments.

Actions: Have students research and compare tuition costs, job placement rates, and return on investment for multiple colleges.

Example:

1. Gather data on tuition, scholarships, and average graduate salaries.
2. Compare schools based on financial burden and earning potential.
3. Identify risks such as student loan debt and job market instability.
4. Weigh long-term financial stability before making a decision.

Pairing 2: Evaluating the Risks of a Part-Time Job During School

Big Skill: Risk Assessment – Weighing potential downsides such as lower grades, stress, and exhaustion.

Small Skill: Data Gathering & Comparison – Researching different job options and their impacts.

Teaching Technique: Guided exploration.

Actions: Have students compare multiple part-time job options to evaluate workload balance.

Example:

1. Research job wages, required hours, and flexibility.
2. Compare how different jobs impact time for school and extracurriculars.
3. Identify risks such as falling grades or lack of sleep.
4. Choose a job based on personal ability to balance responsibilities.

Pairing 3: Assessing the Risks of AP vs. Dual Enrollment Courses

Big Skill: Risk Assessment – Understanding potential academic and college application consequences.

Small Skill: Data Gathering & Comparison – Researching credit transferability, workload, and difficulty.

Teaching Technique: Problem-solving exercises.

Actions: Have students compare the benefits and risks of AP and dual enrollment courses.

Example:

1. Gather information on how AP vs. dual enrollment credits transfer to colleges.
2. Compare workload, exam difficulty, and GPA impact.
3. Identify risks such as failing an AP exam or struggling with a college-level class.

4. Make an informed course selection based on strengths and college goals.

Pairing 4: Deciding Whether to Buy or Lease a Car

Big Skill: Risk Assessment – Analyzing financial risks, long-term commitments, and depreciation.

Small Skill: Data Gathering & Comparison – Collecting data on cost, insurance, and resale value.

Teaching Technique: Observation journals.

Actions: Have students compare leasing vs. buying by researching costs and financial implications.

Example:

1. Research car purchase vs. lease agreements.
2. Compare monthly costs, maintenance, and total expense over five years.
3. Identify risks such as early lease termination fees or repair costs.
4. Decide based on financial stability and long-term plans.

Pairing 5: Choosing a College Major with Career Uncertainty

Big Skill: Risk Assessment – Understanding job market trends and financial stability.

Small Skill: Data Gathering & Comparison – Researching employment rates, salaries, and demand for different fields.

Teaching Technique: Hands-on experiments.

Actions: Have students compare different career paths based on income potential and job security.

Example:

1. Gather job placement rates and average salaries for multiple fields.
2. Compare projected job market growth for each major.
3. Identify risks such as automation replacing jobs.
4. Choose a major that balances passion and career stability.

Pairing 6: Investing in Stocks vs. a Savings Account

Big Skill: Risk Assessment – Understanding short-term and long-term financial risks.

Small Skill: Data Gathering & Comparison – Researching market trends, interest rates, and investment risks.

Teaching Technique: Sensory exploration.

Actions: Have students compare the risks of investing in stocks vs. saving in a bank.

Example:

1. Research interest rates for savings accounts and stock market returns.
2. Compare risks of losing money in volatile markets vs. slow savings growth.
3. Identify benefits of diversification and long-term investment.
4. Choose an investment strategy based on risk tolerance and goals.

Pairing 7: Assessing the Risks of Studying Abroad

Big Skill: Risk Assessment – Evaluating financial, academic, and cultural challenges.

Small Skill: Data Gathering & Comparison – Researching cost, safety, and educational value of different programs.

Teaching Technique: Collaborative group work.

Actions: Have students compare different study-abroad programs and assess potential risks.

Example:

1. Gather data on program costs, scholarships, and visa requirements.
2. Compare safety, quality of education, and language barriers.
3. Identify risks such as homesickness, financial burden, or credit transfer issues.
4. Make a decision based on affordability and personal growth opportunities.

Pairing 8: Choosing Between Online vs. In-Person College Courses

Big Skill: Risk Assessment – Weighing flexibility against learning effectiveness.

Small Skill: Data Gathering & Comparison – Researching tuition differences, student outcomes, and accessibility.

Teaching Technique: Real-world applications.

Actions: Have students compare learning formats and their potential academic impacts.

Example:

1. Research tuition costs, flexibility, and student success rates.
2. Compare engagement, networking, and hands-on learning opportunities.
3. Identify risks such as lack of in-person support or distractions at home.
4. Choose the best learning format based on career goals and personal study habits.

Pairing 9: Starting a Small Business in High School

Big Skill: Risk Assessment – Understanding financial risks, time commitment, and competition.

Small Skill: Data Gathering & Comparison – Researching market demand, startup costs, and business strategies.

Teaching Technique: Cultural exploration.

Actions: Have students evaluate business opportunities and risks before launching a venture.

Example:

1. Research startup costs and market demand for business ideas.
2. Compare competition, pricing strategies, and potential profitability.
3. Identify risks such as high costs, time management issues, or legal concerns.
4. Decide whether to start the business based on financial and personal feasibility.

Pairing 10: Choosing a Safe But Exciting Travel Destination

Big Skill: Risk Assessment – Evaluating safety risks, cultural differences, and budget constraints.

Small Skill: Data Gathering & Comparison – Researching safety reports, cost of travel, and local laws.

Teaching Technique: Hands-on experiments.

Actions: Have students compare multiple destinations based on cost, safety, and experience.

Example:

1. Gather data on crime rates, travel advisories, and cultural norms.
2. Compare costs of flights, lodging, and activities.
3. Identify risks such as language barriers, health risks, or political instability.
4. Select the safest and most enjoyable travel destination.

Decision Making – (Big Skill) Risk Assessment vs. (Small Skill) Justifying Decisions

---

Pairing 1: Choosing a College with Financial Risks

Big Skill: Risk Assessment – Evaluating potential consequences and uncertainties in decisions.

Small Skill: Justifying Decisions – Writing or presenting rational explanations for choices.

Teaching Technique: Hands-on experiments.

Actions: Have students assess the financial risks of different colleges and write a justification for their final choice.

Example:

1. Identify tuition, scholarships, and student loan amounts for each school.
2. Weigh job placement rates and earning potential.

3. Assess risks such as long-term debt or job market instability.
4. Write an essay justifying why one college is the best financial choice.

Pairing 2: Taking a Part-Time Job While in School

Big Skill: Risk Assessment – Balancing income benefits with academic risks.

Small Skill: Justifying Decisions – Defending the decision in a structured format.

Teaching Technique: Guided exploration.

Actions: Have students analyze the pros and cons of working while in high school and present their decision in a debate.

Example:

1. Consider pay, scheduling, and impact on grades.
2. Weigh risks of burnout or lower academic performance.
3. Write a speech or participate in a debate justifying the decision.
4. Adjust argument based on counterpoints raised by peers.

Pairing 3: Deciding Whether to Take AP or Regular Courses

Big Skill: Risk Assessment – Evaluating workload, stress, and long-term academic benefits.

Small Skill: Justifying Decisions – Structuring a logical argument for the chosen course path.

Teaching Technique: Problem-solving exercises.

Actions: Have students weigh the risks and benefits of taking AP courses and write a justification for their choice.

Example:

1. Assess GPA impact, study time, and college credit potential.
2. Compare difficulty level and stress factors.
3. Write a justification essay for taking AP or regular classes.
4. Present and defend the decision in class discussion.

Pairing 4: Investing in Stocks vs. Savings Account

Big Skill: Risk Assessment – Understanding financial risks and potential rewards.

Small Skill: Justifying Decisions – Presenting a logical argument for an investment choice.

Teaching Technique: Observation journals.

Actions: Have students compare investment options and justify their decision in a financial plan.

Example:

1. Research risk factors for stocks vs. savings.
2. Assess long-term potential vs. short-term security.
3. Write a financial analysis explaining the best choice.
4. Present findings in a personal finance discussion.

Pairing 5: Selecting a College Major with Career Uncertainty

Big Skill: Risk Assessment – Evaluating job stability, salary expectations, and passion.

Small Skill: Justifying Decisions – Writing an argument for the major selection.

Teaching Technique: Hands-on experiments.

Actions: Have students research job outlooks and write a defense of their chosen major.

Example:

1. Compare industry demand, salary, and job availability.
2. Assess risk of automation or market saturation.
3. Write a research-backed justification for the major.
4. Present and revise the decision based on peer feedback.

Pairing 6: Studying Abroad vs. Staying Local for College

Big Skill: Risk Assessment – Weighing cultural, financial, and academic risks.

Small Skill: Justifying Decisions – Writing a structured argument for the choice.

Teaching Technique: Sensory exploration.

Actions: Have students compare the risks of studying abroad vs. staying local and justify their decision.

Example:

1. Consider tuition, language barriers, and culture shock.
2. Compare networking opportunities and educational benefits.
3. Write an argument explaining why one choice is better.
4. Present to a peer review panel for discussion.

Pairing 7: Deciding Whether to Join a Competitive Sports Team

Big Skill: Risk Assessment – Balancing time, injury risk, and benefits of teamwork.

Small Skill: Justifying Decisions – Structuring a logical case for joining or not joining.

Teaching Technique: Collaborative group work.

Actions: Have students analyze the pros and cons of joining a competitive sports team and present a persuasive justification.

Example:

1. Compare benefits (fitness, discipline) and risks (injury, time commitment).
2. Assess how sports affect academics and social life.
3. Write a statement of justification explaining the choice.
4. Present in a class debate on time management in high school.

Pairing 8: Choosing Between Online vs. In-Person College Courses

Big Skill: Risk Assessment – Weighing flexibility, networking, and learning effectiveness.

Small Skill: Justifying Decisions – Writing a structured justification for learning format.

Teaching Technique: Real-world applications.

Actions: Have students compare online and in-person education and write a recommendation.

Example:

1. Research student satisfaction and learning outcomes.
2. Compare tuition, accessibility, and networking opportunities.
3. Justify the best learning format based on research.
4. Present findings in a student-led discussion.

Pairing 9: Renting an Apartment vs. Staying in a Dorm

Big Skill: Risk Assessment – Evaluating financial risks, social experience, and responsibilities.

Small Skill: Justifying Decisions – Structuring a financial and lifestyle justification.

Teaching Technique: Cultural exploration.

Actions: Have students assess housing options for college and write a justification.

Example:

1. Compare rent, utilities, and dorm fees.
2. Assess risks like roommate conflicts, safety, and costs.
3. Write a structured argument for their choice.
4. Present to a mock "financial advisor" panel.

Pairing 10: Launching a Personal Business in High School

Big Skill: Risk Assessment – Understanding financial and time risks of entrepreneurship.

Small Skill: Justifying Decisions – Writing a structured business case.

Teaching Technique: Hands-on experiments.

Actions: Have students analyze the feasibility of launching a small business and justify their decision.

Example:

1. Compare start-up costs, profit potential, and competition.
2. Assess time investment and school impact.
3. Write a business proposal justifying the decision.
4. Present the plan to a "Shark Tank" style class panel.

Decision Making – (Big Skill) Logical and Ethical Considerations vs. (Small Skill) Structured Decision Frameworks

---

Pairing 1: Deciding Whether to Report Academic Dishonesty

Big Skill: Logical and Ethical Considerations – Weighing fairness, personal integrity, and potential consequences.

Small Skill: Structured Decision Frameworks – Using a Decision Matrix to assess the best course of action.

Teaching Technique: Hands-on experiments.

Actions: Have students apply a Decision Matrix to weigh the consequences of reporting or ignoring academic dishonesty.

Example:

1. List ethical considerations (integrity vs. peer pressure).
2. Compare possible outcomes for self, peers, and school.
3. Assign values to long-term vs. short-term consequences.
4. Use the matrix to decide the most ethical response.

Pairing 2: Choosing a College with Ethical Values in Mind

Big Skill: Logical and Ethical Considerations – Evaluating a school's policies, inclusivity, and environmental impact.

Small Skill: Structured Decision Frameworks – Applying SWOT analysis to compare ethical and logical factors.

Teaching Technique: Guided exploration.

Actions: Have students analyze potential colleges based on ethics, affordability, and career benefits.

Example:

1. Strengths: Strong academic programs, diversity, sustainability efforts.
2. Weaknesses: High tuition, lack of need-based aid.
3. Opportunities: Scholarships, ethical student organizations.
4. Threats: Misalignment of institutional values with personal beliefs.

Pairing 3: Deciding Whether to Support a Social Cause Online

Big Skill: Logical and Ethical Considerations – Weighing personal beliefs, misinformation risks, and social impact.

Small Skill: Structured Decision Frameworks – Using a Decision Matrix to analyze social media activism.

Teaching Technique: Problem-solving exercises.

Actions: Have students use a Decision Matrix to determine the risks and benefits of supporting a cause online.

Example:

1. Consider impact, credibility, and public perception.
2. Weigh risks of misinformation or negative consequences.
3. Compare alternative ways to support the cause.
4. Justify the final decision based on matrix scores.

Pairing 4: Evaluating the Ethics of AI in Hiring Decisions

Big Skill: Logical and Ethical Considerations – Considering fairness, bias, and efficiency in AI-based hiring.

Small Skill: Structured Decision Frameworks – Applying SWOT analysis to AI in employment decisions.

Teaching Technique: Observation journals.

Actions: Have students evaluate AI hiring systems using SWOT analysis to assess ethical concerns.

Example:

1. Strengths: Efficiency, reduced costs, bias removal.
2. Weaknesses: Potential for discrimination, lack of human judgment.
3. Opportunities: Fairer hiring with proper regulation.
4. Threats: Legal and ethical ramifications of biased AI decisions.

Pairing 5: Deciding Whether to Buy Products from an Unethical Company

Big Skill: Logical and Ethical Considerations – Weighing convenience, affordability, and moral responsibility.

Small Skill: Structured Decision Frameworks – Using a Decision Matrix to evaluate purchasing decisions.

Teaching Technique: Hands-on experiments.

Actions: Have students analyze their shopping habits by weighing ethical business practices against product value.

Example:

1. Compare quality, cost, and ethical concerns.
2. Weigh environmental impact and labor policies.
3. Assign values to ethical vs. practical considerations.
4. Justify the final purchasing decision.

Pairing 6: Choosing a Political Candidate to Support

Big Skill: Logical and Ethical Considerations – Evaluating policies, moral stance, and leadership effectiveness.

Small Skill: Structured Decision Frameworks – Applying SWOT analysis to compare candidates.

Teaching Technique: Sensory exploration.

Actions: Have students research candidates and use SWOT analysis to assess logical and ethical implications.

Example:

1. Strengths: Policy alignment, leadership skills.
2. Weaknesses: History of inconsistency, public perception.
3. Opportunities: Potential for social change, advocacy efforts.
4. Threats: Policy failures, controversies, voter division.

Pairing 7: Deciding Whether to Accept a Job Offer from a Questionable Employer

Big Skill: Logical and Ethical Considerations – Weighing personal values against financial stability.

Small Skill: Structured Decision Frameworks – Using a Decision Matrix to assess risks and rewards.

Teaching Technique: Collaborative group work.

Actions: Have students compare job offers based on ethics, salary, and long-term benefits.

Example:

1. Compare salary vs. ethical concerns about the company.
2. Evaluate long-term job growth and career impact.
3. Weigh risks of working for a company with unethical practices.
4. Make a decision based on logical and ethical considerations.

Pairing 8: Deciding Whether to Confront a Friend About Harmful Behavior

Big Skill: Logical and Ethical Considerations – Balancing honesty, friendship, and long-term well-being.

Small Skill: Structured Decision Frameworks – Using a Decision Matrix to analyze personal conflict.

Teaching Technique: Real-world applications.

Actions: Have students weigh emotional and logical factors in addressing a difficult situation.

Example:

1. Identify benefits of confronting (helping friend, honesty).
2. Assess risks (losing friendship, causing conflict).
3. Compare indirect vs. direct confrontation approaches.
4. Justify the final choice based on values and logic.

Pairing 9: Choosing a Research Topic with Ethical Implications

Big Skill: Logical and Ethical Considerations – Considering fairness, scientific integrity, and public impact.

Small Skill: Structured Decision Frameworks – Applying SWOT analysis to controversial research topics.

Teaching Technique: Cultural exploration.

Actions: Have students compare ethical concerns of different research topics before choosing one.

Example:

1. Strengths: Interest in topic, scientific relevance.
2. Weaknesses: Potential bias or misinformation risks.
3. Opportunities: Advancing ethical knowledge, informing policy.
4. Threats: Controversy, data privacy concerns.

Pairing 10: Evaluating the Ethics of a School Dress Code Policy

Big Skill: Logical and Ethical Considerations – Analyzing fairness, inclusivity, and practicality.

Small Skill: Structured Decision Frameworks – Using a Decision Matrix to assess policy impact.

Teaching Technique: Hands-on experiments.

Actions: Have students apply a Decision Matrix to determine whether a dress code policy is fair and effective.

Example:

1. Consider impact on students' expression and learning environment.
2. Compare enforcement consistency, gender bias concerns.
3. Weigh practicality vs. ethical inclusivity.
4. Justify support or opposition to the policy.

Decision Making – (Big Skill) Logical and Ethical Considerations vs. (Small Skill) Data Gathering & Comparison

---

Pairing 1: Choosing a College Based on Ethical and Academic Considerations

Big Skill: Logical and Ethical Considerations – Weighing personal values, diversity, sustainability, and academic reputation.

Small Skill: Data Gathering & Comparison – Collecting relevant information to compare different options.

Teaching Technique: Hands-on experiments.

Actions: Have students research colleges' ethical policies, sustainability practices, and academic rankings.

Example:

1. Gather data on diversity, tuition, and environmental efforts.
2. Compare ethical commitments, such as financial aid policies or fair admissions.
3. Identify institutions that align with personal values.
4. Make an informed college selection based on both ethics and academics.

Pairing 2: Deciding Whether to Buy Ethically Sourced Products

Big Skill: Logical and Ethical Considerations – Weighing price, quality, and impact on labor rights and sustainability.

Small Skill: Data Gathering & Comparison – Researching brands' ethical standards and sourcing practices.

Teaching Technique: Guided exploration.

Actions: Have students compare product prices, ethical business ratings, and labor policies before purchasing.

Example:

1. Research fair trade vs. conventional products.
2. Compare costs, quality, and sustainability impact.
3. Weigh ethical considerations against affordability.
4. Choose a purchasing strategy that balances ethics and budget.

Pairing 3: Evaluating Whether to Support a Political Candidate

Big Skill: Logical and Ethical Considerations – Weighing policy positions, moral character, and leadership effectiveness.

Small Skill: Data Gathering & Comparison – Researching candidates' past actions, policies, and ethical history.

Teaching Technique: Problem-solving exercises.

Actions: Have students research multiple political candidates and compare their policies using fact-based sources.

Example:

1. Gather data on policy platforms and past voting records.
2. Compare ethical stances on key issues.
3. Evaluate credibility, leadership skills, and moral consistency.
4. Make an informed decision on candidate support.

Pairing 4: Deciding Whether to Use Artificial Intelligence for Hiring

Big Skill: Logical and Ethical Considerations – Analyzing bias, efficiency, and fairness in AI-assisted hiring.

Small Skill: Data Gathering & Comparison – Researching AI bias cases and ethical implications in hiring processes.

Teaching Technique: Observation journals.

Actions: Have students compare the risks and benefits of AI-driven hiring and human-led recruitment.

Example:

1. Research cases of AI bias in employment.
2. Compare AI's efficiency vs. potential for discrimination.
3. Assess how AI aligns with ethical hiring standards.
4. Decide whether AI should play a role in recruitment.

Pairing 5: Choosing News Source for Reliable Information

Big Skill: Logical and Ethical Considerations – Evaluating media bias, factual accuracy, and reliability.

Small Skill: Data Gathering & Comparison – Analyzing different news outlets' credibility and reporting styles.

Teaching Technique: Hands-on experiments.

Actions: Have students compare multiple news sources on the same topic for accuracy and bias.

Example:

1. Select three articles on the same event from different sources.
2. Compare tone, cited sources, and factual consistency.
3. Research each outlet us history and reputation.
4. Choose the most reliable and unbiased source.

Pairing 6: Deciding Whether to Support a Social Media Movement

Big Skill: Logical and Ethical Considerations – Assessing authenticity, potential misinformation, and social impact.

Small Skill: Data Gathering & Comparison – Researching movement origins, claims, and real-world effectiveness.

Teaching Technique: Sensory exploration.

Actions: Have students evaluate online movements by researching their legitimacy and impact.

Example:

1. Gather information on movement leaders and funding.
2. Compare media portrayal and independent reports.
3. Assess risks of misinformation or performative activism.
4. Decide whether to actively support or remain skeptical.

Pairing 7: Determining Whether to Use Animal Testing for Medicine

Big Skill: Logical and Ethical Considerations – Weighing scientific advancement against ethical concerns.

Small Skill: Data Gathering & Comparison – Researching success rates and ethical alternatives.

Teaching Technique: Collaborative group work.

Actions: Have students compare the effectiveness and moral implications of animal testing vs. alternative research methods.

Example:

1. Research medical breakthroughs from animal testing.
2. Compare alternative methods like organ-on-a-chip technology.
3. Weigh ethical concerns about animal suffering.
4. Make an informed argument for or against the practice.

Pairing 8: Choosing an Employer Based on Ethical Practices

Big Skill: Logical and Ethical Considerations – Analyzing fair wages, labor laws, and corporate responsibility.

Small Skill: Data Gathering & Comparison – Researching companies' ethical records and employee reviews.

Teaching Technique: Real-world applications.

Actions: Have students compare job opportunities based on salary, ethical labor policies, and workplace treatment.

Example:

1. Gather data on company treatment of employees.
2. Compare wages, diversity policies, and sustainability efforts.
3. Weigh ethical concerns against financial benefits.
4. Choose the employer with the best ethical balance.

Pairing 9: Deciding Whether to Pursue a Career in a High-Paying but Unethical Industry

Big Skill: Logical and Ethical Considerations – Balancing personal gain with ethical responsibility.

Small Skill: Data Gathering & Comparison – Researching industry practices and ethical concerns.

Teaching Technique: Cultural exploration.

Actions: Have students compare career paths in terms of salary, ethical concerns, and personal values.

Example:

1. Research job prospects in industries with ethical concerns (e.g., fossil fuels, fast fashion).
2. Compare salary potential vs. social impact.
3. Weigh personal values against financial stability.
4. Justify the career path that aligns best with ethical and financial goals.

Pairing 10: Evaluating the Morality of School Dress Codes

Big Skill: Logical and Ethical Considerations – Weighing fairness, freedom of expression, and safety.

Small Skill: Data Gathering & Comparison – Researching policies in different schools and their outcomes.

Teaching Technique: Hands-on experiments.

Actions: Have students research school dress code policies and their impact on different student groups.

Example:

1. Compare dress codes from multiple schools.
2. Research cases of bias or unfair enforcement.
3. Weigh freedom of expression against school culture.
4. Determine whether dress codes are ethical or discriminatory.

Decision Making – (Big Skill) Logical and Ethical Considerations vs. (Small Skill) Justifying Decisions

---

Pairing 1: Deciding Whether to Report Cheating in a Class

Big Skill: Logical and Ethical Considerations – Weighing fairness, personal integrity, and peer relationships.

Small Skill: Justifying Decisions – Writing or presenting rational explanations for choices.

Teaching Technique: Hands-on experiments.

Actions: Have students write a persuasive justification for either reporting or not reporting academic dishonesty.

Example:

1. Consider the impact on personal integrity and fairness.
2. Weigh potential social repercussions.
3. Justify the choice in an argumentative essay.
4. Present in a class debate.

Pairing 2: Choosing a College Based on Ethical and Academic Considerations

Big Skill: Logical and Ethical Considerations – Evaluating a school's diversity, environmental policies, and funding ethics.

Small Skill: Justifying Decisions – Defending the decision in a structured written or oral argument.

Teaching Technique: Guided exploration.

Actions: Have students write a formal justification for choosing a college based on ethical factors.

Example:

1. Research financial aid, sustainability, and diversity policies.
2. Weigh academic benefits against ethical concerns.
3. Write a structured essay explaining the decision.
4. Present justification in a college decision forum.

Pairing 3: Deciding Whether to Support a Political Candidate

Big Skill: Logical and Ethical Considerations – Weighing policy positions, leadership skills, and integrity.

Small Skill: Justifying Decisions – Constructing a well-supported argument for a political choice.

Teaching Technique: Problem-solving exercises.

Actions: Have students research candidates and write a persuasive essay supporting their choice.

Example:

1. Gather data on key policies and leadership history.
2. Weigh ethical and logical concerns.
3. Write a speech justifying candidate support.
4. Deliver in a mock election setting.

Pairing 4: Evaluating Whether to Use AI for Hiring Decisions

Big Skill: Logical and Ethical Considerations – Analyzing bias, fairness, and efficiency.

Small Skill: Justifying Decisions – Writing a well-structured argument in support of or against AI hiring.

Teaching Technique: Observation journals.

Actions: Have students research AI bias in hiring and construct a defense for or against its use.

Example:

1. Compare AI hiring efficiency vs. bias risks.
2. Weigh fairness against corporate benefits.
3. Write an essay justifying the stance.
4. Present the decision to a class panel.

Pairing 5: Deciding Whether to Boycott an Unethical Company

Big Skill: Logical and Ethical Considerations – Weighing economic benefits vs. ethical concerns.

Small Skill: Justifying Decisions – Writing a rationale for participating or not participating in a boycott.

Teaching Technique: Hands-on experiments.

Actions: Have students analyze a company's ethical record and write a justification for supporting or boycotting it.

Example:

1. Research company labor and environmental practices.
2. Compare financial benefits vs. ethical costs.
3. Justify the consumer decision in an argumentative essay.
4. Present arguments in a debate.

Pairing 6: Deciding Whether to Post About a Social Cause on Social Media

Big Skill: Logical and Ethical Considerations – Assessing authenticity, social impact, and risks of misinformation.

Small Skill: Justifying Decisions – Writing a defense for engaging or not engaging in online activism.

Teaching Technique: Sensory exploration.

Actions: Have students analyze an online movement and justify their decision to post or refrain from posting.

Example:

1. Research movement legitimacy and effectiveness.
2. Compare risks of misinformation vs. raising awareness.
3. Write a structured defense of their stance.
4. Present in a mock social media forum.

Pairing 7: Determining Whether to Accept a Job Offer from a Company with Ethical Issues

Big Skill: Logical and Ethical Considerations – Weighing career benefits against moral concerns.

Small Skill: Justifying Decisions – Writing a structured defense for accepting or rejecting the offer.

Teaching Technique: Collaborative group work.

Actions: Have students research ethical concerns and justify their career decision in a professional setting.

Example:

1. Compare salary, benefits, and ethical concerns.
2. Weigh personal values against financial needs.
3. Write a business-style justification letter.
4. Present as part of a mock interview.

Pairing 8: Choosing Whether to Confront a Friend About Harmful Behavior

Big Skill: Logical and Ethical Considerations – Balancing honesty, friendship, and long-term consequences.

Small Skill: Justifying Decisions – Writing a rationale for confrontation or avoidance.

Teaching Technique: Real-world applications.

Actions: Have students write a justification for either addressing or ignoring the issue.

Example:

1. Weigh honesty vs. preserving friendship.
2. Consider possible long-term consequences.
3. Write a personal reflection defending the choice.
4. Present anonymously for class discussion.

Pairing 9: Deciding Whether to Support Genetic Engineering in Medicine

Big Skill: Logical and Ethical Considerations – Balancing medical advancements with ethical concerns.

Small Skill: Justifying Decisions – Writing a structured argument for or against genetic modifications.

Teaching Technique: Cultural exploration.

Actions: Have students research genetic engineering and construct a well-supported defense.

Example:

1. Compare medical benefits vs. potential risks.
2. Consider ethical concerns like human enhancement.
3. Write a policy-style justification.
4. Present in a mock scientific debate.

Pairing 10: Evaluating the Morality of School Dress Codes

Big Skill: Logical and Ethical Considerations – Weighing freedom of expression against school policies.

Small Skill: Justifying Decisions – Writing a structured argument in support of or against dress codes.

Teaching Technique: Hands-on experiments.

Actions: Have students research dress code policies and justify their stance.

Example:

1. Compare enforcement fairness across different student groups.
2. Consider practical benefits vs. ethical concerns.
3. Write a persuasive argument.
4. Present in a mock school policy meeting.

Example Teaching Strategies for Big vs. Small Thinking Skill Pairings of Creative Problem Solving at the High School Level

Here are 90 unique teaching strategy examples for the big thinking and small thinking skill pairings, each described in detail to align with **Big Thinking Skills** (Concepts & Strategies) and **Small Thinking Skills** (Step-by-Step Execution) for the **Creative Problem Solving Skill** taught at the high school level.

Creative Problem Solving

Big Thinking Skills (Concepts & Strategies)

- **Divergent Thinking:** Generating multiple solutions to a problem through brainstorming.
- **Lateral Thinking:** Making unexpected connections between ideas to find innovative solutions.
- **Pattern Recognition in Problems:** Identifying underlying structures in complex issues.

Small Thinking Skills (Step-by-Step Execution)

- **Applying Problem-Solving Techniques:** Using structured methods like mind mapping, Six Thinking Hats, or SCAMPER.
- **Experimenting with Ideas:** Testing different solutions in a controlled manner.
- **Refining Solutions:** Iterating on an idea based on feedback.

Creative Problem Solving – (Big Skill) Divergent Thinking vs. (Small Skill) Applying Problem-Solving Techniques

---

Pairing 1: Finding Ways to Reduce School Waste

Big Skill: Divergent Thinking – Generating multiple solutions to a problem through brainstorming.

Small Skill: Applying Problem-Solving Techniques – Using structured methods like mind mapping, Six Thinking Hats, or SCAMPER.

Teaching Technique: Hands-on experiments.

Actions: Have students use mind mapping to explore innovative ways to reduce waste in their school.

Example:

1. Identify key areas of waste (paper, plastic, food).
2. Use mind mapping to generate creative solutions.
3. Categorize ideas based on feasibility and impact.
4. Develop a school-wide initiative based on the best solutions.

Pairing 2: Designing a More Engaging Classroom Environment

Big Skill: Divergent Thinking – Exploring different ways to improve student engagement.

Small Skill: Applying Problem-Solving Techniques – Using SCAMPER to modify traditional classroom settings.

Teaching Technique: Guided exploration.

Actions: Have students apply SCAMPER to redesign aspects of the classroom for better learning experiences.

Example:

1. Substitute traditional seating with flexible seating.
2. Combine technology with traditional learning.
3. Modify layouts for collaborative learning spaces.
4. Present proposed classroom designs to school administrators.

Pairing 3: Developing Solutions to Reduce Traffic Congestion in the Community

Big Skill: Divergent Thinking – Thinking of multiple ways to solve traffic issues.

Small Skill: Applying Problem-Solving Techniques – Using Six Thinking Hats to approach the issue from different perspectives.

Teaching Technique: Problem-solving exercises.

Actions: Have students apply the Six Thinking Hats technique to explore traffic congestion solutions.

Example:

1. Use the White Hat (facts) to analyze current congestion data.
2. Use the Yellow Hat (positive) to highlight existing solutions.
3. Use the Green Hat (creativity) to propose new transit strategies.
4. Present findings to a local government representative.

Pairing 4: Inventing New Ways to Conserve Water at Home

Big Skill: Divergent Thinking – Brainstorming diverse methods to reduce water waste.

Small Skill: Applying Problem-Solving Techniques – Using mind mapping to explore solutions.

Teaching Technique: Observation journals.

Actions: Have students create a mind map listing home water conservation strategies.

Example:

1. Identify high water-usage activities at home.
2. Brainstorm alternative ways to reduce water use.
3. Categorize ideas by cost, effectiveness, and feasibility.
4. Implement and document results in an observation journal.

Pairing 5: Creating an Innovative Recycling Program

Big Skill: Divergent Thinking – Generating new ways to encourage recycling.

Small Skill: Applying Problem-Solving Techniques – Using SCAMPER to rethink existing recycling models.

Teaching Technique: Hands-on experiments.

Actions: Have students apply SCAMPER to redesign a school recycling program.

Example:

1. Substitute traditional bins with smart recycling stations.
2. Modify incentives to encourage participation.
3. Reverse the current process to make recycling more accessible.
4. Implement a pilot program and track its effectiveness.

Pairing 6: Enhancing the School Lunch Experience

Big Skill: Divergent Thinking – Brainstorming creative ways to improve school lunches.

Small Skill: Applying Problem-Solving Techniques – Using the Six Thinking Hats method to evaluate options.

Teaching Technique: Sensory exploration.

Actions: Have students apply Six Thinking Hats to improve cafeteria food quality and experience.

Example:

1. White Hat: Gather data on student lunch preferences.
2. Black Hat: Identify problems with current meals.
3. Green Hat: Brainstorm healthier, more appealing meal options.
4. Present ideas to the school administration for implementation.

Pairing 7: Designing a Community Service Project for Teens

Big Skill: Divergent Thinking – Exploring multiple ways for students to contribute to the community.

Small Skill: Applying Problem-Solving Techniques – Using mind mapping to organize potential projects.

Teaching Technique: Collaborative group work.

Actions: Have students mind map various service projects that align with student interests.

Example:

1. Identify key community needs (homelessness, education, environment).
2. Brainstorm projects that address these needs creatively.
3. Categorize ideas based on feasibility, impact, and student interest.
4. Launch the most promising initiative.

Pairing 8: Rethinking the School Homework Policy

Big Skill: Divergent Thinking – Generating solutions to make homework more effective.

Small Skill: Applying Problem-Solving Techniques – Using SCAMPER to modify existing policies.

Teaching Technique: Real-world applications.

Actions: Have students apply SCAMPER to propose alternative homework models.

Example:

1. Substitute written assignments with hands-on projects.
2. Modify deadlines to increase flexibility.
3. Reverse the format by using in-class homework and at-home lessons.
4. Propose policy changes to school administrators.

Pairing 9: Developing Strategies to Reduce Screen Time Among Teenagers

Big Skill: Divergent Thinking – Brainstorming multiple ways to reduce excessive screen use.

Small Skill: Applying Problem-Solving Techniques – Using the Six Thinking Hats to evaluate solutions.

Teaching Technique: Cultural exploration.

Actions: Have students use Six Thinking Hats to analyze screen time reduction strategies.

Example:

1. White Hat: Gather statistics on screen time usage.
2. Black Hat: Identify health and social drawbacks.
3. Green Hat: Propose creative alternatives like tech-free zones.
4. Develop and implement an awareness campaign.

Pairing 10: Improving Public Transportation Accessibility

Big Skill: Divergent Thinking – Exploring new ways to make transportation more inclusive.

Small Skill: Applying Problem-Solving Techniques – Using mind mapping to categorize solutions.

Teaching Technique: Hands-on experiments.

Actions: Have students mind map solutions to improve public transport accessibility.

Example:

1. Identify barriers for people with disabilities or low income.
2. Brainstorm technological and policy-based solutions.
3. Categorize ideas based on cost and feasibility.
4. Present solutions to local transportation officials.

Creative Problem Solving – (Big Skill) Divergent Thinking vs. (Small Skill) Experimenting with Ideas

---

Pairing 1: Finding Ways to Reduce Energy Use at School

Big Skill: Divergent Thinking – Brainstorming multiple ways to reduce energy consumption.

Small Skill: Experimenting with Ideas – Testing different solutions in a controlled manner.

Teaching Technique: Hands-on experiments.

Actions: Have students evaluate different energy-saving methods and measure their effectiveness.

Example:

1. Brainstorm energy-saving strategies (LED lights, motion sensors, unplugging devices).
2. Implement changes in different classrooms for a week.
3. Measure energy consumption before and after.
4. Compare results and propose the best long-term solution.

Pairing 2: Creating a More Engaging School Library Space

Big Skill: Divergent Thinking – Generating ideas for improving student use of the library.

Small Skill: Experimenting with Ideas – Testing different space arrangements or activities.

Teaching Technique: Guided exploration.

Actions: Have students rearrange library layouts and test student engagement levels.

Example:

1. Brainstorm changes (seating areas, reading corners, study zones).
2. Rearrange the space and introduce new interactive activities.
3. Measure student feedback and usage over a month.
4. Adjust design based on collected data.

Pairing 3: Reducing Plastic Waste in the Cafeteria

Big Skill: Divergent Thinking – Brainstorming alternative solutions to replace plastic utensils and packaging.

Small Skill: Experimenting with Ideas – Testing the effectiveness of different materials or incentives.

Teaching Technique: Problem-solving exercises.

Actions: Have students experiment with biodegradable alternatives and measure impact.

Example:

1. Brainstorm sustainable alternatives (compostable trays, reusable cutlery).
2. Implement one alternative for a week.
3. Track waste reduction and student participation.
4. Decide on the best long-term solution based on results.

Pairing 4: Increasing Student Participation in School Events

Big Skill: Divergent Thinking – Brainstorming creative ways to boost attendance.

Small Skill: Experimenting with Ideas – Testing different promotional strategies.

Teaching Technique: Observation journals.

Actions: Have students evaluate different event marketing strategies and measure effectiveness.

Example:

1. Brainstorm ways to promote school events (social media, posters, announcements).
2. Implement different strategies for separate events.
3. Track attendance and student engagement.
4. Identify the most successful promotion method for future events.

Pairing 5: Designing a Sustainable School Garden

Big Skill: Divergent Thinking – Exploring multiple approaches to create an eco-friendly garden.

Small Skill: Experimenting with Ideas – Testing different gardening methods.

Teaching Technique: Hands-on experiments.

Actions: Have students evaluate various planting techniques and soil compositions.

Example:

1. Brainstorm eco-friendly gardening methods (hydroponics, composting, vertical farming).
2. Implement and compare the growth rates of different methods.
3. Track sustainability factors like water usage and plant yield.
4. Select the best method for long-term implementation.

Pairing 6: Developing Alternative Transportation for Students

Big Skill: Divergent Thinking – Brainstorming different commuting solutions for students.

Small Skill: Experimenting with Ideas – Testing feasibility through small-scale trials.

Teaching Technique: Sensory exploration.

Actions: Have students evaluate different alternative transport methods.

Example:

1. Brainstorm options (carpooling, biking, school shuttle).
2. Implement carpool incentives or bike-to-school days.
3. Track participation and environmental impact.
4. Determine the most effective method for long-term use.

Pairing 7: Reducing Noise Levels in Classrooms

Big Skill: Divergent Thinking – Generating creative ways to reduce noise disruptions.

Small Skill: Experimenting with Ideas – Testing different noise control techniques.

Teaching Technique: Collaborative group work.

Actions: Have students experiment with noise reduction strategies.

Example:

1. Brainstorm noise reduction methods (white noise, seating rearrangements, quiet zones).
2. Evaluate one method per classroom for a week.

3. Measure decibel levels before and after implementation.
4. Identify the most effective strategy and recommend it to the school.

Pairing 8: Encouraging Healthier Eating Habits Among Students

Big Skill: Divergent Thinking – Brainstorming ways to improve student nutrition.

Small Skill: Experimenting with Ideas – Testing different incentive programs.

Teaching Technique: Real-world applications.

Actions: Have students evaluate various incentive programs for promoting healthy eating.

Example:

1. Brainstorm ideas (discounts on healthy meals, reward points, peer-led challenges).
2. Implement one method for a month.
3. Track student participation and meal selection.
4. Choose the most effective method for long-term use.

Pairing 9: Making Science Labs More Interactive

Big Skill: Divergent Thinking – Exploring creative ways to enhance engagement in science experiments.

Small Skill: Experimenting with Ideas – Testing new teaching techniques and tools.

Teaching Technique: Cultural exploration.

Actions: Have students evaluate alternative lab formats or digital tools.

Example:

1. Brainstorm ways to make experiments more engaging (virtual labs, real-world applications).
2. Implement and compare student engagement levels.
3. Collect feedback and adjust methods accordingly.
4. Recommend the best approach for science teachers to adopt.

Pairing 10: Designing a Student-Run Business for Fundraising

Big Skill: Divergent Thinking – Brainstorming various student business ideas.

Small Skill: Experimenting with Ideas – Testing different products or services.

Teaching Technique: Hands-on experiments.

Actions: Have students pilot different business models to determine the most profitable one.

Example:

1. Brainstorm potential business ideas (tutoring, handmade crafts, food sales).
2. Run small-scale trials to evaluate profitability and student interest.
3. Collect sales and customer feedback data.
4. Launch the most successful model as a permanent fundraising initiative.

Creative Problem Solving – (Big Skill) Divergent Thinking vs. (Small Skill) Refining Solutions

---

Pairing 1: Improving School Recycling Programs

Big Skill: Divergent Thinking – Brainstorming multiple ideas for increasing recycling participation.

Small Skill: Refining Solutions – Iterating on an idea based on student and staff feedback.

Teaching Technique: Hands-on experiments.

Actions: Have students brainstorm recycling improvements, evaluate different strategies, and refine the most effective one.

Example:

1. Generate ideas such as incentive programs, better signage, and composting.
2. Implement a pilot version of selected strategies.
3. Gather feedback from students and custodial staff.
4. Adjust the program based on participation rates and feedback.

Pairing 2: Designing a More Effective Study App

Big Skill: Divergent Thinking – Brainstorming multiple features for a study aid.

Small Skill: Refining Solutions – Iterating on the app based on user feedback.

Teaching Technique: Guided exploration.

Actions: Have students develop a study tool prototype and refine it through user testing.

Example:

1. Brainstorm features such as flashcards, AI tutors, and progress tracking.
2. Create a prototype using available software.
3. Gather feedback from classmates using the app for a week.
4. Modify the app design to enhance usability and effectiveness.

Pairing 3: Developing a Sustainable Water Conservation Plan

Big Skill: Divergent Thinking – Generating creative ways to reduce water waste.

Small Skill: Refining Solutions – Testing and adjusting methods based on results.

Teaching Technique: Problem-solving exercises.

Actions: Have students evaluate different water-saving strategies and refine based on effectiveness.

Example:

1. Brainstorm conservation ideas such as rainwater collection or low-flow systems.
2. Implement a small-scale version of selected ideas.
3. Track water usage data before and after.
4. Adjust the plan based on results and feasibility.

Pairing 4: Improving Traffic Flow Around the School

Big Skill: Divergent Thinking – Brainstorming solutions to reduce congestion.

Small Skill: Refining Solutions – Testing and modifying proposed solutions based on community input.

Teaching Technique: Observation journals.

Actions: Have students analyze traffic patterns and propose solutions, refining based on input.

Example:

1. Brainstorm ideas such as staggered drop-off times, carpool incentives, or bike lanes.
2. Implement one idea on a trial basis.
3. Survey students, parents, and staff on its effectiveness.
4. Adjust based on feedback and traffic flow data.

Pairing 5: Creating a New School Club for Student Well-Being

Big Skill: Divergent Thinking – Generating different club themes and engagement strategies.

Small Skill: Refining Solutions – Adjusting club structure based on student feedback.

Teaching Technique: Hands-on experiments.

Actions: Have students create a pilot club meeting and refine structure based on participation.

Example:

1. Brainstorm club focus areas (mental health, social activism, creativity).
2. Launch a test session with different activity formats.
3. Collect feedback from attendees on engagement.
4. Modify meeting format to improve participation.

Pairing 6: Enhancing School Lunch Options

Big Skill: Divergent Thinking – Exploring multiple ways to improve cafeteria food variety.

Small Skill: Refining Solutions – Testing menu changes and modifying based on student response.

Teaching Technique: Sensory exploration.

Actions: Have students survey peers, suggest new meal options, and refine menu based on feedback.

Example:

1. Brainstorm healthier or more diverse menu items.
2. Introduce a few new options for a trial week.
3. Collect feedback through surveys or focus groups.
4. Adjust menu offerings based on student preferences.

Pairing 7: Designing an Anti-Bullying Awareness Campaign

Big Skill: Divergent Thinking – Brainstorming creative ways to engage students in bullying prevention.

Small Skill: Refining Solutions – Testing campaign elements and improving based on impact.

Teaching Technique: Collaborative group work.

Actions: Have students design a campaign, evaluate it in a small group, and refine based on engagement.

Example:

1. Brainstorm campaign strategies (posters, videos, student testimonials).
2. Run an initial campaign week with selected elements.
3. Collect feedback on student awareness and participation.
4. Modify campaign elements for better engagement.

Pairing 8: Revamping Classroom Seating Arrangements for Better Learning

Big Skill: Divergent Thinking – Exploring different seating layouts to improve engagement.

Small Skill: Refining Solutions – Testing and iterating on seating changes based on student performance.

Teaching Technique: Real-world applications.

Actions: Have students experiment with different classroom layouts and adjust based on effectiveness.

Example:

1. Brainstorm alternative seating configurations (groups, rows, flexible seating).
2. Implement one arrangement for a trial period.
3. Measure participation and student feedback.
4. Adjust based on which layout enhances engagement the most.

Pairing 9: Reducing Social Media Distractions During School Hours

Big Skill: Divergent Thinking – Brainstorming solutions to minimize distractions.

Small Skill: Refining Solutions – Testing different strategies and modifying based on results.

Teaching Technique: Cultural exploration.

Actions: Have students evaluate different approaches and refine based on effectiveness.

Example:

1. Brainstorm solutions (phone-free zones, app blockers, tech-use breaks).
2. Implement a test run of one solution in a controlled setting.
3. Collect feedback on effectiveness from students and teachers.
4. Modify the approach for better results.

Pairing 10: Creating a More Inclusive School Environment

Big Skill: Divergent Thinking – Generating ways to improve inclusivity and representation.

Small Skill: Refining Solutions – Testing initiatives and modifying based on community feedback.

Teaching Technique: Hands on experiments.

Actions: Have students propose and implement initiatives, refining them based on school response.

Example:

1. Brainstorm initiatives (cultural events, diversity panels, inclusive curriculum changes).
2. Implement a pilot event or program.
3. Gather feedback from student groups.
4. Adjust based on effectiveness in fostering inclusivity.

Creative Problem Solving – (Big Skill) Lateral Thinking vs. (Small Skill) Applying Problem-Solving Techniques

---

Pairing 1: Creating a Cross-Disciplinary Learning Approach

Big Skill: Lateral Thinking – Making unexpected connections between different subjects.

Small Skill: Applying Problem-Solving Techniques – Using mind mapping to link ideas across disciplines.

Teaching Technique: Hands-on experiments.

Actions: Have students use mind mapping to connect concepts from different subjects and create a new learning model.

Example:

1. Connect history with technology by exploring how AI could analyze historical events.
2. Link physics with sports to optimize athlete performance.
3. Develop a lesson plan using cross-disciplinary knowledge.
4. Present findings in an interdisciplinary showcase.

Pairing 2: Reinventing School Fundraising Strategies

Big Skill: Lateral Thinking – Finding creative ways to make fundraising more engaging.

Small Skill: Applying Problem-Solving Techniques – Using SCAMPER to modify traditional fundraising methods.

Teaching Technique: Guided exploration.

Actions: Have students apply SCAMPER to redesign fundraising efforts to increase participation.

Example:

1. Substitute bake sales with digital fundraisers (e.g., online auctions).
2. Modify traditional donation drives to be competition-based.
3. Adapt strategies from corporate marketing to promote school events.
4. Pilot the best idea and measure its effectiveness.

Pairing 3: Developing Smart Traffic Solutions for the School Area

Big Skill: Lateral Thinking – Applying unconventional solutions from unrelated fields.

Small Skill: Applying Problem-Solving Techniques – Using Six Thinking Hats to explore different perspectives.

Teaching Technique: Problem-solving exercises.

Actions: Have students analyze traffic issues using Six Thinking Hats to design a more efficient system.

Example:

1. White Hat: Gather school traffic data.
2. Black Hat: Identify congestion challenges.
3. Green Hat: Brainstorm tech-based solutions (e.g., app-based carpooling).

4. Evaluate and present a final recommendation to local authorities.

Pairing 4: Improving School Cafeteria Food Waste Management

Big Skill: Lateral Thinking – Applying restaurant industry waste reduction techniques to schools.

Small Skill: Applying Problem-Solving Techniques – Using mind mapping to categorize causes and solutions.

Teaching Technique: Observation journals.

Actions: Have students analyze cafeteria waste patterns and apply restaurant strategies to reduce waste.

Example:

1. Identify food waste sources (uneaten food, overproduction).
2. Research how professional kitchens minimize waste.
3. Use mind mapping to propose school adaptations (e.g., donation programs, portion control).
4. Implement and track the impact of the best solution.

Pairing 5: Designing an Inclusive Sports Program

Big Skill: Lateral Thinking – Integrating elements from adaptive sports into traditional school sports.

Small Skill: Applying Problem-Solving Techniques – Using SCAMPER to rethink sports programs.

Teaching Technique: Hands-on experiments.

Actions: Have students apply SCAMPER to adapt school sports for broader accessibility.

Example:

1. Substitute traditional scoring systems with goal-based achievements.
2. Modify game rules to accommodate different abilities.
3. Adapt professional adaptive sports innovations.
4. Implement changes in a pilot tournament.

Pairing 6: Creating a Hybrid Learning Environment

Big Skill: Lateral Thinking – Combining physical and digital learning techniques innovatively.

Small Skill: Applying Problem-Solving Techniques – Using Six Thinking Hats to evaluate hybrid learning benefits.

Teaching Technique: Sensory exploration.

Actions: Have students analyze digital and in-person learning benefits using Six Thinking Hats.

Example:

1. Red Hat: Gauge student emotions about hybrid learning.
2. White Hat: Compare success rates of digital vs. in-person methods.
3. Green Hat: Design a flexible classroom model.
4. Evaluate and refine the model based on student feedback.

Pairing 7: Enhancing School Safety with Unexpected Innovations

Big Skill: Lateral Thinking – Applying security strategies from other industries to school settings.

Small Skill: Applying Problem-Solving Techniques – Using mind mapping to brainstorm solutions.

Teaching Technique: Collaborative group work.

Actions: Have students study security innovations in banks, airports, and tech companies, then adapt them for school safety.

Example:

1. Map security challenges in schools (bullying, unauthorized access).
2. Identify industry practices like AI surveillance or RFID entry.
3. Adapt and propose an improved security strategy.
4. Evaluate a small-scale implementation (e.g., visitor management upgrades).

Pairing 8: Redesigning the Public Transportation Experience for Students

Big Skill: Lateral Thinking – Applying hospitality industry customer experience strategies to buses.

Small Skill: Applying Problem-Solving Techniques – Using SCAMPER to reimagine public transit for students.

Teaching Technique: Real-world applications.

Actions: Have students analyze public transit pain points and use SCAMPER to improve them.

Example:

1. Substitute outdated systems with student-friendly transit apps.
2. Modify bus routes based on student demand.
3. Adapt airline customer service techniques for school buses.
4. Implement and track effectiveness.

Pairing 9: Creating an Eco-Friendly School Building Plan

Big Skill: Lateral Thinking – Applying biomimicry and nature-based solutions to school design.

Small Skill: Applying Problem-Solving Techniques – Using mind mapping to develop green school initiatives.

Teaching Technique: Cultural exploration.

Actions: Have students study nature-inspired architecture and apply it to school sustainability.

Example:

1. Research biomimicry in building design (e.g., termite mounds for ventilation).
2. Identify school infrastructure inefficiencies.
3. Mind map ways to integrate nature-inspired solutions.
4. Propose and test one initiative (e.g., green roofs, solar energy collection).

Pairing 10: Rethinking Exam Formats for 21st-Century Learning

Big Skill: Lateral Thinking – Exploring alternative assessment methods inspired by gaming and corporate training.

Small Skill: Applying Problem-Solving Techniques – Using Six Thinking Hats to evaluate new exam formats.

Teaching Technique: Hands-on experiments.

Actions: Have students design and test alternative exam formats inspired by gaming and professional certifications.

Example:

1. White Hat: Research the effectiveness of traditional vs. alternative assessments.
2. Yellow Hat: Highlight benefits of gamified testing.
3. Black Hat: Identify potential challenges (grading consistency).
4. Pilot an alternative exam format and compare outcomes with traditional exams.

Creative Problem Solving – (Big Skill) Lateral Thinking vs. (Small Skill) Experimenting with Ideas

---

Pairing 1: Using Music to Improve Memory in Studying

Big Skill: Lateral Thinking – Exploring how musical rhythms can enhance memory retention.

Small Skill: Experimenting with Ideas – Testing different types of background music in study sessions.

Teaching Technique: Hands-on experiments.

Actions: Have students evaluate how different genres of music impact focus and memory retention.

Example:

1. Select various types of music (classical, jazz, lo-fi, silence).
2. Study the same material under different music conditions.
3. Take quizzes and analyze performance differences.
4. Identify which type of music optimizes memory and concentration.

Pairing 2: Applying Sports Strategy to Improve Debate Team Performance

Big Skill: Lateral Thinking – Using principles from team sports to enhance debate strategies.

Small Skill: Experimenting with Ideas – Testing different debate "plays" inspired by basketball and football.

Teaching Technique: Guided exploration.

Actions: Have students apply offensive and defensive strategies from sports to debate team competitions.

Example:

1. Develop playbooks with strategic argument formations.
2. Evaluate a "fast break" argument vs. a slow, methodical approach.
3. Measure effectiveness in a debate competition.
4. Refine strategies based on performance feedback.

Pairing 3: Designing a Biodegradable Alternative to Plastic Wrappers

Big Skill: Lateral Thinking – Using nature-inspired solutions (biomimicry) to solve environmental issues.

Small Skill: Experimenting with Ideas – Testing different natural materials for durability and decomposition.

Teaching Technique: Problem-solving exercises.

Actions: Have students experiment with edible and biodegradable materials for packaging.

Example:

1. Research plant-based materials like seaweed or banana leaves.
2. Create sample wrappers and test for durability and water resistance.
3. Compare decomposition rates of different materials.

4. Identify the most sustainable alternative.

Pairing 4: Enhancing School Cafeteria Efficiency with Fast-Food Service Models

Big Skill: Lateral Thinking – Applying restaurant industry efficiency models to school lunches.

Small Skill: Experimenting with Ideas – Testing different queue management strategies.

Teaching Technique: Observation journals.

Actions: Have students evaluate various service techniques from the fast-food industry in their cafeteria.

Example:

1. Implement express checkout lines for pre-selected meals.
2. Compare wait times before and after implementation.
3. Survey students about their experience.
4. Adjust the service method based on data.

Pairing 5: Using Gamification to Increase Homework Completion Rates

Big Skill: Lateral Thinking – Applying video game reward systems to schoolwork.

Small Skill: Experimenting with Ideas – Testing different incentive structures.

Teaching Technique: Hands-on experiments.

Actions: Have students design a gamified homework system and evaluate its impact on completion rates.

Example:

1. Introduce point-based rewards for completed assignments.
2. Compare engagement with and without gamification.
3. Survey students on motivation levels.
4. Adjust the system based on effectiveness.

Pairing 6: Implementing VR Technology for History Lessons

Big Skill: Lateral Thinking – Using virtual reality to enhance historical learning experiences.

Small Skill: Experimenting with Ideas – Testing different VR simulations in history classes.

Teaching Technique: Sensory exploration.

Actions: Have students evaluate different VR history experiences and measure their impact on engagement.

Example:

1. Use VR to "visit" historical events.
2. Compare comprehension levels with traditional textbook learning.
3. Gather feedback on student engagement.
4. Modify lesson plans to integrate the most effective VR tools.

Pairing 7: Applying Nature's Self-Cleaning Mechanisms to School Maintenance

Big Skill: Lateral Thinking – Adapting biological self-cleaning techniques for practical applications.

Small Skill: Experimenting with Ideas – Testing different self-cleaning surface treatments in the school.

Teaching Technique: Collaborative group work.

Actions: Have students apply nature-inspired coatings and compare cleanliness levels.

Example:

1. Study how lotus leaves repel water and dirt.
2. Apply similar coatings to classroom surfaces.
3. Measure dirt accumulation over weeks.
4. Evaluate the effectiveness of self-cleaning methods.

Pairing 8: Using Psychological Tricks to Improve Public Speaking Confidence

Big Skill: Lateral Thinking – Incorporating confidence-building strategies from psychology into public speaking.

Small Skill: Experimenting with Ideas – Testing different pre-speech rituals and mindset shifts.

Teaching Technique: Real-world applications.

Actions: Have students experiment with power poses, visualization, and breathing techniques before speeches.

Example:

1. Evaluate different pre-speech routines (visualization, deep breathing, mirror affirmations).
2. Measure self-reported confidence levels before and after.
3. Compare speech delivery scores.
4. Identify the most effective method for reducing anxiety.

Pairing 9: Improving Student Sleep Habits with Military Sleep Techniques

Big Skill: Lateral Thinking – Adapting military sleep strategies for student use.

Small Skill: Experimenting with Ideas – Testing different sleep routines for effectiveness.

Teaching Technique: Cultural exploration.

Actions: Have students evaluate different sleep training methods and compare their impact on restfulness.

Example:

1. Implement military "2-minute sleep" techniques.
2. Track sleep quality and alertness in school.
3. Compare results with traditional sleep schedules.
4. Develop an optimized student sleep routine.

Pairing 10: Using Storytelling in Math to Improve Problem-Solving

Big Skill: Lateral Thinking – Applying narrative techniques to math instruction.

Small Skill: Experimenting with Ideas – Testing different story-driven problem-solving methods.

Teaching Technique: Hands-on experiments.

Actions: Have students apply storytelling techniques to explain complex math concepts and measure comprehension.

Example:

1. Create a narrative around a math problem (e.g., budgeting for a space mission).
2. Compare understanding between narrative-based and traditional teaching.
3. Gather feedback from students.
4. Adjust and refine storytelling methods for math instruction.

Creative Problem Solving – (Big Skill) Lateral Thinking vs. (Small Skill) Refining Solutions

---

Pairing 1: Improving Traffic Flow Using Video Game Pathfinding Strategies

Big Skill: Lateral Thinking – Applying pathfinding algorithms from video games to real-world traffic management.

Small Skill: Refining Solutions – Iterating on the design based on test results and feedback.

Teaching Technique: Hands-on experiments.

Actions: Have students apply AI-based pathfinding strategies to model traffic flow around the school and refine based on real-time feedback.

Example:

1. Analyze congestion areas in school parking lots.
2. Apply video game AI techniques to propose alternate routes.
3. Evaluate solutions using simulation software.
4. Gather feedback and refine the model for real-world application.

Pairing 2: Using Theater Acting Techniques to Improve Job Interview Success

Big Skill: Lateral Thinking – Applying stage presence and improvisation skills to professional settings.

Small Skill: Refining Solutions – Adjusting techniques based on mock interview feedback.

Teaching Technique: Guided exploration.

Actions: Have students use acting methods such as vocal projection, body language, and character immersion to enhance their interview presence.

Example:

1. Train using role-play exercises with scripted and impromptu responses.
2. Conduct mock interviews with different audiences.
3. Gather feedback on tone, confidence, and clarity.
4. Refine responses and body language for real interviews.

Pairing 3: Reducing Food Waste by Repurposing Leftovers into New Meal Creations

Big Skill: Lateral Thinking – Applying culinary fusion techniques to reduce cafeteria waste.

Small Skill: Refining Solutions – Testing recipes and adjusting based on student taste tests.

Teaching Technique: Problem-solving exercises.

Actions: Have students create a new meal plan using cafeteria leftovers and refine based on feedback.

Example:

1. Research global food waste reduction strategies.
2. Develop new recipes using common cafeteria leftovers.
3. Conduct taste tests with students and staff.
4. Refine the best-received recipes and propose them to the school's food program.

Pairing 4: Enhancing Classroom Engagement by Incorporating Escape Room Elements

Big Skill: Lateral Thinking – Integrating puzzle-solving techniques from escape rooms into learning.

Small Skill: Refining Solutions – Adjusting puzzles and engagement methods based on student feedback.

Teaching Technique: Observation journals.

Actions: Have students design an educational escape room challenge and refine it through trials.

Example:

1. Develop puzzles related to course material (e.g., math riddles, historical clues).
2. Implement in a class session and observe engagement levels.
3. Collect feedback on difficulty and learning effectiveness.
4. Modify clues and pacing based on student responses.

Pairing 5: Developing a Multilingual School Announcements System Using AI Translation

Big Skill: Lateral Thinking – Using real-time AI translation tools to increase accessibility.

Small Skill: Refining Solutions – Testing different translation services and improving based on accuracy.

Teaching Technique: Hands-on experiments.

Actions: Have students evaluate multiple AI translation tools for school announcements and refine based on effectiveness.

Example:

1. Compare different AI tools for speed and accuracy.
2. Run trials in different languages and gather feedback from native speakers.
3. Improve phrasing and script delivery based on feedback.
4. Implement the most effective translation method school-wide.

Pairing 6: Using Gamification to Improve School Fundraising Participation

Big Skill: Lateral Thinking – Applying gaming reward mechanics to charity fundraising.

Small Skill: Refining Solutions – Adjusting the mechanics based on participation data.

Teaching Technique: Sensory exploration.

Actions: Have students design a points-based system for fundraising and refine it based on user engagement.

Example:

1. Introduce leaderboard-style incentives for donation goals.
2. Monitor engagement levels in different classes.
3. Gather feedback on rewards and ease of participation.

4. Adjust rules and incentives to optimize engagement.

Pairing 7: Improving Public Speaking by Using Podcasting Techniques

Big Skill: Lateral Thinking – Borrowing techniques from audio storytelling to refine public speaking.

Small Skill: Refining Solutions – Iterating speech structure and vocal delivery based on audience feedback.

Teaching Technique: Collaborative group work.

Actions: Have students record speeches as podcasts, analyze audience engagement, and refine delivery.

Example:

1. Record speeches using podcasting equipment.
2. Compare scripted vs. unscripted delivery styles.
3. Collect listener feedback on clarity, pacing, and engagement.
4. Adjust speech structure based on audience reception.

Pairing 8: Designing an Eco-Friendly School Uniform with Upcycled Materials

Big Skill: Lateral Thinking – Incorporating sustainable fashion practices into school uniforms.

Small Skill: Refining Solutions – Testing materials and improving based on comfort and durability feedback.

Teaching Technique: Real-world applications.

Actions: Have students prototype an upcycled school uniform and refine it based on student comfort testing.

Example:

1. Research sustainable fabrics and recycled materials.
2. Design and construct a test uniform.
3. Gather feedback on comfort, durability, and appearance.
4. Adjust material choices based on wearability tests.

Pairing 9: Improving Student Focus by Using Aromatherapy Techniques

Big Skill: Lateral Thinking – Applying scent-based concentration techniques used in therapy and marketing.

Small Skill: Refining Solutions – Testing different scents and modifying based on student response.

Teaching Technique: Cultural exploration.

Actions: Have students experiment with different essential oils in study environments and refine based on focus improvements.

Example:

1. Select scents commonly associated with focus (e.g., peppermint, lavender).
2. Have students study in rooms with different scents.
3. Measure perceived focus levels and retention.
4. Identify the most effective scent for cognitive performance.

Pairing 10: Enhancing Time Management by Adopting Airline Boarding Techniques for School Transitions

Big Skill: Lateral Thinking – Adapting efficiency models from the airline industry to student movement.

Small Skill: Refining Solutions – Testing different hallway transition methods and improving based on timing data.

Teaching Technique: Hands-on experiments.

Actions: Have students apply airline-style boarding zones to school hallway transitions and adjust based on congestion levels.

Example:

1. Implement staggered class releases based on location.
2. Monitor congestion and student movement efficiency.
3. Collect feedback from students and teachers.
4. Modify the strategy to maximize efficiency without disruption.

Creative Problem Solving – (Big Skill) Pattern Recognition in Problems vs. (Small Skill) Applying Problem-Solving Techniques

---

Pairing 1: Identifying Traffic Congestion Patterns and Optimizing Routes

Big Skill: Pattern Recognition in Problems – Identifying recurring traffic congestion points and peak times.

Small Skill: Applying Problem-Solving Techniques – Using mind mapping to visualize alternative traffic flow strategies.

Teaching Technique: Hands-on experiments.

Actions: Have students observe traffic congestion near the school, map patterns, and propose alternate solutions.

Example:

1. Analyze peak traffic hours and bottlenecks.
2. Use mind mapping to explore alternative traffic solutions (carpooling, staggered release times).
3. Compare simulated outcomes using software.
4. Present a proposal to school administrators.

Pairing 2: Detecting Cybersecurity Threat Patterns in Online Behavior

Big Skill: Pattern Recognition in Problems – Recognizing trends in phishing attacks and data breaches.

Small Skill: Applying Problem-Solving Techniques – Using Six Thinking Hats to evaluate security responses.

Teaching Technique: Guided exploration.

Actions: Have students analyze case studies of cyberattacks and apply structured thinking to prevent future breaches.

Example:

1. Examine common cyber threats (phishing emails, password breaches).
2. Use Six Thinking Hats to analyze perspectives (victims, hackers, IT security teams).
3. Develop best practices to prevent attacks.
4. Present findings as a cybersecurity awareness campaign.

Pairing 3: Recognizing Social Media Trends and Predicting Viral Content

Big Skill: Pattern Recognition in Problems – Identifying common traits in viral posts and online trends.

Small Skill: Applying Problem-Solving Techniques – Using SCAMPER to modify existing content for engagement.

Teaching Technique: Problem-solving exercises.

Actions: Have students analyze viral content trends and apply SCAMPER to create engaging posts.

Example:

1. Study elements of viral videos, memes, and challenges.
2. Apply SCAMPER (e.g., modify existing memes, combine trending ideas).
3. Post content and track engagement metrics.
4. Adjust strategy based on real-time audience feedback.

Pairing 4: Recognizing Climate Change Patterns and Creating Actionable Solutions

Big Skill: Pattern Recognition in Problems – Identifying temperature rise trends and extreme weather occurrences.

Small Skill: Applying Problem-Solving Techniques – Using mind mapping to visualize climate mitigation strategies.

Teaching Technique: Observation journals.

Actions: Have students track climate trends, map connections between human activities, and develop solutions.

Example:

1. Research historical climate data.
2. Use mind mapping to link pollution sources and environmental impact.
3. Develop an awareness campaign or school-wide sustainability initiative.
4. Implement a pilot program (e.g., reducing plastic waste).

Pairing 5: Recognizing Patterns in Student Stress Levels and Academic Performance

Big Skill: Pattern Recognition in Problems – Identifying the link between stress, workload, and performance.

Small Skill: Applying Problem-Solving Techniques – Using Six Thinking Hats to evaluate different stress-reduction strategies.

Teaching Technique: Hands-on experiments.

Actions: Have students survey peers on academic stress and apply structured thinking to design solutions.

Example:

1. Collect data on stress levels across subjects and extracurricular activities.
2. Use Six Thinking Hats to evaluate the impact of different stress-reduction strategies.
3. Implement strategies like mindfulness breaks or workload redistribution.
4. Measure and compare student performance before and after interventions.

Pairing 6: Identifying Crime Rate Patterns and Proposing Preventative Measures

Big Skill: Pattern Recognition in Problems – Recognizing crime hotspots and repeat offenses in urban settings.

Small Skill: Applying Problem-Solving Techniques – Using mind mapping to explore crime prevention strategies.

Teaching Technique: Sensory exploration.

Actions: Have students analyze crime reports, identify trends, and design solutions for community safety.

Example:

1. Research local crime rate statistics.
2. Use mind mapping to categorize causes (economic, social, lack of surveillance).
3. Develop strategies like neighborhood watch programs or tech-based reporting.
4. Present proposals to local law enforcement or city officials.

Pairing 7: Detecting Consumer Spending Patterns and Creating Smart Budgeting Strategies

Big Skill: Pattern Recognition in Problems – Analyzing trends in spending and financial decision-making.

Small Skill: Applying Problem-Solving Techniques – Using SCAMPER to refine personal budgeting methods.

Teaching Technique: Collaborative group work.

Actions: Have students analyze personal spending habits and refine budgeting strategies using SCAMPER.

Example:

1. Track spending over a month and identify unnecessary expenses.
2. Apply SCAMPER to improve budgeting techniques (e.g., automating savings, changing expense priorities).
3. Compare budgeting success before and after adjustments.
4. Share best practices with classmates.

Pairing 8: Recognizing Sleep Patterns and Improving Rest Quality

Big Skill: Pattern Recognition in Problems – Identifying correlations between screen time, study habits, and sleep quality.

Small Skill: Applying Problem-Solving Techniques – Using Six Thinking Hats to explore different sleep improvement strategies.

Teaching Technique: Real-world applications.

Actions: Have students track their sleep habits, recognize disruptions, and develop strategies to enhance rest.

Example:

1. Record pre-bedtime habits (screen usage, diet, relaxation methods).
2. Use Six Thinking Hats to analyze perspectives (science, personal habits, health risks).

3. Implement and test sleep-improvement strategies (e.g., blue light filters, relaxation exercises).
4. Compare sleep quality before and after interventions.

Pairing 9: Recognizing the Root Causes of Student Dropout Rates

Big Skill: Pattern Recognition in Problems – Identifying common risk factors for students leaving school.

Small Skill: Applying Problem-Solving Techniques – Using mind mapping to connect causes and solutions.

Teaching Technique: Cultural exploration.

Actions: Have students analyze dropout rates, find patterns, and propose intervention strategies.

Example:

1. Research risk factors (financial issues, lack of engagement, family circumstances).
2. Use mind mapping to visualize cause-effect relationships.
3. Develop solutions like mentorship programs or alternative education paths.
4. Present recommendations to school administrators.

Pairing 10: Identifying Patterns in Water Usage and Reducing Waste

Big Skill: Pattern Recognition in Problems – Recognizing inefficient water use in schools and households.

Small Skill: Applying Problem-Solving Techniques – Using SCAMPER to modify existing conservation methods.

Teaching Technique: Hands-on experiments.

Actions: Have students track water consumption patterns and apply SCAMPER to create conservation strategies.

Example:

1. Monitor water usage across school bathrooms, kitchens, and sports fields.
2. Apply SCAMPER (e.g., modify faucet designs, introduce auto-shutoff systems, incentivize conservation).
3. Evaluate solutions and track reduction in water waste.
4. Recommend sustainable water usage strategies for the school.

Creative Problem Solving – (Big Skill) Pattern Recognition in Problems vs. (Small Skill) Experimenting with Ideas

---

Pairing 1: Identifying Patterns in Classroom Disruptions and Testing New Seating Arrangements

Big Skill: Pattern Recognition in Problems – Identifying common causes of classroom disruptions.

Small Skill: Experimenting with Ideas – Testing different seating arrangements to improve focus and participation.

Teaching Technique: Hands-on experiments.

Actions: Have students observe classroom behavior patterns and test seating strategies to enhance focus.

Example:

1. Collect data on disruptions by location, time, and student groups.
2. Rearrange seating (e.g., rows, clusters, U-shape) and compare engagement levels.
3. Gather student and teacher feedback on effectiveness.
4. Identify the most effective classroom layout for optimal learning.

Pairing 2: Detecting Patterns in Student Sleep and Testing Sleep Hygiene Techniques

Big Skill: Pattern Recognition in Problems – Identifying connections between sleep habits and academic performance.

Small Skill: Experimenting with Ideas – Testing various sleep improvement strategies to enhance cognitive function.

Teaching Technique: Guided exploration.

Actions: Have students track their sleep habits and experiment with different pre-bedtime routines to find optimal rest strategies.

Example:

1. Record daily sleep hours and alertness levels.
2. Evaluate sleep strategies (reduced screen time, meditation, reading).
3. Compare energy levels and focus after different sleep routines.
4. Determine the best sleep practices for academic performance.

Pairing 3: Recognizing Water Waste Trends and Testing Reduction Methods

Big Skill: Pattern Recognition in Problems – Identifying where the most water is wasted in a school setting.

Small Skill: Experimenting with Ideas – Testing different water conservation techniques.

Teaching Technique: Problem-solving exercises.

Actions: Have students analyze school water usage patterns and implement conservation experiments.

Example:

1. Observe water consumption in restrooms, cafeteria, and sports facilities.
2. Evaluate strategies (low-flow faucets, timed water use, awareness campaigns).
3. Measure water waste reduction before and after implementation.
4. Propose long-term conservation solutions based on findings.

Pairing 4: Identifying Patterns in Food Waste and Testing Sustainable Solutions

Big Skill: Pattern Recognition in Problems – Finding the most commonly wasted foods in school lunches.

Small Skill: Experimenting with Ideas – Testing different waste reduction solutions.

Teaching Technique: Observation journals.

Actions: Have students record patterns of food waste and test various solutions to minimize waste.

Example:

1. Track which food items are discarded most frequently.
2. Implement food-sharing stations or portion control options.
3. Measure the decrease in waste over a week.
4. Adjust strategies based on cafeteria staff and student feedback.

Pairing 5: Recognizing Exercise Habits and Testing Productivity-Boosting Workouts

Big Skill: Pattern Recognition in Problems – Identifying links between physical activity and productivity levels.

Small Skill: Experimenting with Ideas – Testing different short workout routines to improve focus.

Teaching Technique: Hands-on experiments.

Actions: Have students track their focus levels before and after different types of exercise.

Example:

1. Record productivity levels before and after school workouts.
2. Evaluate different activities (stretching, high-intensity training, yoga).
3. Compare focus and alertness between workout types.
4. Recommend the best routine for boosting classroom performance.

Pairing 6: Identifying Public Transportation Delays and Testing Efficiency Solutions

Big Skill: Pattern Recognition in Problems – Analyzing trends in public transportation delays.

Small Skill: Experimenting with Ideas – Testing different route optimization strategies.

Teaching Technique: Sensory exploration.

Actions: Have students monitor transit patterns and test strategies for improving efficiency.

Example:

1. Collect data on peak traffic times and common delays.
2. Experiment with different routes or departure times.
3. Measure differences in arrival punctuality.
4. Develop an optimal school commute plan.

Pairing 7: Recognizing Patterns in Student Stress Levels and Testing Mindfulness Strategies

Big Skill: Pattern Recognition in Problems – Identifying connections between student stress triggers and academic performance.

Small Skill: Experimenting with Ideas – Testing different mindfulness exercises to reduce stress.

Teaching Technique: Collaborative group work.

Actions: Have students track stress patterns and implement mindfulness techniques to observe their effects.

Example:

1. Identify high-stress periods (exams, deadlines).
2. Experiment with different stress-relief methods (breathing exercises, guided meditation, journaling).
3. Track changes in stress levels and academic focus.
4. Determine which method is most effective for student well-being.

Pairing 8: Recognizing Shopping Trends and Testing Consumer Behavior Theories

Big Skill: Pattern Recognition in Problems – Identifying shopping habits and common impulsive purchase triggers.

Small Skill: Experimenting with Ideas – Testing different behavioral psychology techniques to resist unnecessary spending.

Teaching Technique: Real-world applications.

Actions: Have students track their spending habits and experiment with strategies to make better financial decisions.

Example:

1. Analyze personal purchase patterns (frequency, impulse buys).
2. Evaluate strategies like delayed gratification and budgeting apps.
3. Compare spending habits before and after interventions.
4. Identify the most effective method for reducing unnecessary spending.

Pairing 9: Identifying Air Quality Patterns and Testing Purification Solutions

Big Skill: Pattern Recognition in Problems – Identifying pollution trends and air quality fluctuations.

Small Skill: Experimenting with Ideas – Testing different air purification methods.

Teaching Technique: Cultural exploration.

Actions: Have students monitor air quality inside and outside school buildings and test various air purification techniques.

Example:

1. Measure air quality using sensors or DIY indicators.
2. Evaluate air purification techniques (indoor plants, air filters, ventilation adjustments).
3. Track changes in air quality levels over time.
4. Recommend the best purification method for different environments.

Pairing 10: Identifying Patterns in Energy Consumption and Testing Energy-Saving Strategies

Big Skill: Pattern Recognition in Problems – Recognizing trends in electricity usage in classrooms and homes.

Small Skill: Experimenting with Ideas – Testing different energy-efficient strategies to reduce consumption.

Teaching Technique: Hands-on experiments.

Actions: Have students monitor energy use and experiment with conservation strategies.

Example:

1. Track daily energy usage in different locations.
2. Implement changes (LED lighting, unplugging devices, adjusting thermostats).
3. Measure energy consumption before and after interventions.
4. Propose best practices for reducing energy waste.

Creative Problem Solving – (Big Skill) Pattern Recognition in Problems vs. (Small Skill) Refining Solutions

---

Pairing 1: Identifying Student Time Management Issues and Refining Study Schedules

Big Skill: Pattern Recognition in Problems – Identifying patterns in student procrastination and time mismanagement.

Small Skill: Refining Solutions – Adjusting study schedules based on effectiveness and feedback.

Teaching Technique: Hands-on experiments.

Actions: Have students track their study habits, identify ineffective time blocks, and refine their schedules.

Example:

1. Log study times and distractions for a week.
2. Identify peak productivity periods and ineffective study habits.
3. Create an optimized study schedule and implement it.
4. Adjust based on personal performance and feedback.

Pairing 2: Recognizing Patterns in School Cafeteria Waste and Refining Sustainability Programs

Big Skill: Pattern Recognition in Problems – Identifying which types of food are most commonly wasted.

Small Skill: Refining Solutions – Adjusting school lunch menus and waste reduction strategies based on student feedback.

Teaching Technique: Guided exploration.

Actions: Have students track food waste data and refine a school-wide initiative to reduce waste.

Example:

1. Record leftover food items daily.
2. Survey students about meal preferences.
3. Implement a revised meal plan with portion control adjustments.
4. Measure food waste reduction and refine the initiative.

Pairing 3: Detecting Traffic Flow Issues and Refining School Drop-Off/Pick-Up Plans

Big Skill: Pattern Recognition in Problems – Analyzing congestion patterns at school entrances and exits.

Small Skill: Refining Solutions – Modifying traffic routes based on trial results and community input.

Teaching Technique: Problem-solving exercises.

Actions: Have students collect traffic flow data and propose new pickup/drop-off strategies, refining based on trials.

Example:

1. Track peak congestion times and bottlenecks.
2. Propose new routes or staggered pickup schedules.
3. Implement trial solutions for a week and gather driver feedback.
4. Adjust final recommendations based on data.

Pairing 4: Identifying Cybersecurity Threats and Refining Security Protocols

Big Skill: Pattern Recognition in Problems – Recognizing common online threats faced by students.

Small Skill: Refining Solutions – Iterating on security guidelines based on test case results.

Teaching Technique: Observation journals.

Actions: Have students analyze phishing attempts and refine security awareness strategies.

Example:

1. Collect examples of phishing emails and social media scams.
2. Evaluate school-wide awareness by sending mock phishing attempts.
3. Evaluate how many students fall for the traps and revise training materials.
4. Implement a final cybersecurity campaign to improve safety.

Pairing 5: Recognizing Water Waste Patterns and Refining Conservation Initiatives

Big Skill: Pattern Recognition in Problems – Identifying where water waste is most prevalent in the school.

Small Skill: Refining Solutions – Adjusting water conservation efforts based on effectiveness.

Teaching Technique: Hands-on experiments.

Actions: Have students measure water usage, implement conservation strategies, and refine them based on results.

Example:

1. Identify high-consumption areas (bathrooms, kitchens, gym showers).
2. Implement conservation strategies (automatic faucets, awareness posters).
3. Measure water savings after a month.
4. Modify strategies based on data analysis.

Pairing 6: Recognizing Patterns in Student Stress and Refining Wellness Programs

Big Skill: Pattern Recognition in Problems – Identifying major stress triggers in students.

Small Skill: Refining Solutions – Modifying stress-reduction programs based on student feedback.

Teaching Technique: Sensory exploration.

Actions: Have students survey stress levels and refine wellness activities to be more effective.

Example:

1. Identify peak stress periods (exams, assignment deadlines).
2. Implement different stress-relief techniques (yoga, peer support groups).
3. Survey students about effectiveness.
4. Adjust programs to better serve student needs.

Pairing 7: Detecting Patterns in Online Misinformation and Refining Digital Literacy Strategies

Big Skill: Pattern Recognition in Problems – Identifying trends in the spread of fake news and misinformation.

Small Skill: Refining Solutions – Adjusting educational content to improve media literacy.

Teaching Technique: Collaborative group work.

Actions: Have students analyze misinformation sources and refine critical thinking training programs.

Example:

1. Collect misleading news articles and analyze their spread.
2. Evaluate students' ability to detect misinformation before and after training.
3. Gather feedback on which fact-checking methods were most effective.
4. Adjust digital literacy materials accordingly.

Pairing 8: Identifying Patterns in Energy Consumption and Refining Efficiency Plans

Big Skill: Pattern Recognition in Problems – Recognizing peak electricity usage trends at school.

Small Skill: Refining Solutions – Testing energy-saving techniques and improving based on feedback.

Teaching Technique: Real-world applications.

Actions: Have students track energy use and refine conservation initiatives for maximum impact.

Example:

1. Monitor electricity use in different school departments.
2. Implement energy-saving techniques (smart lighting, timers).
3. Compare power consumption before and after changes.
4. Adjust strategies to maximize savings.

Pairing 9: Recognizing Patterns in Student Participation and Refining Extracurricular Activities

Big Skill: Pattern Recognition in Problems – Identifying which extracurricular activities attract the most students.

Small Skill: Refining Solutions – Adjusting club structures based on participation and feedback.

Teaching Technique: Cultural exploration.

Actions: Have students analyze trends in club engagement and refine student activity options.

Example:

1. Track attendance rates across various extracurriculars.
2. Survey students about interests and barriers to participation.
3. Modify activity schedules and promotional strategies.
4. Implement changes and measure increases in participation.

Pairing 10: Recognizing Patterns in Air Pollution and Refining School Greening Efforts

Big Skill: Pattern Recognition in Problems – Identifying how pollution levels fluctuate throughout the day.

Small Skill: Refining Solutions – Testing pollution-reduction methods and improving based on effectiveness.

Teaching Technique: Hands-on experiments.

Actions: Have students monitor air quality levels and refine greening efforts like tree planting and ventilation improvements.

Example:

1. Measure air pollution levels at different times and locations.
2. Implement solutions (plant trees, increase natural airflow).
3. Track changes in air quality over a semester.
4. Adjust efforts based on which strategies were most effective.

Example Teaching Strategies for Big vs. Small Thinking Skill Pairings of Adaptability at the High School Level

Here are 90 unique teaching strategy examples for the big thinking and small thinking skill pairings, each described in detail to align with **Big Thinking Skills** (Concepts & Strategies) and **Small Thinking Skills** (Step-by-Step Execution) for the **Adaptability Skill** taught at the high school level.

Adaptability

Big Thinking Skills (Concepts & Strategies)

- **Flexibility in Thinking:** Recognizing when a change in approach is necessary.
- **Resilience & Growth Mindset:** Learning from failures rather than seeing them as roadblocks.
- **Abstract & Concrete Thinking Balance:** Switching between big-picture strategies and fine details.

Small Thinking Skills (Step-by-Step Execution)

- **Revising & Iterating Plans:** Adjusting strategies based on new information or setbacks.
- **Scenario Planning:** Practicing "what if" exercises to anticipate changes.
- **Stress Testing Decisions:** Evaluating how robust a plan is by considering worst-case scenarios.

Adaptability – (Big Skill) Flexibility in Thinking vs. (Small Skill) Revising & Iterating Plans

---

Pairing 1: Adjusting Study Techniques for a Challenging Subject

Big Skill: Flexibility in Thinking – Recognizing when a study method is not working and being open to alternative strategies.

Small Skill: Revising & Iterating Plans – Testing and refining different learning methods based on results.

Teaching Technique: Hands-on experiments.

Actions: Have students track their study habits, identify ineffective methods, and adjust based on test performance.

Example:

1. Try passive studying (reading notes, highlighting).
2. Take a quiz to measure retention.
3. Switch to active recall and teaching others.
4. Analyze which method works best and refine accordingly.

Pairing 2: Adapting a Business Plan in an Entrepreneurship Class

Big Skill: Flexibility in Thinking – Recognizing when a business idea or strategy needs modification based on market conditions.

Small Skill: Revising & Iterating Plans – Adjusting the business model based on feedback and performance.

Teaching Technique: Guided exploration.

Actions: Have students evaluate a business concept, gather customer feedback, and revise their approach.

Example:

1. Develop a product or service idea.
2. Conduct market testing with peers.
3. Modify pricing, branding, or delivery based on customer input.
4. Relaunch the revised business model.

Pairing 3: Modifying a Fitness Plan for Better Results

Big Skill: Flexibility in Thinking – Understanding when a workout routine is ineffective or causing injury.

Small Skill: Revising & Iterating Plans – Making adjustments to exercise frequency, intensity, or technique.

Teaching Technique: Problem-solving exercises.

Actions: Have students track workout progress, adjust based on results, and refine for optimal performance.

Example:

1. Follow a 4-week workout routine.
2. Measure strength, endurance, or weight change.
3. Identify areas where progress is slow.
4. Adjust exercises or intensity based on results.

Pairing 4: Revising a Debate Strategy Based on Opponent Arguments

Big Skill: Flexibility in Thinking – Recognizing when an argument needs to shift due to new counterpoints.

Small Skill: Revising & Iterating Plans – Adjusting the debate strategy in real time based on opponent responses.

Teaching Technique: Observation journals.

Actions: Have students engage in live debate practice, identify weaknesses, and refine arguments dynamically.

Example:

1. Present an argument in a debate.
2. Note weaknesses based on opponent rebuttals.
3. Modify response tactics for the next round.
4. Refine and implement improved strategy in future debates.

Pairing 5: Adjusting a Science Experiment Based on Unexpected Results

Big Skill: Flexibility in Thinking – Recognizing when a hypothesis is incorrect and adjusting the approach.

Small Skill: Revising & Iterating Plans – Modifying experiment variables based on findings.

Teaching Technique: Hands-on experiments.

Actions: Have students conduct an experiment, analyze results, and refine methodology if needed.

Example:

1. Evaluate a chemical reaction hypothesis.
2. Observe unexpected reaction rates.
3. Adjust temperature, concentration, or method.
4. Compare refined experiment results to the original.

Pairing 6: Changing a Coding Approach for a More Efficient Algorithm

Big Skill: Flexibility in Thinking – Recognizing inefficiencies in a program and trying new approaches.

Small Skill: Revising & Iterating Plans – Debugging and refining the code based on errors and inefficiencies.

Teaching Technique: Sensory exploration.

Actions: Have students develop a program, identify performance bottlenecks, and revise code for efficiency.

Example:

1. Write an algorithm to sort data.
2. Measure execution speed and efficiency.
3. Identify areas where it can be optimized.
4. Implement a refined, faster algorithm.

Pairing 7: Adapting a Speech to Engage a Specific Audience

Big Skill: Flexibility in Thinking – Recognizing when a speech is not resonating with an audience.

Small Skill: Revising & Iterating Plans – Adjusting speech delivery and tone based on audience reaction.

Teaching Technique: Collaborative group work.

Actions: Have students deliver speeches to different audiences and refine content and tone based on feedback.

Example:

1. Present a speech to a classroom audience.
2. Observe audience engagement (facial expressions, responses).
3. Adjust language, examples, or tone.
4. Deliver the refined speech and compare audience reactions.

Pairing 8: Revising a Novel or Essay Based on Peer Feedback

Big Skill: Flexibility in Thinking – Understanding that the first draft is not final and embracing constructive criticism.

Small Skill: Revising & Iterating Plans – Adjusting writing structure, arguments, or themes based on feedback.

Teaching Technique: Real-world applications.

Actions: Have students write an essay, receive peer critiques, and refine content accordingly.

Example:

1. Write a first draft of a persuasive essay.
2. Receive peer feedback on structure and clarity.
3. Revise argument flow and word choice.
4. Submit the final refined version.

Pairing 9: Adapting a Leadership Style in Group Projects

Big Skill: Flexibility in Thinking – Recognizing when a leadership approach is ineffective.

Small Skill: Revising & Iterating Plans – Adjusting leadership strategies based on team dynamics and feedback.

Teaching Technique: Cultural exploration.

Actions: Have students lead group activities, assess leadership effectiveness, and adjust strategies.

Example:

1. Lead a team on a project.
2. Observe communication challenges or disengagement.
3. Switch leadership approach (delegation, collaboration, directive).
4. Measure team productivity before and after changes.

Pairing 10: Modifying a Marketing Strategy for a School Event

Big Skill: Flexibility in Thinking – Recognizing when an event promotion strategy is not attracting enough participation.

Small Skill: Revising & Iterating Plans – Changing marketing approaches based on feedback and engagement levels.

Teaching Technique: Hands-on experiments.

Actions: Have students track event attendance based on different promotional methods and refine strategy accordingly.

Example:

1. Promote an event using posters.
2. Measure sign-ups and engagement.
3. Shift to social media and personal outreach.
4. Compare attendance before and after strategy change.

Adaptability – (Big Skill) Flexibility in Thinking vs. (Small Skill) Scenario Planning

---

Pairing 1: Preparing for Unforeseen Exam Challenges

Big Skill: Flexibility in Thinking – Recognizing when a study approach is not working and adapting to a new method.

Small Skill: Scenario Planning – Practicing "what if" exercises for potential exam challenges.

Teaching Technique: Hands-on experiments.

Actions: Have students prepare for different unexpected exam conditions to build adaptability in test-taking.

Example:

1. Predict possible exam difficulties (tricky questions, time pressure).
2. Develop backup study strategies (quick review techniques, stress management).
3. Simulate different exam conditions (timed, open-ended questions, distractions).
4. Adapt and refine study methods based on experience.

Pairing 2: Adapting a Business Strategy to Economic Changes

Big Skill: Flexibility in Thinking – Recognizing when business models need to shift due to external factors.

Small Skill: Scenario Planning – Running "what if" scenarios to prepare for market changes.

Teaching Technique: Guided exploration.

Actions: Have students create a business plan and evaluate it under different simulated economic conditions.

Example:

1. Develop a startup idea with a pricing model.
2. Evaluate "what if" situations like inflation, supply shortages, or competition.
3. Adjust strategies for each scenario (price adjustments, new revenue streams).
4. Identify the most resilient business plan.

Pairing 3: Adjusting to Unexpected Changes in a Science Experiment

Big Skill: Flexibility in Thinking – Recognizing when an experimental result requires a new approach.

Small Skill: Scenario Planning – Considering different possible outcomes before conducting the experiment.

Teaching Technique: Problem-solving exercises.

Actions: Have students predict multiple experiment results and plan alternative approaches.

Example:

1. Set up a chemistry experiment with multiple possible reactions.
2. Predict "what if" scenarios (chemical imbalance, wrong reaction).
3. Prepare adjustment strategies before beginning the experiment.
4. Implement changes and compare actual results to predictions.

Pairing 4: Preparing for Technological Failures During Presentations

Big Skill: Flexibility in Thinking – Recognizing the need for alternative presentation methods.

Small Skill: Scenario Planning – Practicing backup strategies for tech-related issues.

Teaching Technique: Observation journals.

Actions: Have students develop and test contingency plans for delivering presentations under different conditions.

Example:

1. Prepare a digital slideshow for a school project.
2. Run scenarios where the projector, internet, or laptop fails.
3. Evaluate alternative approaches (paper handouts, verbal delivery, whiteboard use).
4. Evaluate effectiveness of backup strategies.

Pairing 5: Adjusting to Unexpected Opponent Strategies in Debate

Big Skill: Flexibility in Thinking – Modifying arguments in real-time based on opposing viewpoints.

Small Skill: Scenario Planning – Anticipating counterarguments before the debate begins.

Teaching Technique: Hands-on experiments.

Actions: Have students prepare multiple debate strategies and evaluate them against surprise arguments.

Example:

1. Develop a strong argument for a debate topic.
2. Predict "what if" counterpoints from the opposing team.
3. Prepare alternative arguments and rebuttals for each scenario.
4. Implement flexibility in live debate sessions.

Pairing 6: Navigating Career Uncertainty in a Changing Job Market

Big Skill: Flexibility in Thinking – Being open to alternative career paths and adapting to industry trends.

Small Skill: Scenario Planning – Considering different future career shifts and preparing for them.

Teaching Technique: Sensory exploration.

Actions: Have students map out different career paths based on possible industry changes.

Example:

1. Choose an initial career interest.
2. Identify "what if" scenarios (automation replacing jobs, economic downturns).
3. Plan alternative career routes (upskilling, shifting industries).
4. Develop a flexible long-term career roadmap.

Pairing 7: Adjusting Budget Plans Based on Unforeseen Expenses

Big Skill: Flexibility in Thinking – Recognizing the need for financial adaptability.

Small Skill: Scenario Planning – Testing different budget adjustments for unexpected expenses.

Teaching Technique: Collaborative group work.

Actions: Have students simulate a personal or business budget and adjust for financial surprises.

Example:

1. Create a monthly budget with planned expenses.
2. Introduce "what if" events (unexpected car repair, medical bill).
3. Develop strategies to reallocate funds without going into debt.
4. Compare financial stability across different scenarios.

Pairing 8: Adjusting to Rapidly Changing Weather Conditions in Outdoor Activities

Big Skill: Flexibility in Thinking – Understanding when weather conditions require a shift in plans.

Small Skill: Scenario Planning – Preparing alternative strategies for different weather situations.

Teaching Technique: Real-world applications.

Actions: Have students plan an outdoor event or sports competition and prepare for weather disruptions.

Example:

1. Plan a school sports event with a timeline.
2. Predict weather-related "what if" scenarios (rain, extreme heat, wind).
3. Develop contingency plans (indoor alternatives, schedule shifts).
4. Implement the best backup strategy when needed.

Pairing 9: Managing Unexpected Changes in Group Project Team Members

Big Skill: Flexibility in Thinking – Recognizing when a team structure needs to change due to member availability.

Small Skill: Scenario Planning – Preparing for scenarios like losing a group member unexpectedly.

Teaching Technique: Cultural exploration.

Actions: Have students simulate team projects where sudden changes require adjustments in roles.

Example:

1. Form groups for a class project with assigned roles.
2. Introduce a "what if" scenario (a key member leaves).
3. Develop strategies to redistribute tasks and keep the project on track.
4. Implement the best approach and reflect on the process.

Pairing 10: Adapting Marketing Strategies Based on Changing Consumer Preferences

Big Skill: Flexibility in Thinking – Understanding that consumer behavior shifts, and marketing strategies must evolve.

Small Skill: Scenario Planning – Testing different advertising messages for different audience reactions.

Teaching Technique: Hands-on experiments.

Actions: Have students develop marketing campaigns and evaluate them under different consumer preference scenarios.

Example:

1. Create a marketing campaign for a fictional product.
2. Introduce "what if" scenarios (consumer backlash, trend shifts).
3. Adjust messaging and branding to align with changing demands.
4. Compare engagement metrics before and after adjustments.

Adaptability – (Big Skill) Flexibility in Thinking vs. (Small Skill) Stress Testing Decisions

---

Pairing 1: Evaluating College Admissions Plans Under Rejection Scenarios

Big Skill: Flexibility in Thinking – Recognizing when to adjust college application strategies based on results.

Small Skill: Stress Testing Decisions – Analyzing worst-case scenarios in the college admissions process.

Teaching Technique: Hands-on experiments.

Actions: Have students map out alternative college or career paths if their top-choice schools reject them.

Example:

1. Create a primary college application list.
2. Simulate a scenario where top choices reject the application.

3. Develop contingency plans (alternative schools, gap year, community college pathway).
4. Identify the most adaptable plan.

Pairing 2: Assessing Business Startup Risks Under Financial Failure Scenarios

Big Skill: Flexibility in Thinking – Recognizing the need to shift business strategies if revenue falls short.

Small Skill: Stress Testing Decisions – Evaluating how financial losses could impact a startup's survival.

Teaching Technique: Guided exploration.

Actions: Have students simulate financial failure scenarios and modify their business strategy accordingly.

Example:

1. Develop a startup budget with projected revenue and expenses.
2. Introduce stress tests (unexpected competition, economic downturns).
3. Adjust pricing, cost structure, or product offerings to remain viable.
4. Compare outcomes of different crisis-management strategies.

Pairing 3: Planning for Leadership Crises in a School Organization

Big Skill: Flexibility in Thinking – Understanding how to pivot when a team leader is absent.

Small Skill: Stress Testing Decisions – Simulating leadership failures and developing alternative action plans.

Teaching Technique: Problem-solving exercises.

Actions: Have students design a leadership transition plan in case a key team member drops out.

Example:

1. Assign leadership roles in a school club or project.
2. Simulate the sudden loss of the president or key member.
3. Evaluate different ways to delegate responsibilities.
4. Implement and refine the most effective transition strategy.

Pairing 4: Testing Emergency Plans for Natural Disasters at School

Big Skill: Flexibility in Thinking – Adjusting safety procedures for unexpected emergencies.

Small Skill: Stress Testing Decisions – Running drills under different disaster conditions.

Teaching Technique: Observation journals.

Actions: Have students evaluate school emergency procedures and evaluate their effectiveness under multiple disaster scenarios.

Example:

1. Review the school's current emergency response plan.
2. Simulate worst-case scenarios (earthquake, fire, active shooter).
3. Identify weaknesses and propose modifications.
4. Implement and test revised procedures in drills.

Pairing 5: Evaluating Personal Finance Plans for Worst-Case Scenarios

Big Skill: Flexibility in Thinking – Adapting financial strategies when faced with unexpected expenses.

Small Skill: Stress Testing Decisions – Running simulations of financial setbacks.

Teaching Technique: Hands-on experiments.

Actions: Have students create a monthly budget and test how well it holds up under sudden financial stress.

Example:

1. Develop a budget based on expected income and expenses.
2. Introduce financial setbacks (job loss, medical emergency).
3. Adjust the budget to maintain stability.
4. Identify the most resilient financial plan.

Pairing 6: Evaluating Travel Plans for Worst-Case Scenarios

Big Skill: Flexibility in Thinking – Adjusting itineraries based on unexpected travel disruptions.

Small Skill: Stress Testing Decisions – Running simulations of travel delays, lost luggage, or cancellations.

Teaching Technique: Sensory exploration.

Actions: Have students plan a trip and test how their itinerary adapts to different disruptions.

Example:

1. Develop a travel itinerary with flight and lodging details.
2. Simulate setbacks (missed flight, hotel overbooking, lost passport).
3. Adjust plans to accommodate each scenario.
4. Determine the most adaptable itinerary.

Pairing 7: Assessing Election Campaign Strategies for Negative Publicity Scenarios

Big Skill: Flexibility in Thinking – Modifying campaign messaging based on unforeseen public reactions.

Small Skill: Stress Testing Decisions – Simulating crisis communication strategies for political campaigns.

Teaching Technique: Collaborative group work.

Actions: Have students develop a school election campaign and stress test it against negative media coverage.

Example:

1. Create a political campaign with slogans and policies.
2. Simulate "what if" scenarios (false rumors, bad publicity).
3. Develop damage control messaging and strategy shifts.
4. Compare audience reactions before and after adjustments.

Pairing 8: Adapting Sports Team Strategies for Unexpected Game Situations

Big Skill: Flexibility in Thinking – Adjusting sports strategies when a team's usual approach is not working.

Small Skill: Stress Testing Decisions – Running worst-case scenario drills during practice.

Teaching Technique: Real-world applications.

Actions: Have students simulate game situations where key players are injured, or strategies fail.

Example:

1. Design a sports game plan for a school competition.
2. Introduce challenges (injured player, referee bias, opponent strategy change).
3. Adjust gameplay strategy on the spot.
4. Identify the best adaptive approach.

Pairing 9: Evaluating Disaster Preparedness in a Business Setting

Big Skill: Flexibility in Thinking – Adjusting business operations in the face of supply chain disruptions.

Small Skill: Stress Testing Decisions – Simulating different business crises and developing responses.

Teaching Technique: Cultural exploration.

Actions: Have students design a business continuity plan and evaluate its resilience against external disruptions.

Example:

1. Choose a business type (e.g., restaurant, retail, tech company).
2. Introduce stress tests (supply shortages, PR scandals, cyberattacks).
3. Develop contingency plans for each scenario.
4. Compare business survival rates with and without adaptability.

Pairing 10: Preparing for Unexpected Setbacks in a Research Project

Big Skill: Flexibility in Thinking – Recognizing when research findings require a change in direction.

Small Skill: Stress Testing Decisions – Evaluating potential flaws or obstacles in a research study.

Teaching Technique: Hands-on experiments.

Actions: Have students design a research project and test how it holds up under different complications.

Example:

1. Develop a hypothesis and data collection method.
2. Simulate setbacks (equipment failure, lack of data, conflicting results).
3. Adjust research methods to overcome obstacles.
4. Refine and implement the most effective research adjustments.

Adaptability – (Big Skill) Resilience & Growth Mindset vs. (Small Skill) Revising & Iterating Plans

---

Pairing 1: Improving Exam Performance After a Poor Grade

Big Skill: Resilience & Growth Mindset – Viewing a bad grade as a learning opportunity rather than a failure.

Small Skill: Revising & Iterating Plans – Adjusting study techniques based on exam mistakes.

Teaching Technique: Hands-on experiments.

Actions: Have students analyze past exam mistakes, adjust study habits, and retest their knowledge.

Example:

1. Review incorrect answers on an exam.
2. Identify study weaknesses (e.g., poor time management, weak topic understanding).
3. Develop a revised study plan (e.g., active recall, group study).
4. Retake practice tests to measure improvement.

Pairing 2: Recovering from a Failed Business Pitch in an Entrepreneurship Class

Big Skill: Resilience & Growth Mindset – Using rejection as feedback rather than a reason to quit.

Small Skill: Revising & Iterating Plans – Adjusting a business plan based on investor or teacher feedback.

Teaching Technique: Guided exploration.

Actions: Have students revise their pitch based on feedback and present an improved version.

Example:

1. Deliver an initial business pitch to a panel.
2. Gather feedback on weak areas (e.g., unclear value proposition).
3. Refine the business plan and adjust the pitch.
4. Present an improved version and measure investor interest.

Pairing 3: Overcoming Challenges in a Science Experiment

Big Skill: Resilience & Growth Mindset – Recognizing that failed experiments are part of scientific discovery.

Small Skill: Revising & Iterating Plans – Adjusting experimental procedures based on unexpected results.

Teaching Technique: Problem-solving exercises.

Actions: Have students analyze failed experiments and refine their approach.

Example:

1. Conduct a chemistry experiment with an expected reaction.
2. Observe an unexpected result.
3. Adjust variables (e.g., temperature, concentration).
4. Conduct the experiment again and document improvements.

Pairing 4: Improving Athletic Performance After a Loss

Big Skill: Resilience & Growth Mindset – Viewing a sports loss as motivation to improve skills.

Small Skill: Revising & Iterating Plans – Adjusting training techniques based on past performance.

Teaching Technique: Observation journals.

Actions: Have students analyze game footage, identify weaknesses, and refine training.

Example:

1. Watch a recorded sports match to identify mistakes.
2. Adjust training drills to improve weak areas.
3. Implement the revised training plan in practice.
4. Compete again and compare performance.

Pairing 5: Revising a Failed Creative Writing Assignment

Big Skill: Resilience & Growth Mindset – Using writing feedback as a tool for growth rather than discouragement.

Small Skill: Revising & Iterating Plans – Adjusting writing style based on constructive criticism.

Teaching Technique: Hands-on experiments.

Actions: Have students rewrite a creative piece incorporating teacher or peer feedback.

Example:

1. Receive a low grade on an essay due to weak argumentation.
2. Analyze feedback and rewrite sections for clarity and depth.
3. Submit the revised version for reassessment.
4. Compare the original and improved grades.

Pairing 6: Learning from Rejection in College Applications

Big Skill: Resilience & Growth Mindset – Seeing college rejections as opportunities to grow.

Small Skill: Revising & Iterating Plans – Strengthening applications based on rejection feedback.

Teaching Technique: Sensory exploration.

Actions: Have students revise their college applications based on areas for improvement.

Example:

1. Receive a rejection letter from a preferred college.
2. Identify application weaknesses (e.g., essays, test scores, extracurriculars).
3. Strengthen the application and apply to other schools.
4. Track acceptances and compare the revised approach's success.

Pairing 7: Recovering from a Public Speaking Mishap

Big Skill: Resilience & Growth Mindset – Overcoming embarrassment from a poor speech performance.

Small Skill: Revising & Iterating Plans – Practicing and improving delivery based on past mistakes.

Teaching Technique: Collaborative group work.

Actions: Have students analyze past speech mistakes and refine delivery for future presentations.

Example:

1. Deliver a speech with nervousness and stuttering.
2. Review video footage or receive peer feedback.
3. Implement improvements (e.g., pacing, voice modulation, posture).
4. Deliver an improved speech with greater confidence.

Pairing 8: Adjusting a Marketing Plan After Low Consumer Engagement

Big Skill: Resilience & Growth Mindset – Recognizing marketing failures as a step toward success.

Small Skill: Revising & Iterating Plans – Tweaking advertising strategies based on data.

Teaching Technique: Real-world applications.

Actions: Have students analyze the performance of a school marketing campaign and refine it.

Example:

1. Launch a social media campaign with limited engagement.
2. Identify weak points (e.g., poor visuals, unclear messaging).
3. Adjust content strategy (e.g., better images, revised captions).
4. Relaunch and compare engagement data.

Pairing 9: Overcoming Programming Errors in a Coding Project

Big Skill: Resilience & Growth Mindset – Understanding that debugging is an essential part of coding.

Small Skill: Revising & Iterating Plans – Fixing and optimizing code based on test results.

Teaching Technique: Cultural exploration.

Actions: Have students debug and improve a malfunctioning program.

Example:

1. Write a program with logical errors.
2. Identify and analyze the bugs using testing tools.
3. Revise code structure and rerun tests.
4. Compare original and optimized performance.

Pairing 10: Adjusting a Time Management Plan After Procrastination

Big Skill: Resilience & Growth Mindset – Accepting time management struggles as part of the learning process.

Small Skill: Revising & Iterating Plans – Refining productivity techniques based on setbacks.

Teaching Technique: Hands-on experiments.

Actions: Have students track their productivity and refine their time management strategies.

Example:

1. Attempt to follow a time management plan but fail to complete tasks on time.
2. Identify distraction sources and inefficient habits.
3. Implement new techniques (e.g., Pomodoro method, daily planning).
4. Measure productivity before and after adjustments.

Adaptability – (Big Skill) Resilience & Growth Mindset vs. (Small Skill) Scenario Planning

---

Pairing 1: Preparing for Exam Failure and Creating Alternative Study Plans

Big Skill: Resilience & Growth Mindset – Viewing a low test score as an opportunity for learning rather than failure.

Small Skill: Scenario Planning – Practicing "what if" exercises to anticipate different test outcomes.

Teaching Technique: Hands-on experiments.

Actions: Have students plan for different possible exam results and develop adaptable study strategies.

Example:

1. Predict multiple test score scenarios (low, average, high).
2. Develop alternative study plans for different outcomes.
3. If the test score is low, adjust studying techniques for improvement.
4. If the test score is high, analyze what worked and refine methods.

Pairing 2: Responding to College Rejections with Alternative Plans

Big Skill: Resilience & Growth Mindset – Accepting rejection letters as a redirection rather than a failure.

Small Skill: Scenario Planning – Considering different career and education paths in case of rejections.

Teaching Technique: Guided exploration.

Actions: Have students prepare for college rejections and create backup career or education plans.

Example:

1. Apply to multiple colleges with a ranked preference list.
2. Develop "what if" plans (community college, trade school, gap year, alternative universities).
3. Implement alternative plans if initial college applications are rejected.
4. Track long-term outcomes of alternate educational paths.

Pairing 3: Adapting to Job or Internship Rejections

Big Skill: Resilience & Growth Mindset – Learning from job or internship rejection experiences.

Small Skill: Scenario Planning – Preparing different job-seeking strategies based on potential outcomes.

Teaching Technique: Problem-solving exercises.

Actions: Have students develop multiple job search strategies in case of rejection.

Example:

1. Apply to several internships/jobs.
2. Create contingency plans (applying to more companies, networking, skills training).
3. If rejected, analyze feedback and adjust resume/interview skills.
4. Implement an improved strategy and measure success.

Pairing 4: Anticipating and Overcoming Setbacks in Group Projects

Big Skill: Resilience & Growth Mindset – Viewing group conflicts and setbacks as learning experiences.

Small Skill: Scenario Planning – Identifying potential teamwork challenges and preparing solutions.

Teaching Technique: Observation journals.

Actions: Have students role-play group project challenges and develop strategies to overcome them.

Example:

1. Assign roles in a group project.
2. Plan for setbacks (missed deadlines, conflicts, lack of participation).
3. Develop solutions (redistributing work, conflict resolution methods).
4. Implement strategies and evaluate project success.

Pairing 5: Preparing for Public Speaking Mishaps

Big Skill: Resilience & Growth Mindset – Handling nervousness or mistakes during a speech without panic.

Small Skill: Scenario Planning – Practicing ways to recover from common speaking errors.

Teaching Technique: Hands-on experiments.

Actions: Have students prepare for different speech failures and develop adaptive responses.

Example:

1. Prepare a speech on a chosen topic.
2. Simulate potential issues (forgetting lines, losing audience attention, tech failure).
3. Practice alternative recovery strategies (pause, humor, summarization).
4. Deliver an adapted speech based on real-time challenges.

Pairing 6: Adjusting to Unexpected Sports Competition Challenges

Big Skill: Resilience & Growth Mindset – Handling losses or unexpected challenges in sports.

Small Skill: Scenario Planning – Preparing for different game situations and opponent strategies.

Teaching Technique: Sensory exploration.

Actions: Have students simulate different sports scenarios and develop strategic adaptations.

Example:

1. Identify common challenges in their sport (injuries, losing streaks, rule changes).
2. Develop alternative training methods and gameplay strategies.
3. Simulate worst-case scenarios in practice sessions.
4. Apply learned adaptability techniques in real competitions.

Pairing 7: Preparing for Unexpected Expenses in Personal Finance

Big Skill: Resilience & Growth Mindset – Understanding that financial setbacks are learning opportunities.

Small Skill: Scenario Planning – Practicing budgeting strategies for unexpected financial situations.

Teaching Technique: Collaborative group work.

Actions: Have students create a personal budget and adjust for unexpected costs.

Example:

1. Create a monthly budget plan.
2. Simulate emergency expenses (car repair, medical bill).
3. Adjust spending and savings to accommodate new financial needs.
4. Compare the financial outcome before and after adjustments.

Pairing 8: Responding to Environmental or Natural Disaster Changes

Big Skill: Resilience & Growth Mindset – Adapting to environmental challenges rather than panicking.

Small Skill: Scenario Planning – Preparing responses to different types of natural disasters.

Teaching Technique: Real-world applications.

Actions: Have students develop emergency response plans for various disaster scenarios.

Example:

1. Research local environmental risks (hurricanes, wildfires, floods).
2. Create an emergency plan with different levels of severity.
3. Simulate a disaster drill and evaluate the plan.
4. Revise the plan based on weaknesses found in the drill.

Pairing 9: Anticipating Ethical Dilemmas in Decision-Making

Big Skill: Resilience & Growth Mindset – Navigating moral and ethical dilemmas with a problem-solving approach.

Small Skill: Scenario Planning – Practicing ethical decision-making with "what if" scenarios.

Teaching Technique: Cultural exploration.

Actions: Have students role-play ethical dilemmas and develop logical responses.

Example:

1. Identify common ethical dilemmas (cheating, lying, peer pressure).
2. Develop responses based on different perspectives.
3. Evaluate different outcomes by discussing real-life implications.
4. Reflect on which ethical approach was most effective.

Pairing 10: Adjusting Career Goals Based on Industry Changes

Big Skill: Resilience & Growth Mindset – Adapting career aspirations based on industry shifts.

Small Skill: Scenario Planning – Exploring different career paths in case of job market changes.

Teaching Technique: Hands-on experiments.

Actions: Have students research future job market trends and create flexible career paths.

Example:

1. Choose a career interest.
2. Research how automation, technology, or economic shifts might impact that field.
3. Develop alternative career routes (additional skills, certifications).
4. Adjust the career plan based on new industry trends.

Adaptability – (Big Skill) Resilience & Growth Mindset vs. (Small Skill) Stress Testing Decisions

---

Pairing 1: Overcoming Academic Failure with a Stronger Study Plan

Big Skill: Resilience & Growth Mindset – Viewing poor academic performance as a chance to improve rather than a failure.

Small Skill: Stress Testing Decisions – Evaluating study plans by testing how they hold up under worst-case scenarios.

Teaching Technique: Hands-on experiments.

Actions: Have students evaluate different study methods under stressful conditions to find the most effective approach.

Example:

1. Create a study plan with set schedules and methods.
2. Introduce stress tests (shorter prep time, distractions, different test formats).
3. Measure retention and adaptability of study techniques.
4. Revise and strengthen the plan for real-world exams.

Pairing 2: Handling Business Setbacks in an Entrepreneurship Project

Big Skill: Resilience & Growth Mindset – Learning from business losses instead of abandoning the idea.

Small Skill: Stress Testing Decisions – Simulating financial crises and testing business model durability.

Teaching Technique: Guided exploration.

Actions: Have students develop business models and evaluate them against worst-case economic conditions.

Example:

1. Create a business idea with pricing and revenue projections.
2. Simulate setbacks (supply chain issues, market crashes).
3. Adjust business strategy (cost-cutting, alternative revenue streams).
4. Evaluate the revised plan's success under multiple scenarios.

Pairing 3: Adapting Leadership Approaches in a Group Project

Big Skill: Resilience & Growth Mindset – Staying motivated after leadership failures rather than stepping down.

Small Skill: Stress Testing Decisions – Testing different leadership styles under high-pressure team scenarios.

Teaching Technique: Problem-solving exercises.

Actions: Have students simulate leadership crises and evaluate different management approaches.

Example:

1. Assign a student leader for a group project.
2. Introduce worst-case scenarios (team conflicts, missed deadlines).
3. Evaluate different leadership strategies (delegation, assertiveness, conflict resolution).
4. Analyze which approach worked best under pressure.

Pairing 4: Preparing for Athletic Performance Under Unexpected Challenges

Big Skill: Resilience & Growth Mindset – Viewing athletic setbacks as motivation to improve.

Small Skill: Stress Testing Decisions – Simulating high-pressure conditions to build adaptability in performance.

Teaching Technique: Observation journals.

Actions: Have students train under stress conditions to see how they respond to unexpected game scenarios.

Example:

1. Develop a training routine for a sports competition.
2. Simulate challenges (injuries, extreme weather, referee bias).
3. Evaluate performance under different levels of stress.
4. Adapt training methods to manage unpredictable factors.

Pairing 5: Managing Public Speaking Setbacks and Improving Delivery

Big Skill: Resilience & Growth Mindset – Learning from speaking errors instead of avoiding public speaking.

Small Skill: Stress Testing Decisions – Practicing speeches under worst-case scenarios.

Teaching Technique: Hands-on experiments.

Actions: Have students deliver speeches under different high-pressure conditions to develop adaptability.

Example:

1. Prepare a speech on a chosen topic.
2. Simulate challenges (audience distraction, forgotten lines, time cuts).
3. Adjust speaking strategies to maintain composure.
4. Re-test and refine delivery for resilience.

Pairing 6: Building a Financial Safety Net for Unexpected Expenses

Big Skill: Resilience & Growth Mindset – Viewing financial difficulties as learning experiences.

Small Skill: Stress Testing Decisions – Simulating emergency expenses to evaluate budgeting effectiveness.

Teaching Technique: Sensory exploration.

Actions: Have students create a budget and test how well it holds up under financial emergencies.

Example:

1. Develop a monthly budget with savings goals.
2. Introduce unexpected expenses (car repair, medical bill).
3. Assess whether the budget can absorb the financial shock.
4. Adjust financial planning strategies accordingly.

Pairing 7: Adapting to Changes in College or Career Plans

Big Skill: Resilience & Growth Mindset – Remaining open to new career paths if original plans fail.

Small Skill: Stress Testing Decisions – Running "what if" scenarios for career obstacles.

Teaching Technique: Collaborative group work.

Actions: Have students map out career plans and stress-test them under different future conditions.

Example:

1. Choose a desired career path.
2. Simulate obstacles (job automation, industry decline).
3. Develop alternative career pathways (upskilling, side careers).
4. Adjust plans based on evolving job market trends.

Pairing 8: Strengthening Disaster Preparedness Plans for a School or Community

Big Skill: Resilience & Growth Mindset – Viewing crisis preparedness as a skill that improves over time.

Small Skill: Stress Testing Decisions – Evaluating emergency response effectiveness under extreme conditions.

Teaching Technique: Real-world applications.

Actions: Have students design emergency response plans and stress-test them under real-life conditions.

Example:

1. Develop an emergency response plan (fire drill, lockdown procedures).
2. Simulate different disaster intensities (earthquake, school shooter scenario).
3. Evaluate weaknesses in the plan and revise accordingly.
4. Implement an updated plan for increased resilience.

Pairing 9: Handling Failures in Coding and Software Development

Big Skill: Resilience & Growth Mindset – Seeing programming errors as learning opportunities.

Small Skill: Stress Testing Decisions – Running stress tests to identify weaknesses in software programs.

Teaching Technique: Cultural exploration.

Actions: Have students debug software and run stress tests under extreme user conditions.

Example:

1. Develop a simple software application.
2. Simulate high user loads, incorrect inputs, and cyberattacks.
3. Identify and fix system vulnerabilities.
4. Improve coding strategies based on test results.

Pairing 10: Overcoming Failures in Event Planning and Execution

Big Skill: Resilience & Growth Mindset – Learning from event mishaps rather than seeing them as complete failures.

Small Skill: Stress Testing Decisions – Testing the event plan against worst-case logistical failures.

Teaching Technique: Hands-on experiments.

Actions: Have students plan a school event and simulate failures to strengthen contingency planning.

Example:

1. Develop a timeline and checklist for an event (fundraiser, concert, school fair).
2. Introduce worst-case scenarios (canceled venue, low turnout, missing resources).
3. Evaluate alternative solutions to keep the event successful.
4. Implement final strategies based on test results.

Adaptability – (Big Skill) Abstract & Concrete Thinking Balance vs. (Small Skill) Revising & Iterating Plans

---

Pairing 1: Improving a Science Fair Project Based on Trial Runs

Big Skill: Abstract & Concrete Thinking Balance – Understanding the broad scientific concept while refining experiment details.

Small Skill: Revising & Iterating Plans – Adjusting the experimental procedure based on trial results.

Teaching Technique: Hands-on experiments.

Actions: Have students develop a hypothesis, evaluate it, analyze results, and adjust their methodology.

Example:

1. Conduct an experiment on plant growth with a broad goal of sustainable farming.
2. Analyze initial results and identify issues (water inconsistency, incorrect lighting).
3. Revise the experimental method to control variables more precisely.
4. Re-run the experiment with improved methodology.

Pairing 2: Refining a Debate Strategy After a Practice Round

Big Skill: Abstract & Concrete Thinking Balance – Seeing the overarching debate theme while fine-tuning argument structure.

Small Skill: Revising & Iterating Plans – Modifying points and counterarguments based on feedback.

Teaching Technique: Guided exploration.

Actions: Have students engage in a mock debate, assess weak points, and refine their arguments.

Example:

1. Prepare an argument on climate policy using general principles.
2. Engage in a practice debate and identify weak rebuttals.
3. Adjust supporting evidence and refine specific argument details.
4. Reattempt the debate with a stronger, more nuanced approach.

Pairing 3: Adjusting a Business Plan After Market Research

Big Skill: Abstract & Concrete Thinking Balance – Understanding broad industry trends while refining product details.

Small Skill: Revising & Iterating Plans – Adjusting business strategies based on customer feedback.

Teaching Technique: Problem-solving exercises.

Actions: Have students develop a startup idea, evaluate customer responses, and refine their business model.

Example:

1. Develop an online tutoring service with a general mission of academic support.
2. Conduct a survey on student preferences and identify demand gaps.
3. Modify the service structure based on market data (offer group discounts, change pricing).
4. Relaunch with an improved business plan.

Pairing 4: Improving a Novel Draft Based on Peer Reviews

Big Skill: Abstract & Concrete Thinking Balance – Maintaining the overall narrative theme while refining specific details.

Small Skill: Revising & Iterating Plans – Making changes based on constructive feedback.

Teaching Technique: Observation journals.

Actions: Have students write, review, and edit a short story or novel based on feedback.

Example:

1. Write a dystopian story with a broad theme of societal collapse.
2. Receive peer feedback on plot holes and unclear character motivations.
3. Adjust storyline elements while preserving the central theme.
4. Publish a revised and more cohesive version of the story.

Pairing 5: Enhancing a Science Presentation After a Trial Run

Big Skill: Abstract & Concrete Thinking Balance – Communicating complex scientific ideas while simplifying key takeaways.

Small Skill: Revising & Iterating Plans – Improving clarity and engagement based on audience feedback.

Teaching Technique: Hands-on experiments.

Actions: Have students present scientific research, gather audience reactions, and refine their delivery.

Example:

1. Prepare a physics presentation on black holes.
2. Conduct a test run in front of a small audience.
3. Adjust explanations and visuals based on confusion points.
4. Deliver the final improved version at a school science fair.

Pairing 6: Fine-Tuning a Marketing Campaign for a School Event

Big Skill: Abstract & Concrete Thinking Balance – Understanding overall branding strategy while adjusting specific messaging.

Small Skill: Revising & Iterating Plans – Modifying marketing techniques based on engagement data.

Teaching Technique: Sensory exploration.

Actions: Have students develop and test different marketing approaches for an event.

Example:

1. Create a social media campaign for a school fundraiser.
2. Track engagement metrics (likes, shares, event sign-ups).
3. Adjust posting frequency, content format, and messaging for higher reach.
4. Relaunch with a refined strategy that maximizes engagement.

Pairing 7: Enhancing a Coding Project Based on User Testing

Big Skill: Abstract & Concrete Thinking Balance – Structuring an overall software design while fixing detailed coding errors.

Small Skill: Revising & Iterating Plans – Debugging and refining a program based on user feedback.

Teaching Technique: Collaborative group work.

Actions: Have students develop a basic application, evaluate it with users, and refine its usability.

Example:

1. Create a budgeting app with an intuitive interface.
2. Gather user feedback on functionality and ease of use.
3. Fix navigation issues and improve design elements.
4. Relaunch with an improved user experience.

Pairing 8: Adapting a Sports Playbook After a Scrimmage Match

Big Skill: Abstract & Concrete Thinking Balance – Developing an overarching game strategy while adjusting minor details.

Small Skill: Revising & Iterating Plans – Changing tactics based on opponent reactions.

Teaching Technique: Real-world applications.

Actions: Have students analyze game footage, refine strategies, and implement adjustments in practice.

Example:

1. Develop a soccer game strategy focusing on possession play.
2. Evaluate it in a scrimmage match and analyze defensive weaknesses.
3. Adjust player positioning and passing techniques.
4. Implement thc refined strategy in the next match.

Pairing 9: Revising a Budgeting Plan Based on Economic Changes

Big Skill: Abstract & Concrete Thinking Balance – Maintaining long-term financial goals while adjusting short-term expenses.

Small Skill: Revising & Iterating Plans – Adjusting budget plans based on income fluctuations.

Teaching Technique: Cultural exploration.

Actions: Have students develop a financial plan and adjust it for unexpected income changes.

Example:

1. Create a savings plan for a big purchase.
2. Simulate job loss or unexpected expenses.
3. Adjust savings contributions and discretionary spending.
4. Rebalance the budget to meet financial goals.

Pairing 10: Refining an Architectural Design After Structural Testing

Big Skill: Abstract & Concrete Thinking Balance – Combining artistic vision with practical structural constraints.

Small Skill: Revising & Iterating Plans – Modifying designs based on stability tests.

Teaching Technique: Hands-on experiments.

Actions: Have students build architectural models, evaluate their durability, and refine the design.

Example:

1. Construct a model bridge using various materials.
2. Apply stress tests to simulate real-world load capacity.
3. Identify weak points and reinforce structural supports.
4. Implement a revised, more durable design.

Adaptability – (Big Skill) Abstract & Concrete Thinking Balance vs. (Small Skill) Scenario Planning

---

Pairing 1: Preparing for Different Outcomes in College Applications

Big Skill: Abstract & Concrete Thinking Balance – Seeing the long-term impact of a college choice while managing application details.

Small Skill: Scenario Planning – Practicing "what if" exercises for different acceptance scenarios.

Teaching Technique: Hands-on experiments.

Actions: Have students map out different acceptance scenarios and prepare contingency plans.

Example:

1. Apply to a mix of reach, match, and safety schools.
2. Develop "what if" plans (accepted to top choice, waitlisted, rejected).
3. Plan alternative pathways (gap year, community college, job experience).
4. Adapt the final decision based on actual results.

Pairing 2: Adapting Business Strategies in an Entrepreneurship Class

Big Skill: Abstract & Concrete Thinking Balance – Seeing market trends while refining pricing, branding, and logistics.

Small Skill: Scenario Planning – Considering how market changes might affect business success.

Teaching Technique: Guided exploration.

Actions: Have students evaluate different business strategies against hypothetical market shifts.

Example:

1. Develop a business model for a school startup.
2. Simulate "what if" events (economic downturn, new competition).
3. Adjust pricing, advertising, or product features.
4. Select and implement the most adaptable strategy.

Pairing 3: Creating a Flexible Study Plan for Exam Preparation

Big Skill: Abstract & Concrete Thinking Balance – Balancing long-term knowledge retention with short-term memorization.

Small Skill: Scenario Planning – Practicing strategies for different test conditions.

Teaching Technique: Problem-solving exercises.

Actions: Have students evaluate study plans under different simulated conditions.

Example:

1. Develop a study schedule for an upcoming exam.
2. Simulate obstacles (unexpected quiz, reduced study time, mental fatigue).
3. Adjust strategies (focus on key concepts, use active recall, practice under pressure).
4. Identify the most effective approach.

Pairing 4: Adjusting Debate Strategies Based on Opponent Tactics

Big Skill: Abstract & Concrete Thinking Balance – Understanding broad debate themes while refining specific arguments.

Small Skill: Scenario Planning – Anticipating different counterarguments.

Teaching Technique: Observation journals.

Actions: Have students prepare multiple argument strategies and evaluate them against different counterpoints.

Example:

1. Research a controversial topic and develop key arguments.
2. Predict opponent responses and prepare counterarguments.
3. Evaluate different debating styles (aggressive, logical, emotional appeal).
4. Implement the best strategy in the actual debate.

Pairing 5: Planning for Career Path Changes Due to Industry Trends

Big Skill: Abstract & Concrete Thinking Balance – Maintaining a broad career vision while adapting to industry shifts.

Small Skill: Scenario Planning – Considering "what if" scenarios like automation or economic downturns.

Teaching Technique: Hands-on experiments.

Actions: Have students map out career goals and develop alternative routes.

Example:

1. Choose a desired career path.
2. Identify potential disruptions (AI automation, recession, industry decline).
3. Develop alternative career plans (additional certifications, networking).
4. Adjust the plan based on real-world trends.

Pairing 6: Managing School Club Events Under Unpredictable Conditions

Big Skill: Abstract & Concrete Thinking Balance – Organizing large-scale events while focusing on logistics.

Small Skill: Scenario Planning – Preparing for event challenges like low attendance or weather issues.

Teaching Technique: Sensory exploration.

Actions: Have students design an event and prepare for unexpected setbacks.

Example:

1. Plan a fundraiser with marketing, ticket sales, and venue setup.
2. Simulate issues (bad weather, low turnout, technical difficulties).
3. Develop solutions (backup venue, better advertising, live streaming).
4. Implement the revised plan for the actual event.

Pairing 7: Managing Financial Setbacks in Personal Budgeting

Big Skill: Abstract & Concrete Thinking Balance – Maintaining long-term financial goals while managing short-term expenses.

Small Skill: Scenario Planning – Stress-testing budgets with "what if" exercises.

Teaching Technique: Collaborative group work.

Actions: Have students develop a personal budget and evaluate it against financial surprises.

Example:

1. Create a monthly budget with income and expenses.

2. Introduce unexpected costs (car repair, lost job).
3. Adjust spending and saving strategies accordingly.
4. Evaluate the most resilient budgeting method.

Pairing 8: Designing Flexible Travel Itineraries

Big Skill: Abstract & Concrete Thinking Balance – Planning long-term travel goals while adjusting for unexpected events.

Small Skill: Scenario Planning – Preparing for delays, cancellations, and emergencies.

Teaching Technique: Real-world applications.

Actions: Have students create travel plans with backup options.

Example:

1. Plan a trip with destinations, flights, and accommodations.
2. Simulate disruptions (missed flights, bad weather, lost luggage).
3. Develop alternative routes and booking strategies.
4. Adapt travel plans to ensure the trip stays on track.

Pairing 9: Adjusting Scientific Research Plans Based on Experiment Failures

Big Skill: Abstract & Concrete Thinking Balance – Understanding the broader scientific question while refining specific experiments.

Small Skill: Scenario Planning – Preparing for different experimental outcomes.

Teaching Technique: Cultural exploration.

Actions: Have students design experiments and predict alternative results.

Example:

1. Develop a hypothesis and experimental procedure.
2. Consider "what if" results (inconsistent data, no reaction, unexpected variables).
3. Adjust methods (different chemical concentrations, repeated trials).
4. Conduct the improved experiment and analyze findings.

Pairing 10: Preparing for Setbacks in a Creative Project

Big Skill: Abstract & Concrete Thinking Balance – Maintaining an artistic vision while fixing technical issues.

Small Skill: Scenario Planning – Testing different creative approaches when original ideas do not work.

Teaching Technique: Hands-on experiments.

Actions: Have students design and stress-test creative projects under different conditions.

Example:

1. Create a short film or design project with an initial concept.
2. Simulate potential problems (lack of materials, time constraints, equipment failure).
3. Develop alternative approaches (different art mediums, editing techniques).
4. Complete the project using the best backup strategy.

Adaptability – (Big Skill) Abstract & Concrete Thinking Balance vs. (Small Skill) Stress Testing Decisions

---

Pairing 1: Evaluating the Stability of a College Plan

Big Skill: Abstract & Concrete Thinking Balance – Seeing the broad impact of college choices while managing application details.

Small Skill: Stress Testing Decisions – Testing the robustness of the college plan under various challenges.

Teaching Technique: Hands-on experiments.

Actions: Have students identify possible college-related obstacles and develop alternative strategies.

Example:

1. Create a college application plan, considering tuition, location, and career goals.
2. Simulate setbacks (financial aid rejection, housing issues, admission waitlist).
3. Develop alternative solutions (scholarships, community college, different schools).
4. Adapt the plan based on real-world outcomes.

Pairing 2: Evaluating a Business Model for Market Crashes

Big Skill: Abstract & Concrete Thinking Balance – Understanding economic trends while refining business details.

Small Skill: Stress Testing Decisions – Simulating different financial crises to evaluate business resilience.

Teaching Technique: Guided exploration.

Actions: Have students develop a business idea and analyze how it performs under economic downturns.

Example:

1. Develop a small business plan for a startup.
2. Introduce stress scenarios (inflation, competitor undercutting, supply chain disruption).

3. Adjust pricing, branding, and customer engagement strategies.
4. Evaluate resilience with a modified business model.

Pairing 3: Preparing for Emergency Situations in Event Planning

Big Skill: Abstract & Concrete Thinking Balance – Managing the big-picture vision of an event while managing logistics.

Small Skill: Stress Testing Decisions – Identifying weak points in event plans and simulating worst-case scenarios.

Teaching Technique: Problem-solving exercises.

Actions: Have students plan an event and test how well it holds up against unexpected disruptions.

Example:

1. Develop a school concert plan with logistics, marketing, and budget.
2. Simulate worst-case scenarios (weather issues, low ticket sales, equipment failure).
3. Modify backup plans to keep the event functional.
4. Execute the event with contingency strategies in place.

Pairing 4: Stress-Testing a Debate Strategy Against Aggressive Opponents

Big Skill: Abstract & Concrete Thinking Balance – Developing strong overarching themes while adjusting responses on the fly.

Small Skill: Stress Testing Decisions – Simulating high-pressure debate scenarios to assess argument adaptability.

Teaching Technique: Observation journals.

Actions: Have students prepare their stance on a topic and stress-test it with rapid opposition arguments.

Example:

1. Develop a structured argument for a debate competition.
2. Simulate opponent strategies that attack weak points.
3. Adjust counterarguments to maintain logical consistency.
4. Compete with an improved, stress-tested debate approach.

Pairing 5: Ensuring Financial Stability in Personal Budgeting

Big Skill: Abstract & Concrete Thinking Balance – Managing overall financial goals while adjusting to daily expenses.

Small Skill: Stress Testing Decisions – Testing the sustainability of a personal budget under financial stress.

Teaching Technique: Hands-on experiments.

Actions: Have students simulate financial challenges and adjust their budget accordingly.

Example:

1. Develop a monthly financial plan, including savings and expenses.
2. Introduce "what if" events (unexpected medical expenses, job loss).
3. Adjust spending priorities and emergency savings plans.
4. Identify financial habits that ensure long-term stability.

Pairing 6: Evaluating the Durability of a Science Experiment Design

Big Skill: Abstract & Concrete Thinking Balance – Balancing theoretical scientific concepts with real-world execution.

Small Skill: Stress Testing Decisions – Running experiments under extreme conditions to assess robustness.

Teaching Technique: Sensory exploration.

Actions: Have students evaluate and refine their experiment designs under varying conditions.

Example:

1. Design a chemistry experiment for a research fair.
2. Simulate extreme conditions (higher temperatures, different reactant ratios).
3. Modify the experiment based on observed inconsistencies.
4. Present a refined experiment with stronger reliability.

Pairing 7: Evaluating a Sports Team's Adaptability to Unexpected Setbacks

Big Skill: Abstract & Concrete Thinking Balance – Maintaining overall team strategies while adjusting for game situations.

Small Skill: Stress Testing Decisions – Running game scenarios where key players are absent.

Teaching Technique: Collaborative group work.

Actions: Have students practice different gameplay strategies under adverse conditions.

Example:

1. Develop a playbook strategy for a soccer team.
2. Simulate setbacks (injured key player, referee bias, bad weather).
3. Adjust team formation and tactics to maintain competitive edge.
4. Implement revised strategies in a live game.

Pairing 8: Evaluating the Strength of a Software Program Under Heavy Usage

Big Skill: Abstract & Concrete Thinking Balance – Managing a broad software vision while debugging small issues.

Small Skill: Stress Testing Decisions – Running performance tests under maximum load conditions.

Teaching Technique: Real-world applications.

Actions: Have students develop a program and evaluate its limits under stress conditions.

Example:

1. Code a web application for school use.
2. Simulate user overload (multiple logins, simultaneous data processing).
3. Identify failure points and optimize code.
4. Relaunch with improved performance and efficiency.

Pairing 9: Evaluating Crisis Responses in Leadership Roles

Big Skill: Abstract & Concrete Thinking Balance – Balancing leadership responsibilities while managing emergencies.

Small Skill: Stress Testing Decisions – Practicing leadership adaptability under high-pressure scenarios.

Teaching Technique: Cultural exploration.

Actions: Have students role-play leadership crisis situations and develop adaptive strategies.

Example:

1. Assume a leadership position in a class project or school club.
2. Introduce crisis scenarios (team conflict, lack of funding, sudden policy change).
3. Develop solutions that maintain team cohesion and success.
4. Implement lessons learned into real-world leadership roles.

Pairing 10: Strengthening an Environmental Policy Proposal for Sustainability

Big Skill: Abstract & Concrete Thinking Balance – Crafting sustainable environmental policies with practical implementation.

Small Skill: Stress Testing Decisions – Identifying potential failures in sustainability proposals.

Teaching Technique: Hands-on experiments.

Actions: Have students evaluate their environmental policy proposals under various stress conditions.

Example:

1. Design a school-wide sustainability initiative.
2. Introduce challenges (lack of participation, budget cuts, opposing viewpoints).
3. Modify the proposal to ensure feasibility and effectiveness.
4. Implement the strongest version of the plan in a school or community setting.

Example Teaching Strategies for Big vs. Small Thinking Skill Pairings of Systems Thinking Mastery at the College/Technical School Level

Here are 90 unique teaching strategy examples for the big thinking and small thinking skill pairings, each described in detail to align with **Big Thinking Skills** (Concepts & Strategies) and **Small Thinking Skills** (Step-by-Step Execution) for the **Systems Thinking Mastery Skill** taught at the college/technical school level.

Systems Thinking Mastery

Big Thinking Skills (Concepts & Strategies)

- **Understanding Interconnected Systems:** Recognizing how components interact in fields like economics, ecosystems, and organizational structures.
- **Identifying Systemic Patterns:** Recognizing feedback loops, delays, and cascading effects.
- **Holistic Problem-Solving:** Considering multiple perspectives when analyzing complex systems.

Small Thinking Skills (Step-by-Step Execution)

- **Mapping System Interactions:** Using flowcharts, causal loop diagrams, and network models to visualize systems.
- **Simulating System Changes:** Running models to predict the effects of small changes on an entire system.
- **Applying Systems Thinking Tools:** Using methodologies like stock-and-flow diagrams and leverage point identification.

Systems Thinking Mastery – (Big Skill) Understanding Interconnected Systems vs. (Small Skill) Mapping System Interactions

---

Pairing 1: Mapping Global Supply Chains and Their Disruptions

Big Skill: Understanding Interconnected Systems – Analyzing how global trade networks impact supply and demand.

Small Skill: Mapping System Interactions – Using flowcharts to visualize supply chain dependencies.

Teaching Technique: Hands-on experiments.

Actions: Have students trace how raw materials move through different countries before reaching consumers.

Example:

1. Choose a product (smartphones, coffee, clothing).
2. Map its supply chain from raw material sourcing to consumer sales.
3. Introduce disruptions (COVID-19, trade wars, natural disasters).
4. Identify vulnerabilities and propose solutions for resilience.

Pairing 2: Analyzing Climate Change Feedback Loops

Big Skill: Understanding Interconnected Systems – Exploring how environmental factors influence climate change.

Small Skill: Mapping System Interactions – Using causal loop diagrams to model feedback cycles.

Teaching Technique: Guided exploration.

Actions: Have students create diagrams showing how greenhouse gas emissions affect climate patterns.

Example:

1. Identify key components (deforestation, CO_2 emissions, ice melt, rising temperatures).
2. Map causal loops showing reinforcing cycles (e.g., melting ice reduces reflectivity, increasing warming).
3. Analyze intervention points for slowing climate change.
4. Compare natural vs. human-driven feedback loops.

Pairing 3: Visualizing Economic Inflation and Interest Rate Effects

Big Skill: Understanding Interconnected Systems – Examining how economic policies affect businesses and individuals.

Small Skill: Mapping System Interactions – Creating network models of financial decision-making.

Teaching Technique: Problem-solving exercises.

Actions: Have students trace the effects of interest rate changes on inflation, loans, and consumer spending.

Example:

1. Identify key economic factors (Federal Reserve rates, borrowing, business growth).
2. Map interactions (higher interest rates → reduced borrowing → slowed inflation).
3. Simulate policy changes and predict long-term outcomes.

4. Discuss how central banks use this knowledge to stabilize economies.

Pairing 4: Modeling Disease Spread in a Population

Big Skill: Understanding Interconnected Systems – Recognizing how individual actions influence public health.

Small Skill: Mapping System Interactions – Using network models to simulate disease transmission.

Teaching Technique: Observation journals.

Actions: Have students build models predicting how different policies affect disease spread.

Example:

1. Choose an infectious disease (flu, COVID-19, measles).
2. Map transmission pathways (close contact, airborne, vaccination rates).
3. Simulate different interventions (social distancing, vaccines, masks).
4. Evaluate the most effective strategies for controlling outbreaks.

Pairing 5: Mapping Social Media Influence on Public Opinion

Big Skill: Understanding Interconnected Systems – Investigating how digital platforms shape beliefs and behaviors.

Small Skill: Mapping System Interactions – Using network diagrams to track information flows.

Teaching Technique: Hands-on experiments.

Actions: Have students analyze how news spreads through social networks and impacts real-world decisions.

Example:

1. Select a recent news event or viral trend.
2. Map interactions (who shares, where misinformation spreads, what platforms dominate).
3. Evaluate how algorithms amplify certain narratives.
4. Suggest policies to balance freedom of speech and misinformation control.

Pairing 6: Understanding Urban Traffic Flow and Congestion

Big Skill: Understanding Interconnected Systems – Examining how infrastructure, population growth, and policies affect traffic.

Small Skill: Mapping System Interactions – Creating flowcharts showing road networks and congestion points.

Teaching Technique: Sensory exploration.

Actions: Have students simulate how changes in road systems impact traffic patterns.

Example:

1. Choose a busy urban area or highway system.
2. Map road networks, public transportation routes, and bottlenecks.
3. Introduce changes (new roads, tolls, public transit expansion).
4. Predict how these changes affect congestion and commuter behavior.

Pairing 7: Examining Power Structures in Government and Policy-Making

Big Skill: Understanding Interconnected Systems – Investigating how government branches and agencies interact.

Small Skill: Mapping System Interactions – Using flowcharts to model decision-making processes.

Teaching Technique: Collaborative group work.

Actions: Have students trace how policies move through legislative, executive, and judicial branches.

Example:

1. Select a policy (healthcare reform, environmental regulation).
2. Map how it moves from proposal to implementation.
3. Identify key influencers (lobbyists, media, voters).
4. Discuss where public engagement has the most impact.

Pairing 8: Understanding How AI Algorithms Shape Online Experiences

Big Skill: Understanding Interconnected Systems – Exploring how AI-driven recommendations influence behavior.

Small Skill: Mapping System Interactions – Using network models to show decision trees in AI algorithms.

Teaching Technique: Real-world applications.

Actions: Have students trace how recommendation systems (YouTube, TikTok, Spotify) personalize content.

Example:

1. Pick a platform and analyze how content is recommended.
2. Map how user data feeds into AI algorithms.
3. Evaluate potential biases and their social impact.
4. Suggest ethical improvements to algorithmic decision-making.

Pairing 9: Mapping Energy Flow in Ecosystems

Big Skill: Understanding Interconnected Systems – Studying how energy moves through food chains and biomes.

Small Skill: Mapping System Interactions – Using ecological flowcharts to represent trophic levels.

Teaching Technique: Cultural exploration.

Actions: Have students visualize how changes in one species population affect entire ecosystems.

Example:

1. Select an ecosystem (rainforest, ocean, grassland).
2. Map energy transfer from primary producers to apex predators.
3. Introduce disruptions (species extinction, habitat destruction).
4. Predict ecosystem-wide consequences.

Pairing 10: Understanding the Psychological and Social Impact of School Policies

Big Skill: Understanding Interconnected Systems – Evaluating how school rules impact student well-being and academic performance.

Small Skill: Mapping System Interactions – Using causal loop diagrams to visualize policy effects.

Teaching Technique: Hands-on experiments.

Actions: Have students analyze how different school policies (grading systems, disciplinary measures) influence student behavior.

Example:

1. Choose a policy (attendance requirements, dress codes, mental health programs).
2. Map interactions between students, teachers, and administration.
3. Introduce changes (more mental health resources, flexible grading).
4. Predict how these changes impact school culture and outcomes.

Systems Thinking Mastery – (Big Skill) Understanding Interconnected Systems vs. (Small Skill) Simulating System Changes

---

Pairing 1: Simulating the Effects of a Minimum Wage Increase on an Economy

Big Skill: Understanding Interconnected Systems – Examining how economic policies impact businesses, employment, and consumer spending.

Small Skill: Simulating System Changes – Running models to predict how wage adjustments affect the entire economy.

Teaching Technique: Hands-on experiments.

Actions: Have students develop a model to track the effects of raising the minimum wage on small businesses, inflation, and employment.

Example:

1. Create a model that represents businesses, workers, and consumers.
2. Increase the minimum wage and observe effects on employment and prices.
3. Compare different scenarios (gradual vs. sudden wage hikes).
4. Draw conclusions on the trade-offs of wage policies.

Pairing 2: Predicting the Impact of Removing a Predator from an Ecosystem

Big Skill: Understanding Interconnected Systems – Analyzing how species interact within an ecosystem.

Small Skill: Simulating System Changes – Modeling the consequences of removing a keystone species.

Teaching Technique: Guided exploration.

Actions: Have students simulate what happens to a food web if a key predator disappears.

Example:

1. Select an ecosystem (ocean, savannah, rainforest).
2. Remove a predator (wolves, sharks, lions) and observe effects on prey.
3. Track cascading changes in vegetation, smaller predators, and overall biodiversity.
4. Assess human intervention strategies to restore balance.

Pairing 3: Evaluating the Effect of Social Media Bans on Public Opinion

Big Skill: Understanding Interconnected Systems – Investigating how social networks shape public discourse.

Small Skill: Simulating System Changes – Running models on how limiting social media access affects information spread.

Teaching Technique: Problem-solving exercises.

Actions: Have students simulate how removing major social media platforms affects political discussions and misinformation.

Example:

1. Model how news spreads through different platforms.
2. Remove or limit access to one platform (e.g., Twitter, TikTok).

3. Observe shifts in information flow and alternative communication methods.
4. Analyze how this impacts political engagement and knowledge gaps.

Pairing 4: Modeling the Effects of Deforestation on Carbon Sequestration

Big Skill: Understanding Interconnected Systems – Examining how forests regulate atmospheric carbon.

Small Skill: Simulating System Changes – Running climate models to measure deforestation impact.

Teaching Technique: Observation journals.

Actions: Have students simulate different deforestation rates and measure their impact on global temperatures.

Example:

1. Create a carbon cycle model with forests, oceans, and atmosphere.
2. Remove sections of forest and track CO_2 levels over time.
3. Simulate tree regrowth and compare carbon absorption rates.
4. Predict long-term climate effects and propose sustainable solutions.

Pairing 5: Analyzing the Impact of Traffic Policy Changes on Congestion

Big Skill: Understanding Interconnected Systems – Studying how urban policies affect transportation flow.

Small Skill: Simulating System Changes – Testing different traffic management strategies.

Teaching Technique: Hands-on experiments.

Actions: Have students run a simulation where traffic laws, public transport, or road expansions change.

Example:

1. Model a city's road network.
2. Simulate changes like toll roads, new bus lanes, or speed limit adjustments.
3. Track congestion, commute times, and accident rates.
4. Identify the most effective transportation solutions.

Pairing 6: Evaluating the Consequences of a National Healthcare Policy Change

Big Skill: Understanding Interconnected Systems – Investigating how government policies affect public health.

Small Skill: Simulating System Changes – Running models to predict healthcare system reactions to reforms.

Teaching Technique: Sensory exploration.

Actions: Have students evaluate how different policy changes affect hospital capacity, insurance costs, and patient care.

Example:

1. Select a healthcare reform (universal care, privatization, increased funding).
2. Simulate effects on wait times, treatment quality, and costs.
3. Compare models in different socioeconomic groups.
4. Evaluate which policy leads to the best overall outcomes.

Pairing 7: Examining the Impact of Renewable Energy Adoption on an Electrical Grid

Big Skill: Understanding Interconnected Systems – Studying how energy sources interact in a national grid.

Small Skill: Simulating System Changes – Testing how increased wind and solar power impact grid stability.

Teaching Technique: Collaborative group work.

Actions: Have students simulate different energy mix scenarios to evaluate grid reliability and efficiency.

Example:

1. Model an energy grid powered by coal, natural gas, solar, and wind.
2. Increase renewable energy use and measure fluctuations in power supply.
3. Add backup storage (batteries) and analyze system stability.
4. Determine the best balance between reliability and sustainability.

Pairing 8: Predicting the Economic Impact of a Major Technology Breakthrough

Big Skill: Understanding Interconnected Systems – Analyzing how technological advancements affect industries.

Small Skill: Simulating System Changes – Running models to measure job losses, productivity shifts, and market adaptation.

Teaching Technique: Real-world applications.

Actions: Have students simulate how AI, automation, or biotech breakthroughs impact the economy.

Example:

1. Select a technology (self-driving cars, AI-driven healthcare, quantum computing).
2. Model workforce disruptions and new job creation.
3. Assess industry shifts (who benefits, who loses).
4. Propose policies to support displaced workers.

Pairing 9: Forecasting the Impact of a Currency Collapse on Global Trade

Big Skill: Understanding Interconnected Systems – Exploring how financial markets interact.

Small Skill: Simulating System Changes – Modeling how a major currency failure affects global markets.

Teaching Technique: Cultural exploration.

Actions: Have students simulate what happens when a major currency (USD, Euro, Yen) rapidly devalues.

Example:

1. Model global trade flows with a stable currency.
2. Introduce a sudden devaluation and track shifts in import/export costs.
3. Analyze inflation, stock markets, and government interventions.
4. Predict how businesses and nations would react.

Pairing 10: Measuring the Effects of a Mass Migration Event on a City

Big Skill: Understanding Interconnected Systems – Studying how population changes affect urban infrastructure.

Small Skill: Simulating System Changes – Testing how different cities respond to large refugee or migration influxes.

Teaching Technique: Hands-on experiments.

Actions: Have students simulate how an unexpected population increase impacts housing, jobs, and public services.

Example:

1. Model a city's population, housing, and job market.
2. Introduce an influx of 100,000 new residents.
3. Track changes in job availability, housing costs, and school enrollment.
4. Develop urban planning solutions to accommodate rapid growth.

Systems Thinking Mastery – (Big Skill) Understanding Interconnected Systems vs. (Small Skill) Applying Systems Thinking Tools

---

Pairing 1: Mapping Water Usage and Conservation Strategies

Big Skill: Understanding Interconnected Systems – Analyzing how water sources, consumption, and policies interact in an urban setting.

Small Skill: Applying Systems Thinking Tools – Using stock-and-flow diagrams to track water movement and leverage points for conservation.

Teaching Technique: Hands-on experiments.

Actions: Have students create a model of a city's water supply and explore different conservation methods.

Example:

1. Map water sources (reservoirs, aquifers, desalination plants).
2. Track stock-and-flow rates of consumption, waste, and replenishment.
3. Identify leverage points (reducing household waste, improving infrastructure).
4. Propose policies that maximize water conservation.

Pairing 2: Analyzing the Impact of Carbon Emissions on Climate Change

Big Skill: Understanding Interconnected Systems – Studying how different sectors contribute to climate change.

Small Skill: Applying Systems Thinking Tools – Using stock-and-flow models to analyze carbon accumulation in the atmosphere.

Teaching Technique: Guided exploration.

Actions: Have students develop models showing CO_2 emissions, absorption, and accumulation trends.

Example:

1. Identify key carbon sources (factories, transportation, agriculture).
2. Use stock-and-flow models to measure how carbon builds up.
3. Determine leverage points (switching to renewables, reforestation).
4. Assess the impact of policy changes on emission reduction.

Pairing 3: Understanding the Flow of Money in a Circular Economy

Big Skill: Understanding Interconnected Systems – Exploring how money circulates within local and global markets.

Small Skill: Applying Systems Thinking Tools – Using stock-and-flow diagrams to track economic transactions.

Teaching Technique: Problem-solving exercises.

Actions: Have students model how businesses, consumers, and governments exchange money.

Example:

1. Create a diagram with stocks (savings, investments, wages).
2. Track flows between businesses, banks, and households.

3. Identify leverage points (tax changes, stimulus checks).
4. Predict how policy changes impact economic stability.

Pairing 4: Evaluating the Effectiveness of Public Health Policies

Big Skill: Understanding Interconnected Systems – Studying how health systems respond to policy changes.

Small Skill: Applying Systems Thinking Tools – Using leverage point identification to improve public health outcomes.

Teaching Technique: Observation journals.

Actions: Have students evaluate vaccination programs and how small interventions affect disease control.

Example:

1. Identify key variables (vaccine distribution, public awareness, infection rates).
2. Use leverage points to model interventions (public campaigns, mandates).
3. Analyze which policies reduce infection rates fastest.
4. Recommend policies based on data-driven insights.

Pairing 5: Designing Sustainable Food Supply Chains

Big Skill: Understanding Interconnected Systems – Recognizing how food production, transportation, and consumption interact.

Small Skill: Applying Systems Thinking Tools – Using stock-and-flow models to optimize sustainability.

Teaching Technique: Hands-on experiments.

Actions: Have students map food production systems and find ways to reduce waste.

Example:

1. Track food from farms to consumers.
2. Identify where waste occurs (harvest loss, packaging, expiration).
3. Find leverage points (improving storage, reducing overproduction).
4. Propose solutions to make supply chains more sustainable.

Pairing 6: Mapping the Growth and Regulation of Social Media Platforms

Big Skill: Understanding Interconnected Systems – Examining how social networks evolve and affect users.

Small Skill: Applying Systems Thinking Tools – Using network analysis to find leverage points for regulation.

Teaching Technique: Sensory exploration.

Actions: Have students study social media platform growth and develop strategies for ethical regulation.

Example:

1. Map user engagement trends (likes, shares, ad revenue).
2. Identify stock (active users) and flow (content creation).
3. Determine leverage points (algorithm changes, misinformation control).
4. Suggest ethical improvements to social media policies.

Pairing 7: Evaluating Renewable Energy Transitions in Power Grids

Big Skill: Understanding Interconnected Systems – Exploring how different energy sources interact in national grids.

Small Skill: Applying Systems Thinking Tools – Using stock-and-flow models to simulate transitions.

Teaching Technique: Collaborative group work.

Actions: Have students analyze the challenges of integrating solar, wind, and hydroelectric power.

Example:

1. Model fossil fuel dependency vs. renewable adoption.
2. Track how energy supply and demand fluctuate.
3. Identify leverage points (battery storage, grid updates).
4. Develop policies for a smooth energy transition.

Pairing 8: Investigating Urban Transportation and Traffic Congestion

Big Skill: Understanding Interconnected Systems – Studying how city transportation networks influence efficiency and pollution.

Small Skill: Applying Systems Thinking Tools – Using causal loop diagrams to find congestion solutions.

Teaching Technique: Real-world applications.

Actions: Have students map traffic flows and test different policies for reducing congestion.

Example:

1. Identify stock (number of vehicles) and flow (commuters entering/leaving cities).
2. Introduce different policies (tolls, bike lanes, public transit expansion).
3. Use leverage points to simulate reductions in traffic.
4. Recommend the most effective strategies for urban planners.

Pairing 9: Analyzing Corporate Environmental Responsibility

Big Skill: Understanding Interconnected Systems – Examining how businesses impact global sustainability efforts.

Small Skill: Applying Systems Thinking Tools – Using leverage points to drive sustainable corporate decisions.

Teaching Technique: Cultural exploration.

Actions: Have students assess how companies can balance profit and environmental responsibility.

Example:

1. Track resource extraction, production, and waste output.
2. Identify leverage points (green supply chains, carbon offsets).
3. Simulate changes in sustainability policies.
4. Recommend corporate strategies for reducing ecological footprints.

Pairing 10: Studying How Education Policies Affect Student Success

Big Skill: Understanding Interconnected Systems – Investigating how curriculum, funding, and teacher quality impact learning.

Small Skill: Applying Systems Thinking Tools – Using stock-and-flow diagrams to measure education outcomes.

Teaching Technique: Hands-on experiments.

Actions: Have students model how different education investments affect student performance.

Example:

1. Map student progress based on school funding levels.
2. Introduce policy changes (teacher training, standardized testing).
3. Measure effects on graduation rates and knowledge retention.
4. Identify leverage points for improving educational access.

Systems Thinking Mastery – (Big Skill) Identifying Systemic Patterns vs. (Small Skill) Mapping System Interactions

---

Pairing 1: Understanding Economic Recessions and Recovery Cycles

Big Skill: Identifying Systemic Patterns – Recognizing feedback loops in economic growth, recessions, and inflation cycles.

Small Skill: Mapping System Interactions – Using causal loop diagrams to visualize how financial crises unfold.

Teaching Technique: Hands-on experiments.

Actions: Have students map the economic cycle, identifying key causes and effects.

Example:

1. Identify key factors: consumer spending, business investment, unemployment.
2. Create a causal loop diagram showing how recession spirals occur.
3. Introduce policy changes (stimulus, interest rate cuts) and observe effects.
4. Predict the long-term recovery timeline based on different interventions.

Pairing 2: Analyzing Climate Change Feedback Loops

Big Skill: Identifying Systemic Patterns – Recognizing reinforcing and balancing loops in global warming.

Small Skill: Mapping System Interactions – Using flowcharts to model the interactions between emissions, ice melt, and rising temperatures.

Teaching Technique: Guided exploration.

Actions: Have students visualize how small temperature increases create accelerating environmental effects.

Example:

1. Map out major feedback loops (melting ice → reduced reflectivity → more heat absorption).
2. Identify leverage points (reducing emissions, afforestation).
3. Simulate different intervention strategies.
4. Compare short-term vs. long-term environmental impacts.

Pairing 3: Tracking the Spread of Misinformation in Social Networks

Big Skill: Identifying Systemic Patterns – Recognizing cascading effects of misinformation online.

Small Skill: Mapping System Interactions – Using network models to track how false information spreads.

Teaching Technique: Problem-solving exercises.

Actions: Have students analyze real-world misinformation cases and model their spread.

Example:

1. Choose a viral misinformation case (e.g., vaccine myths, election fraud claims).
2. Map its spread across different platforms.

3. Identify amplification mechanisms (bot networks, influencers).
4. Develop strategies for slowing misinformation spread.

Pairing 4: Understanding Urban Traffic Congestion and Solutions

Big Skill: Identifying Systemic Patterns – Recognizing how traffic congestion builds through feedback loops.

Small Skill: Mapping System Interactions – Using causal loop diagrams to analyze road network strain.

Teaching Technique: Observation journals.

Actions: Have students track how small changes in traffic policy affect congestion.

Example:

1. Identify main congestion factors (population growth, limited public transit).
2. Map interactions between road capacity, rush hour patterns, and commute delays.
3. Simulate solutions (bike lanes, tolls, ride-sharing incentives).
4. Predict long-term congestion trends under different policies.

Pairing 5: Examining How Stock Markets React to Investor Behavior

Big Skill: Identifying Systemic Patterns – Recognizing reinforcing and balancing loops in stock market fluctuations.

Small Skill: Mapping System Interactions – Using network diagrams to model investor confidence cycles.

Teaching Technique: Hands-on experiments.

Actions: Have students analyze how investor panic and speculation create market booms and crashes.

Example:

1. Track how stock prices rise due to positive news cycles.
2. Identify feedback loops (rising prices → more investors → further rises).
3. Simulate negative scenarios (mass sell-offs, economic downturns).
4. Predict long-term stabilization mechanisms (government regulations, market corrections).

Pairing 6: Mapping Food Supply Chain Disruptions

Big Skill: Identifying Systemic Patterns – Recognizing how delays and shortages cascade through global supply chains.

Small Skill: Mapping System Interactions – Using flowcharts to trace food production, transportation, and delivery networks.

Teaching Technique: Sensory exploration.

Actions: Have students analyze how one disruption (e.g., drought, pandemic) affects the whole supply chain.

Example:

1. Choose a key food product (wheat, beef, coffee).
2. Track its journey from farms to consumers.
3. Introduce disruptions (drought, labor strikes, war) and observe cascading effects.
4. Identify leverage points for resilience (local sourcing, alternative transport).

Pairing 7: Evaluating Feedback Loops in Political Campaigns

Big Skill: Identifying Systemic Patterns – Understanding how political messaging reinforces voter perceptions.

Small Skill: Mapping System Interactions – Using causal loop diagrams to track campaign strategies.

Teaching Technique: Collaborative group work.

Actions: Have students study how campaign ads and debates shift public opinion.

Example:

1. Identify key campaign components (media coverage, debates, social media).
2. Map how negative ads affect voter engagement.
3. Simulate different strategies (positive messaging, grassroots movements).
4. Analyze election outcome predictions based on different campaign choices.

Pairing 8: Analyzing Crime Rates and Law Enforcement Policies

Big Skill: Identifying Systemic Patterns – Recognizing how crime rates fluctuate based on social conditions.

Small Skill: Mapping System Interactions – Using flowcharts to show the impact of policing strategies.

Teaching Technique: Real-world applications.

Actions: Have students model how increased policing, economic opportunities, and education affect crime rates.

Example:

1. Map out influences on crime (poverty, employment, school funding).
2. Introduce policy changes (community policing, youth programs).

3. Simulate short-term and long-term effects.
4. Evaluate which policies create the most sustainable reductions in crime.

Pairing 9: Modeling Disease Spread and Healthcare Responses

Big Skill: Identifying Systemic Patterns – Recognizing exponential growth patterns in disease outbreaks.

Small Skill: Mapping System Interactions – Using network models to track infection rates and intervention effects.

Teaching Technique: Cultural exploration.

Actions: Have students analyze the spread of infectious diseases using system mapping.

Example:

1. Choose a disease (COVID-19, flu, Ebola).
2. Map transmission pathways (airborne, contact, vaccination rates).
3. Introduce intervention strategies (lockdowns, social distancing).
4. Predict outcomes based on different levels of compliance.

Pairing 10: Understanding Renewable Energy Transition Challenges

Big Skill: Identifying Systemic Patterns – Recognizing delays and feedback loops in energy system transitions.

Small Skill: Mapping System Interactions – Using causal loop diagrams to track the shift from fossil fuels to renewables.

Teaching Technique: Hands-on experiments.

Actions: Have students analyze the challenges of switching energy grids to sustainable sources.

Example:

1. Map energy sources (coal, wind, solar, nuclear).
2. Identify barriers to transition (grid updates, storage limitations, costs).
3. Simulate different investment strategies and timelines.
4. Predict long-term sustainability impacts based on different policy approaches.

Systems Thinking Mastery – (Big Skill) Identifying Systemic Patterns vs. (Small Skill) Simulating System Changes

---

Pairing 1: Simulating Economic Inflation and Its Long-Term Effects

Big Skill: Identifying Systemic Patterns – Recognizing feedback loops between wages, prices, and purchasing power.

Small Skill: Simulating System Changes – Running models to evaluate how price increases impact an economy over time.

Teaching Technique: Hands-on experiments.

Actions: Have students adjust inflation rates in a simulation to observe how they affect businesses and consumers.

Example:

1. Model inflation at 2%, 5%, and 10% growth rates.
2. Observe effects on wages, consumer spending, and business pricing.
3. Introduce policy changes (interest rate hikes, wage increases).
4. Predict long-term economic stability under different scenarios.

Pairing 2: Predicting the Spread of Wildfires Based on Climate Variables

Big Skill: Identifying Systemic Patterns – Recognizing how drought, wind, and deforestation increase fire risk.

Small Skill: Simulating System Changes – Modeling how altering variables changes fire intensity and spread.

Teaching Technique: Guided exploration.

Actions: Have students create wildfire spread models to evaluate fire prevention strategies.

Example:

1. Set baseline conditions (temperature, wind speed, vegetation density).
2. Simulate a small fire ignition under different conditions.
3. Adjust firebreaks, controlled burns, and reforestation efforts.
4. Identify the most effective mitigation strategies.

Pairing 3: Modeling the Impact of Social Media Algorithms on Public Opinion

Big Skill: Identifying Systemic Patterns – Understanding how algorithms create echo chambers and misinformation loops.

Small Skill: Simulating System Changes – Running models that tweak algorithmic weightings.

Teaching Technique: Problem-solving exercises.

Actions: Have students modify engagement-based ranking algorithms to evaluate how they affect information spread.

Example:

1. Model an algorithm that prioritizes engagement vs. accuracy.
2. Measure changes in misinformation spread over time.
3. Adjust transparency settings (fact-checking labels, AI moderation).
4. Evaluate the balance between free speech and responsible content curation.

Pairing 4: Testing Renewable Energy Integration into Power Grids

Big Skill: Identifying Systemic Patterns – Recognizing delays and inefficiencies in switching to clean energy.

Small Skill: Simulating System Changes – Running energy grid models with different renewable energy sources.

Teaching Technique: Observation journals.

Actions: Have students simulate how different levels of wind and solar power affect energy reliability.

Example:

1. Start with an energy grid powered by fossil fuels.
2. Introduce 20%, 50%, and 80% renewable energy replacements.
3. Observe changes in energy reliability and battery storage needs.
4. Identify the best energy mix for sustainability and stability.

Pairing 5: Simulating Urban Development and Traffic Congestion

Big Skill: Identifying Systemic Patterns – Recognizing how city growth affects transportation efficiency.

Small Skill: Simulating System Changes – Modeling how changes in road networks and public transit impact congestion.

Teaching Technique: Hands-on experiments.

Actions: Have students modify city layouts in a simulation to evaluate traffic solutions.

Example:

1. Start with a baseline city traffic system.
2. Add population growth and new businesses.
3. Evaluate interventions like new highways, bike lanes, and subway expansions.
4. Determine which strategies most effectively reduce congestion.

Pairing 6: Modeling Disease Spread and Public Health Responses

Big Skill: Identifying Systemic Patterns – Recognizing exponential infection growth and healthcare strain.

Small Skill: Simulating System Changes – Testing the effects of different interventions on an outbreak.

Teaching Technique: Sensory exploration.

Actions: Have students simulate pandemic responses to compare different policies.

Example:

1. Set an initial infection rate and population density.
2. Evaluate interventions (mask mandates, social distancing, vaccines).
3. Measure reductions in infection spread for each scenario.
4. Identify which policies are most effective at slowing disease outbreaks.

Pairing 7: Simulating Agricultural Changes Due to Climate Shifts

Big Skill: Identifying Systemic Patterns – Recognizing how climate change disrupts food production.

Small Skill: Simulating System Changes – Running models to evaluate different farming adaptation strategies.

Teaching Technique: Collaborative group work.

Actions: Have students experiment with different crops, irrigation, and soil treatments under climate variability.

Example:

1. Model a farm under stable conditions.
2. Introduce temperature increases and rainfall variability.
3. Evaluate adaptation methods (drought-resistant crops, better irrigation).
4. Predict which strategies ensure the highest yield sustainability.

Pairing 8: Evaluating Economic Effects of Universal Basic Income (UBI)

Big Skill: Identifying Systemic Patterns – Recognizing how cash payments influence work incentives and economic cycles.

Small Skill: Simulating System Changes – Running models that introduce UBI and measure economic effects.

Teaching Technique: Real-world applications.

Actions: Have students simulate UBI implementation at different income levels.

Example:

1. Set baseline economic indicators (employment, consumer spending).
2. Introduce monthly UBI payments to all citizens.
3. Observe changes in work participation, savings, and investments.
4. Assess long-term sustainability of UBI programs.

Pairing 9: Predicting the Effects of Deforestation on Biodiversity

Big Skill: Identifying Systemic Patterns – Recognizing cascading ecological effects of habitat loss.

Small Skill: Simulating System Changes – Running ecosystem models to evaluate reforestation and conservation strategies.

Teaching Technique: Cultural exploration.

Actions: Have students compare biodiversity levels under different deforestation rates.

Example:

1. Model a rainforest ecosystem with existing species balance.
2. Remove sections of forest and track species population changes.
3. Evaluate reforestation efforts and their effects on ecosystem recovery.
4. Determine the most effective conservation methods.

Pairing 10: Simulating Economic Effects of AI-Driven Job Automation

Big Skill: Identifying Systemic Patterns – Recognizing delays and disruptions in job markets due to automation.

Small Skill: Simulating System Changes – Running models on how AI adoption affects employment trends.

Teaching Technique: Hands-on experiments.

Actions: Have students simulate different AI adoption rates in various industries.

Example:

1. Select an industry (manufacturing, finance, retail).
2. Introduce AI-driven automation at different speeds.
3. Observe changes in employment rates, wages, and business profits.
4. Identify policies (retraining programs, tax incentives) that mitigate disruption.

Systems Thinking Mastery – (Big Skill) Identifying Systemic Patterns vs. (Small Skill) Applying Systems Thinking Tools

---

Pairing 1: Modeling Economic Boom-and-Bust Cycles

Big Skill: Identifying Systemic Patterns – Recognizing feedback loops in economic expansions and recessions.

Small Skill: Applying Systems Thinking Tools – Using stock-and-flow diagrams to analyze investment and spending patterns.

Teaching Technique: Hands-on experiments.

Actions: Have students map out economic cycles and test different fiscal policies.

Example:

1. Identify factors (consumer confidence, bank lending, inflation).
2. Use a stock-and-flow diagram to track investment growth and contractions.
3. Introduce government interventions (stimulus packages, tax cuts).
4. Predict long-term stability under different economic strategies.

Pairing 2: Analyzing the Carbon Cycle and Climate Change

Big Skill: Identifying Systemic Patterns – Recognizing how carbon moves through natural and human systems.

Small Skill: Applying Systems Thinking Tools – Using stock-and-flow models to track CO_2 emissions and absorption.

Teaching Technique: Guided exploration.

Actions: Have students simulate carbon accumulation and identify leverage points for reduction.

Example:

1. Identify sources (fossil fuels, deforestation) and sinks (oceans, forests).
2. Model stock-and-flow dynamics of CO_2 in the atmosphere.
3. Introduce policy interventions (carbon taxes, reforestation).
4. Evaluate the effectiveness of different mitigation strategies.

Pairing 3: Examining Supply Chain Resilience in Global Trade

Big Skill: Identifying Systemic Patterns – Recognizing delays and bottlenecks in supply chains.

Small Skill: Applying Systems Thinking Tools – Using leverage point identification to improve logistics.

Teaching Technique: Problem-solving exercises.

Actions: Have students analyze disruptions (COVID-19, labor shortages) and propose solutions.

Example:

1. Track goods from production to consumer.
2. Identify points where delays cause cascading failures.
3. Model improvements (local sourcing, automation).
4. Predict how changes enhance supply chain resilience.

Pairing 4: Understanding Energy Grid Stability and Blackouts

Big Skill: Identifying Systemic Patterns – Recognizing feedback loops in energy demand and production.

Small Skill: Applying Systems Thinking Tools – Using stock-and-flow diagrams to track energy storage and consumption.

Teaching Technique: Observation journals.

Actions: Have students simulate how energy demand spikes lead to blackouts.

Example:

1. Identify power sources (coal, renewables, nuclear).
2. Track fluctuations in electricity demand.
3. Model leverage points (battery storage, grid updates).
4. Propose policies for a more resilient energy system.

Pairing 5: Predicting the Impact of Social Media on Political Polarization

Big Skill: Identifying Systemic Patterns – Recognizing self-reinforcing echo chambers.

Small Skill: Applying Systems Thinking Tools – Using network models to track information flow.

Teaching Technique: Hands-on experiments.

Actions: Have students map online conversations and identify misinformation feedback loops.

Example:

1. Identify key platforms and their algorithms.
2. Map stock (users) and flow (information shared).
3. Introduce leverage points (fact-checking, algorithm changes).
4. Predict changes in polarization with different interventions.

Pairing 6: Evaluating Healthcare System Capacity and Strain

Big Skill: Identifying Systemic Patterns – Recognizing delays in treatment availability and hospital capacity.

Small Skill: Applying Systems Thinking Tools – Using stock-and-flow diagrams to analyze patient inflow and outflow.

Teaching Technique: Sensory exploration.

Actions: Have students simulate hospital overcrowding and test policy interventions.

Example:

1. Identify key variables (hospital beds, staffing levels, patient admissions).
2. Map stock (available beds) and flow (patient turnover).
3. Introduce leverage points (preventive care, telemedicine).
4. Determine which policies reduce hospital strain.

Pairing 7: Investigating Biodiversity Loss and Ecosystem Stability

Big Skill: Identifying Systemic Patterns – Recognizing cascading effects of species extinction.

Small Skill: Applying Systems Thinking Tools – Using network models to visualize species interactions.

Teaching Technique: Collaborative group work.

Actions: Have students simulate the removal of keystone species from an ecosystem.

Example:

1. Identify trophic levels in an ecosystem.
2. Map how species depend on one another for survival.
3. Introduce human impact scenarios (deforestation, overfishing).
4. Predict biodiversity shifts under different conservation efforts.

Pairing 8: Forecasting Urban Population Growth and Infrastructure Demand

Big Skill: Identifying Systemic Patterns – Recognizing how city expansion impacts housing, transit, and public services.

Small Skill: Applying Systems Thinking Tools – Using stock-and-flow diagrams to track population and resource use.

Teaching Technique: Real-world applications.

Actions: Have students model how rapid urbanization stresses municipal systems.

Example:

1. Map urban growth over time.
2. Track stock (housing supply) and flow (migration rates).
3. Identify leverage points (zoning laws, public transport investment).
4. Predict outcomes of different city planning approaches.

Pairing 9: Analyzing the Effects of Interest Rate Changes on Consumer Behavior

Big Skill: Identifying Systemic Patterns – Recognizing how financial incentives shape spending habits.

Small Skill: Applying Systems Thinking Tools – Using stock-and-flow models to examine savings and borrowing trends.

Teaching Technique: Cultural exploration.

Actions: Have students simulate how interest rate adjustments affect mortgage markets and investments.

Example:

1. Model stock (savings) and flow (loan approvals).
2. Track consumer reactions to rate hikes vs. cuts.
3. Identify leverage points (government stimulus, tax policies).
4. Predict financial stability under different interest rate scenarios.

Pairing 10: Studying the Feedback Loops in Public Education Reform

Big Skill: Identifying Systemic Patterns – Recognizing how funding, student outcomes, and teacher retention interact.

Small Skill: Applying Systems Thinking Tools – Using leverage point analysis to improve school performance.

Teaching Technique: Hands-on experiments.

Actions: Have students simulate how different education policies impact graduation rates.

Example:

1. Identify key components (funding, curriculum, teacher salaries).
2. Model stock (student retention) and flow (graduation rates).
3. Introduce policy changes (smaller class sizes, standardized testing).
4. Predict long-term effects on educational achievement.

Systems Thinking Mastery – (Big Skill) Holistic Problem-Solving vs. (Small Skill) Mapping System Interactions

---

Pairing 1: Analyzing Global Water Scarcity and Management

Big Skill: Holistic Problem-Solving – Considering economic, environmental, and social factors in water conservation.

Small Skill: Mapping System Interactions – Using causal loop diagrams to track water distribution, consumption, and shortages.

Teaching Technique: Hands-on experiments.

Actions: Have students map out how various industries, governments, and consumers contribute to water scarcity.

Example:

1. Identify key players (agriculture, urban areas, policymakers).
2. Create a flowchart tracking water use and replenishment.
3. Introduce different conservation policies and predict their effects.
4. Assess long-term sustainability under different climate conditions.

Pairing 2: Examining the Interplay Between Climate Change and Migration

Big Skill: Holistic Problem-Solving – Analyzing the economic, social, and political impacts of climate-induced displacement.

Small Skill: Mapping System Interactions – Using network models to visualize migration flows and policy responses.

Teaching Technique: Guided exploration.

Actions: Have students model how environmental degradation leads to migration and geopolitical shifts.

Example:

1. Identify climate stressors (rising sea levels, drought, extreme weather).
2. Map migration pathways and policy responses.
3. Introduce different adaptation strategies (infrastructure investment, relocation plans).
4. Evaluate the feasibility of proposed solutions.

Pairing 3: Mapping the Effects of Urbanization on Biodiversity

Big Skill: Holistic Problem-Solving – Balancing human development with environmental conservation.

Small Skill: Mapping System Interactions – Using causal loop diagrams to analyze urban expansion effects.

Teaching Technique: Problem-solving exercises.

Actions: Have students study how urban growth impacts ecosystems and propose mitigation strategies.

Example:

1. Map key urbanization factors (deforestation, pollution, housing expansion).
2. Track how habitat destruction affects local species.
3. Simulate policy interventions (green spaces, wildlife corridors).
4. Predict biodiversity levels under different urban planning models.

Pairing 4: Understanding the Complexity of Food Supply Chains

Big Skill: Holistic Problem-Solving – Evaluating economic, ethical, and logistical challenges in food distribution.

Small Skill: Mapping System Interactions – Using flowcharts to track food production from farms to consumers.

Teaching Technique: Observation journals.

Actions: Have students analyze disruptions (pandemics, trade policies) and propose resilient systems.

Example:

1. Identify major food suppliers and distribution networks.
2. Map disruptions caused by labor shortages or climate change.
3. Introduce leverage points (local farming, improved logistics).
4. Compare the efficiency of different supply chain strategies.

Pairing 5: Investigating the Relationship Between Mental Health and Social Media

Big Skill: Holistic Problem-Solving – Balancing digital connectivity with psychological well-being.

Small Skill: Mapping System Interactions – Using network models to track engagement patterns and mental health impacts.

Teaching Technique: Hands-on experiments.

Actions: Have students examine how social media algorithms shape behavior and propose ethical guidelines.

Example:

1. Identify feedback loops (dopamine triggers, validation cycles).
2. Map the relationship between usage time and mental health indicators.
3. Introduce different platform policies (screen time limits, content moderation).
4. Evaluate the effects of these interventions.

Pairing 6: Evaluating the Intersection of Economic Policies and Income Inequality

Big Skill: Holistic Problem-Solving – Weighing fiscal strategies against social and economic consequences.

Small Skill: Mapping System Interactions – Using causal loop diagrams to analyze tax structures and wealth distribution.

Teaching Technique: Sensory exploration.

Actions: Have students simulate different tax policies and their effects on income disparity.

Example:

1. Identify economic forces affecting income distribution.
2. Map how tax breaks and social programs influence wealth gaps.
3. Introduce policy scenarios (universal basic income, progressive taxes).
4. Predict which strategies reduce inequality most effectively.

Pairing 7: Understanding the Impact of Public Transportation on Sustainability

Big Skill: Holistic Problem-Solving – Analyzing infrastructure, economic feasibility, and environmental benefits.

Small Skill: Mapping System Interactions – Using flowcharts to compare public vs. private transportation models.

Teaching Technique: Collaborative group work.

Actions: Have students map traffic flow, emissions, and cost-benefit analysis for transit systems.

Example:

1. Identify public transit factors (cost, accessibility, ridership).
2. Map carbon emissions for different transportation modes.
3. Introduce policies (bike lanes, subsidies, congestion charges).
4. Predict environmental and social outcomes.

Pairing 8: Examining Global Health and Vaccine Distribution Challenges

Big Skill: Holistic Problem-Solving – Considering logistical, political, and ethical factors in vaccine distribution.

Small Skill: Mapping System Interactions – Using network models to visualize supply chains and accessibility.

Teaching Technique: Real-world applications.

Actions: Have students analyze how vaccines are distributed globally and propose fairer solutions.

Example:

1. Identify factors influencing vaccine availability (manufacturing, patents, logistics).
2. Map distribution inequities among different countries.
3. Introduce policy changes (waived patents, public-private partnerships).
4. Predict which strategies increase global vaccine access.

Pairing 9: Mapping the Relationship Between Renewable Energy Adoption and Job Markets

Big Skill: Holistic Problem-Solving – Balancing economic shifts with sustainable energy goals.

Small Skill: Mapping System Interactions – Using causal loop diagrams to show energy policy effects.

Teaching Technique: Cultural exploration.

Actions: Have students analyze workforce shifts in response to green energy policies.

Example:

1. Identify jobs linked to fossil fuels and renewables.
2. Map employment shifts as energy transitions accelerate.
3. Introduce leverage points (job retraining, government incentives).
4. Predict employment trends under different policy approaches.

Pairing 10: Evaluating the Effects of Globalization on Local Cultures

Big Skill: Holistic Problem-Solving – Weighing economic benefits against cultural preservation.

Small Skill: Mapping System Interactions – Using flowcharts to show how global influences affect local traditions.

Teaching Technique: Hands-on experiments.

Actions: Have students trace the spread of global trends and their impact on Indigenous cultures.

Example:

1. Identify key industries driving globalization (media, fashion, food).
2. Map cultural shifts in response to international influences.
3. Introduce policy scenarios (cultural preservation laws, economic tariffs).
4. Predict long-term cultural adaptation trends.

Systems Thinking Mastery – (Big Skill) Holistic Problem-Solving vs. (Small Skill) Simulating System Changes

---

Pairing 1: Simulating the Effects of Raising the Minimum Wage on the Economy

Big Skill: Holistic Problem-Solving – Considering worker benefits, business costs, and inflationary pressures.

Small Skill: Simulating System Changes – Running models to analyze how wage increases impact employment, prices, and productivity.

Teaching Technique: Hands-on experiments.

Actions: Have students modify wage policies in a simulated economy and observe the effects.

Example:

1. Start with a base economy where wages remain constant.
2. Increase the minimum wage by 10%, 20%, and 30%.
3. Measure effects on job creation, consumer spending, and business closures.
4. Identify optimal wage levels that balance economic growth with worker well-being.

Pairing 2: Predicting the Impact of Renewable Energy Transition on Electricity Grids

Big Skill: Holistic Problem-Solving – Balancing sustainability goals, grid reliability, and economic feasibility.

Small Skill: Simulating System Changes – Testing how different renewable energy sources impact grid stability.

Teaching Technique: Guided exploration.

Actions: Have students simulate the gradual phase-out of fossil fuels in a power grid.

Example:

1. Start with a fossil fuel-dominated power grid.
2. Introduce varying percentages of solar, wind, and hydro power.
3. Analyze how energy supply fluctuates and how battery storage compensates.
4. Determine the most efficient renewable energy mix for grid stability.

Pairing 3: Modeling the Impact of Social Media Restrictions on Misinformation Spread

Big Skill: Holistic Problem-Solving – Considering freedom of speech, public safety, and misinformation control.

Small Skill: Simulating System Changes – Running models to measure how different content moderation policies affect information dissemination.

Teaching Technique: Problem-solving exercises.

Actions: Have students implement different moderation policies and track misinformation spread.

Example:

1. Simulate a social media platform where misinformation spreads unchecked.
2. Introduce content warnings, fact-checking labels, or algorithm adjustments.
3. Observe changes in user engagement and misinformation circulation.
4. Identify which strategy effectively balances free speech and public awareness.

Pairing 4: Analyzing the Long-Term Effects of Universal Healthcare Implementation

Big Skill: Holistic Problem-Solving – Considering economic, social, and political perspectives on healthcare access.

Small Skill: Simulating System Changes – Modeling how universal healthcare affects public health and government budgets.

Teaching Technique: Observation journals.

Actions: Have students run different healthcare models to compare costs and benefits.

Example:

1. Simulate a healthcare system where private insurance dominates.
2. Introduce government-funded universal healthcare.
3. Track changes in wait times, disease prevention, and tax burdens.
4. Identify which model best balances affordability and service quality.

Pairing 5: Simulating the Consequences of Deforestation on Local and Global Ecosystems

Big Skill: Holistic Problem-Solving – Weighing economic benefits of deforestation against environmental degradation.

Small Skill: Simulating System Changes – Modeling the effects of tree loss on biodiversity, climate, and local economies.

Teaching Technique: Hands-on experiments.

Actions: Have students analyze different rates of deforestation and their consequences.

Example:

1. Start with a model of a rainforest ecosystem.
2. Gradually increase deforestation and track changes in carbon sequestration, wildlife populations, and soil erosion.
3. Introduce reforestation efforts and observe recovery rates.
4. Identify policies that promote both economic growth and sustainability.

Pairing 6: Examining the Impact of AI Automation on Job Markets

Big Skill: Holistic Problem-Solving – Considering efficiency gains, job losses, and workforce retraining needs.

Small Skill: Simulating System Changes – Running models to measure employment trends and economic shifts.

Teaching Technique: Sensory exploration.

Actions: Have students evaluate different AI adoption rates in various industries.

Example:

1. Select industries (manufacturing, finance, healthcare) for automation.
2. Simulate workforce reductions and new job opportunities.
3. Introduce retraining programs and measure economic recovery.
4. Evaluate policies that balance automation benefits with job security.

Pairing 7: Predicting the Effects of Mass Migration on Urban Infrastructure

Big Skill: Holistic Problem-Solving – Balancing humanitarian needs with urban planning and economic impacts.

Small Skill: Simulating System Changes – Testing how different migration policies affect city resources.

Teaching Technique: Collaborative group work.

Actions: Have students analyze population influx scenarios in urban environments.

Example:

1. Choose a city with an initial stable population.
2. Introduce an influx of 100,000 migrants.
3. Observe strain on housing, healthcare, and employment.
4. Identify policies that promote successful urban integration.

Pairing 8: Simulating the Effects of High-Speed Rail on Regional Economies

Big Skill: Holistic Problem-Solving – Evaluating infrastructure investments, environmental impact, and economic benefits.

Small Skill: Simulating System Changes – Running models to analyze travel patterns and business growth.

Teaching Technique: Real-world applications.

Actions: Have students simulate regional economic growth with and without high-speed rail.

Example:

1. Map a region with major urban centers.
2. Introduce a high-speed rail network.
3. Track changes in business investment, tourism, and commuting patterns.
4. Predict long-term economic and environmental benefits.

Pairing 9: Evaluating the Impact of a Plastic Ban on Global Supply Chains

Big Skill: Holistic Problem-Solving – Considering environmental benefits, corporate adaptation, and consumer behavior.

Small Skill: Simulating System Changes – Modeling how industries and economies respond to a plastic ban.

Teaching Technique: Cultural exploration.

Actions: Have students simulate the effects of banning single-use plastics across different industries.

Example:

1. Select key industries (packaging, retail, fast food).
2. Implement a plastic banning and track alternative materials adoption.
3. Measure economic costs, waste reduction, and consumer response.
4. Assess long-term feasibility of phasing out plastics.

Pairing 10: Forecasting the Consequences of Ocean Acidification on Marine Life and Fisheries

Big Skill: Holistic Problem-Solving – Considering environmental, economic, and food security impacts of ocean changes.

Small Skill: Simulating System Changes – Running models on pH level shifts and marine biodiversity.

Teaching Technique: Hands-on experiments.

Actions: Have students evaluate how ocean acidification affects coral reefs and fish populations.

Example:

1. Set baseline ocean pH levels and marine biodiversity data.
2. Simulate CO_2 increases and measure acidification effects.
3. Observe changes in marine life (coral bleaching, fish migration).
4. Propose policies to mitigate damage (emission reductions, marine conservation).

Systems Thinking Mastery – (Big Skill) Holistic Problem-Solving vs. (Small Skill) Applying Systems Thinking Tools

---

Pairing 1: Examining the Global Energy Transition

Big Skill: Holistic Problem-Solving – Considering environmental, economic, and political factors in moving away from fossil fuels.

Small Skill: Applying Systems Thinking Tools – Using stock-and-flow diagrams to analyze energy production and consumption.

Teaching Technique: Hands-on experiments.

Actions: Have students model how different energy policies affect supply and demand.

Example:

1. Identify energy sources (coal, nuclear, wind, solar).
2. Use a stock-and-flow diagram to track energy generation and storage.
3. Introduce leverage points (government subsidies, carbon taxes).
4. Predict how various policies impact the speed of energy transitions.

Pairing 2: Analyzing the Economic Impact of Automation on Jobs

Big Skill: Holistic Problem-Solving – Weighing efficiency, employment rates, and societal adaptation to AI and automation.

Small Skill: Applying Systems Thinking Tools – Using leverage point identification to analyze labor market shifts.

Teaching Technique: Guided exploration.

Actions: Have students model employment trends with different levels of automation adoption.

Example:

1. Identify industries at high risk of automation (manufacturing, logistics, retail).
2. Use stock-and-flow diagrams to track job displacement and retraining efforts.
3. Introduce leverage points (universal basic income, workforce reskilling programs).
4. Predict how different strategies mitigate unemployment.

Pairing 3: Understanding the Spread of Infectious Diseases

Big Skill: Holistic Problem-Solving – Considering social behavior, healthcare capacity, and public policy in disease outbreaks.

Small Skill: Applying Systems Thinking Tools – Using stock-and-flow models to track infection rates and interventions.

Teaching Technique: Problem-solving exercises.

Actions: Have students simulate disease spread and test containment strategies.

Example:

1. Model an outbreak with variables (population density, transmission rate).
2. Introduce containment measures (mask mandates, vaccination campaigns).
3. Identify leverage points (early intervention, public awareness).
4. Predict long-term health system impacts.

Pairing 4: Mapping the Water Crisis in Urban and Rural Areas

Big Skill: Holistic Problem-Solving – Addressing economic, environmental, and infrastructure challenges in water distribution.

Small Skill: Applying Systems Thinking Tools – Using causal loop diagrams to track supply, demand, and waste.

Teaching Technique: Observation journals.

Actions: Have students analyze water consumption patterns and propose solutions.

Example:

1. Identify water sources (reservoirs, aquifers, desalination plants).
2. Use a stock-and-flow model to track usage and replenishment.
3. Introduce leverage points (recycling systems, agricultural efficiency).
4. Predict how conservation policies affect long-term availability.

Pairing 5: Evaluating the Impact of Climate Change on Food Security

Big Skill: Holistic Problem-Solving – Considering weather patterns, supply chain disruptions, and food distribution.

Small Skill: Applying Systems Thinking Tools – Using stock-and-flow models to track agricultural production.

Teaching Technique: Hands-on experiments.

Actions: Have student mappings show how climate changes alter food supply.

Example:

1. Track agricultural productivity under different climate scenarios.
2. Use leverage points to evaluate adaptive farming techniques (drought-resistant crops, precision irrigation).
3. Identify global trade influences on food availability.
4. Predict which strategies ensure the most stable food supply.

Pairing 6: Investigating the Effects of Urbanization on Traffic Congestion

Big Skill: Holistic Problem-Solving – Balancing infrastructure growth, environmental sustainability, and economic expansion.

Small Skill: Applying Systems Thinking Tools – Using causal loop diagrams to analyze traffic flow.

Teaching Technique: Sensory exploration.

Actions: Have students evaluate different city planning models for reducing congestion.

Example:

1. Identify congestion causes (population growth, limited public transit).
2. Map stock-and-flow dynamics of vehicle numbers and road capacity.
3. Introduce leverage points (bike lanes, high-speed rail, congestion pricing).
4. Predict how each strategy improves urban mobility.

Pairing 7: Assessing the Impact of Deforestation on Biodiversity

Big Skill: Holistic Problem-Solving – Weighing economic benefits of land development against ecological damage.

Small Skill: Applying Systems Thinking Tools – Using stock-and-flow models to track forest loss and species decline.

Teaching Technique: Collaborative group work.

Actions: Have students model deforestation effects and mitigation efforts.

Example:

1. Identify major drivers of deforestation (logging, agriculture, urban expansion).
2. Track species population changes due to habitat loss.
3. Introduce leverage points (reforestation programs, conservation policies).
4. Predict biodiversity trends under different scenarios.

Pairing 8: Examining the Impact of Consumer Behavior on Plastic Waste

Big Skill: Holistic Problem-Solving – Addressing corporate responsibility, government regulations, and public awareness.

Small Skill: Applying Systems Thinking Tools – Using stock-and-flow diagrams to analyze waste generation and recycling.

Teaching Technique: Real-world applications.

Actions: Have students evaluate policies that reduce plastic waste.

Example:

1. Track plastic production, consumption, and disposal rates.
2. Identify leverage points (biodegradable alternatives, bans on single-use plastics).
3. Introduce economic incentives for recycling programs.
4. Predict long-term waste reduction trends.

Pairing 9: Evaluating the Role of Political Polarization in Policy-Making

Big Skill: Holistic Problem-Solving – Weighing ideological divisions, media influence, and legislative effectiveness.

Small Skill: Applying Systems Thinking Tools – Using causal loop diagrams to model political discourse and policy outcomes.

Teaching Technique: Cultural exploration.

Actions: Have students analyze how different information sources shape public opinion.

Example:

1. Identify key influences (news media, social networks, political leaders).
2. Map how polarization affects voter behavior and policy decisions.
3. Introduce leverage points (bipartisan initiatives, media literacy education).
4. Predict the effectiveness of depolarization strategies.

Pairing 10: Analyzing the Relationship Between Education Funding and Student Performance

Big Skill: Holistic Problem-Solving – Balancing financial investment, curriculum effectiveness, and social equity.

Small Skill: Applying Systems Thinking Tools – Using stock-and-flow diagrams to track school resources and student outcomes.

Teaching Technique: Hands-on experiments.

Actions: Have students evaluate different funding models and their effects on education quality.

Example:

1. Identify key education funding sources (local, state, federal).
2. Track stock-and-flow dynamics of resources per student.
3. Introduce leverage points (teacher training, technology access).
4. Predict long-term performance trends under different funding scenarios.

Example Teaching Strategies for Big vs. Small Thinking Skill Pairings of Ethical Decision Making at the College/Technical School Level

Here are 90 unique teaching strategy examples for the big thinking and small thinking skill pairings, each described in detail to align with **Big Thinking Skills** (Concepts & Strategies) and **Small Thinking Skills** (Step-by-Step Execution) for the **Ethical Decision-Making Skill** taught at the college/technical school level.

Ethical Decision Making

Big Thinking Skills (Concepts & Strategies)

- **Applying Ethical Frameworks:** Evaluating moral and societal implications using models like utilitarianism, deontology, and virtue ethics.
- **Assessing Stakeholder Impact:** Understanding how decisions affect individuals, organizations, and society at large.
- **Weighing Competing Values:** Balancing business goals, social responsibility, and personal ethics.

Small Thinking Skills (Step-by-Step Execution)

- **Breaking Down Ethical Dilemmas:** Identifying key stakeholders, conflicts, and possible resolutions.
- **Applying Case Study Analysis:** Evaluating real-world ethical situations and extracting lessons.
- **Justifying Ethical Decisions:** Forming structured arguments using established ethical reasoning models.

Ethical Decision Making – (Big Skill) Applying Ethical Frameworks vs. (Small Skill) Breaking Down Ethical Dilemmas

---

Pairing 1: The Ethics of AI in Hiring Practices

Big Skill: Applying Ethical Frameworks – Evaluating fairness, bias, and privacy in AI-driven hiring using utilitarianism, deontology, and virtue ethics.

Small Skill: Breaking Down Ethical Dilemmas – Identifying how automated hiring affects employers, job applicants, and marginalized communities.

Teaching Technique: Case study analysis.

Actions: Have students examine real-world AI hiring cases and apply ethical reasoning models.

Example:

1. Identify stakeholders (HR departments, job seekers, minority applicants).
2. Map ethical concerns (bias in algorithms, data privacy, fairness).
3. Apply utilitarianism (maximizing efficiency vs. harm), deontology (fair treatment), and virtue ethics (just hiring).
4. Develop ethical hiring guidelines for AI use.

Pairing 2: Genetic Editing and Designer Babies

Big Skill: Applying Ethical Frameworks – Evaluating bioethics through the lens of consequentialist, duty-based, and virtue perspectives.

Small Skill: Breaking Down Ethical Dilemmas – Identifying conflicts between scientific advancement, parental rights, and societal equity.

Teaching Technique: Debate and structured argumentation.

Actions: Have students analyze the ethics of CRISPR and gene editing for non-therapeutic enhancements.

Example:

1. Identify stakeholders (scientists, future children, policymakers, society).
2. Discuss dilemmas (access inequality, unintended consequences, human identity).
3. Apply ethical models (utilitarianism: eliminating disease, deontology: respecting human dignity, virtue ethics: moral character in genetic choices).
4. Argue for or against designer babies from different ethical perspectives.

Pairing 3: Data Privacy in Social Media

Big Skill: Applying Ethical Frameworks – Assessing corporate responsibility in protecting user data through ethical lenses.

Small Skill: Breaking Down Ethical Dilemmas – Identifying the conflict between user autonomy, business interests, and national security.

Teaching Technique: Scenario-based learning.

Actions: Have students evaluate different social media privacy scandals and propose ethical reforms.

Example:

1. Identify key stakeholders (users, companies, governments).
2. Map ethical issues (informed consent, targeted advertising, surveillance).
3. Apply ethical frameworks (utilitarianism: balancing privacy with public safety, deontology: user consent, virtue ethics: corporate responsibility).

4. Design a privacy policy for a hypothetical social media platform.

Pairing 4: Whistleblowing in Corporate Corruption

Big Skill: Applying Ethical Frameworks – Weighing the moral duty to report unethical practices against potential retaliation.

Small Skill: Breaking Down Ethical Dilemmas – Identifying key players, motivations, and consequences in corporate whistleblowing.

Teaching Technique: Role-playing ethical dilemmas.

Actions: Have students role-play whistleblowing cases and apply different ethical models.

Example:

1. Identify stakeholders (employees, companies, regulators, public).
2. Discuss conflicts (job security vs. public interest, legal vs. ethical obligations).
3. Apply ethical frameworks (deontology: duty to report, utilitarianism: harm vs. benefit, virtue ethics: integrity).
4. Simulate a boardroom discussion on a real whistleblowing case.

Pairing 5: Ethical AI in Warfare

Big Skill: Applying Ethical Frameworks – Evaluating just war theory and ethical AI use in military technology.

Small Skill: Breaking Down Ethical Dilemmas – Identifying key moral conflicts in autonomous weapons and civilian protection.

Teaching Technique: Ethical scenario simulations.

Actions: Have students assess drone warfare and AI-targeting systems.

Example:

1. Identify affected parties (soldiers, civilians, governments, AI developers).
2. Analyze dilemmas (autonomy in decision-making, accountability, collateral damage).
3. Apply ethical models (deontology: banning autonomous weapons, utilitarianism: minimizing casualties, virtue ethics: humane conduct in war).
4. Propose ethical AI policies for warfare.

Pairing 6: Labor Exploitation in Fast Fashion

Big Skill: Applying Ethical Frameworks – Weighing corporate profit motives against workers' rights and sustainability.

Small Skill: Breaking Down Ethical Dilemmas – Identifying global supply chain conflicts in labor conditions.

Teaching Technique: Research-based discussion.

Actions: Have students analyze fast fashion companies and propose ethical alternatives.

Example:

1. Identify stakeholders (workers, companies, consumers, policymakers).
2. Map dilemmas (cheap production vs. fair wages, sustainability vs. profit).
3. Apply ethical models (virtue ethics: corporate responsibility, deontology: fair labor laws, utilitarianism: economic benefits vs. worker harm).
4. Develop an ethical fashion certification model.

Pairing 7: AI-Generated Art and Copyright Issues

Big Skill: Applying Ethical Frameworks – Evaluating intellectual property rights and artistic integrity in AI-generated works.

Small Skill: Breaking Down Ethical Dilemmas – Identifying the rights of artists, consumers, and AI developers.

Teaching Technique: Ethical case study analysis.

Actions: Have students examine legal and ethical implications of AI in creative industries.

Example:

1. Identify stakeholders (human artists, AI developers, consumers).
2. Discuss conflicts (fair use, originality, AI vs. human labor).
3. Apply ethical frameworks (deontology: respecting creators' rights, utilitarianism: accessibility of AI art, virtue ethics: valuing human creativity).
4. Draft an AI copyright policy.

Pairing 8: The Ethics of Organ Transplant Allocation

Big Skill: Applying Ethical Frameworks – Assessing fairness in distributing limited medical resources.

Small Skill: Breaking Down Ethical Dilemmas – Identifying moral concerns in patient priority rankings.

Teaching Technique: Case-based analysis.

Actions: Have students simulate decision-making for organ transplant recipients.

Example:

1. Identify stakeholders (patients, doctors, policymakers, families).
2. Map dilemmas (age vs. need, ability to pay vs. fairness, first-come-first-served vs. best chance of survival).
3. Apply ethical models (utilitarianism: maximizing life-years, deontology: fair process, virtue ethics: compassion in medicine).

4. Develop ethical transplant policies.

Pairing 9: The Ethics of Censorship in Digital Content

Big Skill: Applying Ethical Frameworks – Balancing freedom of speech with harm prevention.

Small Skill: Breaking Down Ethical Dilemmas – Identifying government, corporate, and societal interests in content moderation.

Teaching Technique: Policy evaluation.

Actions: Have students assess real-world cases of censorship and propose balanced policies.

Example:

1. Identify stakeholders (governments, tech companies, content creators, users).
2. Discuss conflicts (hate speech vs. free speech, misinformation vs. open dialogue).
3. Apply ethical frameworks (utilitarianism: public safety, deontology: right to express, virtue ethics: responsibility).
4. Design a fair digital content policy.

Pairing 10: Self-Driving Cars and Moral Decision-Making

Big Skill: Applying Ethical Frameworks – Evaluating how AI should make life-or-death decisions in autonomous vehicles.

Small Skill: Breaking Down Ethical Dilemmas – Identifying passenger safety, pedestrian rights, and liability concerns.

Teaching Technique: Moral dilemma simulations.

Actions: Have students design ethical decision-making algorithms for self-driving cars.

Example:

1. Identify affected groups (drivers, pedestrians, AI developers).
2. Analyze ethical conflicts (protect passengers vs. minimize overall harm).
3. Apply ethical models (utilitarianism: minimizing casualties, deontology: equal protection, virtue ethics: designing ethical AI).
4. Draft an AI-driven car ethics code.

Ethical Decision Making – (Big Skill) Applying Ethical Frameworks vs. (Small Skill) Applying Case Study Analysis

---

Pairing 1: Facebook and the Cambridge Analytica Data Scandal

Big Skill: Applying Ethical Frameworks – Evaluating privacy violations and corporate responsibility using utilitarianism, deontology, and virtue ethics.

Small Skill: Applying Case Study Analysis – Examining the impact of data misuse in political influence.

Teaching Technique: Case study deconstruction.

Actions: Have students analyze Facebook's handling of user data and apply ethical reasoning models.

Example:

1. Identify ethical concerns (user consent, political manipulation, corporate responsibility).
2. Apply **utilitarianism** (greatest good vs. harm caused by misinformation), **deontology** (duty to protect data privacy), and **virtue ethics** (honest corporate behavior).
3. Debate whether Facebook acted ethically or if regulations should have been stricter.
4. Propose a corporate ethics policy to prevent similar violations.

Pairing 2: Theranos and Deceptive Medical Technology

Big Skill: Applying Ethical Frameworks – Evaluating fraudulent business practices and public health risks.

Small Skill: Applying Case Study Analysis – Investigating how misleading claims about medical technology led to ethical violations.

Teaching Technique: Role-playing ethical perspectives.

Actions: Have students assess the ethical responsibility of business leaders and investors.

Example:

1. Identify ethical failures (deception, investor fraud, patient risk).
2. Apply **virtue ethics** (honesty in business), **utilitarianism** (risk-benefit for investors and patients), and **deontology** (duty to provide truthful health information).
3. Simulate an ethics board discussion on whistleblower accountability.
4. Develop an ethical checklist for medical startups.

Pairing 3: Apple and Ethical Sourcing of Materials

Big Skill: Applying Ethical Frameworks – Evaluating labor conditions and environmental responsibility.

Small Skill: Applying Case Study Analysis – Investigating Apple's supply chain and use of conflict minerals.

Teaching Technique: Policy critique.

Actions: Have students analyze Apple's supplier policies and propose ethical sourcing reforms.

Example:

1. Identify ethical concerns (child labor, unsafe conditions, resource exploitation).
2. Apply **utilitarianism** (balancing affordable tech with fair labor), **deontology** (corporate duty to workers), and **virtue ethics** (ethical leadership).
3. Assess Apple's existing policies and gaps in accountability.
4. Propose an ethical certification program for tech supply chains.

Pairing 4: Volkswagen Emissions Scandal

Big Skill: Applying Ethical Frameworks – Evaluating corporate deception and environmental ethics.

Small Skill: Applying Case Study Analysis – Reviewing how Volkswagen manipulated emissions tests.

Teaching Technique: Group discussions and ethical ranking.

Actions: Have students assess Volkswagen's decisions and consequences for stakeholders.

Example:

1. Identify ethical dilemmas (false advertising, environmental harm, public trust).
2. Apply **deontology** (company's duty to comply with regulations), **utilitarianism** (balancing business interests with pollution impact), and **virtue ethics** (corporate honesty).
3. Rank Volkswagen's ethical failures and discuss corrective measures.
4. Develop a transparent emissions policy for the auto industry.

Pairing 5: The Ethics of Self-Driving Cars and Decision Algorithms

Big Skill: Applying Ethical Frameworks – Evaluating responsibility in AI decision-making.

Small Skill: Applying Case Study Analysis – Reviewing real-life cases of self-driving car accidents and liability.

Teaching Technique: Simulation of ethical dilemmas.

Actions: Have students analyze different decision-making models for self-driving AI.

Example:

1. Identify ethical concerns (passenger safety, pedestrian protection, legal liability).
2. Apply **utilitarianism** (minimizing casualties), **deontology** (ensuring fair treatment in crash scenarios), and **virtue ethics** (building moral AI).
3. Evaluate different ethical models for AI responses in unavoidable accidents.
4. Develop an ethical decision-making framework for autonomous vehicles.

Pairing 6: Pharmaceutical Pricing and Access to Life-Saving Drugs

Big Skill: Applying Ethical Frameworks – Evaluating corporate responsibility and access to essential medicine.

Small Skill: Applying Case Study Analysis – Investigating cases where pharmaceutical companies set high drug prices.

Teaching Technique: Debate on corporate and societal responsibility.

Actions: Have students assess the justification for high drug prices using different ethical theories.

Example:

1. Identify ethical concerns (public health, profit motives, accessibility).
2. Apply **utilitarianism** (weighing R&D costs vs. public benefit), **deontology** (right to affordable healthcare), and **virtue ethics** (corporate integrity).
3. Debate whether drug companies should have price limits.
4. Develop an ethical pricing policy for pharmaceuticals.

Pairing 7: Google's Censorship and Government Compliance in China

Big Skill: Applying Ethical Frameworks – Evaluating ethical trade-offs between market expansion and free speech.

Small Skill: Applying Case Study Analysis – Reviewing Google's compliance with Chinese internet censorship laws.

Teaching Technique: Policy assessment.

Actions: Have students analyze Google's choice to censor search results.

Example:

1. Identify ethical issues (freedom of speech vs. business interests).
2. Apply **deontology** (upholding freedom of information), **utilitarianism** (weighing economic growth vs. government suppression), and **virtue ethics** (corporate moral leadership).
3. Propose ethical guidelines for multinational corporations in authoritarian countries.
4. Discuss potential global consequences of compliance.

Pairing 8: Nike and Ethical Manufacturing Practices

Big Skill: Applying Ethical Frameworks – Evaluating labor conditions and corporate responsibility.

Small Skill: Applying Case Study Analysis – Investigating Nike's past use of sweatshops.

Teaching Technique: Ethical impact assessment.

Actions: Have students analyze how public backlash influenced Nike's ethical reforms.

Example:

1. Identify stakeholders (workers, consumers, corporate leadership).
2. Apply **utilitarianism** (affordable goods vs. worker exploitation), **deontology** (workers' rights), and **virtue ethics** (ethical sourcing).
3. Assess Nike's changes and discuss areas for further improvement.
4. Create an ethical compliance plan for manufacturing companies.

Pairing 9: Ethical Issues in Cryptocurrency and Financial Fraud

Big Skill: Applying Ethical Frameworks – Evaluating financial decentralization and fraud risks.

Small Skill: Applying Case Study Analysis – Investigating cryptocurrency scams like FTX and Ponzi schemes.

Teaching Technique: Risk and ethics mapping.

Actions: Have students examine cryptocurrency fraud cases and propose ethical regulations.

Example:

1. Identify ethical dilemmas (decentralization, lack of oversight, investor losses).
2. Apply **utilitarianism** (democratized finance vs. fraud risk), **deontology** (fair financial practices), and **virtue ethics** (transparency in markets).
3. Develop ethical guidelines for cryptocurrency markets.
4. Debate whether governments should regulate decentralized finance.

Pairing 10: Amazon's Warehouse Working Conditions

Big Skill: Applying Ethical Frameworks – Evaluating labor rights, worker treatment, and corporate responsibility.

Small Skill: Applying Case Study Analysis – Examining ethical concerns about warehouse labor conditions.

Teaching Technique: Employee perspective simulation.

Actions: Have students analyze Amazon's fulfillment center policies and propose ethical reforms.

Example:

1. Identify ethical issues (workplace stress, surveillance, injury risks).
2. Apply **utilitarianism** (fast shipping vs. worker well-being), **deontology** (duty to provide fair labor conditions), and **virtue ethics** (corporate empathy).
3. Rank ethical concerns and propose improvements.
4. Draft a fair labor policy for e-commerce companies.

Ethical Decision Making – (Big Skill) Applying Ethical Frameworks vs. (Small Skill) Justifying Ethical Decisions

---

Pairing 1: Ethical AI in Hiring Practices

Big Skill: Applying Ethical Frameworks – Evaluating fairness, bias, and privacy concerns in AI-driven hiring using utilitarianism, deontology, and virtue ethics.

Small Skill: Justifying Ethical Decisions – Forming structured arguments for and against AI decision-making in employment.

Teaching Technique: Debate-based ethical analysis.

Actions: Have students argue whether AI hiring tools should be regulated or banned.

Example:

1. Identify ethical concerns (bias in algorithms, fairness, discrimination).
2. Apply **utilitarianism** (efficiency vs. risk of bias), **deontology** (equal treatment obligations), and **virtue ethics** (corporate integrity).
3. Justify an ethical stance on regulating AI hiring.
4. Draft an ethical hiring policy for AI-based recruitment tools.

Pairing 2: Data Privacy and Surveillance by Governments

Big Skill: Applying Ethical Frameworks – Evaluating state surveillance in national security vs. individual privacy rights.

Small Skill: Justifying Ethical Decisions – Structuring arguments on whether mass surveillance is ethical.

Teaching Technique: Structured ethical debate.

Actions: Have students defend or oppose government data collection policies.

Example:

1. Identify ethical dilemmas (public safety vs. privacy rights).
2. Apply **deontology** (duty to protect citizens vs. right to privacy), **utilitarianism** (security trade-offs), and **virtue ethics** (trust in government).
3. Justify whether surveillance is necessary or unethical.
4. Draft an ethical framework for government surveillance programs.

Pairing 3: Autonomous Vehicles and Liability in Accidents

Big Skill: Applying Ethical Frameworks – Evaluating responsibility when self-driving cars cause accidents.

Small Skill: Justifying Ethical Decisions – Arguing who should be held accountable (manufacturers, users, or AI developers).

Teaching Technique: Role-playing ethical perspectives.

Actions: Have students analyze real-world self-driving car incidents and argue liability.

Example:

1. Identify stakeholders (passengers, pedestrians, companies).
2. Apply **deontology** (holding AI accountable), **utilitarianism** (minimizing casualties), and **virtue ethics** (moral responsibility).
3. Justify whether liability should fall on carmakers or human users.
4. Develop an ethical liability policy for autonomous vehicles.

Pairing 4: Pharmaceutical Pricing and Access to Medicine

Big Skill: Applying Ethical Frameworks – Evaluating corporate pricing strategies for life-saving medications.

Small Skill: Justifying Ethical Decisions – Forming ethical arguments on whether drug prices should be capped.

Teaching Technique: Case study evaluation.

Actions: Have students analyze drug pricing cases and justify price controls or free-market policies.

Example:

1. Identify ethical conflicts (profitability vs. affordability).
2. Apply **utilitarianism** (greatest good for most people), **deontology** (right to healthcare), and **virtue ethics** (corporate responsibility).
3. Justify if governments should regulate drug prices.
4. Draft a fair pricing policy for pharmaceuticals.

Pairing 5: Ethical Considerations in Genetic Engineering

Big Skill: Applying Ethical Frameworks – Weighing the benefits and risks of gene editing for humans.

Small Skill: Justifying Ethical Decisions – Developing structured arguments on whether genetic modifications should be legal.

Teaching Technique: Ethical scenario simulations.

Actions: Have students assess the ethical implications of CRISPR-based gene editing.

Example:

1. Identify stakeholders (scientists, patients, regulators, society).

2. Apply **deontology** (duty to prevent harm), **utilitarianism** (eliminating disease), and **virtue ethics** (moral responsibility in science).
3. Justify whether gene editing should be restricted or allowed.
4. Develop an ethical policy for medical genetic modification.

Pairing 6: Workplace Ethics and Whistleblowing

Big Skill: Applying Ethical Frameworks – Evaluating employee responsibilities vs. company loyalty.

Small Skill: Justifying Ethical Decisions – Structuring ethical arguments on when whistleblowing is morally justified.

Teaching Technique: Role-playing ethical conflicts.

Actions: Have students defend or criticize whistleblowers in real-world cases.

Example:

1. Identify ethical concerns (loyalty vs. exposing harm).
2. Apply **virtue ethics** (integrity), **deontology** (duty to report wrongdoing), and **utilitarianism** (balancing harm and benefits).
3. Justify if whistleblowing should be legally protected.
4. Draft an ethical policy for corporate transparency.

Pairing 7: Artificial Intelligence and Creative Copyright Laws

Big Skill: Applying Ethical Frameworks – Assessing AI's role in artistic creation and intellectual property rights.

Small Skill: Justifying Ethical Decisions – Structuring arguments for AI-generated content ownership.

Teaching Technique: Legal and ethical case analysis.

Actions: Have students analyze AI art cases and justify whether AI creations should have copyright protections.

Example:

1. Identify key issues (ownership, originality, economic impact).
2. Apply **deontology** (artists' rights), **utilitarianism** (public benefit vs. artist compensation), and **virtue ethics** (ethical innovation).
3. Justify an ethical stance on AI-generated art ownership.
4. Propose legal guidelines for AI-created content.

Pairing 8: Ethical Dilemmas in Organ Transplant Allocation

Big Skill: Applying Ethical Frameworks – Evaluating fairness in distributing limited medical resources.

Small Skill: Justifying Ethical Decisions – Arguing whether age, wealth, or social contribution should impact transplant eligibility.

Teaching Technique: Structured argumentation.

Actions: Have students debate different allocation models for scarce medical resources.

Example:

1. Identify ethical concerns (equal access vs. prioritization).
2. Apply **deontology** (equal treatment), **utilitarianism** (maximizing life-years saved), and **virtue ethics** (compassion in medicine).
3. Justify an allocation model for organ transplants.
4. Draft an ethical policy for patient selection.

Pairing 9: The Ethics of War and Drone Strikes

Big Skill: Applying Ethical Frameworks – Evaluating just war theory and collateral damage concerns.

Small Skill: Justifying Ethical Decisions – Arguing the morality of autonomous drone warfare.

Teaching Technique: Ethical case study analysis.

Actions: Have students analyze military drone policies and justify their ethical stance.

Example:

1. Identify key ethical dilemmas (civilian harm, combatant safety, national security).
2. Apply **utilitarianism** (minimizing soldier casualties), **deontology** (strict rules of engagement), and **virtue ethics** (moral warfare conduct).
3. Justify the use or prohibition of drone warfare.
4. Propose ethical guidelines for military AI.

Pairing 10: The Morality of Capital Punishment

Big Skill: Applying Ethical Frameworks – Assessing the justice and deterrence effects of the death penalty.

Small Skill: Justifying Ethical Decisions – Structuring ethical arguments for or against capital punishment.

Teaching Technique: Policy evaluation and debate.

Actions: Have students defend or oppose the death penalty using ethical reasoning.

Example:

1. Identify ethical concerns (fair trials, deterrence vs. retribution).
2. Apply **deontology** (right to life), **utilitarianism** (deterring crime vs. wrongful executions), and **virtue ethics** (justice and rehabilitation).
3. Justify whether capital punishment aligns with ethical principles.
4. Draft an ethical policy on sentencing for serious crimes.

Ethical Decision Making – (Big Skill) Assessing Stakeholder Impact vs. (Small Skill) Breaking Down Ethical Dilemmas

---

Pairing 1: The Ethics of AI in Healthcare Diagnostics

Big Skill: Assessing Stakeholder Impact – Understanding how AI-driven healthcare decisions affect patients, doctors, and insurance providers.

Small Skill: Breaking Down Ethical Dilemmas – Identifying the trade-offs between accuracy, accessibility, and human oversight.

Teaching Technique: Case study analysis.

Actions: Have students evaluate how AI-driven diagnosis impacts patient outcomes, medical ethics, and liability.

Example:

1. Identify stakeholders (patients, doctors, AI developers, regulators).
2. Analyze conflicts (accuracy vs. bias, cost vs. accessibility, human oversight vs. automation).
3. Discuss possible resolutions (human-AI collaboration, policy changes).
4. Develop an ethical framework for AI in medical diagnostics.

Pairing 2: The Impact of Autonomous Vehicles on Employment and Safety

Big Skill: Assessing Stakeholder Impact – Weighing the benefits of autonomous driving against job losses and liability concerns.

Small Skill: Breaking Down Ethical Dilemmas – Identifying conflicts between industry innovation and workforce displacement.

Teaching Technique: Debate on societal impact.

Actions: Have students discuss who should be held responsible for self-driving accidents and how to manage job losses.

Example:

1. Identify stakeholders (drivers, tech companies, insurance firms, lawmakers).
2. Analyze conflicts (safety improvements vs. job loss, legal responsibility).

3. Discuss potential solutions (reskilling programs, new liability laws).
4. Propose an ethical transition strategy for self-driving cars.

Pairing 3: Data Privacy in Social Media Advertising

Big Skill: Assessing Stakeholder Impact – Evaluating how companies, consumers, and governments are affected by data tracking.

Small Skill: Breaking Down Ethical Dilemmas – Identifying conflicts between business growth and consumer privacy rights.

Teaching Technique: Ethical role-play.

Actions: Have students argue from the perspectives of social media companies, consumers, and privacy advocates.

Example:

1. Identify stakeholders (advertisers, social media users, regulators).
2. Analyze conflicts (personalized ads vs. user tracking, revenue vs. privacy).
3. Explore solutions (opt-in policies, stronger data laws).
4. Develop ethical advertising policies that balance profit and privacy.

Pairing 4: The Ethical Implications of Fast Fashion

Big Skill: Assessing Stakeholder Impact – Analyzing how low-cost clothing production affects workers, consumers, and the environment.

Small Skill: Breaking Down Ethical Dilemmas – Identifying conflicts between affordability, labor rights, and sustainability.

Teaching Technique: Policy evaluation.

Actions: Have students examine the long-term effects of fast fashion and propose ethical alternatives.

Example:

1. Identify stakeholders (workers, brands, consumers, environmental groups).
2. Analyze conflicts (cheap production vs. fair wages, convenience vs. sustainability).
3. Explore solutions (ethical sourcing, consumer awareness campaigns).
4. Draft a sustainable business model for fashion brands.

Pairing 5: The Morality of Gig Economy Work Conditions

Big Skill: Assessing Stakeholder Impact – Evaluating the balance between worker flexibility and labor rights.

Small Skill: Breaking Down Ethical Dilemmas – Identifying conflicts between fair pay, worker protections, and business models.

Teaching Technique: Ethical impact assessment.

Actions: Have students analyze how gig platforms affect worker stability and company profits.

Example:

1. Identify stakeholders (gig workers, platform companies, government agencies).
2. Analyze conflicts (flexibility vs. worker exploitation, company profit vs. fair wages).
3. Explore solutions (minimum pay standards, better benefits).
4. Develop an ethical labor policy for gig economy companies.

Pairing 6: The Ethics of Pharmaceutical Pricing and Accessibility

Big Skill: Assessing Stakeholder Impact – Considering how drug pricing affects patients, pharmaceutical companies, and governments.

Small Skill: Breaking Down Ethical Dilemmas – Identifying conflicts between profit motives and healthcare accessibility.

Teaching Technique: Structured ethical debate.

Actions: Have students argue for and against government intervention in drug pricing.

Example:

1. Identify stakeholders (patients, healthcare providers, drug manufacturers).
2. Analyze conflicts (research investment vs. affordability, regulation vs. innovation).
3. Explore solutions (government subsidies, price caps, alternative pricing models).
4. Develop an ethical framework for pharmaceutical pricing.

Pairing 7: The Role of AI in Predictive Policing

Big Skill: Assessing Stakeholder Impact – Evaluating how predictive policing affects communities, law enforcement, and policymakers.

Small Skill: Breaking Down Ethical Dilemmas – Identifying conflicts between crime prevention, privacy, and potential bias.

Teaching Technique: Scenario-based learning.

Actions: Have students examine predictive policing cases and recommend ethical policies.

Example:

1. Identify stakeholders (police, community members, tech developers).
2. Analyze conflicts (public safety vs. bias, crime prevention vs. privacy concerns).
3. Explore solutions (transparency in AI models, oversight committees).
4. Propose ethical policies for AI-driven policing.

Pairing 8: The Impact of Renewable Energy Projects on Local Communities

Big Skill: Assessing Stakeholder Impact – Balancing economic, environmental, and community interests in energy development.

Small Skill: Breaking Down Ethical Dilemmas – Identifying conflicts between sustainability goals and displacement concerns.

Teaching Technique: Community impact study.

Actions: Have students evaluate real cases of wind and solar farm expansions.

Example:

1. Identify stakeholders (residents, energy companies, environmental groups).
2. Analyze conflicts (green energy vs. land use, economic benefits vs. community displacement).
3. Explore solutions (compensation for landowners, community integration plans).
4. Develop an ethical policy for sustainable energy projects.

Pairing 9: The Ethics of Cancel Culture and Free Speech

Big Skill: Assessing Stakeholder Impact – Understanding how social media accountability affects individuals, companies, and society.

Small Skill: Breaking Down Ethical Dilemmas – Identifying conflicts between accountability, free speech, and reputational harm.

Teaching Technique: Case study deconstruction.

Actions: Have students debate the ethical limits of online shaming and accountability.

Example:

1. Identify stakeholders (canceled individuals, activists, corporations, media platforms).
2. Analyze conflicts (accountability vs. public shaming, consequences vs. redemption).
3. Explore solutions (ethical social media policies, restorative justice).
4. Develop guidelines for ethical digital discourse.

Pairing 10: Ethical Challenges in Space Exploration and Colonization

Big Skill: Assessing Stakeholder Impact – Considering how space colonization affects humanity, private companies, and ethical responsibility.

Small Skill: Breaking Down Ethical Dilemmas – Identifying conflicts between scientific progress, commercial interests, and planetary protection.

Teaching Technique: Futuristic policy simulation.

Actions: Have students assess the ethical considerations of human expansion into space.

Example:

1. Identify stakeholders (space agencies, private companies, future settlers, global governments).
2. Analyze conflicts (resource exploitation vs. planetary preservation, accessibility vs. monopolization).
3. Explore solutions (global space treaties, fair colonization ethics).
4. Develop an ethical framework for space exploration governance.

Ethical Decision Making – (Big Skill) Assessing Stakeholder Impact vs. (Small Skill) Applying Case Study Analysis

---

Pairing 1: Facebook and the Cambridge Analytica Scandal

Big Skill: Assessing Stakeholder Impact – Understanding how data privacy violations affected users, political integrity, and corporate accountability.

Small Skill: Applying Case Study Analysis – Evaluating Facebook's ethical responsibility in handling user data.

Teaching Technique: Case study deconstruction.

Actions: Have students analyze the impact of data misuse and propose ethical guidelines for tech companies.

Example:

1. Identify stakeholders (users, governments, advertisers, Facebook executives).
2. Analyze ethical conflicts (privacy violations, election manipulation).
3. Assess outcomes (public backlash, government fines, policy changes).
4. Develop an ethical data protection policy for tech companies.

Pairing 2: The Volkswagen Emissions Scandal

Big Skill: Assessing Stakeholder Impact – Understanding how corporate deception affected consumers, environmental policies, and the automotive industry.

Small Skill: Applying Case Study Analysis – Reviewing Volkswagen's ethical failure in manipulating emissions tests.

Teaching Technique: Ethical impact assessment.

Actions: Have students evaluate Volkswagen's ethical breaches and suggest accountability measures.

Example:

1. Identify stakeholders (car buyers, environmental agencies, Volkswagen employees).
2. Analyze ethical dilemmas (corporate profit vs. environmental responsibility).

3. Assess consequences (legal fines, loss of public trust, stricter regulations).
4. Develop an ethical compliance strategy for automakers.

Pairing 3: Amazon's Warehouse Working Conditions

Big Skill: Assessing Stakeholder Impact – Evaluating the effects of Amazon's work culture on employees, consumers, and corporate profits.

Small Skill: Applying Case Study Analysis – Investigating reports of worker exploitation, injury rates, and labor policies.

Teaching Technique: Role-playing ethical perspectives.

Actions: Have students examine labor conditions and propose reforms.

Example:

1. Identify stakeholders (warehouse workers, management, customers, unions).
2. Analyze ethical conflicts (fast shipping vs. worker well-being).
3. Assess changes (wage increases, improved safety measures, unionization efforts).
4. Develop a sustainable labor policy for e-commerce companies.

Pairing 4: The Boeing 737 MAX Crisis

Big Skill: Assessing Stakeholder Impact – Understanding how corporate decisions on safety affected passengers, engineers, and regulatory bodies.

Small Skill: Applying Case Study Analysis – Evaluating the Boeing 737 MAX crashes and the company's response.

Teaching Technique: Policy evaluation.

Actions: Have students analyze Boeing's ethical failures and propose safety regulations.

Example:

1. Identify stakeholders (pilots, passengers, Boeing executives, regulatory agencies).
2. Analyze ethical conflicts (profit motives vs. passenger safety).
3. Assess impact (fatal crashes, lawsuits, regulatory scrutiny).
4. Develop a corporate transparency and safety protocol.

Pairing 5: The Theranos Fraud Scandal

Big Skill: Assessing Stakeholder Impact – Evaluating the effects of fraudulent medical technology claims on investors, patients, and regulators.

Small Skill: Applying Case Study Analysis – Reviewing how Theranos misled stakeholders and caused harm.

Teaching Technique: Structured ethical debate.

Actions: Have students debate whether stricter regulations could have prevented the fraud.

Example:

1. Identify stakeholders (patients, doctors, investors, Theranos leadership).
2. Analyze ethical issues (misrepresentation of technology, patient safety risks).
3. Assess impact (financial losses, criminal charges, industry skepticism).
4. Develop ethical guidelines for medical startups.

Pairing 6: Nike and Sweatshop Labor Controversy

Big Skill: Assessing Stakeholder Impact – Analyzing how Nike's production model affected workers, consumers, and corporate branding.

Small Skill: Applying Case Study Analysis – Investigating past allegations of child labor and unfair wages in Nike factories.

Teaching Technique: Ethical policy critique.

Actions: Have students assess Nike's response and suggest ethical sourcing policies.

Example:

1. Identify stakeholders (factory workers, Nike executives, consumers, labor rights groups).
2. Analyze ethical conflicts (low-cost production vs. fair wages and working conditions).
3. Assess corporate responses (factory audits, wage increases, reputation management).
4. Propose an ethical labor sourcing framework.

Pairing 7: Google's Censorship and Compliance in China

Big Skill: Assessing Stakeholder Impact – Weighing the balance between corporate profit and freedom of expression.

Small Skill: Applying Case Study Analysis – Reviewing Google's decision to comply with China's censorship laws.

Teaching Technique: Policy and ethics debate.

Actions: Have students analyze whether tech companies should comply with authoritarian governments.

Example:

1. Identify stakeholders (Google users, the Chinese government, free speech advocates).
2. Analyze ethical conflicts (business expansion vs. restricting information access).
3. Assess corporate consequences (criticism from human rights organizations, profit loss in Western markets).

4. Propose ethical guidelines for multinational corporations in restricted media environments.

Pairing 8: The Ethics of Self-Driving Car Decision-Making

Big Skill: Assessing Stakeholder Impact – Evaluating who bears responsibility for autonomous vehicle accidents.

Small Skill: Applying Case Study Analysis – Examining real-world cases of self-driving car failures.

Teaching Technique: Scenario-based analysis.

Actions: Have students debate liability in self-driving accidents.

Example:

1. Identify stakeholders (passengers, pedestrians, AI developers, car manufacturers).
2. Analyze ethical dilemmas (AI decision-making in life-threatening situations).
3. Assess impact (legal frameworks, consumer trust, corporate liability).
4. Develop ethical policies for autonomous vehicle regulations.

Pairing 9: Pharmaceutical Companies and Opioid Crisis Responsibility

Big Skill: Assessing Stakeholder Impact – Understanding the role of pharmaceutical companies in the opioid epidemic.

Small Skill: Applying Case Study Analysis – Investigating Purdue Pharma's role in opioid addiction.

Teaching Technique: Public health ethics discussion.

Actions: Have students analyze corporate responsibility and public health consequences.

Example:

1. Identify stakeholders (patients, doctors, pharmaceutical executives, government regulators).
2. Analyze ethical conflicts (profit motives vs. public health risks).
3. Assess outcomes (legal settlements, policy changes, addiction crisis).
4. Develop corporate responsibility guidelines for pharmaceutical marketing.

Pairing 10: The Ethics of Space Commercialization by Private Companies

Big Skill: Assessing Stakeholder Impact – Evaluating the consequences of privatized space exploration on society and global equity.

Small Skill: Applying Case Study Analysis – Examining SpaceX, Blue Origin, and other private space ventures.

Teaching Technique: Future policy simulation.

Actions: Have students assess ethical challenges and propose international space regulations.

Example:

1. Identify stakeholders (space agencies, private companies, global governments).
2. Analyze ethical dilemmas (access to space, environmental concerns, commercialization).
3. Assess corporate impact (privileged access, militarization concerns).
4. Develop an ethical framework for private space exploration governance.

Ethical Decision Making – (Big Skill) Assessing Stakeholder Impact vs. (Small Skill) Justifying Ethical Decisions

---

Pairing 1: The Ethics of AI in Hiring Decisions

Big Skill: Assessing Stakeholder Impact – Understanding how AI-driven hiring affects job seekers, employers, and workplace diversity.

Small Skill: Justifying Ethical Decisions – Using ethical models to argue for or against AI's role in hiring.

Teaching Technique: Structured debate.

Actions: Have students argue whether AI-based hiring systems are fair or discriminatory.

Example:

1. Identify stakeholders (job seekers, HR managers, AI developers, regulators).
2. Analyze ethical conflicts (efficiency vs. bias, cost-cutting vs. diversity).
3. Apply ethical models:
    - **Utilitarianism**: AI increases efficiency but may exclude marginalized groups.
    - **Deontology**: Hiring should be based on fair, transparent processes.
    - **Virtue Ethics**: Ethical hiring requires integrity and inclusivity.
4. Justify whether AI should be regulated in hiring.

Pairing 2: Corporate Responsibility in Climate Change

Big Skill: Assessing Stakeholder Impact – Understanding how corporate environmental decisions affect society and the economy.

Small Skill: Justifying Ethical Decisions – Constructing arguments on corporate responsibility for climate action.

Teaching Technique: Ethical policy evaluation.

Actions: Have students analyze a corporation's carbon footprint and argue for ethical policies.

Example:

1. Identify stakeholders (corporations, consumers, environmental groups, future generations).
2. Analyze ethical conflicts (profit vs. environmental sustainability).
3. Apply ethical models:
   - **Utilitarianism**: Reducing emissions benefits the planet but may raise costs.
   - **Deontology**: Companies have a duty to act sustainably.
   - **Virtue Ethics**: Ethical leaders should prioritize sustainability.
4. Justify corporate environmental policies.

Pairing 3: Consumer Privacy vs. Targeted Advertising

Big Skill: Assessing Stakeholder Impact – Understanding the effects of personalized ads on consumers, businesses, and data privacy.

Small Skill: Justifying Ethical Decisions – Defending or opposing the ethical use of consumer data.

Teaching Technique: Role-playing different perspectives.

Actions: Have students take on the roles of advertisers, consumers, and regulators in a debate.

Example:

1. Identify stakeholders (consumers, advertisers, tech companies, regulators).
2. Analyze ethical dilemmas (privacy invasion vs. free services, consumer choice vs. manipulation).
3. Apply ethical models:
   - **Utilitarianism**: Personalized ads benefit consumers but compromise privacy.
   - **Deontology**: Users should have control over their data.
   - **Virtue Ethics**: Companies should be transparent and fair.
4. Justify a balanced approach to targeted advertising.

Pairing 4: Ethical AI in Predictive Policing

Big Skill: Assessing Stakeholder Impact – Evaluating how predictive policing affects communities, law enforcement, and civil rights.

Small Skill: Justifying Ethical Decisions – Arguing whether predictive policing is ethical.

Teaching Technique: Scenario-based ethical reasoning.

Actions: Have students analyze case studies where predictive policing was used and argue its ethical standing.

Example:

1. Identify stakeholders (law enforcement, civilians, policy makers, AI developers).
2. Analyze ethical dilemmas (reducing crime vs. racial profiling, efficiency vs. fairness).
3. Apply ethical models:
   - **Utilitarianism**: Reducing crime is good, but bias could harm communities.
   - **Deontology**: Law enforcement must act impartially.
   - **Virtue Ethics**: Policing should be just and accountable.
4. Justify whether predictive policing should be adopted.

Pairing 5: The Ethics of Gig Economy Work

Big Skill: Assessing Stakeholder Impact – Understanding how gig work affects workers, businesses, and consumers.

Small Skill: Justifying Ethical Decisions – Arguing whether gig companies should provide worker benefits.

Teaching Technique: Policy analysis and ethical debate.

Actions: Have students evaluate gig work policies and propose ethical labor reforms.

Example:

1. Identify stakeholders (gig workers, companies, consumers, labor unions).
2. Analyze ethical dilemmas (flexibility vs. stability, fair pay vs. business costs).
3. Apply ethical models:
   - **Utilitarianism**: Gig work offers jobs, but workers lack security.
   - **Deontology**: Companies should treat workers fairly.
   - **Virtue Ethics**: Ethical leadership values worker well-being.
4. Justify ethical labor standards for gig economy platforms.

Pairing 6: The Role of AI in Content Moderation

Big Skill: Assessing Stakeholder Impact – Understanding the effects of AI-driven censorship on free speech, misinformation, and platform responsibility.

Small Skill: Justifying Ethical Decisions – Arguing the ethical limits of AI moderation.

Teaching Technique: Case study analysis.

Actions: Have students assess real cases of AI moderation failures and justify ethical solutions.

Example:

1. Identify stakeholders (social media users, platform owners, regulators).
2. Analyze ethical dilemmas (protecting speech vs. preventing harm).

3. Apply ethical models:

   - **Utilitarianism**: Censorship can prevent harm but may stifle debate.
   - **Deontology**: Speech rights should be upheld.
   - **Virtue Ethics**: Platforms should be responsible yet fair.

4. Justify ethical AI content moderation guidelines.

Pairing 7: Genetic Editing and Human Enhancement

Big Skill: Assessing Stakeholder Impact – Understanding the implications of gene editing on individuals and society.

Small Skill: Justifying Ethical Decisions – Arguing the ethics of modifying human genetics.

Teaching Technique: Structured ethical discussion.

Actions: Have students examine CRISPR technology and justify limits on genetic modification.

Example:

1. Identify stakeholders (patients, doctors, future generations, governments).
2. Analyze ethical dilemmas (curing diseases vs. designer babies, accessibility vs. privilege).
3. Apply ethical models:

   - **Utilitarianism**: Reducing disease is good, but inequality could arise.
   - **Deontology**: Genetic modification must follow ethical principles.
   - **Virtue Ethics**: Science should prioritize moral responsibility.

4. Justify ethical genetic engineering policies.

Pairing 8: The Ethics of Autonomous Weapons in Warfare

Big Skill: Assessing Stakeholder Impact – Evaluating how AI weapons affect military strategy, civilians, and global security.

Small Skill: Justifying Ethical Decisions – Arguing whether autonomous weapons should be banned.

Teaching Technique: Ethical impact mapping.

Actions: Have students analyze military case studies and debate AI weapon policies.

Example:

1. Identify stakeholders (militaries, civilians, AI developers, policymakers).
2. Analyze ethical dilemmas (reduced soldier casualties vs. loss of human control).
3. Apply ethical models:

- **Utilitarianism**: AI reduces troop deaths but may escalate conflicts.
- **Deontology**: War decisions should not be left to AI.
- **Virtue Ethics**: Ethical war conduct requires human judgment.

4. Justify international policies on AI weapons.

Pairing 9: Ethical Considerations in Organ Transplant Prioritization

Big Skill: Assessing Stakeholder Impact – Understanding how transplant allocation affects patients, doctors, and policymakers.

Small Skill: Justifying Ethical Decisions – Arguing ethical organ distribution models.

Teaching Technique: Medical ethics debate.

Actions: Have students evaluate allocation methods and justify ethical selection criteria.

Example:

1. Identify stakeholders (patients, doctors, organ donors, policymakers).
2. Analyze ethical dilemmas (age vs. need, wealth vs. fairness).
3. Apply ethical models:

   - **Utilitarianism**: Maximizing life-years.
   - **Deontology**: Equal access to organs.
   - **Virtue Ethics**: Compassionate medical ethics.

4. Justify fair organ transplant policies.

Pairing 10: The Ethics of Corporate Influence in Political Campaigns

Big Skill: Assessing Stakeholder Impact – Understanding how corporate donations affect democracy, elections, and policymaking.

Small Skill: Justifying Ethical Decisions – Forming structured arguments on whether corporations should be allowed to finance political campaigns.

Teaching Technique: Policy analysis and ethical debate.

Actions: Have students analyze campaign finance laws and justify ethical reforms.

Example:

1. Identify stakeholders (corporations, politicians, voters, advocacy groups).
2. Analyze ethical dilemmas (corporate free speech vs. electoral fairness, public influence vs. political corruption).
3. Apply ethical models:

   - **Utilitarianism**: Political donations help candidates but may bias laws toward corporate interests.
   - **Deontology**: Elections should be free from undue influence.

- **Virtue Ethics**: Ethical democracy should prioritize transparency and fairness.

4. Justify ethical campaign finance policies and propose reforms (e.g., donation limits, transparency laws, public funding models).

Ethical Decision Making – (Big Skill) Weighing Competing Values vs. (Small Skill) Breaking Down Ethical Dilemmas

---

Pairing 1: Corporate Profits vs. Environmental Responsibility

Big Skill: Weighing Competing Values – Balancing business profitability with environmental sustainability.

Small Skill: Breaking Down Ethical Dilemmas – Identifying conflicts between shareholder returns and ecological impact.

Teaching Technique: Case study analysis.

Actions: Have students evaluate a corporation's decision to use cheaper but environmentally harmful materials.

Example:

1. Identify stakeholders (corporate executives, consumers, environmental groups, regulators).
2. Analyze conflicts (higher profits vs. long-term sustainability, corporate image vs. operational costs).
3. Explore solutions (sustainable production methods, government incentives, consumer education).
4. Develop a policy balancing profit and sustainability for companies.

Pairing 2: Employee Well-being vs. Competitive Business Demands

Big Skill: Weighing Competing Values – Balancing worker productivity with fair labor conditions.

Small Skill: Breaking Down Ethical Dilemmas – Identifying conflicts between corporate efficiency and worker health.

Teaching Technique: Ethical impact assessment.

Actions: Have students analyze whether companies should implement a four-day workweek despite potential profit reduction.

Example:

1. Identify stakeholders (employees, business owners, customers, labor unions).
2. Analyze conflicts (work-life balance vs. productivity, labor costs vs. worker satisfaction).

3. Explore solutions (trial periods, productivity assessments, flexible schedules).
4. Develop a company policy that optimizes both employee well-being and business performance.

Pairing 3: Free Speech vs. Content Moderation on Social Media

Big Skill: Weighing Competing Values – Balancing free expression with responsible digital content oversight.

Small Skill: Breaking Down Ethical Dilemmas – Identifying conflicts between open discourse and the prevention of harmful content.

Teaching Technique: Role-playing different perspectives.

Actions: Have students debate whether social media platforms should regulate controversial content.

Example:

1. Identify stakeholders (platform users, social media companies, regulators, advocacy groups).
2. Analyze conflicts (free speech vs. misinformation, platform liability vs. user rights).
3. Explore solutions (clearer guidelines, AI-driven moderation, user-driven reporting).
4. Develop a balanced content moderation policy for social media platforms.

Pairing 4: Drug Pricing and Access to Healthcare

Big Skill: Weighing Competing Values – Balancing pharmaceutical innovation with affordable drug pricing.

Small Skill: Breaking Down Ethical Dilemmas – Identifying conflicts between research investment, drug costs, and patient access.

Teaching Technique: Structured debate.

Actions: Have students evaluate whether life-saving medications should be patent-protected or made publicly available.

Example:

1. Identify stakeholders (patients, pharmaceutical companies, insurance providers, governments).
2. Analyze conflicts (high R&D costs vs. affordability, profit-driven vs. humanitarian approaches).
3. Explore solutions (government subsidies, generic drug access, public-private partnerships).
4. Develop ethical pricing policies for essential medications.

Pairing 5: Artificial Intelligence in Hiring Decisions

Big Skill: Weighing Competing Values – Balancing efficiency and fairness in AI-driven hiring.

Small Skill: Breaking Down Ethical Dilemmas – Identifying conflicts between automation, bias, and human oversight.

Teaching Technique: Case study analysis.

Actions: Have students assess whether AI should replace human recruiters in hiring.

Example:

1. Identify stakeholders (job seekers, companies, HR departments, AI developers).
2. Analyze conflicts (efficiency vs. potential bias, cost-saving vs. fairness).
3. Explore solutions (AI-human collaboration, bias audits, transparency policies).
4. Develop ethical AI guidelines for hiring processes.

Pairing 6: Corporate Donations and Political Influence

Big Skill: Weighing Competing Values – Balancing corporate free speech with fair democratic processes.

Small Skill: Breaking Down Ethical Dilemmas – Identifying conflicts between business interests and public trust in elections.

Teaching Technique: Policy analysis.

Actions: Have students debate whether corporations should fund political campaigns.

Example:

1. Identify stakeholders (voters, political candidates, corporate leaders, lawmakers).
2. Analyze conflicts (freedom to donate vs. undue influence, corporate interests vs. public trust).
3. Explore solutions (donation limits, public campaign funding, transparency laws).
4. Develop an ethical framework for corporate political contributions.

Pairing 7: Ethical Considerations in Workplace Surveillance

Big Skill: Weighing Competing Values – Balancing workplace security with employee privacy.

Small Skill: Breaking Down Ethical Dilemmas – Identifying conflicts between monitoring productivity and respecting worker rights.

Teaching Technique: Scenario-based discussion.

Actions: Have students evaluate whether companies should use AI to monitor employees.

Example:

1. Identify stakeholders (employees, managers, HR departments, tech developers).

2. Analyze conflicts (security vs. privacy, productivity tracking vs. trust).
3. Explore solutions (transparent policies, opt-in monitoring, performance-based evaluations).
4. Develop ethical surveillance guidelines for companies.

Pairing 8: The Ethics of Fast Fashion

Big Skill: Weighing Competing Values – Balancing consumer demand for affordable clothing with fair labor practices.

Small Skill: Breaking Down Ethical Dilemmas – Identifying conflicts between low-cost production, worker conditions, and environmental impact.

Teaching Technique: Ethical supply chain analysis.

Actions: Have students assess the ethical impact of fast fashion brands and propose alternatives.

Example:

1. Identify stakeholders (consumers, garment workers, fashion brands, environmentalists).
2. Analyze conflicts (affordable clothing vs. worker exploitation, industry growth vs. sustainability).
3. Explore solutions (ethical sourcing, sustainable materials, fair-trade labor).
4. Develop a corporate responsibility plan for fashion retailers.

Pairing 9: Privacy vs. National Security in Government Surveillance

Big Skill: Weighing Competing Values – Balancing public safety with personal privacy rights.

Small Skill: Breaking Down Ethical Dilemmas – Identifying conflicts between national security interests and civil liberties.

Teaching Technique: Case study debate.

Actions: Have students analyze real-world surveillance programs and propose ethical policies.

Example:

1. Identify stakeholders (citizens, government agencies, cybersecurity experts, advocacy groups).
2. Analyze conflicts (public safety vs. government overreach, data privacy vs. threat prevention).
3. Explore solutions (oversight committees, legal safeguards, transparency).
4. Develop ethical national security policies balancing privacy and protection.

Pairing 10: The Ethics of Autonomous Weapons in Warfare

Big Skill: Weighing Competing Values – Balancing military efficiency with moral responsibility.

Small Skill: Breaking Down Ethical Dilemmas – Identifying conflicts between technological warfare and ethical combat practices.

Teaching Technique: Ethical impact simulation.

Actions: Have students assess the use of autonomous weapons and propose regulations.

Example:

1. Identify stakeholders (militaries, civilians, policymakers, AI developers).
2. Analyze conflicts (reducing human casualties vs. dehumanizing warfare, strategic advantages vs. accountability).
3. Explore solutions (human oversight, UN regulations, restricted deployment).
4. Develop ethical military policies regarding AI weaponry.

Ethical Decision Making – (Big Skill) Weighing Competing Values vs. (Small Skill) Applying Case Study Analysis

---

Pairing 1: Facebook and User Data Monetization

Big Skill: Weighing Competing Values – Balancing business profitability with consumer privacy and digital ethics.

Small Skill: Applying Case Study Analysis – Evaluating how Facebook's data collection practices impact users and society.

Teaching Technique: Case study analysis.

Actions: Have students assess the ethical implications of Facebook selling user data to third parties.

Example:

1. Identify key stakeholders (users, advertisers, regulators, Facebook executives).
2. Analyze ethical dilemmas (monetization vs. privacy, targeted ads vs. manipulation).
3. Evaluate lessons learned (GDPR regulations, consumer trust).
4. Develop ethical data policies for social media platforms.

Pairing 2: The Boeing 737 MAX Scandal

Big Skill: Weighing Competing Values – Balancing corporate profit with passenger safety.

Small Skill: Applying Case Study Analysis – Investigating how Boeing's rushed production led to fatal design flaws.

Teaching Technique: Ethical impact assessment.

Actions: Have students analyze Boeing's decision to prioritize cost savings over additional safety measures.

Example:

1. Identify key stakeholders (Boeing engineers, airlines, passengers, regulators).
2. Analyze ethical dilemmas (profits vs. passenger safety, speed-to-market vs. quality control).
3. Evaluate consequences (loss of trust, lawsuits, FAA scrutiny).
4. Develop corporate transparency guidelines for aviation safety.

Pairing 3: Amazon's Labor Practices in Warehouses

Big Skill: Weighing Competing Values – Balancing worker well-being with operational efficiency and consumer demands.

Small Skill: Applying Case Study Analysis – Examining ethical concerns in Amazon's warehouse working conditions.

Teaching Technique: Role-playing different perspectives.

Actions: Have students assess whether Amazon should implement stronger labor protections despite cost increases.

Example:

1. Identify key stakeholders (Amazon executives, warehouse workers, consumers, labor unions).
2. Analyze ethical dilemmas (fast shipping vs. worker well-being, automation vs. job security).
3. Evaluate past corporate actions (wage increases, unionization efforts).
4. Develop ethical labor policies that balance efficiency with fair working conditions.

Pairing 4: The Theranos Fraud Case

Big Skill: Weighing Competing Values – Balancing medical innovation with honesty and consumer protection.

Small Skill: Applying Case Study Analysis – Reviewing how Theranos misled investors and patients.

Teaching Technique: Structured ethical debate.

Actions: Have students examine Elizabeth Holmes' decisions and their impact on stakeholders.

Example:

1. Identify key stakeholders (investors, patients, regulators, Theranos executives).

2. Analyze ethical dilemmas (funding growth vs. scientific integrity, secrecy vs. transparency).
3. Evaluate regulatory failures (lack of oversight, whistleblower suppression).
4. Develop ethical leadership guidelines for biotech startups.

Pairing 5: The Rana Plaza Factory Collapse

Big Skill: Weighing Competing Values – Balancing global supply chain efficiency with worker safety and human rights.

Small Skill: Applying Case Study Analysis – Examining the ethical failings that led to a tragic factory collapse in Bangladesh.

Teaching Technique: Ethical supply chain assessment.

Actions: Have students evaluate corporate responsibility in overseas manufacturing.

Example:

1. Identify key stakeholders (factory workers, fast fashion brands, consumers, regulators).
2. Analyze ethical dilemmas (low-cost production vs. fair wages, safety vs. profit).
3. Evaluate responses (corporate audits, sustainability pledges).
4. Develop ethical manufacturing guidelines for global brands.

Pairing 6: Google's Censorship Agreement with China

Big Skill: Weighing Competing Values – Balancing access to foreign markets with free speech and human rights.

Small Skill: Applying Case Study Analysis – Investigating Google's ethical decision to comply with Chinese censorship laws.

Teaching Technique: Policy and ethics debate.

Actions: Have students assess whether Google should prioritize profit over free expression.

Example:

1. Identify key stakeholders (Chinese users, Google executives, international governments, free speech advocates).
2. Analyze ethical dilemmas (economic expansion vs. supporting censorship).
3. Evaluate lessons learned (withdrawal from China, ethical tech policies).
4. Develop ethical globalization strategies for multinational companies.

Pairing 7: The Volkswagen Emissions Scandal

Big Skill: Weighing Competing Values – Balancing corporate competitiveness with environmental sustainability and regulatory compliance.

Small Skill: Applying Case Study Analysis – Reviewing Volkswagen's ethical violations in emissions testing.

Teaching Technique: Case study deconstruction.

Actions: Have students examine the impact of Volkswagen's deception on various stakeholders.

Example:

1. Identify key stakeholders (consumers, regulatory agencies, Volkswagen executives, environmental groups).
2. Analyze ethical dilemmas (performance vs. emissions compliance, profit vs. environmental impact).
3. Evaluate corporate consequences (fines, reputation damage, legal actions).
4. Develop an ethical compliance strategy for automakers.

Pairing 8: AI and Facial Recognition Surveillance

Big Skill: Weighing Competing Values – Balancing security advancements with privacy rights and civil liberties.

Small Skill: Applying Case Study Analysis – Investigating ethical concerns in government and corporate facial recognition use.

Teaching Technique: Ethical scenario simulations.

Actions: Have students analyze cases where facial recognition was misused and propose ethical solutions.

Example:

1. Identify key stakeholders (citizens, law enforcement, tech companies, privacy advocates).
2. Analyze ethical dilemmas (crime prevention vs. racial bias, security vs. personal privacy).
3. Evaluate policy responses (bans, regulations, AI transparency requirements).
4. Develop a responsible AI usage framework for security applications.

Pairing 9: SpaceX and the Ethics of Commercial Space Travel

Big Skill: Weighing Competing Values – Balancing scientific progress, privatization, and global equity in space travel.

Small Skill: Applying Case Study Analysis – Examining the ethical challenges of private-sector space exploration.

Teaching Technique: Future policy simulation.

Actions: Have students evaluate whether space colonization should be private, government-controlled, or internationally regulated.

Example:

1. Identify key stakeholders (governments, private space companies, scientists, developing nations).
2. Analyze ethical dilemmas (profit-driven exploration vs. fair access, environmental impact of launches).
3. Evaluate past industry trends (NASA partnerships, monopolization concerns).
4. Develop ethical space policies for commercial exploration.

Pairing 10: The Opioid Crisis and Pharmaceutical Ethics

Big Skill: Weighing Competing Values – Balancing pharmaceutical profits with public health and social responsibility.

Small Skill: Applying Case Study Analysis – Investigating Purdue Pharma's role in the opioid epidemic.

Teaching Technique: Public health ethics discussion.

Actions: Have students analyze corporate responsibility and propose regulations for preventing future drug crises.

Example:

1. Identify key stakeholders (patients, doctors, pharmaceutical executives, regulators).
2. Analyze ethical dilemmas (profit-driven marketing vs. public health, addiction risks vs. pain management).
3. Evaluate regulatory failures (lack of oversight, deceptive marketing tactics).
4. Develop corporate responsibility frameworks for pharmaceutical companies.

Ethical Decision Making – (Big Skill) Weighing Competing Values vs. (Small Skill) Justifying Ethical Decisions

---

Pairing 1: The Ethics of Artificial Intelligence in Hiring

Big Skill: Weighing Competing Values – Balancing efficiency, fairness, and bias in AI-driven hiring processes.

Small Skill: Justifying Ethical Decisions – Using ethical models to evaluate whether AI should be used in recruitment.

Teaching Technique: Structured debate.

Actions: Have students argue whether AI-driven hiring is ethical or if human oversight is necessary.

Example:

1. Identify stakeholders (job seekers, employers, AI developers, policymakers).
2. Analyze conflicts (efficiency vs. fairness, cost-saving vs. discrimination).
3. Apply ethical models:
   - **Utilitarianism:** AI streamlines hiring but may reinforce biases.
   - **Deontology:** Hiring should be based on ethical, transparent criteria.
   - **Virtue Ethics:** Employers should prioritize fairness over profit.
4. Justify ethical hiring policies balancing automation and human decision-making.

Pairing 2: Corporate Social Responsibility vs. Profit Maximization

Big Skill: Weighing Competing Values – Balancing corporate financial success with societal well-being.

Small Skill: Justifying Ethical Decisions – Defending or opposing corporate initiatives for social responsibility.

Teaching Technique: Policy evaluation.

Actions: Have students assess whether corporations should prioritize social initiatives over profit.

Example:

1. Identify stakeholders (shareholders, employees, consumers, communities).
2. Analyze conflicts (higher profits vs. sustainability, short-term gains vs. long-term responsibility).
3. Apply ethical models:
   - **Utilitarianism:** CSR benefits society but may hurt profitability.
   - **Deontology:** Companies have a duty to act ethically, regardless of profit.
   - **Virtue Ethics:** Ethical businesses build long-term consumer trust.
4. Justify CSR policies that balance ethics and profit.

Pairing 3: Fast Fashion and Labor Ethics

Big Skill: Weighing Competing Values – Balancing affordability, profitability, and fair labor practices.

Small Skill: Justifying Ethical Decisions – Using ethical reasoning to defend or criticize fast fashion.

Teaching Technique: Ethical supply chain analysis.

Actions: Have students evaluate whether fashion companies should raise prices to improve worker conditions.

Example:

1. Identify stakeholders (factory workers, consumers, fashion brands, environmental groups).
2. Analyze conflicts (low prices vs. fair wages, fast production vs. ethical sourcing).
3. Apply ethical models:
    - **Utilitarianism:** Low-cost clothing benefits many, but at workers' expense.
    - **Deontology:** Workers' rights should be respected regardless of business costs.
    - **Virtue Ethics:** Ethical fashion should prioritize human dignity.
4. Justify ethical sourcing policies for global fashion brands.

Pairing 4: Consumer Privacy vs. Targeted Advertising

Big Skill: Weighing Competing Values – Balancing business growth with data privacy rights.

Small Skill: Justifying Ethical Decisions – Forming structured arguments for or against data collection.

Teaching Technique: Role-playing different perspectives.

Actions: Have students debate whether targeted advertising is ethical.

Example:

1. Identify stakeholders (consumers, advertisers, tech companies, regulators).
2. Analyze conflicts (personalization vs. surveillance, profit vs. privacy).
3. Apply ethical models:
    - **Utilitarianism:** Ads benefit users but erode privacy.
    - **Deontology:** Privacy should not be violated for profit.
    - **Virtue Ethics:** Ethical companies should be transparent.
4. Justify privacy policies balancing user rights and business needs.

Pairing 5: The Ethics of Autonomous Vehicles

Big Skill: Weighing Competing Values – Balancing safety, innovation, and job displacement.

Small Skill: Justifying Ethical Decisions – Evaluating the impact of self-driving cars using ethical reasoning.

Teaching Technique: Ethical impact assessment.

Actions: Have students argue for or against self-driving cars based on ethical principles.

Example:

1. Identify stakeholders (drivers, tech companies, consumers, policymakers).
2. Analyze conflicts (fewer accidents vs. job losses, AI control vs. human oversight).

3. Apply ethical models:
    - **Utilitarianism:** Fewer accidents benefit society, but truckers lose jobs.
    - **Deontology:** Human drivers should retain control over machines.
    - **Virtue Ethics:** Companies should transition workers ethically.
4. Justify policies for responsible self-driving car adoption.

Pairing 6: Pharmaceutical Pricing Ethics

Big Skill: Weighing Competing Values – Balancing research investment, accessibility, and corporate profit.

Small Skill: Justifying Ethical Decisions – Defending or challenging high drug prices.

Teaching Technique: Structured ethical debate.

Actions: Have students evaluate whether life-saving medications should be patented or publicly funded.

Example:

1. Identify stakeholders (patients, pharmaceutical companies, regulators, insurance providers).
2. Analyze conflicts (high R&D costs vs. affordability, patents vs. access).
3. Apply ethical models:
    - **Utilitarianism:** High prices fund innovation but harm the poor.
    - **Deontology:** Health should not be commodified.
    - **Virtue Ethics:** Ethical pharma companies prioritize access.
4. Justify ethical pricing policies that balance innovation and affordability.

Pairing 7: Workplace Diversity Initiatives

Big Skill: Weighing Competing Values – Balancing diversity, meritocracy, and business goals.

Small Skill: Justifying Ethical Decisions – Using ethical reasoning to evaluate diversity programs.

Teaching Technique: Policy analysis.

Actions: Have students assess whether diversity hiring quotas are ethical.

Example:

1. Identify stakeholders (job applicants, companies, employees, advocacy groups).
2. Analyze conflicts (equal opportunity vs. preferential treatment).
3. Apply ethical models:
    - **Utilitarianism:** Diversity strengthens businesses but may create tension.

- **Deontology:** All hiring should be fair and unbiased.
- **Virtue Ethics:** Ethical businesses value inclusivity.

4. Justify diversity hiring policies balancing fairness and effectiveness.

Pairing 8: AI in Predictive Policing

Big Skill: Weighing Competing Values – Balancing crime prevention, civil liberties, and racial justice.

Small Skill: Justifying Ethical Decisions – Evaluating AI's ethical role in law enforcement.

Teaching Technique: Scenario-based learning.

Actions: Have students debate whether predictive policing should be used.

Example:

1. Identify stakeholders (law enforcement, civilians, policymakers, AI developers).
2. Analyze conflicts (crime reduction vs. racial bias, efficiency vs. fairness).
3. Apply ethical models:
   - **Utilitarianism:** AI prevents crime but may target minorities unfairly.
   - **Deontology:** Policing must be impartial.
   - **Virtue Ethics:** Fairness and justice should guide policing.
4. Justify ethical policies for AI in law enforcement.

Pairing 9: Corporate Influence in Politics

Big Skill: Weighing Competing Values – Balancing free speech, corporate interests, and democratic fairness.

Small Skill: Justifying Ethical Decisions – Arguing for or against corporate political donations.

Teaching Technique: Ethical impact mapping.

Actions: Have students analyze corporate influence on elections and propose ethical reforms.

Example:

1. Identify stakeholders (corporations, politicians, voters, advocacy groups).
2. Analyze conflicts (free speech vs. corruption, business interests vs. democracy).
3. Apply ethical models:
   - **Utilitarianism:** Donations support policies but distort democracy.
   - **Deontology:** Elections should be fair, not bought.
   - **Virtue Ethics:** Politicians should act ethically, not for donors.
4. Justify campaign finance policies balancing fairness and free speech.

Pairing 10: The Ethics of Biotechnology and Human Genetic Engineering

Big Skill: Weighing Competing Values – Balancing medical advancements, human dignity, and ethical responsibility.

Small Skill: Justifying Ethical Decisions – Using structured ethical reasoning to assess the moral implications of genetic modifications.

Teaching Technique: Bioethics case study analysis.

Actions: Have students evaluate whether genetic editing technologies like CRISPR should be used to modify human embryos.

Example:

1. Identify stakeholders (scientists, medical professionals, future generations, regulators, ethicists).
2. Analyze conflicts (curing diseases vs. designer babies, accessibility vs. socioeconomic inequality).
3. Apply ethical models:
    - **Utilitarianism:** Genetic engineering could eliminate diseases but might widen social inequality.
    - **Deontology:** Modifying human life raises ethical concerns about playing "God."
    - **Virtue Ethics:** Responsible science should focus on human well-being and fairness.
4. Justify bioethical policies that balance medical progress with ethical considerations, such as regulating genetic modifications only for severe medical conditions.

Example Teaching Strategies for Big vs. Small Thinking Skill Pairings of Leadership Thinking at the College/Technical School Level

Here are 90 unique teaching strategy examples for the big thinking and small thinking skill pairings, each described in detail to align with **Big Thinking Skills** (Concepts & Strategies) and **Small Thinking Skills** (Step-by-Step Execution) for the **Leadership Thinking Skill** taught at the college/technical school level.

Leadership Thinking

Big Thinking Skills (Concepts & Strategies)

- **Visionary & Strategic Alignment:** Understanding how to align team goals with broader organizational objectives.
- **Influence & Decision-Making:** Developing the ability to inspire, guide, and persuade others.
- **Situational Adaptability:** Adjusting leadership style based on team needs, crises, or strategic shifts.

Small Thinking Skills (Step-by-Step Execution)

- **Delegating & Prioritizing Tasks:** Breaking down objectives into actionable steps for team members.
- **Facilitating Team Communication:** Leading discussions, resolving conflicts, and ensuring clear messaging.
- **Tracking Leadership Effectiveness:** Using performance metrics and feedback loops to assess and improve leadership impact.

Leadership Thinking – (Big Skill) Visionary & Strategic Alignment vs. (Small Skill) Delegating & Prioritizing Tasks

---

Pairing 1: Launching a New Product in a Tech Startup

Big Skill: Visionary & Strategic Alignment – Aligning product development with the company's mission and market needs.

Small Skill: Delegating & Prioritizing Tasks – Assigning key roles in engineering, marketing, and sales.

Teaching Technique: Strategic project simulation.

Actions: Have students create a launch strategy, delegate tasks among teams, and prioritize deadlines.

Example:

1. Define strategic goals (market penetration, user engagement).
2. Break down tasks (engineers build MVP, marketing sets outreach plan, sales identify key accounts).
3. Delegate responsibilities (team leads own their domains).
4. Prioritize actions (focus first on MVP, then marketing, then sales).

Pairing 2: Leading a Corporate Rebranding Initiative

Big Skill: Visionary & Strategic Alignment – Ensuring that the rebranding reflects long-term business strategy.

Small Skill: Delegating & Prioritizing Tasks – Assigning responsibilities to design, communications, and legal teams.

Teaching Technique: Role-based leadership exercise.

Actions: Have students plan and execute a fictional company rebranding.

Example:

1. Set vision (modernized branding for new audience).
2. Assign roles (graphic designers for logo, marketers for messaging, legal for trademarks).
3. Prioritize steps (market research → design → testing → rollout).
4. Evaluate alignment (does the new brand match company values?).

Pairing 3: Managing a Nonprofit Fundraising Campaign

Big Skill: Visionary & Strategic Alignment – Aligning fundraising efforts with the nonprofit is mission.

Small Skill: Delegating & Prioritizing Tasks – Assigning responsibilities to fundraising, outreach, and finance teams.

Teaching Technique: Case study planning.

Actions: Have students design a fundraising campaign and allocate resources.

Example:

1. Define strategic goal ($100K fundraising target for education programs).
2. Assign roles (event planners for fundraising gala, marketing for donor outreach, finance for budgeting).
3. Prioritize actions (sponsorship outreach → event planning → donor engagement).
4. Track performance (adjust efforts based on donor responses).

Pairing 4: Organizing a Large-Scale Corporate Training Program

Big Skill: Visionary & Strategic Alignment – Ensuring that employee training aligns with company growth goals.

Small Skill: Delegating & Prioritizing Tasks – Assigning tasks to HR, department leads, and training facilitators.

Teaching Technique: Leadership scenario role-play.

Actions: Have students develop a training roadmap and assign responsibilities.

Example:

1. Set leadership vision (upskilling employees for digital transformation).
2. Break down tasks (HR for logistics, department heads for content, trainers for execution).
3. Prioritize execution (pilot program → company-wide rollout → feedback assessment).
4. Monitor alignment (does training improve key performance areas?).

Pairing 5: Expanding a Business into International Markets

Big Skill: Visionary & Strategic Alignment – Aligning expansion strategy with long-term growth objectives.

Small Skill: Delegating & Prioritizing Tasks – Assigning market research, compliance, and operational setup.

Teaching Technique: International business simulation.

Actions: Have students plan an expansion strategy, delegate responsibilities, and set priorities.

Example:

1. Set strategic goal (expand into European market in one year).
2. Assign roles (research team for market analysis, legal for regulations, operations for supply chain).
3. Prioritize execution (regulatory approvals → partnerships → marketing launch).
4. Track success (adjust approach based on early market feedback).

Pairing 6: Crisis Management in a Corporate Setting

Big Skill: Visionary & Strategic Alignment – Aligning crisis response with company reputation management.

Small Skill: Delegating & Prioritizing Tasks – Assigning PR, legal, and operational crisis teams.

Teaching Technique: Crisis response simulation

Actions: Have students develop a crisis response plan and delegate key roles.

Example:

1. Define crisis objectives (mitigate reputation damage, ensure transparency).
2. Assign responsibilities (PR handles communication, legal assesses liability, ops fixes root cause).
3. Prioritize actions (internal briefing → public statement → corrective measures).
4. Evaluate outcomes (does response align with company values?).

Pairing 7: Leading a Sustainability Initiative for a Corporation

Big Skill: Visionary & Strategic Alignment – Ensuring environmental sustainability aligns with business operations.

Small Skill: Delegating & Prioritizing Tasks – Assigning sustainability research, policy implementation, and reporting.

Teaching Technique: ESG (Environmental, Social, Governance) case study.

Actions: Have students create a corporate sustainability plan with clear delegation.

Example:

1. Set vision (net-zero carbon emissions by 2035).
2. Assign teams (research for energy alternatives, HR for sustainability culture, compliance for reporting).
3. Prioritize phases (energy efficiency upgrades → supply chain changes → carbon offset programs).
4. Monitor progress (KPIs for emissions reduction).

Pairing 8: Managing a High-Performance Sales Team

Big Skill: Visionary & Strategic Alignment – Aligning sales strategies with broader revenue targets.

Small Skill: Delegating & Prioritizing Tasks – Assigning sales goals, client segmentation, and marketing strategies.

Teaching Technique: Competitive strategy simulation.

Actions: Have students design a sales strategy and delegate roles accordingly.

Example:

1. Set revenue target ($10M in new contracts in 12 months).
2. Assign roles (lead gen team for prospecting, sales reps for conversion, marketers for brand positioning).
3. Prioritize approach (target high-value clients first → leverage referral network → expand to mid-market).
4. Track performance (realign strategy based on sales data).

Pairing 9: Leading a Digital Transformation Initiative

Big Skill: Visionary & Strategic Alignment – Aligning digital adoption with business efficiency goals.

Small Skill: Delegating & Prioritizing Tasks – Assigning tech implementation, employee training, and process redesign.

Teaching Technique: Technology adoption case study.

Actions: Have students develop a digital transformation roadmap and allocate responsibilities.

Example:

1. Define transformation goal (migrate all operations to cloud by 2025).
2. Assign teams (IT for cloud infrastructure, HR for employee training, finance for budgeting).
3. Prioritize implementation (pilot program → company-wide rollout → feedback integration).
4. Measure success (efficiency gains, employee adoption rates).

Pairing 10: Merging Two Corporate Cultures Post-Acquisition

Big Skill: Visionary & Strategic Alignment – Ensuring organizational alignment during a company merger.

Small Skill: Delegating & Prioritizing Tasks – Assigning HR integration, policy standardization, and leadership roles.

Teaching Technique: Organizational change simulation.

Actions: Have students develop an integration plan and prioritize key tasks.

Example:

1. Set vision (seamless cultural and operational integration within 12 months).
2. Assign responsibilities (HR for employee engagement, legal for policy alignment, executives for vision-setting).
3. Prioritize efforts (leadership alignment → employee onboarding → operational merging).
4. Monitor cultural fit (employee surveys, retention rates).

Leadership Thinking – (Big Skill) Visionary & Strategic Alignment vs. (Small Skill) Facilitating Team Communication

---

Pairing 1: Aligning a Cross-Functional Team for Product Development

Big Skill: Visionary & Strategic Alignment – Ensuring product goals align with the company's market strategy.

Small Skill: Facilitating Team Communication – Encouraging collaboration between design, engineering, and marketing teams.

Teaching Technique: Agile project simulation.

Actions: Have students lead a simulated product development meeting, ensuring clear communication and resolving conflicts.

Example:

1. Set vision (develop an eco-friendly smart device in six months).
2. Facilitate team meetings (ensure engineers, designers, and marketers communicate priorities).
3. Resolve conflicts (mediate between cost-efficient designs and high-end features).
4. Ensure alignment (use a product roadmap to keep teams on track).

Pairing 2: Leading a Corporate Strategy Session for a Business Expansion

Big Skill: Visionary & Strategic Alignment – Aligning a company's expansion goals with market trends.

Small Skill: Facilitating Team Communication – Ensuring executives, analysts, and operations teams communicate clearly.

Teaching Technique: Business case analysis.

Actions: Have students role-play as executives leading an expansion strategy meeting.

Example:

1. Define expansion objectives (launch in two new international markets).
2. Facilitate discussions (finance team addresses budget, marketing ensures localization).
3. Resolve conflicts (risk analysis vs. growth projections).
4. Summarize decisions (create an executive report for stakeholders).

Pairing 3: Managing a Team During Organizational Change

Big Skill: Visionary & Strategic Alignment – Ensuring teams remain motivated during a company restructuring.

Small Skill: Facilitating Team Communication – Leading open discussions and addressing employee concerns.

Teaching Technique: Change management workshop.

Actions: Have students simulate a leadership role during a restructuring announcement.

Example:

1. Communicate strategic goals (explain why the restructuring is happening).
2. Hold town hall meetings (address employee fears and uncertainties).
3. Encourage open dialogue (let employees voice concerns and provide feedback).
4. Provide clear next steps (send follow-up documentation on changes).

Pairing 4: Leading a Crisis Response Team

Big Skill: Visionary & Strategic Alignment – Aligning crisis response strategies with company values.

Small Skill: Facilitating Team Communication – Managing emergency communications internally and externally.

Teaching Technique: Crisis response simulation.

Actions: Have students develop a communication plan during a simulated PR crisis.

Example:

1. Identify the crisis (data breach affecting customers).
2. Coordinate with teams (IT for solutions, PR for messaging, legal for compliance).
3. Hold daily briefings (keep employees informed and aligned).
4. Issue a public statement (ensure transparency and accountability).

Pairing 5: Leading a Diversity & Inclusion Initiative

Big Skill: Visionary & Strategic Alignment – Aligning company culture with diversity, equity, and inclusion goals.

Small Skill: Facilitating Team Communication – Leading discussions about inclusivity and policy improvements.

Teaching Technique: Ethical leadership forum.

Actions: Have students facilitate a discussion on improving diversity initiatives within a company.

Example:

1. Define strategic DEI goals (increase leadership diversity by 20% in five years).
2. Encourage employee input (gather feedback on workplace inclusivity).
3. Address conflicts (resolve misunderstandings about DEI initiatives).
4. Develop a DEI action plan (publish policies and progress updates).

Pairing 6: Overseeing a Remote Team's Performance

Big Skill: Visionary & Strategic Alignment – Ensuring remote teams stay engaged and aligned with company objectives.

Small Skill: Facilitating Team Communication – Establishing effective virtual communication channels.

Teaching Technique: Remote leadership simulation.

Actions: Have students develop a virtual communication strategy for a remote team.

Example:

1. Define remote work policies (set clear performance metrics).
2. Host regular check-ins (weekly video meetings to align goals).
3. Resolve communication gaps (clarify expectations across time zones).
4. Implement collaboration tools (use project management software to track progress).

Pairing 7: Running a Company-Wide Innovation Challenge

Big Skill: Visionary & Strategic Alignment – Encouraging employees to contribute to company-wide innovation.

Small Skill: Facilitating Team Communication – Organizing brainstorming sessions and cross-team discussions.

Teaching Technique: Innovation lab exercise.

Actions: Have students design an internal hackathon where teams pitch new ideas.

Example:

1. Set the innovation goal (develop a cost-saving initiative or new product idea).
2. Encourage idea sharing (use breakout groups for brainstorming).
3. Moderate discussions (ensure balanced participation).
4. Summarize insights (present winning ideas to leadership).

Pairing 8: Managing Interdepartmental Collaboration for a Major Project

Big Skill: Visionary & Strategic Alignment – Ensuring different departments work cohesively on a shared objective.

Small Skill: Facilitating Team Communication – Leading cross-functional meetings and resolving conflicts.

Teaching Technique: Role-playing conflict resolution.

Actions: Have students mediate a disagreement between IT, marketing, and finance teams.

Example:

1. Define shared project goals (launch a new enterprise software).
2. Facilitate interdepartmental meetings (align goals between departments).
3. Resolve miscommunication (clarify overlapping responsibilities).
4. Document decisions (send follow-ups to ensure accountability).

Pairing 9: Leading a Sales Strategy Discussion

Big Skill: Visionary & Strategic Alignment – Aligning sales goals with company growth objectives.

Small Skill: Facilitating Team Communication – Coaching sales teams on key messaging and negotiation tactics.

Teaching Technique: Sales leadership workshop.

Actions: Have students practice leading a sales team meeting and refining messaging strategies.

Example:

1. Set sales targets (increase revenue by 15% in one year).
2. Facilitate knowledge-sharing (sales reps discuss top-performing strategies).
3. Resolve objections (train team on handling customer concerns).
4. Refine messaging (ensure consistency across all sales pitches).

Pairing 10: Driving Corporate Social Responsibility (CSR) Initiatives

Big Skill: Visionary & Strategic Alignment – Aligning CSR efforts with company brand identity and mission.

Small Skill: Facilitating Team Communication – Leading community engagement discussions and project coordination.

Teaching Technique: Ethical business case study.

Actions: Have students plan a CSR project and coordinate communication among stakeholders.

Example:

1. Set CSR goals (reduce carbon footprint by 25% over five years).
2. Facilitate internal discussions (gather input from employees and executives).
3. Engage external stakeholders (communicate with nonprofit partners).
4. Publish impact reports (track and share progress with the public).

Leadership Thinking – (Big Skill) Visionary & Strategic Alignment vs. (Small Skill) Tracking Leadership Effectiveness

---

Pairing 1: Leading a New Market Expansion Strategy

Big Skill: Visionary & Strategic Alignment – Ensuring a company's expansion plan aligns with long-term business goals.

Small Skill: Tracking Leadership Effectiveness – Using key performance indicators (KPIs) to measure leadership impact.

Teaching Technique: Market strategy simulation.

Actions: Have students develop an expansion strategy, set performance benchmarks, and track leadership effectiveness.

Example:

1. Set strategic vision (expand into three new regions in five years).
2. Define success metrics (market share growth, customer acquisition rates).
3. Implement tracking systems (monthly leadership review meetings).
4. Refine strategy (adjust expansion based on market response).

Pairing 2: Leading a Company-Wide Digital Transformation

Big Skill: Visionary & Strategic Alignment – Ensuring digital adoption aligns with operational efficiency goals.

Small Skill: Tracking Leadership Effectiveness – Measuring employee adoption and impact on productivity.

Teaching Technique: Digital change management exercise.

Actions: Have students design a technology adoption plan and track leadership impact using feedback loops.

Example:

1. Set transformation goals (100% digital transition within two years).
2. Monitor employee engagement (survey team on training effectiveness).
3. Measure efficiency improvements (track cost savings and work productivity).
4. Adjust leadership approach (if resistance is high, provide additional training).

Pairing 3: Managing a Remote Workforce Effectively

Big Skill: Visionary & Strategic Alignment – Aligning remote work policies with business continuity plans.

Small Skill: Tracking Leadership Effectiveness – Using performance dashboards and engagement surveys.

Teaching Technique: Virtual leadership case study.

Actions: Have students develop remote leadership strategies and measure impact through data analytics.

Example:

1. Define remote work objectives (maintain pre-pandemic productivity levels).
2. Implement tracking tools (use software to measure task completion rates).
3. Conduct employee feedback sessions (assess engagement and well-being).
4. Refine policies (adjust based on productivity insights and satisfaction scores).

Pairing 4: Driving a Sustainability Initiative in a Corporation

Big Skill: Visionary & Strategic Alignment – Ensuring corporate sustainability goals align with brand and operational efficiency.

Small Skill: Tracking Leadership Effectiveness – Measuring environmental impact and leadership progress.

Teaching Technique: ESG (Environmental, Social, Governance) leadership analysis.

Actions: Have students develop sustainability benchmarks and measure leadership effectiveness.

Example:

1. Set sustainability targets (reduce carbon footprint by 30% in five years).
2. Track metrics (emission reduction, energy savings, cost impact).
3. Evaluate leadership influence (track executive buy-in and employee participation).
4. Refine strategy (adjust sustainability initiatives based on KPI results).

Pairing 5: Leading a Diversity, Equity, and Inclusion (DEI) Strategy

Big Skill: Visionary & Strategic Alignment – Ensuring DEI policies align with company culture and business goals.

Small Skill: Tracking Leadership Effectiveness – Measuring diversity metrics and leadership inclusivity ratings.

Teaching Technique: DEI impact assessment.

Actions: Have students develop a diversity initiative and track its effectiveness over time.

Example:

1. Set DEI goals (increase leadership diversity by 20% over five years).
2. Track hiring metrics (monitor demographic changes in leadership roles).
3. Measure inclusivity feedback (use employee surveys to assess company culture).
4. Adjust leadership training (implement workshops to address gaps in inclusion).

Pairing 6: Overseeing a Mergers & Acquisitions (M&A) Integration

Big Skill: Visionary & Strategic Alignment – Ensuring smooth integration of corporate cultures and processes.

Small Skill: Tracking Leadership Effectiveness – Measuring employee retention, morale, and integration success.

Teaching Technique: Business integration case study.

Actions: Have students develop an M&A leadership strategy and track its success using feedback loops.

Example:

1. Define integration objectives (retain 90% of employees post-merger).
2. Monitor transition challenges (track turnover rates and team morale).
3. Conduct pulse surveys (measure employee satisfaction with leadership).
4. Refine leadership messaging (address cultural conflicts in team meetings).

Pairing 7: Scaling a Startup from Growth to Maturity

Big Skill: Visionary & Strategic Alignment – Aligning startup expansion with long-term market positioning.

Small Skill: Tracking Leadership Effectiveness – Using revenue growth, employee engagement, and innovation tracking.

Teaching Technique: Growth-stage business simulation.

Actions: Have students plan a startup growth strategy and measure leadership effectiveness.

Example:

1. Define scale-up targets (reach $5M in revenue within three years).
2. Monitor key business metrics (profit margins, customer retention).
3. Measure leadership impact (track how leadership decisions affect employee retention and culture).
4. Adjust strategy (if growth slows, pivot market approach).

Pairing 8: Leading a Global Expansion Project

Big Skill: Visionary & Strategic Alignment – Aligning business internationalization with market needs.

Small Skill: Tracking Leadership Effectiveness – Measuring global team collaboration and operational success.

Teaching Technique: Global business expansion exercise.

Actions: Have students develop an international market entry strategy and track leadership performance.

Example:

1. Set expansion goals (launch operations in three countries).
2. Monitor cross-cultural communication (track engagement between regional offices).
3. Assess market adaptation (use customer feedback to measure effectiveness).
4. Refine global leadership approach (adjust regional leadership styles).

Pairing 9: Running a Large-Scale Corporate Innovation Initiative

Big Skill: Visionary & Strategic Alignment – Ensuring innovation efforts support long-term business growth.

Small Skill: Tracking Leadership Effectiveness – Measuring idea generation, execution rates, and team engagement.

Teaching Technique: Innovation leadership challenge.

Actions: Have students develop an innovation strategy and track its effectiveness using KPIs.

Example:

1. Set innovation goals (develop three new product lines in five years).
2. Monitor R&D investment ROI (track cost vs. market adoption rates).
3. Measure leadership influence (assess participation in ideation programs).
4. Refine initiatives (eliminate low-impact projects and scale high-potential ideas).

Pairing 10: Managing a Customer-Centric Transformation

Big Skill: Visionary & Strategic Alignment – Ensuring business strategy shifts to focus on customer experience.

Small Skill: Tracking Leadership Effectiveness – Using customer satisfaction, NPS (Net Promoter Score), and service benchmarks.

Teaching Technique: Customer experience leadership analysis.

Actions: Have students design a customer-first strategy and track leadership effectiveness through customer metrics.

Example:

1. Set customer experience goals (increase NPS by 25% in two years).
2. Monitor satisfaction scores (gather customer feedback regularly).
3. Track employee engagement (assess frontline team effectiveness).
4. Refine service strategies (adjust based on customer and employee insights).

Leadership Thinking – (Big Skill) Influence & Decision-Making vs. (Small Skill) Delegating & Prioritizing Tasks

---

Pairing 1: Leading a High-Stakes Project for a Corporate Client

Big Skill: Influence & Decision-Making – Inspiring confidence and guiding a team through complex challenges.

Small Skill: Delegating & Prioritizing Tasks – Breaking the project into clear milestones and assigning ownership.

Teaching Technique: Client project simulation.

Actions: Have students develop a strategy to complete a high-stakes project efficiently by delegating tasks.

Example:

1. Set client expectations (ensure alignment on deliverables and timeline).
2. Assign responsibilities (project manager oversees execution, marketing creates branding, finance tracks costs).
3. Prioritize critical tasks (client research → initial draft → revisions → final presentation).
4. Monitor progress (use project tracking tools to ensure deadlines are met).

Pairing 2: Running a Political Campaign for Student Government Elections

Big Skill: Influence & Decision-Making – Persuading voters, leading a campaign team, and managing public image.

Small Skill: Delegating & Prioritizing Tasks – Assigning campaign roles and focusing efforts on key voter groups.

Teaching Technique: Campaign leadership role-play.

Actions: Have students develop and execute a student government campaign strategy.

Example:

1. Define key messaging (focus on student concerns like tuition fees or campus safety).
2. Assign roles (communications handles PR, outreach mobilizes volunteers, data team analyzes polling).
3. Prioritize outreach efforts (social media → campus events → direct voter engagement).
4. Adjust strategy based on feedback (shift focus if a particular issue gains traction).

Pairing 3: Leading a Product Launch in a Startup Environment

Big Skill: Influence & Decision-Making – Convincing stakeholders and guiding a team to meet market demands.

Small Skill: Delegating & Prioritizing Tasks – Ensuring clear division of responsibilities among R&D, marketing, and sales.

Teaching Technique: Startup simulation.

Actions: Have students develop a go-to-market strategy, delegate tasks, and execute launch activities.

Example:

1. Define launch objectives (secure 10K pre-orders within three months).
2. Assign teams (product team refines the prototype, marketing builds buzz, sales secures retail partnerships).
3. Prioritize phases (finalize product → build hype → execute launch event → monitor customer feedback).
4. Evaluate success metrics (adjust strategies based on sales performance).

Pairing 4: Managing a Crisis Response for a PR Scandal

Big Skill: Influence & Decision-Making – Leading public perception management and maintaining company credibility.

Small Skill: Delegating & Prioritizing Tasks – Assigning crisis response roles and ensuring swift resolution.

Teaching Technique: Crisis simulation.

Actions: Have students develop a response plan to manage a corporate scandal effectively.

Example:

1. Identify the crisis (data breach affecting customer privacy).
2. Assign response teams (legal for compliance, PR for media messaging, IT for security patches).
3. Prioritize actions (internal briefing → customer notification → public statement → corrective measures).
4. Track impact (monitor media and public reactions to refine the response).

Pairing 5: Overseeing a Charity Fundraiser with a Large Volunteer Base

Big Skill: Influence & Decision-Making – Motivating volunteers, securing donations, and driving community engagement.

Small Skill: Delegating & Prioritizing Tasks – Coordinating fundraising efforts through structured teams.

Teaching Technique: Philanthropy leadership challenge.

Actions: Have students organize a charity event, assign volunteer roles, and track success metrics.

Example:

1. Define fundraising goals ($50K for a community cause).
2. Assign volunteer teams (event logistics, donor outreach, social media promotion).
3. Prioritize activities (corporate sponsorships → event marketing → donation collection).
4. Evaluate impact (adjust future strategies based on donor feedback).

Pairing 6: Spearheading a Corporate Sustainability Initiative

Big Skill: Influence & Decision-Making – Gaining executive buy-in and leading sustainability efforts.

Small Skill: Delegating & Prioritizing Tasks – Assigning research, implementation, and compliance monitoring.

Teaching Technique: ESG (Environmental, Social, Governance) strategic planning.

Actions: Have students develop an actionable sustainability roadmap and delegate key initiatives.

Example:

1. Set sustainability targets (reduce carbon footprint by 30% over five years).
2. Assign task forces (engineering for energy efficiency, HR for green culture, finance for cost analysis).
3. Prioritize execution (supply chain review → process optimizations → employee education).
4. Track sustainability impact (use environmental KPIs to adjust plans).

Pairing 7: Leading a Major Workforce Restructuring in an Organization

Big Skill: Influence & Decision-Making – Managing workforce morale and making tough decisions on layoffs or role changes.

Small Skill: Delegating & Prioritizing Tasks – Organizing HR teams, communication plans, and transition programs.

Teaching Technique: Organizational change simulation.

Actions: Have students design and implement a workforce restructuring strategy.

Example:

1. Define restructuring objectives (streamline operations while minimizing layoffs).
2. Assign key responsibilities (HR for employee support, PR for communication, finance for severance packages).
3. Prioritize execution (internal announcement → role transition support → external messaging).
4. Monitor leadership effectiveness (adjust support programs based on employee feedback).

Pairing 8: Driving a Digital Marketing Strategy for a Global Brand

Big Skill: Influence & Decision-Making – Persuading leadership and executing a high-impact marketing campaign.

Small Skill: Delegating & Prioritizing Tasks – Dividing responsibilities among content creators, advertisers, and analysts.

Teaching Technique: Marketing strategy simulation.

Actions: Have students develop a multi-channel marketing campaign and delegate execution roles.

Example:

1. Set marketing goals (increase brand awareness by 50% within a year).
2. Assign roles (creative team for visuals, ad buyers for paid media, data team for analytics).
3. Prioritize platforms (social media → influencer partnerships → search engine marketing).
4. Adjust based on performance data (reallocate budget to high-performing channels).

Pairing 9: Leading a Team in an Engineering R&D Project

Big Skill: Influence & Decision-Making – Encouraging collaboration and guiding technical innovation.

Small Skill: Delegating & Prioritizing Tasks – Structuring the research workflow among engineers and scientists.

Teaching Technique: Research & development simulation.

Actions: Have students lead an R&D team and distribute tasks for effective problem-solving.

Example:

1. Define project scope (develop a prototype for a new AI algorithm).
2. Assign work streams (data scientists for model training, engineers for software integration).
3. Prioritize milestones (research → testing → prototype refinement → deployment).
4. Assess progress (adjust research direction based on test results).

Pairing 10: Managing a High-Profile Entertainment or Sports Event

Big Skill: Influence & Decision-Making – Coordinating multiple stakeholders, sponsors, and media.

Small Skill: Delegating & Prioritizing Tasks – Structuring event planning and execution logistics.

Teaching Technique: Event management leadership exercise.

Actions: Have students design a large-scale event strategy with clear delegation.

Example:

1. Set event goals (sell out a 50K-seat stadium event).
2. Assign team leaders (ticketing, media relations, stage production, sponsorships).
3. Prioritize pre-event logistics (venue setup → marketing push → rehearsals).
4. Monitor execution (adjust scheduling based on real-time logistics feedback).

Leadership Thinking – (Big Skill) Influence & Decision-Making vs. (Small Skill) Facilitating Team Communication

---

Pairing 1: Leading a Cross-Functional Team to Solve a Business Challenge

Big Skill: Influence & Decision-Making – Inspiring diverse teams to align and collaborate toward a common business goal.

Small Skill: Facilitating Team Communication – Ensuring all voices are heard and conflict is managed effectively.

Teaching Technique: Cross-functional team workshop.

Actions: Have students role-play as department leaders resolving a business challenge collaboratively.

Example:

1. Define the challenge (improving customer retention by 20%).
2. Facilitate open discussion (allow marketing, sales, and customer service to share insights).
3. Resolve conflicts (mediate between marketing pushing promotions and sales pushing personal outreach).
4. Summarize strategy (document and distribute final action plan).

Pairing 2: Managing a High-Stakes Corporate Negotiation

Big Skill: Influence & Decision-Making – Using persuasion techniques to negotiate terms favorable to the organization.

Small Skill: Facilitating Team Communication – Ensuring internal alignment before and after negotiations.

Teaching Technique: Negotiation role-play exercise.

Actions: Have students simulate a contract negotiation between a supplier and a corporate procurement team.

Example:

1. Set negotiation objectives (secure a cost reduction without compromising quality).
2. Facilitate team discussions (align finance, legal, and operations on acceptable terms).
3. Mediate disagreements (ensure all internal concerns are addressed before finalizing terms).
4. Close negotiation successfully (document key agreements and next steps).

Pairing 3: Leading a Nonprofit Organization Through a Fundraising Campaign

Big Skill: Influence & Decision-Making – Persuading donors and rallying a team around a shared mission.

Small Skill: Facilitating Team Communication – Keeping internal fundraising teams coordinated and motivated.

Teaching Technique: Community leadership challenge.

Actions: Have students develop a strategic fundraising plan and lead team discussions.

Example:

1. Set fundraising target ($500,000 for educational scholarships).
2. Align communication between departments (outreach, donor relations, finance).
3. Resolve team conflicts (prioritize funding strategies based on data).
4. Motivate the team (celebrate milestones and refine tactics based on progress).

Pairing 4: Overseeing an International Business Expansion

Big Skill: Influence & Decision-Making – Aligning global expansion with corporate strategy while navigating cultural differences.

Small Skill: Facilitating Team Communication – Managing multinational teams and bridging communication gaps.

Teaching Technique: Global business case study.

Actions: Have students simulate the leadership of a company expanding into three new countries.

Example:

1. Define expansion goals (open regional offices in Asia, Europe, and South America).
2. Facilitate multinational discussions (ensure all regional leads understand their roles).
3. Manage cultural conflicts (train teams on cultural sensitivity in communication).
4. Summarize global alignment strategy (document local strategies while maintaining corporate values).

Pairing 5: Leading a Corporate Social Responsibility (CSR) Initiative

Big Skill: Influence & Decision-Making – Guiding stakeholders to align business goals with social impact.

Small Skill: Facilitating Team Communication – Keeping all stakeholders aligned and resolving conflicting priorities.

Teaching Technique: Ethical leadership simulation.

Actions: Have students develop and present a CSR proposal while managing team discussions.

Example:

1. Define CSR goals (reduce plastic waste by 50% in five years).
2. Facilitate discussions (bring in operations, marketing, and supply chain leaders).
3. Resolve resistance (mitigate cost concerns from finance teams).
4. Finalize CSR commitment (publish an internal and external action plan).

Pairing 6: Managing a Virtual Team in a Remote Work Environment

Big Skill: Influence & Decision-Making – Keeping virtual teams motivated and aligned despite physical distance.

Small Skill: Facilitating Team Communication – Creating clear messaging structures and ensuring transparency.

Teaching Technique: Remote leadership case study.

Actions: Have students design a virtual communication plan and implement it in a simulation.

Example:

1. Define remote work policies (establish expectations for work hours and deliverables).
2. Facilitate virtual meetings (encourage participation and use structured agendas).
3. Resolve communication issues (address misalignment through real-time feedback tools).
4. Measure success (track employee engagement and adjust communication practices).

Pairing 7: Running an Organizational Restructuring Process

Big Skill: Influence & Decision-Making – Leading employees through a major organizational shift.

Small Skill: Facilitating Team Communication – Ensuring clear messaging and reducing employee uncertainty.

Teaching Technique: Change management leadership exercise.

Actions: Have students develop a restructuring communication strategy and lead discussions.

Example:

1. Define restructuring goals (streamline departments while minimizing layoffs).
2. Facilitate employee discussions (hold Q&A sessions to clarify changes).
3. Address resistance (develop transition support programs).
4. Ensure alignment (track employee sentiment and adjust messaging).

Pairing 8: Leading a Political Debate Preparation Team

Big Skill: Influence & Decision-Making – Training a candidate to persuasively present their platform.

Small Skill: Facilitating Team Communication – Coordinating researchers, speechwriters, and media strategists.

Teaching Technique: Political leadership role-play.

Actions: Have students simulate a political campaign debate strategy session.

Example:

1. Define key messaging (focus on economic and social policies).
2. Facilitate communication (ensure speechwriters and strategists are aligned).
3. Resolve conflicting opinions (mediate between different messaging approaches).
4. Finalize debate strategy (deliver a cohesive and compelling argument).

Pairing 9: Leading an Agile Development Team in a Tech Company

Big Skill: Influence & Decision-Making – Guiding software teams through iterative product development.

Small Skill: Facilitating Team Communication – Ensuring clear communication between engineers, designers, and product managers.

Teaching Technique: Agile methodology workshop.

Actions: Have students lead an agile stand-up meeting and practice resolving team bottlenecks.

Example:

1. Define sprint objectives (launch beta version of an app within two weeks).
2. Facilitate daily check-ins (track blockers and adjust priorities).
3. Resolve team bottlenecks (balance workload between frontend and backend developers).
4. Ensure successful sprint completion (review retrospectives and improve processes).

Pairing 10: Managing a Hollywood Film Production Team

Big Skill: Influence & Decision-Making – Ensuring the creative vision aligns with budget and schedule constraints.

Small Skill: Facilitating Team Communication – Keeping directors, producers, and crew aligned.

Teaching Technique: Film production leadership simulation.

Actions: Have students lead a simulated pre-production meeting, aligning all departments.

Example:

1. Define film production goals (complete shooting in 90 days while staying on budget).
2. Facilitate team alignment (ensure cast, cinematographers, and editors understand the vision).
3. Resolve creative conflicts (balance artistic vision with financial constraints).
4. Execute production plan (track progress and make adjustments as needed).

Leadership Thinking – (Big Skill) Influence & Decision-Making vs. (Small Skill) Tracking Leadership Effectiveness

---

Pairing 1: Leading a Corporate Cultural Transformation

Big Skill: Influence & Decision-Making – Guiding a company through cultural shifts to improve performance and morale.

Small Skill: Tracking Leadership Effectiveness – Measuring changes in employee engagement and productivity.

Teaching Technique: Leadership impact analysis.

Actions: Have students develop and implement a cultural transformation plan, then measure its impact.

Example:

1. Define cultural goals (increase employee engagement by 30% in two years).
2. Implement initiatives (employee recognition programs, leadership coaching).
3. Measure effectiveness (track employee surveys, retention rates).
4. Adjust strategy (refine leadership approach based on feedback data).

Pairing 2: Overseeing a High-Stakes Mergers & Acquisitions (M&A) Deal

Big Skill: Influence & Decision-Making – Leading negotiations and ensuring alignment between merging companies.

Small Skill: Tracking Leadership Effectiveness – Using integration success metrics to assess leadership decisions.

Teaching Technique: Business acquisition case study.

Actions: Have students simulate an M&A process and assess leadership impact.

Example:

1. Define success criteria (maintain 90% employee retention post-merger).
2. Ensure smooth integration (align policies, cultures, and processes).
3. Monitor leadership impact (track transition efficiency and employee morale).
4. Refine post-merger leadership strategy (adjust based on KPI results).

Pairing 3: Leading a Political or Public Policy Campaign

Big Skill: Influence & Decision-Making – Persuading voters, shaping public opinion, and driving legislative initiatives.

Small Skill: Tracking Leadership Effectiveness – Measuring voter support, campaign reach, and policy impact.

Teaching Technique: Political campaign simulation.

Actions: Have students design and execute a political campaign strategy while tracking public response.

Example:

1. Set campaign objectives (increase voter approval ratings by 15%).
2. Implement outreach strategies (social media, debates, public speeches).
3. Analyze leadership influence (polling data, media coverage).
4. Refine messaging (adjust based on demographic engagement metrics).

Pairing 4: Managing a Corporate Crisis Response

Big Skill: Influence & Decision-Making – Leading a company through crisis communication and damage control.

Small Skill: Tracking Leadership Effectiveness – Using public sentiment analysis and media coverage metrics.

Teaching Technique: Crisis response workshop.

Actions: Have students develop and execute a crisis response plan while tracking leadership effectiveness.

Example:

1. Identify crisis goals (minimize brand damage, restore trust).
2. Implement communication strategies (press releases, CEO statements, media Q&A).
3. Measure public response (social media trends, customer feedback, stock impact).
4. Refine approach (adjust messaging based on crisis impact reports).

Pairing 5: Leading a Global Expansion Strategy

Big Skill: Influence & Decision-Making – Convincing stakeholders to invest in new international markets.

Small Skill: Tracking Leadership Effectiveness – Measuring market penetration, team adaptation, and ROI.

Teaching Technique: International business strategy simulation.

Actions: Have students design and lead a global expansion initiative while tracking leadership impact.

Example:

1. Define expansion targets (launch in three international markets within two years).
2. Ensure team alignment (manage international teams, adjust strategies per region).
3. Track leadership success (market share growth, cultural adaptation effectiveness).
4. Refine strategy (adjust based on early market performance).

Pairing 6: Driving a Digital Transformation Initiative

Big Skill: Influence & Decision-Making – Leading an organization's shift to new technologies and digital workflows.

Small Skill: Tracking Leadership Effectiveness – Measuring employee adoption, efficiency gains, and cost savings.

Teaching Technique: Digital transformation case study.

Actions: Have students implement a digital adoption plan and track leadership performance.

Example:

1. Set digital goals (100% transition to cloud-based operations in two years).
2. Implement change management strategies (employee training, phased rollout).
3. Monitor adoption rates (track system usage, employee satisfaction).
4. Refine leadership strategies (adjust training and communication tactics).

Pairing 7: Leading a Social Justice or Activism Movement

Big Skill: Influence & Decision-Making – Mobilizing people and persuading organizations to support social causes.

Small Skill: Tracking Leadership Effectiveness – Measuring community impact, policy changes, and engagement levels.

Teaching Technique: Social movement strategy workshop.

Actions: Have students plan a social justice initiative while tracking effectiveness.

Example:

1. Set advocacy goals (secure legislative change or corporate policy revision).
2. Mobilize support (protests, petitions, social media campaigns).
3. Measure engagement (track policy responses, media reach, public sentiment).
4. Adjust campaign efforts (modify messaging based on real-time impact data).

Pairing 8: Managing a Fortune 500 Leadership Transition

Big Skill: Influence & Decision-Making – Leading an executive transition while maintaining organizational stability.

Small Skill: Tracking Leadership Effectiveness – Using performance reviews, stock price changes, and employee sentiment.

Teaching Technique: Leadership succession planning.

Actions: Have students simulate an executive leadership transition and measure impact.

Example:

1. Define leadership change objectives (ensure stability, maintain stock value).
2. Develop transition plans (internal communications, leadership handover).
3. Monitor leadership impact (track stock price trends, employee sentiment).
4. Refine transition plan (adjust support based on initial post-transition feedback).

Pairing 9: Spearheading a Public-Private Partnership

Big Skill: Influence & Decision-Making – Negotiating between government agencies and private sector leaders.

Small Skill: Tracking Leadership Effectiveness – Measuring stakeholder satisfaction and project ROI.

Teaching Technique: Government-business collaboration simulation.

Actions: Have students develop a partnership proposal and track leadership influence.

Example:

1. Set partnership goals (launch public infrastructure with private funding).
2. Secure stakeholder buy-in (negotiate terms between government and business leaders).
3. Track effectiveness (analyze cost efficiency, service improvements).
4. Refine collaboration approach (adjust based on citizen and investor feedback).

Pairing 10: Leading a Fortune 500 Diversity & Inclusion Initiative

Big Skill: Influence & Decision-Making – Driving corporate cultural change to improve diversity and inclusion.

Small Skill: Tracking Leadership Effectiveness – Using workforce data, promotion rates, and employee feedback.

Teaching Technique: Diversity leadership analysis.

Actions: Have students develop a corporate diversity initiative and measure its leadership impact.

Example:

1. Set diversity objectives (increase leadership diversity by 20% in five years).
2. Implement hiring and training policies (unconscious bias training, mentorship programs).
3. Track key diversity metrics (promotion rates, employee sentiment surveys).
4. Adjust policies based on progress (refine hiring strategies based on data).

Leadership Thinking – (Big Skill) Situational Adaptability vs. (Small Skill) Delegating & Prioritizing Tasks

---

Pairing 1: Leading a Crisis Response Team During a Cyberattack

Big Skill: Situational Adaptability – Adjusting leadership to remain calm and decisive during high-stakes emergencies.

Small Skill: Delegating & Prioritizing Tasks – Assigning responsibilities efficiently to contain and resolve the crisis.

Teaching Technique: Crisis simulation exercise.

Actions: Have students lead a simulated cyberattack response team and delegate tasks based on urgency.

Example:

1. Assess the crisis (identify scope of data breach).
2. Prioritize immediate responses (IT investigates, PR prepares public statement, legal assesses liability).
3. Ensure continuous communication (hold rapid-response team meetings).
4. Monitor and adjust response (analyze impact and update strategy accordingly).

Pairing 2: Managing a Team During a Sudden Leadership Transition

Big Skill: Situational Adaptability – Shifting leadership approach to stabilize a team after an unexpected leadership change.

Small Skill: Delegating & Prioritizing Tasks – Ensuring continuity of operations by distributing leadership responsibilities.

Teaching Technique: Leadership transition role-play.

Actions: Have students simulate taking over a team after a CEO or manager unexpectedly leaves.

Example:

1. Assess immediate needs (identify urgent business priorities).
2. Distribute leadership responsibilities (assign interim leads for key departments).
3. Provide stability (communicate transition plan to employees).
4. Refine leadership approach (adjust based on team morale and operational needs).

Pairing 3: Adjusting Leadership in a Multicultural Global Team

Big Skill: Situational Adaptability – Modifying leadership style to accommodate cultural differences in an international setting.

Small Skill: Delegating & Prioritizing Tasks – Assigning work while considering time zones, communication styles, and team dynamics.

Teaching Technique: Global leadership simulation.

Actions: Have students lead a multinational team project and adjust leadership style based on cultural norms.

Example:

1. Understand cultural differences (research communication styles in each region).
2. Assign tasks based on strengths (match roles to expertise and working preferences).
3. Facilitate cross-cultural collaboration (ensure seamless communication across time zones).
4. Evaluate leadership effectiveness (adjust delegation approach based on team feedback).

Pairing 4: Leading a Team During a Sudden Market Shift

Big Skill: Situational Adaptability – Quickly modifying business strategies to respond to unexpected market changes.

Small Skill: Delegating & Prioritizing Tasks – Ensuring tasks are realigned to new business priorities.

Teaching Technique: Market shift response simulation.

Actions: Have students react to a major competitor's unexpected move and realign their team.

Example:

1. Analyze the market shift (identify competitor's advantage).
2. Reassign resources (focus on counter-strategy: pricing, product features, marketing).
3. Ensure fast execution (assign rapid response teams to priority tasks).
4. Monitor results and adjust (track performance and iterate based on market feedback).

Pairing 5: Adjusting Leadership in a High-Growth Startup

Big Skill: Situational Adaptability – Transitioning from hands-on leadership to delegation as a startup scales.

Small Skill: Delegating & Prioritizing Tasks – Assigning leadership roles to new managers while focusing on strategy.

Teaching Technique: Startup growth leadership challenge.

Actions: Have students transition from a small startup team to a structured leadership hierarchy.

Example:

1. Recognize growth phase (identify when the startup requires structured management).
2. Delegate key roles (appoint department heads to scale operations).
3. Shift focus to vision (founder transitions from daily execution to strategic growth).
4. Adjust leadership as needed (restructure roles based on performance and business needs).

Pairing 6: Leading a Manufacturing Team Through a Supply Chain Crisis

Big Skill: Situational Adaptability – Adjusting leadership style to manage unexpected supply chain disruptions.

Small Skill: Delegating & Prioritizing Tasks – Assigning problem-solving tasks to ensure minimal operational impact.

Teaching Technique: Operations crisis simulation.

Actions: Have students navigate a supply chain crisis and make rapid delegation decisions.

Example:

1. Assess supply chain impact (identify delays, shortages, alternative suppliers).
2. Reprioritize production (focus on high-demand or high-margin products).
3. Assign crisis roles (logistics finds alternate suppliers, finance negotiates new terms).
4. Monitor outcomes and adjust (review supplier performance and refine future risk strategies).

Pairing 7: Adjusting Leadership During a High-Pressure Sales Quarter

Big Skill: Situational Adaptability – Modifying motivation techniques to boost sales team performance.

Small Skill: Delegating & Prioritizing Tasks – Ensuring sales efforts are focused on the highest-yield opportunities.

Teaching Technique: Sales performance case study.

Actions: Have students lead a sales team during an intense quarter and adjust motivation tactics.

Example:

1. Identify key sales drivers (prioritize high-value customer segments).
2. Reassign sales reps (focus best closers on high-profile accounts).
3. Adjust incentives (introduce rewards for hitting short-term goals).
4. Monitor team morale and adjust (use feedback loops to refine motivation strategies).

Pairing 8: Managing a Team During an Industry Disruption

Big Skill: Situational Adaptability – Navigating industry shifts (e.g., AI replacing traditional roles).

Small Skill: Delegating & Prioritizing Tasks – Assigning new training, innovation, and transition tasks to adapt.

Teaching Technique: Industry disruption workshop.

Actions: Have students lead a company through an industry disruption and delegate transition tasks.

Example:

1. Identify disruption risks (e.g., automation reducing jobs).
2. Reallocate workforce efforts (transition employees to training or new roles).
3. Ensure knowledge transfer (document institutional knowledge before shifting processes).
4. Track and adjust effectiveness (refine strategies based on transition success).

Pairing 9: Adapting Leadership in a Creative Team Under Tight Deadlines

Big Skill: Situational Adaptability – Adjusting between directive and collaborative leadership in creative teams.

Small Skill: Delegating & Prioritizing Tasks – Assigning time-sensitive and creativity-focused tasks accordingly.

Teaching Technique: Creative project sprint simulation.

Actions: Have students lead a high-pressure creative project while balancing team autonomy and structure.

Example:

1. Set non-negotiable deadlines (ensure timely delivery of client projects).
2. Delegate execution tasks (technical designers on details, creatives on concepts).
3. Adjust creative freedom (provide structure or flexibility depending on team progress).
4. Monitor and refine approach (adjust leadership style to maintain balance).

Pairing 10: Managing a Disaster Recovery Effort

Big Skill: Situational Adaptability – Leading disaster relief teams with shifting priorities.

Small Skill: Delegating & Prioritizing Tasks – Assigning immediate, mid-term, and long-term relief efforts.

Teaching Technique: Disaster response leadership exercise.

Actions: Have students coordinate relief operations after a simulated disaster.

Example:

1. Assess immediate needs (rescue, medical aid, food distribution).
2. Assign rapid response teams (emergency responders, logistics, communications).
3. Plan long-term recovery (infrastructure rebuilding, economic assistance).
4. Refine response strategy (adjust based on evolving crisis data).

Leadership Thinking – (Big Skill) Situational Adaptability vs. (Small Skill) Facilitating Team Communication

---

Pairing 1: Adapting Leadership in a Remote Work Environment

Big Skill: Situational Adaptability – Adjusting leadership style to manage a remote team effectively.

Small Skill: Facilitating Team Communication – Ensuring clear messaging, transparency, and engagement across virtual channels.

Teaching Technique: Remote team leadership simulation.

Actions: Have students develop a remote leadership strategy and facilitate virtual team discussions.

Example:

1. Set communication norms (establish meeting cadences and response expectations).
2. Facilitate virtual discussions (use video calls, project management tools, and asynchronous messaging).
3. Resolve communication barriers (address time zone challenges, cultural differences).
4. Adjust leadership approach (offer more structure for less experienced employees, more flexibility for senior team members).

Pairing 2: Managing a Crisis Response Team After a PR Scandal

Big Skill: Situational Adaptability – Leading a team through reputational damage with a flexible approach.

Small Skill: Facilitating Team Communication – Ensuring crisis teams communicate effectively and respond swiftly.

Teaching Technique: PR crisis management exercise.

Actions: Have students simulate managing a corporate PR crisis and coordinating internal and external communication.

Example:

1. Identify key messages (maintain transparency while controlling the narrative).
2. Coordinate internal messaging (align PR, legal, and executive teams).
3. Engage with stakeholders (customers, media, regulators).
4. Monitor sentiment and adjust strategy (track public response and refine messaging).

Pairing 3: Leading a High-Stakes Negotiation Between Two Departments

Big Skill: Situational Adaptability – Adapting negotiation style based on team dynamics and priorities.

Small Skill: Facilitating Team Communication – Ensuring that both parties articulate concerns and reach a fair resolution.

Teaching Technique: Business negotiation role-play.

Actions: Have students mediate between two conflicting departments and facilitate constructive dialogue.

Example:

1. Define negotiation goals (resolve resource allocation conflict).
2. Facilitate structured discussions (ensure each team presents its perspective).
3. Mediate compromise solutions (prioritize company goals over department interests).
4. Ensure ongoing communication (set follow-ups to track the resolution's effectiveness).

Pairing 4: Adapting Leadership in a Startup Scaling from 10 to 100 Employees

Big Skill: Situational Adaptability – Transitioning from a hands-on leadership style to a structured management approach.

Small Skill: Facilitating Team Communication – Establishing clear messaging as team size increases.

Teaching Technique: Startup growth leadership challenge.

Actions: Have students develop a communication strategy for a rapidly scaling company.

Example:

1. Define new leadership roles (establish middle management for efficiency).
2. Standardize the communication (introduce weekly company-wide meetings).

3. Maintain cultural cohesion (adjust leadership style to ensure engagement at scale).
4. Assess communication effectiveness (track employee feedback and refine strategies).

Pairing 5: Leading an Interdepartmental Collaboration for Product Development

Big Skill: Situational Adaptability – Modifying leadership approach based on team needs in different project phases.

Small Skill: Facilitating Team Communication – Ensuring smooth coordination between engineering, marketing, and finance.

Teaching Technique: Cross-functional team workshop.

Actions: Have students simulate a product development process, balancing technical and business priorities.

Example:

1. Facilitate kickoff meetings (align team on product vision and market needs).
2. Adapt leadership style (directive in early design, collaborative in testing).
3. Resolve conflicts (mediate disagreements between technical feasibility and marketability).
4. Ensure continuous feedback loops (use retrospectives to refine communication).

Pairing 6: Managing an Event with Unexpected Changes

Big Skill: Situational Adaptability – Adjusting leadership to manage last-minute event disruptions.

Small Skill: Facilitating Team Communication – Coordinating real-time updates to vendors, teams, and attendees.

Teaching Technique: Event crisis management simulation.

Actions: Have students manage an event with unpredictable changes and communicate solutions.

Example:

1. Assess crisis impact (venue change, guest cancellations, technical failures).
2. Assign quick response teams (logistics, media, contingency planning).
3. Keep all stakeholders informed (use centralized messaging for updates).
4. Adjust event execution (make real-time changes while ensuring a seamless experience).

Pairing 7: Leading a Workforce During a Layoff or Restructuring

Big Skill: Situational Adaptability – Navigating organizational changes while maintaining morale.

Small Skill: Facilitating Team Communication – Delivering transparent and empathetic messaging.

Teaching Technique: Change management case study.

Actions: Have students create and present a restructuring communication plan.

Example:

1. Frame the message (ensure clarity on reasons and next steps).
2. Address team concerns (host open forums for Q&A).
3. Support affected employees (provide career transition resources).
4. Monitor workforce morale (adjust messaging based on employee sentiment).

Pairing 8: Leading a Diverse Team Across Generations

Big Skill: Situational Adaptability – Modifying leadership style to engage Baby Boomers, Millennials, and Gen Z employees.

Small Skill: Facilitating Team Communication – Bridging generational communication gaps.

Teaching Technique: Multigenerational workforce training.

Actions: Have students design a strategy to unify a multigenerational team.

Example:

1. Understand generational preferences (Baby Boomers prefer structured meetings, Gen Z prefers digital collaboration).
2. Customize communication methods (blend formal emails, Slack, and video calls).
3. Encourage knowledge sharing (reverse mentorship programs).
4. Adjust leadership approach (balance traditional authority with inclusivity).

Pairing 9: Managing a Research & Development (R&D) Team Under Pressure

Big Skill: Situational Adaptability – Switching between directive and participative leadership styles in research.

Small Skill: Facilitating Team Communication – Ensuring knowledge transfer and transparent discussion among scientists, engineers, and executives.

Teaching Technique: R&D leadership workshop.

Actions: Have students lead a simulated R&D breakthrough process while managing communication.

Example:

1. Set innovation goals (develop a prototype for an AI-powered product).
2. Facilitate collaboration (ensure engineers, designers, and executives align).
3. Address conflicts between speed and accuracy (adjust leadership style accordingly).

4. Communicate findings effectively (translate technical results for stakeholders).

Pairing 10: Leading a Mission-Critical Military or Emergency Response Team

Big Skill: Situational Adaptability – Adjusting leadership approach in life-or-death situations.

Small Skill: Facilitating Team Communication – Ensuring rapid, precise messaging in high-stakes environments.

Teaching Technique: Tactical leadership simulation.

Actions: Have students lead a simulated emergency response team in a disaster scenario.

Example:

1. Assess situational urgency (evaluate risk levels and mission objectives).
2. Establish command hierarchy (ensure clear chain of communication).
3. Adapt orders dynamically (switch from structured to agile leadership when needed).
4. Ensure post-mission debriefing (review effectiveness of leadership communication).

Leadership Thinking – (Big Skill) Situational Adaptability vs. (Small Skill) Tracking Leadership Effectiveness

---

Pairing 1: Leading a Company Through a Market Disruption

Big Skill: Situational Adaptability – Adjusting leadership style to pivot strategies when unexpected market changes occur.

Small Skill: Tracking Leadership Effectiveness – Measuring how well strategic shifts improve financial stability and employee performance.

Teaching Technique: Market disruption case study.

Actions: Have students analyze a real-world example of a company adapting to market changes and track leadership effectiveness.

Example:

1. Assess market impact (identify threats like a competitor launch or economic downturn).
2. Adjust leadership style (shift from aggressive expansion to cost control).
3. Measure effectiveness (track financials, employee morale, and market share).
4. Refine approach (adjust based on performance feedback and industry trends).

Pairing 2: Managing a Leadership Transition in an Organization

Big Skill: Situational Adaptability – Modifying leadership approach when stepping into a new leadership role.

Small Skill: Tracking Leadership Effectiveness – Using performance evaluations, employee feedback, and transition KPIs.

Teaching Technique: Leadership transition simulation.

Actions: Have students develop and implement a transition plan for a new leader and track success metrics.

Example:

1. Set transition goals (maintain productivity while earning team trust).
2. Adjust management style (observe and modify leadership based on team feedback).
3. Measure leadership impact (employee satisfaction surveys, retention rates).
4. Refine transition approach (use data-driven adjustments to solidify leadership style).

Pairing 3: Adapting Leadership in a Fast-Growing Startup

Big Skill: Situational Adaptability – Transitioning from a hands-on leadership style to scalable management.

Small Skill: Tracking Leadership Effectiveness – Measuring productivity, employee engagement, and business growth metrics.

Teaching Technique: Growth leadership challenge.

Actions: Have students simulate leading a startup scaling from 10 to 200 employees.

Example:

1. Define growth challenges (loss of close team dynamics, delegation struggles).
2. Adapt leadership approach (shift from direct management to empowering middle leaders).
3. Monitor effectiveness (use engagement scores, turnover rates, and revenue tracking).
4. Refine strategy (adjust leadership structure based on team needs).

Pairing 4: Leading a Crisis Response Team in a Natural Disaster

Big Skill: Situational Adaptability – Adjusting leadership style to maintain control and ensure effective action in a high-stakes environment.

Small Skill: Tracking Leadership Effectiveness – Using response time, casualty rates, and relief distribution efficiency as metrics.

Teaching Technique: Emergency response simulation.

Actions: Have students coordinate a disaster relief plan and track key leadership performance metrics.

Example:

1. Assess crisis priorities (rescue operations, resource allocation, public safety).

2. Modify leadership as needed (directive for immediate action, collaborative for long-term rebuilding).
3. Track effectiveness (response speed, coordination success, public trust).
4. Refine crisis management strategies (adjust based on debrief reports).

Pairing 5: Leading a Hybrid or Remote Team in a Corporate Environment

Big Skill: Situational Adaptability – Adjusting leadership to manage remote and in-person teams effectively.

Small Skill: Tracking Leadership Effectiveness – Using productivity metrics, employee satisfaction scores, and communication analytics.

Teaching Technique: Remote leadership workshop.

Actions: Have students lead a remote team for a simulated project and track leadership impact.

Example:

1. Establish communication norms (set expectations for meetings, response times).
2. Adjust management style (use data to balance autonomy with accountability).
3. Measure leadership success (track project deadlines, employee feedback, digital tool usage).
4. Refine approach (modify meeting frequency, engagement techniques based on insights).

Pairing 6: Managing a Large-Scale Corporate Restructuring

Big Skill: Situational Adaptability – Leading a company through major structural changes (e.g., downsizing, mergers, department reorganization).

Small Skill: Tracking Leadership Effectiveness – Using financial performance, employee engagement, and operational KPIs.

Teaching Technique: Change management role-play.

Actions: Have students simulate restructuring a corporation and measure leadership success.

Example:

1. Develop restructuring plan (set goals, ensure transparency).
2. Adjust leadership style (top-down for layoffs, collaborative for morale rebuilding).
3. Track effectiveness (monitor employee retention, stock value, department efficiency).
4. Adapt approach (implement additional employee support initiatives if necessary).

Pairing 7: Overseeing a Cross-Cultural Leadership Initiative

Big Skill: Situational Adaptability – Adapting leadership to navigate diverse cultural expectations and working styles.

Small Skill: Tracking Leadership Effectiveness – Using cultural engagement metrics, feedback surveys, and team productivity indicators.

Teaching Technique: Global leadership case study.

Actions: Have students develop a cross-cultural leadership plan and track adaptation effectiveness.

Example:

1. Assess cultural differences (adjust communication styles based on region).
2. Adapt delegation methods (some cultures prefer direct authority, others value consensus).
3. Measure success (track cross-team collaboration, survey cultural satisfaction).
4. Refine leadership practices (adjust strategies based on team feedback).

Pairing 8: Leading an R&D Team in a High-Tech Industry

Big Skill: Situational Adaptability – Balancing structured leadership with creative autonomy for innovation.

Small Skill: Tracking Leadership Effectiveness – Measuring R&D progress, patent filings, and employee satisfaction.

Teaching Technique: Innovation leadership simulation.

Actions: Have students lead an R&D team with shifting priorities and measure success.

Example:

1. Define innovation objectives (increase new patent filings by 20%).
2. Adjust leadership approach (directive in execution, collaborative in brainstorming).
3. Track team performance (use KPIs for creativity, development speed, failure rates).
4. Refine strategy (adjust deadlines or resources based on effectiveness).

Pairing 9: Managing a Political or Advocacy Campaign

Big Skill: Situational Adaptability – Shifting campaign leadership strategies based on public sentiment and opposition tactics.

Small Skill: Tracking Leadership Effectiveness – Using voter engagement, media coverage, and fundraising success metrics.

Teaching Technique: Campaign strategy exercise.

Actions: Have students develop a political campaign, adjust leadership, and measure effectiveness.

Example:

1. Set campaign goals (increase support in key demographics).

2. Adapt messaging strategy (based on polling trends and competitor tactics).
3. Measure impact (use social media analytics, fundraising results, voter surveys).
4. Refine campaign approach (adjust focus areas based on data insights).

Pairing 10: Leading a Fortune 500 Company Through a CEO Change

Big Skill: Situational Adaptability – Adjusting leadership expectations and strategies after a major leadership transition.

Small Skill: Tracking Leadership Effectiveness – Using stock performance, investor confidence, and employee morale metrics.

Teaching Technique: Corporate leadership case study.

Actions: Have students analyze a major CEO transition and track leadership effectiveness metrics.

Example:

1. Assess transition impact (evaluate internal and external expectations).
2. Adjust leadership messaging (reassure investors, engage employees).
3. Monitor leadership effectiveness (stock fluctuations, leadership approval ratings).
4. Refine transition strategy (adjust based on stakeholder response).

Example Teaching Strategies for Big vs. Small Thinking Skill Pairings of Data Analysis and Interpretation at the College/Technical School Level

Here are 90 unique teaching strategy examples for the big thinking and small thinking skill pairings, each described in detail to align with **Big Thinking Skills** (Concepts & Strategies) and **Small Thinking Skills** (Step-by-Step Execution) for the **Data Analysis and Interpretation Skill** taught at the college/technical school level.

Data Analysis and Interpretation

Big Thinking Skills (Concepts & Strategies)

- **Extracting Meaning from Data:** Understanding how to derive insights from raw numbers and trends.
- **Integrating Data into Decision-Making:** Using analytics to guide strategic choices.
- **Cross-Disciplinary Synthesis:** Combining data from multiple fields (e.g., economics + sociology) for holistic analysis.

Small Thinking Skills (Step-by-Step Execution)

- **Using Analytical Tools:** Applying Excel, Python, R, or business intelligence software for data processing.
- **Performing Statistical Analysis:** Calculating measures such as mean, median, correlation, and regression analysis.
- **Visualizing Data:** Creating dashboards, charts, and reports to present findings effectively.

Data Analysis and Interpretation – (Big Skill) Extracting Meaning from Data vs. (Small Skill) Using Analytical Tools

---

Pairing 1: Analyzing Customer Purchase Trends for Retail Optimization

Big Skill: Extracting Meaning from Data – Identifying buying patterns, peak sales periods, and customer preferences.

Small Skill: Using Analytical Tools – Applying Excel pivot tables, SQL queries, and Power BI dashboards.

Teaching Technique: Retail sales data analysis case study.

Actions: Have students analyze real or simulated sales data to find trends and optimize product placement.

Example:

1. Import sales data into Excel.
2. Use pivot tables to analyze seasonal demand.
3. Visualize insights with Power BI to highlight purchasing behaviors.
4. Recommend inventory adjustments based on findings.

Pairing 2: Evaluating Marketing Campaign Performance

Big Skill: Extracting Meaning from Data – Understanding which marketing strategies generate the highest engagement.

Small Skill: Using Analytical Tools – Implementing Python for data cleaning and Matplotlib for visualization.

Teaching Technique: Marketing A/B testing analysis.

Actions: Have students compare two digital ad campaigns and evaluate their effectiveness using Python.

Example:

1. Load ad performance data into Python Pandas.
2. Analyze engagement rates and conversion metrics.
3. Plot trends using Matplotlib to visualize ROI.
4. Suggest budget reallocation for the better-performing ad.

Pairing 3: Identifying Fraudulent Transactions in Banking

Big Skill: Extracting Meaning from Data – Spotting irregular patterns that indicate fraud.

Small Skill: Using Analytical Tools – Applying R for statistical anomaly detection.

Teaching Technique: Financial fraud detection simulation.

Actions: Have students use R to identify suspicious transactions in a dataset.

Example:

1. Load transaction data into R.
2. Use statistical models (e.g., outlier detection via standard deviation).
3. Flag anomalies based on transaction frequency and amount.
4. Develop fraud alerts for the bank's security team.

Pairing 4: Predicting Employee Turnover with HR Analytics

Big Skill: Extracting Meaning from Data – Identifying factors that influence workforce retention.

Small Skill: Using Analytical Tools – Using Excel regression analysis and Power BI for HR dashboards.

Teaching Technique: Workforce analytics project.

Actions: Have students analyze HR datasets to predict employee turnover risks.

Example:

1. Load HR data (employee tenure, salary, department, job satisfaction).
2. Run regression analysis in Excel to identify top risk factors.
3. Create an interactive Power BI dashboard for HR managers.
4. Suggest retention strategies based on insights.

Pairing 5: Forecasting Stock Market Trends

Big Skill: Extracting Meaning from Data – Finding correlations between stock movements and economic indicators.

Small Skill: Using Analytical Tools – Using Python libraries like Pandas and Scikit-learn for predictive modeling.

Teaching Technique: Stock market forecasting exercise.

Actions: Have students develop a predictive model for stock price trends.

Example:

1. Gather historical stock data from Yahoo Finance API.
2. Clean data using Pandas (remove null values, normalize features).
3. Use machine learning models (linear regression or random forests) to predict stock movements.
4. Evaluate model accuracy and refine parameters.

Pairing 6: Analyzing Hospital Patient Readmission Rates

Big Skill: Extracting Meaning from Data – Determining factors leading to hospital readmissions.

Small Skill: Using Analytical Tools – Applying R for statistical testing and data visualization.

Teaching Technique: Healthcare analytics study.

Actions: Have students use R to assess readmission risk based on patient demographics and conditions.

Example:

1. Load patient records into R.
2. Perform chi-square tests to find correlations between diagnosis and readmission.
3. Visualize patterns with ggplot2.
4. Develop recommendations for hospital policy adjustments.

Pairing 7: Evaluating Supply Chain Efficiency in Manufacturing

Big Skill: Extracting Meaning from Data – Identifying bottlenecks and inefficiencies in production.

Small Skill: Using Analytical Tools – Using Excel macros and Power BI for real-time monitoring.

Teaching Technique: Supply chain performance dashboard.

Actions: Have students analyze logistics data and propose improvements.

Example:

1. Import supply chain data into Power BI.
2. Identify slow-moving inventory using trend analysis.
3. Create real-time dashboards to track shipping times and delays.
4. Recommend process optimizations to reduce waste and delays.

Pairing 8: Measuring Social Media Sentiment for Brand Analysis

Big Skill: Extracting Meaning from Data – Understanding customer sentiment from social media interactions.

Small Skill: Using Analytical Tools – Using Python's Natural Language Processing (NLP) libraries.

Teaching Technique: Social media sentiment analysis project.

Actions: Have students scrape Twitter data and analyze sentiment toward a brand.

Example:

1. Collect tweets using Tweepy API.
2. Use NLP techniques (tokenization, stopword removal, sentiment scoring).
3. Visualize sentiment trends (positive, neutral, negative).
4. Suggest improvements for customer engagement strategies.

Pairing 9: Evaluating Academic Performance Metrics in Education

Big Skill: Extracting Meaning from Data – Identifying factors affecting student performance.

Small Skill: Using Analytical Tools – Applying Excel and Power BI for educational data visualization.

Teaching Technique: Student success data analysis.

Actions: Have students analyze academic records to determine key success indicators.

Example:

1. Load student performance data into Power BI.
2. Analyze correlations between attendance, study time, and grades.
3. Develop interactive reports for school administrators.
4. Recommend interventions to improve student outcomes.

Pairing 10: Analyzing Website Traffic and User Behavior

Big Skill: Extracting Meaning from Data – Identifying trends in website visits and conversion rates.

Small Skill: Using Analytical Tools – Using Google Analytics and SQL for web traffic analysis.

Teaching Technique: Website performance analysis.

Actions: Have students track user behavior and optimize website performance.

Example:

1. Import website data into SQL (page views, bounce rates, session duration).
2. Run queries to segment data by traffic source (organic, paid, referral).
3. Use Google Analytics reports to identify high-exit pages.
4. Suggest website design changes to increase conversions.

Data Analysis and Interpretation – (Big Skill) Extracting Meaning from Data vs. (Small Skill) Performing Statistical Analysis

---

Pairing 1: Analyzing Employee Productivity Across Departments

Big Skill: Extracting Meaning from Data – Understanding how department-wise productivity varies.

Small Skill: Performing Statistical Analysis – Calculating mean, standard deviation, and correlation between hours worked and output.

Teaching Technique: Workforce efficiency analysis.

Actions: Have students analyze employee work-hour data to identify productivity trends.

Example:

1. Collect productivity metrics from different teams.
2. Calculate averages and standard deviations of work hours vs. output.
3. Find correlations between hours worked and overall performance.
4. Provide insights to management on work-hour policies.

Pairing 2: Evaluating Student Test Scores for Curriculum Effectiveness

Big Skill: Extracting Meaning from Data – Determining whether a new curriculum improves student outcomes.

Small Skill: Performing Statistical Analysis – Conducting a t-test to compare pre- and post-curriculum scores.

Teaching Technique: Education effectiveness study.

Actions: Have students analyze test score data before and after curriculum changes.

Example:

1. Gather test scores from two semesters.
2. Compute means and standard deviations for both groups.
3. Perform a t-test to check for statistically significant differences.
4. Recommend curriculum changes based on findings.

Pairing 3: Identifying Sales Trends in a Retail Chain

Big Skill: Extracting Meaning from Data – Understanding seasonal variations and sales spikes.

Small Skill: Performing Statistical Analysis – Using regression analysis to predict future sales.

Teaching Technique: Sales forecasting project.

Actions: Have students analyze historical sales data to predict peak demand periods.

Example:

1. Compile monthly sales records over three years.
2. Calculate moving averages to smooth out fluctuations.
3. Apply linear regression to forecast future trends.
4. Suggest inventory stocking strategies based on findings.

Pairing 4: Analyzing Social Media Engagement Metrics

Big Skill: Extracting Meaning from Data – Understanding what type of content performs best on social platforms.

Small Skill: Performing Statistical Analysis – Computing correlation coefficients between post frequency and engagement rates.

Teaching Technique: Social media impact study.

Actions: Have students analyze engagement data from different posting schedules.

Example:

1. Collect social media metrics (likes, shares, comments).
2. Calculate average engagement per post type (video, image, text).
3. Find correlation between posting frequency and interaction levels.
4. Recommend optimal posting strategies for businesses.

Pairing 5: Assessing Patient Recovery Times in Healthcare

Big Skill: Extracting Meaning from Data – Identifying factors influencing patient recovery rates.

Small Skill: Performing Statistical Analysis – Conducting ANOVA tests to compare different treatment methods.

Teaching Technique: Medical data analysis simulation.

Actions: Have students examine patient recovery data from different treatment groups.

Example:

1. Gather patient recovery-time data for three treatments.
2. Calculate means and variances for each group.
3. Perform an ANOVA test to check if differences are significant.
4. Recommend best treatment approach based on statistical results.

Pairing 6: Evaluating Manufacturing Defect Rates

Big Skill: Extracting Meaning from Data – Identifying defect patterns in production lines.

Small Skill: Performing Statistical Analysis – Computing standard deviations and using Six Sigma principles.

Teaching Technique: Quality control analysis.

Actions: Have students examine defect rate data to improve manufacturing efficiency.

Example:

1. Collect defect counts from multiple production batches.
2. Calculate mean and standard deviation for defect rates.
3. Compare against Six Sigma benchmarks for process control.
4. Suggest process improvements to minimize variability.

Pairing 7: Analyzing Consumer Preferences for Product Development

Big Skill: Extracting Meaning from Data – Understanding market trends to refine product offerings.

Small Skill: Performing Statistical Analysis – Using cluster analysis to segment customer preferences.

Teaching Technique: Market segmentation exercise.

Actions: Have students group customers based on survey responses and purchase behavior.

Example:

1. Compile survey data on product preferences.
2. Compute means and standard deviations for different demographics.
3. Use cluster analysis to identify distinct consumer segments.
4. Recommend marketing strategies tailored to each segment.

Pairing 8: Predicting Loan Default Rates in Banking

Big Skill: Extracting Meaning from Data – Identifying risk factors for loan defaults.

Small Skill: Performing Statistical Analysis – Applying logistic regression to classify risk levels.

Teaching Technique: Financial risk modeling exercise.

Actions: Have students analyze past loan data to build a risk model.

Example:

1. Import financial records of borrowers.
2. Calculate mean default rates for different credit score ranges.
3. Apply logistic regression to classify high-risk borrowers.
4. Suggest lending policies based on findings.

Pairing 9: Measuring Advertising ROI for Digital Marketing Campaigns

Big Skill: Extracting Meaning from Data – Understanding how ad spend impacts revenue.

Small Skill: Performing Statistical Analysis – Computing regression analysis between ad spend and revenue growth.

Teaching Technique: Digital marketing case study.

Actions: Have students analyze past ad spend and revenue data to evaluate campaign effectiveness.

Example:

1. Collect advertising and revenue data for multiple campaigns.
2. Compute correlation coefficients between ad spend and revenue.
3. Perform a regression analysis to determine ROI.
4. Optimize future ad budgets based on findings.

Pairing 10: Investigating Traffic Accidents and Road Safety Data

Big Skill: Extracting Meaning from Data – Identifying accident-prone locations and risk factors.

Small Skill: Performing Statistical Analysis – Using chi-square tests to compare accident rates across intersections.

Teaching Technique: Public safety data project.

Actions: Have students analyze traffic data to improve road safety measures.

Example:

1. Compile accident reports for different road intersections.
2. Calculate frequencies of accidents under different conditions (rain, nighttime, rush hour).
3. Apply chi-square tests to detect significant risk factors.
4. Propose traffic regulation changes to improve safety.

Data Analysis and Interpretation – (Big Skill) Extracting Meaning from Data vs. (Small Skill) Visualizing Data

---

Pairing 1: Analyzing E-commerce Sales Performance

Big Skill: Extracting Meaning from Data – Identifying customer purchasing trends and revenue growth patterns.

Small Skill: Visualizing Data – Creating dashboards in Power BI or Tableau to track key metrics.

Teaching Technique: E-commerce performance dashboard project.

Actions: Have students develop an interactive sales dashboard with dynamic filters.

Example:

1. Load sales data from an e-commerce store.
2. Create visualizations (bar charts for best-selling products, line charts for revenue trends).
3. Use interactive filters (category, location, time period).
4. Present insights on product demand fluctuations.

Pairing 2: Tracking COVID-19 Infection Rates by Region

Big Skill: Extracting Meaning from Data – Understanding how the pandemic spreads in different areas.

Small Skill: Visualizing Data – Using heat maps and line graphs to depict trends over time.

Teaching Technique: Public health data visualization exercise.

Actions: Have students visualize infection rates and compare regional differences.

Example:

1. Import COVID-19 case data into Excel or Python.
2. Create a heat map showing infection intensity across locations.
3. Use a line graph to depict case trends over months.
4. Analyze surges and declines to suggest policy interventions.

Pairing 3: Examining Social Media Engagement Metrics

Big Skill: Extracting Meaning from Data – Identifying trends in user interactions across platforms.

Small Skill: Visualizing Data – Using stacked bar charts and scatter plots to compare post engagement.

Teaching Technique: Social media analytics project.

Actions: Have students analyze and visualize engagement across different content types.

Example:

1. Extract social media data (likes, shares, comments).
2. Compare engagement trends using a stacked bar chart.
3. Plot a scatter graph for post frequency vs. engagement rate.
4. Identify content types with the highest audience interaction.

Pairing 4: Visualizing Stock Market Trends

Big Skill: Extracting Meaning from Data – Understanding patterns in stock price fluctuations.

Small Skill: Visualizing Data – Using candlestick charts and moving averages in Python or Excel.

Teaching Technique: Stock market data visualization exercise.

Actions: Have students develop a financial dashboard tracking market performance.

Example:

1. Load stock price data from Yahoo Finance API.
2. Plot a candlestick chart showing price fluctuations.

3. Overlay moving averages to smooth trends.
4. Analyze stock performance over different timeframes.

Pairing 5: Creating a Crime Rate Dashboard for Urban Planning

Big Skill: Extracting Meaning from Data – Identifying high-crime areas and patterns over time.

Small Skill: Visualizing Data – Mapping crime density with geospatial visualizations.

Teaching Technique: Crime data heat map project.

Actions: Have students develop an interactive crime rate visualization for a city.

Example:

1. Collect crime reports categorized by location.
2. Use GIS software to create a heat map.
3. Overlay trends (time of day, seasonality).
4. Provide policy recommendations to law enforcement.

Pairing 6: Understanding Customer Churn in Subscription Services

Big Skill: Extracting Meaning from Data – Identifying why customers cancel subscriptions.

Small Skill: Visualizing Data – Creating funnel charts and retention curves.

Teaching Technique: Customer behavior analysis.

Actions: Have students visualize churn rates using historical subscription data.

Example:

1. Gather retention data (monthly active users, cancellations).
2. Create a funnel chart displaying customer retention steps.
3. Overlay trend lines for churn over time.
4. Suggest retention strategies based on findings.

Pairing 7: Monitoring Energy Consumption in Smart Cities

Big Skill: Extracting Meaning from Data – Understanding electricity usage across urban regions.

Small Skill: Visualizing Data – Using line charts and geospatial visualizations to depict consumption.

Teaching Technique: Smart city energy tracking project.

Actions: Have students analyze power grid usage and develop a real-time dashboard.

Example:

1. Import energy consumption data by city zone.
2. Plot hourly/daily usage trends using a line graph.
3. Use a heat map to show high-consumption areas.
4. Recommend efficiency measures for peak hours.

Pairing 8: Examining Academic Performance Trends

Big Skill: Extracting Meaning from Data – Identifying how student performance varies by subject and demographic factors.

Small Skill: Visualizing Data – Creating scatter plots and comparative bar charts.

Teaching Technique: Education data analysis.

Actions: Have students develop a student performance dashboard to track academic trends.

Example:

1. Load student grades by subject and school district.
2. Compare test score distributions using bar charts.
3. Plot scatter graphs for study time vs. performance.
4. Suggest curriculum changes to address gaps.

Pairing 9: Tracking Traffic Patterns for Urban Mobility Planning

Big Skill: Extracting Meaning from Data – Understanding congestion trends and vehicle movement.

Small Skill: Visualizing Data – Creating flow diagrams and congestion heat maps.

Teaching Technique: Transportation data dashboard project.

Actions: Have students analyze road traffic data and create a visual representation.

Example:

1. Collect GPS-based traffic data.
2. Plot vehicles counted per road segment.
3. Create time-based traffic flow maps.
4. Suggest traffic management improvements.

Pairing 10: Analyzing Weather Data for Climate Research

Big Skill: Extracting Meaning from Data – Identifying temperature and precipitation trends over decades.

Small Skill: Visualizing Data – Using histograms and trend lines to present climate patterns.

Teaching Technique: Climate change data visualization.

Actions: Have students develop a climate dashboard to showcase weather trends.

Example:

1. Gather historical temperature and rainfall data.
2. Create a histogram showing frequency of extreme weather events.
3. Plot a time-series graph to observe climate shifts.
4. Predict future weather trends based on historical patterns.

Data Analysis and Interpretation – (Big Skill) Integrating Data into Decision-Making vs. (Small Skill) Using Analytical Tools

---

Pairing 1: Optimizing Retail Inventory Management

Big Skill: Integrating Data into Decision-Making – Using sales and demand data to manage stock levels efficiently.

Small Skill: Using Analytical Tools – Applying Excel for sales forecasting and Power BI for dashboard reporting.

Teaching Technique: Retail inventory optimization project.

Actions: Have students analyze past sales data and optimize stock levels using predictive models.

Example:

1. Import sales data from multiple store locations.
2. Use Excel's regression tool to predict future demand.
3. Develop a Power BI dashboard tracking real-time inventory levels.
4. Recommend stock reallocation for underperforming stores.

Pairing 2: Allocating Budget for Marketing Campaigns

Big Skill: Integrating Data into Decision-Making – Using ROI analysis to determine the most effective marketing channels.

Small Skill: Using Analytical Tools – Applying Python for data cleaning and Pandas for ad performance analysis.

Teaching Technique: Marketing budget allocation exercise.

Actions: Have students analyze marketing spend vs. conversion rates and adjust budgets accordingly.

Example:

1. Load marketing spend and performance data into Python.
2. Use Pandas to calculate ROI for different advertising channels.

3. Visualize performance trends with Matplotlib.
4. Recommend budget shifts to maximize return on ad spend.

Pairing 3: Enhancing Customer Satisfaction through Data-Driven Insights

Big Skill: Integrating Data into Decision-Making – Using customer feedback and survey data to improve service quality.

Small Skill: Using Analytical Tools – Using R for sentiment analysis and visualization.

Teaching Technique: Customer service analytics case study.

Actions: Have students analyze customer feedback data to identify areas for improvement.

Example:

1. Extract survey responses and online reviews.
2. Apply sentiment analysis in R to categorize positive, neutral, and negative feedback.
3. Create a word cloud visualization to highlight common issues.
4. Recommend service changes based on data insights.

Pairing 4: Improving Healthcare Outcomes with Predictive Analytics

Big Skill: Integrating Data into Decision-Making – Using patient history data to predict readmission risks.

Small Skill: Using Analytical Tools – Applying machine learning in Python to classify risk levels.

Teaching Technique: Predictive healthcare analytics project.

Actions: Have students develop a predictive model for hospital readmission rat**es.**

Example:

1. Load patient records into a Pandas data frame.
2. Use Scikit-learn to train a logistic regression model on past readmissions.
3. Evaluate model accuracy and fine-tune hyperparameters.
4. Develop a hospital decision dashboard to flag high-risk patients.

Pairing 5: Reducing Employee Turnover with HR Analytics

Big Skill: Integrating Data into Decision-Making – Using workforce data to identify key factors influencing retention.

Small Skill: Using Analytical Tools – Utilizing Power BI to track HR metrics and SQL for data queries.

Teaching Technique: Workforce analytics case study.

Actions: Have students analyze HR data to develop strategies for improving retention.

Example:

1. Import employee records (job satisfaction, performance, tenure).
2. Use SQL queries to identify turnover trends by department.
3. Create Power BI dashboards for HR decision-makers.
4. Recommend retention strategies (salary adjustments, training programs).

Pairing 6: Optimizing Supply Chain Efficiency with Data Modeling

Big Skill: Integrating Data into Decision-Making – Using real-time logistics data to enhance supply chain operations.

Small Skill: Using Analytical Tools – Applying Excel Solver for optimization and Power BI for visualization.

Teaching Technique: Supply chain simulation exercise.

Actions: Have students use supply chain data to reduce delivery times and costs.

Example:

1. Gather logistics data (shipping times, supplier delays).
2. Use Excel Solver to minimize transportation costs.
3. Develop a Power BI dashboard tracking on-time deliveries.
4. Adjust supplier contracts based on data insights.

Pairing 7: Improving Fraud Detection in Financial Transactions

Big Skill: Integrating Data into Decision-Making – Using transaction patterns to flag potential fraud.

Small Skill: Using Analytical Tools – Implementing anomaly detection in Python using Scikit-learn.

Teaching Technique: Financial fraud detection simulation.

Actions: Have students develop a fraud detection model using historical transaction data.

Example:

1. Load transaction records into a Pandas data frame.
2. Use machine learning (random forests) to classify fraudulent vs. normal transactions.
3. Evaluate model precision and recall for effectiveness.
4. Suggest fraud prevention measures for banks.

Pairing 8: Forecasting Energy Consumption for Smart Cities

Big Skill: Integrating Data into Decision-Making – Using energy usage data to optimize consumption.

Small Skill: Using Analytical Tools – Applying time series forecasting in R or Python.

Teaching Technique: Smart city energy optimization project.

Actions: Have students analyze city-wide electricity usage and predict future demand.

Example:

1. Import energy consumption data by region.
2. Use time series analysis (ARIMA models) to forecast demand.
3. Create interactive Power BI reports for city planners.
4. Suggest energy-saving strategies (smart grids, off-peak usage incentives).

Pairing 9: Enhancing Manufacturing Productivity Through IoT Data

Big Skill: Integrating Data into Decision-Making – Using real-time factory sensor data to improve efficiency.

Small Skill: Using Analytical Tools – Using SQL databases and Power BI for monitoring.

Teaching Technique: IoT data analysis project.

Actions: Have students track factory machine performance using sensor data.

Example:

1. Collect IoT data from production lines.
2. Use SQL queries to identify downtime trends.
3. Develop Power BI visualizations to monitor performance.
4. Optimize maintenance schedules based on sensor readings.

Pairing 10: Personalizing Online Learning Platforms with Data Insights

Big Skill: Integrating Data into Decision-Making – Using learning analytics to improve student engagement.

Small Skill: Using Analytical Tools – Applying Python and machine learning for personalized content recommendations.

Teaching Technique: Educational data science project.

Actions: Have students analyze student interaction data to enhance learning experiences.

Example:

1. Extract course interaction logs from an online learning platform.
2. Use clustering algorithms to group students based on learning patterns.

3. Develop personalized course recommendations using machine learning models.
4. Present insights to educators for curriculum improvements.

Data Analysis and Interpretation – (Big Skill) Integrating Data into Decision-Making vs. (Small Skill) Performing Statistical Analysis

---

Pairing 1: Forecasting Sales Growth for Business Strategy

Big Skill: Integrating Data into Decision-Making – Using past sales data to guide business expansion.

Small Skill: Performing Statistical Analysis – Applying regression analysis to predict future sales trends.

Teaching Technique: Business forecasting case study.

Actions: Have students analyze past sales data and predict revenue growth for the next quarter.

Example:

1. Collect historical sales data for multiple product categories.
2. Use linear regression to model future sales trends.
3. Calculate correlation coefficients between sales and marketing spend.
4. Make strategic recommendations on where to expand sales efforts.

Pairing 2: Evaluating Employee Performance Trends

Big Skill: Integrating Data into Decision-Making – Using performance metrics to identify training needs.

Small Skill: Performing Statistical Analysis – Calculating mean, median, and standard deviation of employee evaluations.

Teaching Technique: Workforce analytics simulation.

Actions: Have students analyze performance review data to identify high and low performers.

Example:

1. Import employee evaluation scores from HR records.
2. Calculate average performance ratings for different departments.
3. Analyze distribution patterns using standard deviation.
4. Recommend customized training programs based on findings.

Pairing 3: Measuring Customer Satisfaction for Service Improvements

Big Skill: Integrating Data into Decision-Making – Using customer feedback to refine service offerings.

Small Skill: Performing Statistical Analysis – Conducting t-tests to compare satisfaction levels before and after policy changes.

Teaching Technique: Customer service data analysis.

Actions: Have students analyze survey data to evaluate the impact of service modifications.

Example:

1. Collect customer satisfaction scores from pre- and post-service changes.
2. Perform a t-test to assess statistical significance of improvements.
3. Calculate mean and standard deviation to compare distributions.
4. Recommend further service adjustments based on statistical evidence.

Pairing 4: Identifying Student Success Factors in Higher Education

Big Skill: Integrating Data into Decision-Making – Using student performance data to guide academic policies.

Small Skill: Performing Statistical Analysis – Calculating correlation between study habits and GPA.

Teaching Technique: Educational analytics case study.

Actions: Have students analyze academic data to identify key predictors of student success.

Example:

1. Collect data on student study hours, test scores, and GPA.
2. Calculate correlation coefficients to find key relationships.
3. Perform regression analysis to model GPA predictors.
4. Suggest academic interventions to improve student outcomes.

Pairing 5: Improving Healthcare Outcomes Using Data Analysis

Big Skill: Integrating Data into Decision-Making – Using medical records to optimize patient care strategies.

Small Skill: Performing Statistical Analysis – Using ANOVA to compare effectiveness of different treatments.

Teaching Technique: Healthcare data analysis project.

Actions: Have students analyze medical treatment success rates and recommend best practices.

Example:

1. Gather patient recovery data from multiple treatment plans.
2. Perform an ANOVA test to check statistical significance of differences.
3. Calculate standard deviation to measure outcome variability.
4. Recommend treatment optimizations for hospitals.

Pairing 6: Assessing Financial Risk in Investment Portfolios

Big Skill: Integrating Data into Decision-Making – Using financial data to inform investment strategies.

Small Skill: Performing Statistical Analysis – Calculating standard deviation and risk correlation in asset portfolios.

Teaching Technique: Investment risk assessment case study.

Actions: Have students evaluate the risk and return of different stock portfolios.

Example:

1. Collect historical stock price data for multiple asset classes.
2. Compute portfolio standard deviation to measure volatility.
3. Analyze correlations between asset performance and market trends.
4. Recommend portfolio adjustments based on risk tolerance levels.

Pairing 7: Optimizing Inventory for Supply Chain Efficiency

Big Skill: Integrating Data into Decision-Making – Using demand forecasting to improve inventory management.

Small Skill: Performing Statistical Analysis – Conducting time-series analysis to predict stock demand.

Teaching Technique: Supply chain forecasting simulation.

Actions: Have students analyze product demand variations and propose stocking strategies.

Example:

1. Import historical inventory data from a retail store.
2. Perform moving average calculations to detect seasonal demand shifts.
3. Use regression models to predict future stock requirements.
4. Recommend optimal reorder points to minimize shortages and overstock.

Pairing 8: Evaluating Air Quality for Environmental Policy Decisions

Big Skill: Integrating Data into Decision-Making – Using pollution metrics to implement environmental regulations.

Small Skill: Performing Statistical Analysis – Calculating mean pollution levels and correlation with health data.

Teaching Technique: Environmental data analysis project.

Actions: Have students assess air quality data and its impact on public health.

Example:

1. Collect air pollution data from city monitoring stations.
2. Calculate mean and median pollutant concentrations.
3. Analyze correlation between pollution levels and respiratory illnesses.
4. Recommend policy interventions to reduce pollution impact.

Pairing 9: Measuring Effectiveness of Online Advertisements

Big Skill: Integrating Data into Decision-Making – Using engagement metrics to optimize digital marketing campaigns.

Small Skill: Performing Statistical Analysis – Computing correlation between ad spend and conversion rates.

Teaching Technique: Digital marketing performance analysis.

Actions: Have students analyze ad campaign data to maximize ROI.

Example:

1. Collect advertising spend and sales conversion data.
2. Calculate mean conversion rates for different ad formats.
3. Use regression analysis to determine spend-to-sales impact.
4. Recommend ad budget allocation based on statistical findings.

Pairing 10: Detecting Fraud in Banking Transactions

Big Skill: Integrating Data into Decision-Making – Using transaction monitoring to identify fraudulent activities.

Small Skill: Performing Statistical Analysis – Analyzing anomaly detection using Z-scores and standard deviations.

Teaching Technique: Financial fraud detection case study.

Actions: Have students evaluate banking transaction data for irregular activity.

Example:

1. Import transaction records from a financial institution.
2. Calculate mean and standard deviation of transaction amounts.
3. Use Z-score analysis to detect outliers.
4. Recommend fraud prevention measures based on findings.

Data Analysis and Interpretation – (Big Skill) Integrating Data into Decision-Making vs. (Small Skill) Visualizing Data

---

Pairing 1: Optimizing Business Revenue Streams

Big Skill: Integrating Data into Decision-Making – Using revenue analytics to prioritize high-performing business segments.

Small Skill: Visualizing Data – Creating interactive dashboards in Power BI or Tableau.

Teaching Technique: Business revenue analysis project.

Actions: Have students analyze company revenue data and create a visual report for leadership.

Example:

1. Gather sales revenue data by product and region.
2. Develop Power BI charts showing revenue trends over time.
3. Use heat maps to highlight top-performing locations.
4. Recommend strategy shifts to maximize profitability.

Pairing 2: Tracking Consumer Trends in E-commerce

Big Skill: Integrating Data into Decision-Making – Identifying customer purchase behavior patterns to guide marketing.

Small Skill: Visualizing Data – Creating sales funnel charts and customer heatmaps.

Teaching Technique: E-commerce analytics simulation.

Actions: Have students develop an interactive e-commerce performance dashboard.

Example:

1. Extract customer purchase history.
2. Create funnel visualizations to analyze drop-off points.
3. Use heatmaps to track product page engagement.
4. Suggest UI/UX improvements to enhance sales conversions.

Pairing 3: Monitoring Financial Market Performance

Big Skill: Integrating Data into Decision-Making – Using stock market data to guide investment decisions.

Small Skill: Visualizing Data – Creating real-time financial dashboards using Python's Matplotlib and Seaborn.

Teaching Technique: Stock market data visualization.

Actions: Have students track stock performance trends and present investment insights.

Example:

1. Collect historical stock data.
2. Generate line charts for price movement analysis.
3. Use comparative bar charts for sector performance.
4. Recommend investment strategies based on visualized trends.

Pairing 4: Evaluating Hospital Patient Outcomes

Big Skill: Integrating Data into Decision-Making – Using healthcare analytics to improve patient care strategies.

Small Skill: Visualizing Data – Creating bar graphs and pie charts for patient recovery rates.

Teaching Technique: Healthcare analytics case study.

Actions: Have students visualize patient recovery data to assess treatment effectiveness.

Example:

1. Import patient recovery statistics.
2. Create bar graphs comparing different treatments.
3. Use pie charts to show recovery rate distributions.
4. Suggest policy changes to enhance patient care.

Pairing 5: Measuring Digital Marketing Campaign Success

Big Skill: Integrating Data into Decision-Making – Using engagement analytics to refine digital marketing efforts.

Small Skill: Visualizing Data – Creating comparative campaign performance reports using Google Data Studio.

Teaching Technique: Digital marketing data tracking project.

Actions: Have students analyze ad campaign data and present findings visually.

Example:

1. Pull ad engagement metrics (click-through rates, conversions).
2. Develop a comparison dashboard for different campaigns.
3. Use scatter plots to identify the most effective ad formats.
4. Optimize future ad spend based on insights.

Pairing 6: Optimizing Supply Chain Logistics

Big Skill: Integrating Data into Decision-Making – Using logistics data to streamline distribution networks.

Small Skill: Visualizing Data – Creating route efficiency dashboards using Power BI.

Teaching Technique: Supply chain visualization project.

Actions: Have students map supplier routes and identify cost-saving opportunities.

Example:

1. Import shipment data.
2. Use geospatial maps to visualize delivery routes.
3. Develop cost vs. time efficiency charts.
4. Recommend alternative logistics strategies.

Pairing 7: Enhancing College Student Retention Strategies

Big Skill: Integrating Data into Decision-Making – Using enrollment trends to improve student engagement.

Small Skill: Visualizing Data – Creating student retention heatmaps and line charts.

Teaching Technique: Higher education data visualization.

Actions: Have students analyze dropout rates and present solutions visually.

Example:

1. Gather student retention statistics.
2. Use heatmaps to identify at-risk demographics.
3. Develop trend lines showing retention over semesters.
4. Propose engagement initiatives to increase retention.

Pairing 8: Analyzing Energy Consumption for Sustainability Efforts

Big Skill: Integrating Data into Decision-Making – Using energy analytics to optimize sustainability initiatives.

Small Skill: Visualizing Data – Creating real-time dashboards with interactive graphs.

Teaching Technique: Smart energy management project.

Actions: Have students monitor energy usage and identify inefficiencies.

Example:

1. Extract electricity consumption data.
2. Create time-series graphs for peak usage trends.
3. Develop sector-wise comparison charts.
4. Suggest energy reduction strategies.

Pairing 9: Forecasting Weather Patterns for Climate Research

Big Skill: Integrating Data into Decision-Making – Using climate data to predict extreme weather events.

Small Skill: Visualizing Data – Creating time-lapse charts and 3D visualizations of climate shifts.

Teaching Technique: Climate trend visualization exercise.

Actions: Have students track global temperature shifts and visualize findings.

Example:

1. Import historical climate data.
2. Use line graphs to depict temperature fluctuations.
3. Develop 3D contour maps for precipitation changes.
4. Provide policy recommendations for climate adaptation.

Pairing 10: Tracking Crime Trends for Law Enforcement Resource Allocation

Big Skill: Integrating Data into Decision-Making – Using crime analytics to guide public safety policies.

Small Skill: Visualizing Data – Creating heatmaps and comparative charts for crime rates.

Teaching Technique: Public safety data visualization.

Actions: Have students analyze crime reports and develop data-driven policing strategies.

Example:

1. Collect crime data by location and time of day.
2. Use GIS mapping tools to visualize high-risk areas.
3. Compare historical trends with line graphs.
4. Propose targeted law enforcement strategies.

Data Analysis and Interpretation – (Big Skill) Cross-Disciplinary Synthesis vs. (Small Skill) Using Analytical Tools

---

Pairing 1: Analyzing the Impact of Economic Policies on Public Health

Big Skill: Cross-Disciplinary Synthesis – Combining economic trends and healthcare data to assess policy outcomes.

Small Skill: Using Analytical Tools – Applying Python (Pandas, Matplotlib) for data integration and visualization.

Teaching Technique: Policy impact analysis project.

Actions: Have students evaluate how GDP changes affect healthcare accessibility.

Example:

1. Gather economic growth and healthcare funding data.
2. Merge datasets in Pandas to compare spending with public health outcomes.
3. Use Matplotlib to visualize trends in mortality rates vs. economic stability.
4. Develop policy recommendations based on integrated findings.

Pairing 2: Investigating the Relationship Between Education Levels and Crime Rates

Big Skill: Cross-Disciplinary Synthesis – Merging education statistics and law enforcement data for social impact analysis.

Small Skill: Using Analytical Tools – Using R for correlation analysis and geospatial visualization.

Teaching Technique: Social research data synthesis.

Actions: Have students analyze how educational attainment impacts crime rates.

Example:

1. Collect data on literacy rates and crime statistics.
2. Use R to perform correlation tests between education levels and crime rates.
3. Visualize geographic trends using ggplot2.
4. Recommend education-based crime prevention strategies.

Pairing 3: Assessing the Effects of Climate Change on Agriculture

Big Skill: Cross-Disciplinary Synthesis – Integrating climate science with agricultural yield data for sustainability insights.

Small Skill: Using Analytical Tools – Utilizing Excel for time-series forecasting and Tableau for visualization.

Teaching Technique: Climate and agriculture impact study.

Actions: Have students compare temperature fluctuations with crop yield changes.

Example:

1. Import historical temperature and crop production data.
2. Use Excel's regression tools to analyze the impact of rising temperatures on yields.
3. Visualize results with Tableau dashboards.
4. Suggest adaptive farming techniques based on trends.

Pairing 4: Evaluating the Role of Social Media in Political Movements

Big Skill: Cross-Disciplinary Synthesis – Integrating communication studies with political science data.

Small Skill: Using Analytical Tools – Applying sentiment analysis in Python (NLTK, TextBlob).

Teaching Technique: Political trend analysis project.

Actions: Have students analyze the correlation between social media activity and political engagement.

Example:

1. Scrape Twitter data for major political movements.
2. Use NLTK for sentiment analysis of tweets.
3. Compare social media trends with voter turnout data.
4. Propose strategies for digital civic engagement.

Pairing 5: Studying the Economic Impact of Tourism on Urban Development

Big Skill: Cross-Disciplinary Synthesis – Combining urban development statistics with tourism revenue analysis.

Small Skill: Using Analytical Tools – Using Power BI for data integration and visual storytelling.

Teaching Technique: Urban planning and tourism analysis.

Actions: Have students assess how tourism growth affects infrastructure and local economies.

Example:

1. Collect city development and tourism revenue data.
2. Use Power BI to merge datasets and identify trends.
3. Create visual reports showing infrastructure development linked to visitor spending.
4. Recommend sustainable tourism policies.

Pairing 6: Exploring the Link Between Nutrition and Cognitive Performance

Big Skill: Cross-Disciplinary Synthesis – Integrating nutritional science with psychological studies.

Small Skill: Using Analytical Tools – Applying Excel for statistical comparisons and Python for trend modeling.

Teaching Technique: Health and cognitive function study.

Actions: Have students analyze how diet affects student test performance.

Example:

1. Gather student dietary logs and test scores.
2. Use Excel for t-tests comparing nutrition levels with academic outcomes.
3. Use Python's Seaborn to visualize patterns.
4. Develop dietary guidelines for academic performance enhancement.

Pairing 7: Assessing the Role of Income Disparity in Mental Health Issues

Big Skill: Cross-Disciplinary Synthesis – Merging economic inequality data with mental health statistics.

Small Skill: Using Analytical Tools – Using R (ggplot2, dplyr) for data cleaning and visualization.

Teaching Technique: Economic and mental health impact study.

Actions: Have students analyze income brackets against mental health survey data.

Example:

1. Load economic disparity data from census reports.
2. Merge with mental health survey datasets.
3. Use ggplot2 to create heatmaps showing correlations.
4. Propose policy interventions to address disparities.

Pairing 8: Measuring the Economic Benefits of Renewable Energy Adoption

Big Skill: Cross-Disciplinary Synthesis – Integrating environmental science with financial analysis.

Small Skill: Using Analytical Tools – Applying regression analysis in Python (Scikit-learn).

Teaching Technique: Energy economics research project.

Actions: Have students compare energy costs of fossil fuels vs. renewables.

Example:

1. Collect energy cost data for renewables vs. traditional sources.
2. Run linear regression models predicting long-term savings.
3. Visualize cost trends with Matplotlib.
4. Provide policy recommendations for green energy investments.

Pairing 9: Investigating the Relationship Between Housing Prices and Crime Rates

Big Skill: Cross-Disciplinary Synthesis – Merging real estate trends with public safety statistics.

Small Skill: Using Analytical Tools – Using Power BI to create interactive maps.

Teaching Technique: Real estate and crime correlation study.

Actions: Have students analyze crime statistics alongside property values.

Example:

1. Import housing market data by neighborhood.
2. Overlay crime rate data for the same areas.
3. Use Power BI for spatial analysis.
4. Present findings on crime impact on home values.

Pairing 10: Examining the Link Between Workplace Diversity and Financial Performance

Big Skill: Cross-Disciplinary Synthesis – Integrating HR diversity metrics with corporate financial reports.

Small Skill: Using Analytical Tools – Using Excel PivotTables and Python's Pandas for data analysis.

Teaching Technique: Workforce diversity financial impact project.

Actions: Have students compare company diversity levels with profit margins.

Example:

1. Collect workforce demographic data and corporate earnings reports.
2. Use Excel PivotTables to compare diversity levels across industries.
3. Apply correlation analysis in Python to measure financial impact.
4. Provide recommendations for diversity-focused business strategies.

Data Analysis and Interpretation – (Big Skill) Cross-Disciplinary Synthesis vs. (Small Skill) Performing Statistical Analysis

---

Pairing 1: Evaluating the Link Between Economic Growth and Public Health

Big Skill: Cross-Disciplinary Synthesis – Merging economic indicators with healthcare outcomes for policy assessment.

Small Skill: Performing Statistical Analysis – Conducting correlation and regression analysis on GDP vs. health metrics.

Teaching Technique: Economic and health policy case study.

Actions: Have students analyze how economic growth affects life expectancy and healthcare access.

Example:

1. Collect GDP and healthcare expenditure data from multiple countries.

2. Use correlation analysis to measure the strength of the relationship.
3. Apply regression analysis to predict how increased GDP affects health outcomes.
4. Propose policy changes based on the findings.

Pairing 2: Investigating the Impact of Education on Crime Rates

Big Skill: Cross-Disciplinary Synthesis – Combining education statistics and crime data for social analysis.

Small Skill: Performing Statistical Analysis – Using correlation analysis to measure relationships between literacy and crime.

Teaching Technique: Education policy impact study.

Actions: Have students analyze regional education levels and crime rates to identify patterns.

Example:

1. Collect data on literacy rates and crime incidents per region.
2. Calculate correlation coefficients to see if higher education levels reduce crime.
3. Use regression models to predict crime rate changes with increased education funding.
4. Develop policy recommendations for improving education as a crime deterrent.

Pairing 3: Measuring the Economic Effects of Climate Change on Agriculture

Big Skill: Cross-Disciplinary Synthesis – Integrating environmental science with agricultural economics.

Small Skill: Performing Statistical Analysis – Using time-series analysis to track climate impact on crop yields.

Teaching Technique: Climate-agriculture impact study.

Actions: Have students examine how temperature fluctuations and rainfall changes affect crop production.

Example:

1. Import historical climate and crop yield data.
2. Perform regression analysis to identify climate variables affecting production.
3. Calculate correlation coefficients between temperature shifts and agricultural output.
4. Suggest adaptive farming strategies based on data trends.

Pairing 4: Analyzing the Effect of Social Media on Political Participation

Big Skill: Cross-Disciplinary Synthesis – Merging communication studies with political science.

Small Skill: Performing Statistical Analysis – Using regression analysis to predict voter turnout from online activity.

Teaching Technique: Political media impact project.

Actions: Have students compare social media engagement with voter participation rates.

Example:

1. Collect data on political social media engagement and voter turnout by region.
2. Use correlation analysis to measure the strength of the relationship.
3. Perform multiple regression to determine the most influential social media metrics.
4. Develop strategies for increasing political engagement through digital media.

Pairing 5: Assessing the Relationship Between Tourism and Urban Development

Big Skill: Cross-Disciplinary Synthesis – Integrating tourism revenue data with infrastructure development.

Small Skill: Performing Statistical Analysis – Using ANOVA to compare urban growth rates between tourist-heavy and low-tourist areas.

Teaching Technique: Urban planning and tourism analysis.

Actions: Have students analyze how tourism influences economic growth and city development.

Example:

1. Gather data on tourism revenue and infrastructure investment per city.
2. Perform ANOVA tests to check for statistical differences between tourism-heavy and non-tourism cities.
3. Calculate growth correlations to measure long-term impact.
4. Propose policies for sustainable urban development.

Pairing 6: Investigating the Role of Income Inequality in Mental Health Trends

Big Skill: Cross-Disciplinary Synthesis – Integrating economic inequality data with psychological health research.

Small Skill: Performing Statistical Analysis – Using regression analysis to explore socioeconomic determinants of mental health.

Teaching Technique: Economic and mental health study.

Actions: Have students analyze how income disparity impacts rates of depression and anxiety.

Example:

1. Collect economic data (income levels) and mental health survey results.
2. Run correlation analysis to identify key trends.

3. Apply multiple regression to determine the most significant economic predictors of mental health issues.
4. Suggest interventions for mental health support based on economic conditions.

Pairing 7: Understanding the Relationship Between Nutrition and Academic Performance

Big Skill: Cross-Disciplinary Synthesis – Combining nutritional data with educational performance metrics.

Small Skill: Performing Statistical Analysis – Using t-tests to compare performance across different diet groups.

Teaching Technique: Nutrition and academic performance study.

Actions: Have students analyze dietary habits and student GPA data to identify key relationships.

Example:

1. Gather data on student diets and academic performance.
2. Perform t-tests to compare average GPA among different nutritional groups.
3. Calculate correlation coefficients to see which nutrients impact cognitive function.
4. Develop dietary recommendations to enhance learning outcomes.

Pairing 8: Exploring the Economic Benefits of Renewable Energy Adoption

Big Skill: Cross-Disciplinary Synthesis – Merging environmental science with economic growth metrics.

Small Skill: Performing Statistical Analysis – Using regression modeling to assess renewable energy's impact on job creation.

Teaching Technique: Energy and economic impact study.

Actions: Have students examine how transitioning to renewables affects employment rates and energy costs.

Example:

1. Collect energy sector job data from renewable and fossil fuel industries.
2. Use multiple regression models to predict employment growth in green energy sectors.
3. Compare cost savings for businesses switching to renewables.
4. Recommend policies for incentivizing green energy investments.

Pairing 9: Evaluating the Link Between Housing Prices and Local Crime Rates

Big Skill: Cross-Disciplinary Synthesis – Integrating real estate trends with law enforcement data.

Small Skill: Performing Statistical Analysis – Using correlation and regression analysis to predict home values.

Teaching Technique: Real estate and crime impact study.

Actions: Have students analyze how crime rates influence housing prices.

Example:

1. Collect housing market and local crime statistics.
2. Use correlation analysis to measure the impact of crime on property values.
3. Apply regression models to predict price fluctuations in different crime zones.
4. Propose strategies for community revitalization.

Pairing 10: Examining the Relationship Between Workplace Diversity and Company Profitability

Big Skill: Cross-Disciplinary Synthesis – Integrating HR diversity data with corporate financial performance.

Small Skill: Performing Statistical Analysis – Using ANOVA and regression analysis to evaluate the impact of diversity on profit margins.

Teaching Technique: Diversity and business impact study.

Actions: Have students compare corporate diversity levels with financial performance indicators.

Example:

1. Collect workforce demographic data from multiple corporations.
2. Use ANOVA to compare profit margins across companies with varying diversity levels.
3. Run regression models to evaluate if diversity is a predictor of financial success.
4. Provide business strategies to enhance workforce inclusiveness and profitability.

Data Analysis and Interpretation – (Big Skill) Cross-Disciplinary Synthesis vs. (Small Skill) Visualizing Data

---

Pairing 1: Understanding the Economic and Social Impact of Unemployment

Big Skill: Cross-Disciplinary Synthesis – Merging economic and sociological data to analyze unemployment trends.

Small Skill: Visualizing Data – Creating dashboards with interactive charts comparing unemployment rates and social factors.

Teaching Technique: Labor market impact study.

Actions: Have students create interactive dashboards comparing unemployment rates across demographics.

Example:

1. Collect data on unemployment rates, crime rates, and mental health trends.
2. Use Power BI to create comparative bar charts showing economic downturn effects.
3. Develop a heatmap of high-unemployment regions correlated with crime rates.
4. Propose policy recommendations for workforce development programs.

Pairing 2: Examining the Relationship Between Air Pollution and Public Health

Big Skill: Cross-Disciplinary Synthesis – Combining environmental and healthcare data to assess the impact of pollution on respiratory diseases.

Small Skill: Visualizing Data – Creating geospatial visualizations of pollution levels and health outcomes.

Teaching Technique: Environmental health visualization project.

Actions: Have students map pollution levels against respiratory illness rates in different cities.

Example:

1. Collect air quality index (AQI) and hospitalization data for respiratory illnesses.
2. Use Tableau to create a map visualization of pollution hotspots.
3. Overlay hospital admission rates for asthma-related cases.
4. Develop a report recommending pollution control measures.

Pairing 3: Analyzing the Influence of Education on Income Levels

Big Skill: Cross-Disciplinary Synthesis – Combining education statistics with income distribution data to study economic mobility.

Small Skill: Visualizing Data – Using histograms and scatter plots to compare education levels and earnings.

Teaching Technique: Economic mobility research project.

Actions: Have students develop interactive charts showing income variations by education level.

Example:

1. Gather data on median income by education attainment.
2. Create a scatter plot in Python (Matplotlib) to visualize trends.
3. Use box plots to compare income disparity across education groups.
4. Suggest policies to bridge the earnings gap.

Pairing 4: Exploring the Connection Between Social Media Trends and Political Engagement

Big Skill: Cross-Disciplinary Synthesis – Merging communication studies with political science data.

Small Skill: Visualizing Data – Developing a timeline visualization of social media influence on voter turnout.

Teaching Technique: Political trend analysis project.

Actions: Have students track political discourse on social media and compare it to voter turnout rates.

Example:

1. Scrape Twitter and Facebook data for political hashtags.
2. Use Google Data Studio to create a time-series visualization of trending topics.
3. Overlay voter turnout rates for corresponding election periods.
4. Recommend strategies for digital civic engagement.

Pairing 5: Investigating the Relationship Between Tourism and Economic Growth

Big Skill: Cross-Disciplinary Synthesis – Combining tourism revenue with local economic data to analyze growth patterns.

Small Skill: Visualizing Data – Creating comparative charts for GDP growth in tourism-dependent regions.

Teaching Technique: Tourism and economy analysis.

Actions: Have students develop interactive dashboards showcasing the economic benefits of tourism.

Example:

1. Import tourism revenue and GDP growth data.
2. Use Power BI to create bar charts comparing tourism-dependent vs. non-tourism regions.
3. Develop a trend line visualization showing long-term economic benefits.
4. Recommend policies to maximize sustainable tourism revenue.

Pairing 6: Assessing the Link Between Food Deserts and Public Health

Big Skill: Cross-Disciplinary Synthesis – Integrating geographical and health data to study food accessibility.

Small Skill: Visualizing Data – Using GIS mapping tools to display food deserts and their impact on obesity rates.

Teaching Technique: Food accessibility analysis.

Actions: Have students analyze food desert locations and their correlation with health conditions.

Example:

1. Gather census data on grocery store locations and obesity rates.
2. Use QGIS to create a map visualization of food deserts.
3. Overlay obesity rate data to identify at-risk communities.
4. Suggest initiatives to improve food accessibility.

Pairing 7: Evaluating the Impact of Renewable Energy on Job Creation

Big Skill: Cross-Disciplinary Synthesis – Merging environmental sustainability data with economic employment reports.

Small Skill: Visualizing Data – Creating bar charts comparing employment rates in renewable vs. fossil fuel industries.

Teaching Technique: Green energy economic study.

Actions: Have students analyze job growth in renewable energy sectors and compare it to fossil fuel industries.

Example:

1. Collect employment data from both energy sectors.
2. Use Excel to create comparative bar graphs of job creation trends.
3. Visualize salary differences between fossil fuel and green energy jobs.
4. Develop recommendations for transitioning workers to renewable industries.

Pairing 8: Understanding the Effect of Income Inequality on Mental Health

Big Skill: Cross-Disciplinary Synthesis – Integrating economic disparity statistics with mental health survey data.

Small Skill: Visualizing Data – Using scatter plots and heatmaps to display the correlation between income levels and mental health conditions.

Teaching Technique: Socioeconomic health analysis.

Actions: Have students explore how income inequality correlates with depression and anxiety rates.

Example:

1. Gather economic and mental health survey data.
2. Use Python's Seaborn to generate heatmaps of mental health trends across income brackets.
3. Create scatter plots to compare per capita income with reported stress levels.
4. Recommend policies for mental health support in low-income areas.

Pairing 9: Forecasting the Relationship Between Climate Change and Migration Patterns

Big Skill: Cross-Disciplinary Synthesis – Combining climate and demographic data to analyze climate-driven displacement.

Small Skill: Visualizing Data – Using animated maps to track migration shifts due to climate events.

Teaching Technique: Climate migration forecasting.

Actions: Have students analyze how extreme weather events correlate with migration trends.

Example:

1. Collect climate disaster and migration data.
2. Use Tableau to create a dynamic migration map.
3. Visualize time-series trends showing climate disaster frequency and displacement patterns.
4. Suggest policy solutions for climate refugees.

Pairing 10: Analyzing the Effect of Healthcare Accessibility on Life Expectancy

Big Skill: Cross-Disciplinary Synthesis – Merging healthcare infrastructure data with demographic trends.

Small Skill: Visualizing Data – Creating comparative bar charts and regional distribution maps.

Teaching Technique: Healthcare access and life expectancy study.

Actions: Have students analyze how hospital density and medical resources correlate with life expectancy.

Example:

1. Collect data on hospital locations and mortality rates.
2. Use Excel PivotTables to compare healthcare accessibility across regions.
3. Create bar charts showing variations in life expectancy.
4. Recommend investments in underserved areas.

Example Teaching Strategies for Big vs. Small Thinking Skill Pairings of Visionary Thinking at the Business Professional Level

Here are 63 unique teaching strategy examples for the big thinking and small thinking skill pairings, each described in detail to align with **Big Thinking Skills** (Concepts & Strategies) and **Small Thinking Skills** (Step-by-Step Execution) for the **Visionary Thinking Skill** taught at the business professional level.

Visionary Thinking

Big Thinking Skills (Concepts & Strategies)

- **Strategic Visioning:** Developing a long-term vision that aligns with business goals and market trends.
- **Inspirational Leadership:** Communicating a vision in a way that motivates and unifies teams.
- **Foresight & Innovation:** Anticipating industry shifts and proactively shaping strategies.

Small Thinking Skills (Step-by-Step Execution)

- **Translating Vision into Action:** Breaking down broad strategic goals into clear operational tasks.
- **Crafting Persuasive Messaging:** Writing and delivering effective vision statements and presentations.
- **Stakeholder Alignment:** Ensuring that different teams understand and commit to the vision.

Visionary Thinking – (Big Skill) Strategic Visioning vs. (Small Skill) Translating Vision into Action

---

Pairing 1: Defining Market Leadership Through Strategic Vision

Big Skill: Strategic Visioning – Developing a long-term vision that aligns with business goals and market trends.

Small Skill: Translating Vision into Action – Breaking down broad strategic goals into clear operational tasks.

Teaching Technique: Scenario planning and vision articulation.

Actions: Guide professionals in developing a 5-year strategic vision for becoming an industry leader, then breaking it into quarterly action steps.

Example:

1. Define a vision to become the top sustainable fashion brand.
2. Assign R&D, marketing, and supply chain teams specific milestones.
3. Develop a three-phase sustainability plan (material sourcing, production, distribution).
4. Implement a brand awareness campaign on sustainability impact.

Pairing 2: Digital Transformation Strategy Implementation

Big Skill: Strategic Visioning – Anticipating how digital trends impact business evolution.

Small Skill: Translating Vision into Action – Identifying necessary technology upgrades, training programs, and workflow shifts.

Teaching Technique: Digital roadmap creation.

Actions: Have leaders map out a digital transformation strategy by defining short-term IT upgrades, mid-term AI integration, and long-term automation goals.

Example:

1. Identify key digital tools (AI, automation, cloud solutions).
2. Implement AI-driven customer service chatbots in Phase 1.
3. Train employees on new data analytics software.
4. Roll out an enterprise-wide digital workflow automation system.

Pairing 3: Creating a Customer-Centric Business Model

Big Skill: Strategic Visioning – Shaping a long-term strategy focused on customer needs and experience.

Small Skill: Translating Vision into Action – Structuring customer service initiatives and product adjustments to align with the vision.

Teaching Technique: Customer journey mapping.

Actions: Have teams create an end-to-end customer journey map and define step-by-step actions to enhance each touchpoint.

Example:

1. Research customer pain points and expectations.
2. Develop a strategy for seamless omnichannel customer interactions.
3. Implement AI-powered chat support and real-time tracking systems.
4. Measure customer satisfaction and iterate based on feedback.

Pairing 4: Expanding into New Global Markets

Big Skill: Strategic Visioning – Forecasting global expansion opportunities.

Small Skill: Translating Vision into Action – Defining the necessary logistics, partnerships, and regulatory considerations.

Teaching Technique: Market entry planning.

Actions: Guide leaders through a structured approach to identify viable markets, develop partnerships, and outline operational tasks for expansion.

Example:

1. Conduct market research to identify potential regions for expansion.
2. Develop localization strategies for branding and customer service.
3. Secure compliance with regional data laws and regulations.
4. Build a regional distribution network and establish a sales team.

Pairing 5: Leading Sustainable Business Growth

Big Skill: Strategic Visioning – Incorporating sustainability into long-term corporate growth.

Small Skill: Translating Vision into Action – Creating a roadmap for sustainable operations and ethical sourcing.

Teaching Technique: ESG (Environmental, Social, Governance) planning.

Actions: Have leaders outline clear environmental and ethical policies and assign responsibilities across departments.

Example:

1. Establish sustainability KPIs for carbon emissions and resource efficiency.
2. Transition to 50% renewable energy use within three years.
3. Partner with ethical suppliers and implement transparent sourcing.
4. Launch a corporate sustainability report to track progress.

Pairing 6: Building a Culture of Innovation

Big Skill: Strategic Visioning – Establishing a vision for continuous innovation within the company.

Small Skill: Translating Vision into Action – Defining processes and resources for fostering innovation.

Teaching Technique: Innovation pipeline development.

Actions: Guide leaders in developing structured processes for ideation, prototyping, and implementation, assigning ownership at each stage.

Example:

1. Set up an internal innovation hub for employee-led projects.
2. Implement a quarterly innovation challenge with rewards.
3. Develop a cross-functional team to assess and refine ideas.
4. Partner with universities for research collaboration.

Pairing 7: Transforming Company Culture to Enhance Employee Engagement

Big Skill: Strategic Visioning – Creating a long-term vision for a thriving, motivated workforce.

Small Skill: Translating Vision into Action – Implementing leadership development programs, performance incentives, and workplace flexibility.

Teaching Technique: Change management strategy.

Actions: Have HR and leadership develop engagement initiatives and define key milestones to foster a high-performance culture.

Example:

1. Conduct employee engagement surveys to identify key areas for improvement.
2. Implement hybrid work policies with flexible hours.
3. Establish structured career progression plans with mentorship.
4. Launch an employee recognition program with performance incentives.

Visionary Thinking – (Big Skill) Strategic Visioning vs. (Small Skill) Crafting Persuasive Messaging

---

Pairing 1: Defining a Market-Leading Vision Through Compelling Messaging

Big Skill: Strategic Visioning – Developing a long-term vision that aligns with business goals and market trends.

Small Skill: Crafting Persuasive Messaging – Writing and delivering effective vision statements and presentations.

Teaching Technique: Vision statement workshop.

Actions: Guide professionals in crafting a clear and inspiring vision statement that aligns with industry trends and company strengths.

Example:

1. Research competitors' vision statements for differentiation.
2. Draft a concise, aspirational vision statement.
3. Deliver the vision in a company-wide presentation with supporting visuals.
4. Gather employee feedback and refine messaging for clarity.

Pairing 2: Aligning Vision with Investor Messaging

Big Skill: Strategic Visioning – Communicating long-term business direction to investors.

Small Skill: Crafting Persuasive Messaging – Structuring compelling investor presentations.

Teaching Technique: Investor pitch simulation.

Actions: Have executives create and refine an investor pitch that explains the company's vision, projected growth, and industry positioning.

Example:

1. Develop a pitch deck with financial projections and key milestones.
2. Use storytelling techniques to showcase market opportunities.
3. Practice delivery with feedback from mentors or advisors.
4. Present the final version to potential investors.

Pairing 3: Inspiring Internal Teams with a Vision Narrative

Big Skill: Strategic Visioning – Developing a vision that unites employees under a common goal.

Small Skill: Crafting Persuasive Messaging – Translating vision into a relatable and motivating story.

Teaching Technique: Leadership communication training.

Actions: Train leaders to articulate the vision in ways that engage and energize employees, using storytelling and real world examples.

Example:

1. Define key themes that resonate with employees.
2. Craft an internal speech that links the vision to individual roles.
3. Present the vision at a company town hall.
4. Measure employee engagement through surveys and discussions.

Pairing 4: Using Data to Strengthen Vision Communication

Big Skill: Strategic Visioning – Backing vision with market and performance data.

Small Skill: Crafting Persuasive Messaging – Creating data-driven presentations for stakeholders.

Teaching Technique: Data storytelling workshop.

Actions: Teach professionals to integrate industry trends, consumer insights, and financial data into vision messaging.

Example:

1. Collect market research on industry trends.
2. Design a visual report with key takeaways.
3. Present findings in a leadership meeting to align strategic goals.
4. Use the insights to refine the company's long-term vision.

Pairing 5: Launching a Rebranding Initiative with Vision-Driven Messaging

Big Skill: Strategic Visioning – Defining a refreshed company direction through branding.

Small Skill: Crafting Persuasive Messaging – Translating brand vision into marketing content.

Teaching Technique: Brand messaging framework.

Actions: Guide marketing teams in creating messaging that reflects the company's future direction while maintaining brand authenticity.

Example:

1. Develop a new tagline that embodies the company's future.
2. Craft key messaging for website relaunch and social media campaigns.
3. Train customer service teams on new brand language.
4. Roll out vision-aligned branding materials internally and externally.

Pairing 6: Gaining Executive Buy-in for Strategic Change

Big Skill: Strategic Visioning – Presenting a compelling case for transformation.

Small Skill: Crafting Persuasive Messaging – Framing vision in a way that aligns with executive priorities.

Teaching Technique: Executive alignment strategy.

Actions: Have professionals tailor vision communication to resonate with different executive stakeholders, emphasizing ROI and strategic impact.

Example:

1. Identify key decision-makers and their priorities.
2. Craft personalized messaging that aligns with financial and operational goals.
3. Present vision proposals in executive meetings with supporting evidence.
4. Secure commitment for phased implementation of the vision.

Pairing 7: Using Vision Messaging to Drive Customer Engagement

Big Skill: Strategic Visioning – Positioning the company as an industry leader through thought leadership.

Small Skill: Crafting Persuasive Messaging – Communicating vision in customer-facing content.

Teaching Technique: Content marketing strategy.

Actions: Have teams create vision-driven messaging for websites, blogs, and social media to build brand authority and customer loyalty.

Example:

1. Develop blog content that educates customers on industry trends.
2. Incorporate vision statements into website headers and company materials.
3. Host a webinar explaining the company's innovation roadmap.
4. Measure customer engagement through analytics and feedback.

Visionary Thinking – (Big Skill) Strategic Visioning vs. (Small Skill) Stakeholder Alignment

---

Pairing 1: Aligning Cross-Departmental Goals with Strategic Vision

Big Skill: Strategic Visioning – Developing a long-term vision that aligns with business goals and market trends.

Small Skill: Stakeholder Alignment – Ensuring that different teams understand and commit to the vision.

Teaching Technique: Cross-functional alignment workshops.

Actions: Facilitate structured sessions where leadership and department heads align on shared priorities, ensuring each team's objectives contribute to the broader vision.

Example:

1. Conduct a vision-sharing workshop with all department heads.
2. Break down long-term goals into department-specific objectives.
3. Create a shared accountability dashboard to track progress.
4. Review alignment quarterly and adjust where necessary.

Pairing 2: Translating Strategic Vision into Departmental Action Plans

Big Skill: Strategic Visioning – Ensuring business strategy is effectively translated at all levels.

Small Skill: Stakeholder Alignment – Helping departments tailor the vision to their specific roles.

Teaching Technique: Departmental goal-setting framework.

Actions: Assist each department in defining their contribution to the company vision and outlining quarterly action items that align with leadership's strategic plan.

Example:

1. The CEO defines a five-year growth vision.
2. Sales set revenue growth targets aligned with this vision.
3. Marketing develops customer acquisition strategies supporting revenue goals.
4. Operations optimizes supply chain processes to meet demand increases.

Pairing 3: Securing Leadership Buy-In for Strategic Initiatives

Big Skill: Strategic Visioning – Crafting long-term business objectives that require executive approval.

Small Skill: Stakeholder Alignment – Gaining leadership commitment for successful execution.

Teaching Technique: Executive vision alignment meetings.

Actions: Develop a structured approach to communicating strategic goals to C-suite executives, ensuring each leader understands their role in implementation.

Example:

1. Present a vision document with financial and operational projections.
2. Hold one-on-one meetings with key executives to tailor messaging.
3. Address concerns by demonstrating ROI and competitive advantage.
4. Secure sign-off and establish a leadership-driven implementation plan.

Pairing 4: Driving Employee Engagement in the Company Vision

Big Skill: Strategic Visioning – Creating a compelling vision that motivates employees.

Small Skill: Stakeholder Alignment – Ensuring employees at all levels understand and support the vision.

Teaching Technique: Internal communication and culture-building strategy.

Actions: Implement structured communication strategies, such as company town halls, internal newsletters, and storytelling to reinforce the vision.

Example:

1. Leadership hosts quarterly town halls to discuss progress.
2. Create an internal podcast featuring employees contributing to the vision.
3. Develop a recognition program celebrating teams aligned with the vision.
4. Establish a feedback loop for employees to voice concerns and ideas.

Pairing 5: Aligning Investors with Long-Term Strategic Growth

Big Skill: Strategic Visioning – Defining a future-oriented investment strategy.

Small Skill: Stakeholder Alignment – Ensuring investor confidence in the vision's feasibility.

Teaching Technique: Investor roadshows and strategic briefings.

Actions: Build a compelling investor narrative and use performance data to reinforce the company's vision for future growth.

Example:

1. Develop an investor presentation highlighting key growth areas.
2. Demonstrate long-term profitability with financial modeling.
3. Conduct targeted investor Q&A sessions to address concerns.
4. Share quarterly updates to maintain alignment and trust.

Pairing 6: Aligning Mergers & Acquisitions with the Strategic Vision

Big Skill: Strategic Visioning – Ensuring that corporate growth through M&A aligns with company direction.

Small Skill: Stakeholder Alignment – Uniting leadership, employees, and customers around a shared post-merger vision.

Teaching Technique: M&A integration workshops.

Actions: Develop a structured transition plan, ensuring both companies' stakeholders understand and embrace the new vision.

Example:

1. Establish an M&A task force for integration planning.
2. Align branding and messaging to reflect the new entity's vision.
3. Conduct joint employee engagement sessions to foster unity.
4. Ensure customers understand the benefits of the merger through targeted communication.

Pairing 7: Creating an Industry Leadership Position Through Stakeholder Buy-In

Big Skill: Strategic Visioning – Positioning the company as an industry leader.

Small Skill: Stakeholder Alignment – Gaining support from regulators, industry partners, and key influencers.

Teaching Technique: Thought leadership and strategic partnerships.

Actions: Develop initiatives that establish the company's leadership in the industry and ensure key stakeholders contribute to this effort.

Example:

1. Publish industry white papers highlighting future trends.
2. Form strategic partnerships with leading institutions.
3. Participate in global conferences to reinforce the vision.

4. Launch an advisory board with key stakeholders to shape future industry direction.

Visionary Thinking – (Big Skill) Inspirational Leadership vs. (Small Skill) Translating Vision into Action

---

Pairing 1: Motivating Teams with a Clear Vision for Innovation

Big Skill: Inspirational Leadership – Communicating a vision in a way that motivates and unifies teams.

Small Skill: Translating Vision into Action – Breaking down broad strategic goals into clear operational tasks.

Teaching Technique: Vision-driven team workshops.

Actions: Leaders will communicate an inspiring vision for innovation and break it down into actionable team objectives, ensuring clarity and alignment.

Example:

1. Define an innovation goal, such as "Becoming the leading AI-driven company in the industry."
2. Assign R&D to explore AI applications in customer service.
3. Establish a timeline for launching AI-powered features.
4. Set measurable milestones for implementation and employee upskilling.

Pairing 2: Inspiring High Performance Through a Purpose-Driven Vision

Big Skill: Inspirational Leadership – Creating an emotional connection to the company vision.

Small Skill: Translating Vision into Action – Defining actionable team roles to support the vision.

Teaching Technique: Employee motivation strategy.

Actions: Have leaders connect business vision to individual employee roles, showing how their work contributes to a larger purpose.

Example:

1. Define a company vision focused on sustainability.
2. Link employee contributions (e.g., procurement team prioritizing eco-friendly materials).
3. Set sustainability targets for each department.
4. Celebrate progress through company-wide recognition programs.

Pairing 3: Aligning Teams with a Customer-Centric Vision

Big Skill: Inspirational Leadership – Driving engagement by focusing on customer impact.

Small Skill: Translating Vision into Action – Creating a customer-first operational roadmap.

Teaching Technique: Customer journey mapping.

Actions: Guide teams in identifying customer pain points and structuring operational improvements to enhance customer experience.

Example:

1. Set a vision to become the most customer-friendly brand in the industry.
2. Task teams with redesigning customer support processes.
3. Implement a real-time customer feedback loop.
4. Train employees on proactive service strategies.

Pairing 4: Leading Organizational Change Through Inspirational Messaging

Big Skill: Inspirational Leadership – Encouraging buy-in for large-scale transformations.

Small Skill: Translating Vision into Action – Structuring change management plans with clear steps.

Teaching Technique: Change management communication training.

Actions: Teach leaders to articulate the reason for change, outline steps, and keep teams engaged throughout the process.

Example:

1. Announce a strategic shift from traditional to digital services.
2. Establish training programs to upskill employees in digital tools.
3. Create a step-by-step transition timeline.
4. Measure engagement through feedback sessions.

Pairing 5: Driving Employee Engagement in Corporate Vision

Big Skill: Inspirational Leadership – Strengthening company culture through a compelling vision.

Small Skill: Translating Vision into Action – Defining core initiatives that reinforce company values.

Teaching Technique: Cultural transformation workshops.

Actions: Guide teams in aligning workplace culture with company vision, ensuring consistent messaging and behaviors.

Example:

1. Establish a vision for a more inclusive and diverse workforce.
2. Implement training programs on inclusive leadership.
3. Adjust hiring practices to reflect diversity goals.
4. Track progress through employee engagement surveys.

Pairing 6: Unifying Global Teams Under a Shared Vision

Big Skill: Inspirational Leadership – Creating a vision that resonates across diverse teams.

Small Skill: Translating Vision into Action – Localizing strategic initiatives to fit regional needs.

Teaching Technique: Vision adaptation workshops.

Actions: Develop regionalized approaches to align global teams with corporate strategy while considering local market dynamics.

Example:

1. Communicate the company's global expansion vision.
2. Task regional teams with adapting brand messaging for local markets.
3. Align regional product development with corporate goals.
4. Hold cross-regional meetings to ensure cohesion.

Pairing 7: Inspiring Resilience in Times of Crisis

Big Skill: Inspirational Leadership – Keeping teams motivated through challenges.

Small Skill: Translating Vision into Action – Structuring short-term crisis response plans.

Teaching Technique: Crisis leadership simulations.

Actions: Train leaders to maintain morale and provide clear action plans during disruptions.

Example:

1. Announce a long-term recovery plan during an economic downturn.
2. Reprioritize short-term goals to maintain stability.
3. Establish regular check-ins to provide transparency.
4. Implement resilience-building initiatives like stress management training.

Visionary Thinking – (Big Skill) Inspirational Leadership vs. (Small Skill) Crafting Persuasive Messaging

---

Pairing 1: Inspiring Organizational Change with a Vision Statement

Big Skill: Inspirational Leadership – Communicating a vision in a way that motivates and unifies teams.

Small Skill: Crafting Persuasive Messaging – Writing and delivering effective vision statements and presentations.

Teaching Technique: Vision statement development workshop.

Actions: Guide leaders in drafting a compelling vision statement that resonates with employees and stakeholders, then refine messaging for clarity and impact.

Example:

1. Define a vision of becoming a sustainability-driven company.
2. Craft a concise and inspiring statement to rally employees.
3. Develop an internal campaign to reinforce the vision.
4. Deliver an executive speech unveiling the vision.

Pairing 2: Driving Employee Engagement Through Vision Storytelling

Big Skill: Inspirational Leadership – Creating an emotional connection to the company vision.

Small Skill: Crafting Persuasive Messaging – Using storytelling techniques to make the vision relatable.

Teaching Technique: Leadership storytelling training.

Actions: Teach leaders to use narratives, customer success stories, and company history to make the vision more engaging.

Example:

1. Use real customer success stories to show the vision in action.
2. Structure a company-wide address with a compelling opening, middle, and call to action.
3. Train managers to reinforce this vision in team meetings.
4. Measure engagement through feedback and participation.

Pairing 3: Aligning Teams Through Executive Town Halls

Big Skill: Inspirational Leadership – Ensuring teams understand and support the company vision.

Small Skill: Crafting Persuasive Messaging – Delivering engaging and persuasive town hall presentations.

Teaching Technique: Interactive town hall preparation.

Actions: Equip leaders with structured frameworks to communicate vision and company direction effectively.

Example:

1. Prepare a quarterly town hall presentation highlighting company progress.
2. Incorporate visual storytelling with graphs and key achievements.
3. Allow employees to ask questions and provide feedback.
4. Reinforce key vision themes in follow-up communications.

Pairing 4: Winning Investor Confidence Through a Vision Pitch

Big Skill: Inspirational Leadership – Inspiring confidence in the company's long-term direction.

Small Skill: Crafting Persuasive Messaging – Structuring investor presentations for maximum impact.

Teaching Technique: Investor pitch training.

Actions: Teach leaders to create compelling investor narratives that highlight growth potential, competitive advantages, and financial stability.

Example:

1. Develop a high-impact pitch deck with vision-aligned growth plans.
2. Use compelling financial and market data to build credibility.
3. Practice persuasive delivery with executive coaching.
4. Secure investor buy-in for funding and strategic partnerships.

Pairing 5: Motivating Employees with a Vision-Driven Performance Culture

Big Skill: Inspirational Leadership – Communicating the "why" behind company goals to drive motivation.

Small Skill: Crafting Persuasive Messaging – Reinforcing the vision in employee communications.

Teaching Technique: Internal branding and vision reinforcement.

Actions: Develop messaging strategies that consistently tie team goals and achievements back to the company vision.

Example:

1. Create an internal campaign that links KPIs to vision outcomes.
2. Feature employee success stories in company newsletters.
3. Develop a reward program recognizing those who embody the vision.
4. Conduct leadership roadshows to reinforce messaging across teams.

Pairing 6: Leading a Rebrand with Vision-Driven Messaging

Big Skill: Inspirational Leadership – Ensuring brand evolution aligns with long-term company goals.

Small Skill: Crafting Persuasive Messaging – Translating the rebrand into compelling customer-facing content.

Teaching Technique: Brand messaging workshops.

Actions: Guide marketing and leadership teams in refining brand messaging to reflect company values, mission, and vision.

Example:

1. Craft a new brand tagline that captures the company's vision.
2. Update website and marketing materials with vision-driven language.
3. Train sales and customer service teams on how to communicate the new brand positioning.
4. Launch an external campaign reinforcing the company's evolution.

Pairing 7: Unifying Global Teams Through a Shared Vision

Big Skill: Inspirational Leadership – Creating a company-wide sense of purpose.

Small Skill: Crafting Persuasive Messaging – Customizing vision messaging for diverse global teams.

Teaching Technique: Cultural adaptation workshops.

Actions: Develop messaging frameworks that maintain consistency while adapting to regional and cultural nuances.

Example:

1. Develop a universal vision statement with localized translations.
2. Host regional meetings to discuss how the vision applies to each market.
3. Encourage leaders to communicate this vision in culturally relevant ways.
4. Create region-specific messaging toolkits for managers.

Visionary Thinking – (Big Skill) Inspirational Leadership vs. (Small Skill) Stakeholder Alignment

---

Pairing 1: Aligning Cross-Functional Teams with the Vision

Big Skill: Inspirational Leadership – Communicating a vision in a way that motivates and unifies teams.

Small Skill: Stakeholder Alignment – Ensuring that different teams understand and commit to the vision.

Teaching Technique: Cross-functional vision workshops.

Actions: Host collaborative sessions where leaders from various departments translate the company vision into specific objectives for their teams.

Example:

1. Organize a leadership summit to clarify the company's five-year vision.
2. Develop department-specific implementation strategies.
3. Assign accountability roles to track progress toward vision execution.
4. Establish regular check-ins for alignment across teams.

Pairing 2: Securing Executive Buy-In for Visionary Initiatives

Big Skill: Inspirational Leadership – Gaining leadership support for transformative strategies.

Small Skill: Stakeholder Alignment – Aligning executive priorities with the long-term vision.

Teaching Technique: Executive briefing sessions.

Actions: Develop and present a high-level strategy document that links business vision to tangible executive objectives.

Example:

1. Conduct one-on-one executive briefings.
2. Demonstrate how the vision aligns with market trends and financial goals.
3. Address potential concerns with data-driven justifications.
4. Gain approval and establish accountability structures for execution.

Pairing 3: Driving Vision Alignment Across Global Teams

Big Skill: Inspirational Leadership – Ensuring vision consistency across international teams.

Small Skill: Stakeholder Alignment – Adapting the vision for different markets while maintaining core principles.

Teaching Technique: Regional vision adaptation strategy.

Actions: Create a structured communication framework that tailors the vision to local teams while ensuring alignment with global objectives.

Example:

1. Define core company values applicable across all regions.
2. Collaborate with regional leaders to adapt implementation strategies.
3. Create localized messaging for internal and external communications.
4. Conduct periodic global leadership summits to reinforce alignment.

Pairing 4: Motivating Employees with a Unified Vision

Big Skill: Inspirational Leadership – Communicating the company's mission in an engaging way.

Small Skill: Stakeholder Alignment – Ensuring employees understand how their roles contribute to the vision.

Teaching Technique: Vision-driven employee engagement programs.

Actions: Develop interactive initiatives like storytelling campaigns and internal branding efforts to reinforce the company's vision.

Example:

1. Launch a company-wide storytelling initiative featuring employees embodying this vision.
2. Implement vision-aligned performance recognition programs.
3. Use town halls to showcase employee contributions to long-term goals.
4. Track engagement metrics to assess understanding and alignment.

Pairing 5: Strengthening Stakeholder Commitment Through Vision Roadmaps

Big Skill: Inspirational Leadership – Creating a compelling roadmap that outlines the vision's execution.

Small Skill: Stakeholder Alignment – Mapping stakeholders to key milestones and roles.

Teaching Technique: Vision roadmap development.

Actions: Develop a structured roadmap that clearly defines stakeholder responsibilities and tracks progress toward key objectives.

Example:

1. Identify key stakeholders and their role in the vision execution.
2. Develop a step-by-step implementation roadmap with measurable outcomes.
3. Conduct periodic updates to maintain alignment and motivation.
4. Use digital dashboards to provide real-time progress tracking.

Pairing 6: Aligning Customers and Partners with a Shared Vision

Big Skill: Inspirational Leadership – Communicating a customer-centric vision that builds trust and engagement.

Small Skill: Stakeholder Alignment – Ensuring partners and clients align with company values and direction.

Teaching Technique: Vision-driven customer experience strategy.

Actions: Craft customer-facing messaging that ties the company's vision to client needs and industry trends.

Example:

1. Develop a brand narrative that reinforces the company's purpose.
2. Train customer service and sales teams on vision-driven engagement.
3. Align marketing content to reflect customer-centered innovation.
4. Build long-term partnerships by emphasizing shared goals and values.

Pairing 7: Ensuring Vision Consistency During Organizational Change

Big Skill: Inspirational Leadership – Keeping teams aligned during business transformations.

Small Skill: Stakeholder Alignment – Managing change by reinforcing commitment to the vision.

Teaching Technique: Change management communication strategy.

Actions: Design structured messaging and engagement plans to ensure that stakeholders remain aligned and motivated during major transitions.

Example:

1. Communicate the rationale for organizational changes through internal messaging.
2. Create a phased transition plan to minimize uncertainty.
3. Provide regular updates and open Q&A forums for employees.
4. Implement a feedback system to address concerns proactively.

Visionary Thinking – (Big Skill) Foresight & Innovation vs. (Small Skill) Translating Vision into Action

---

Pairing 1: Predicting Market Trends and Developing Actionable Strategies

Big Skill: Foresight & Innovation – Anticipating industry shifts and proactively shaping strategies.

Small Skill: Translating Vision into Action – Breaking down broad strategic goals into clear operational tasks.

Teaching Technique: Trend analysis and strategic execution framework.

Actions: Conduct market research to identify emerging trends, develop strategic objectives, and break them into actionable steps.

Example:

1. Identify the shift toward AI-driven customer service.
2. Set a vision for integrating AI into customer interactions.

3. Develop a phased implementation plan with clear milestones.
4. Assign teams to pilot AI chatbots and measure impact.

Pairing 2: Shaping Future Business Models Based on Industry Disruptions

Big Skill: Foresight & Innovation – Identifying potential disruptions and adapting proactively.

Small Skill: Translating Vision into Action – Structuring innovation initiatives into achievable steps.

Teaching Technique: Business model innovation labs.

Actions: Guide teams in evaluating disruption risks and designing forward-thinking business models with clear execution plans.

Example:

1. Anticipate the impact of blockchain on financial transactions.
2. Define a vision for secure, decentralized payment solutions.
3. Develop an internal task force to explore blockchain applications.
4. Establish pilot programs for testing blockchain integrations.

Pairing 3: Future-Proofing Workforce Skills through Proactive Training Plans

Big Skill: Foresight & Innovation – Predicting skill gaps and technological advancements.

Small Skill: Translating Vision into Action – Implementing workforce development programs.

Teaching Technique: Skills mapping and training roadmap.

Actions: Identify emerging skills needed in the industry and develop structured learning programs to prepare employees.

Example:

1. Predict the rise of machine learning in marketing analytics.
2. Design a vision for an AI-augmented marketing department.
3. Develop a three-year upskilling plan for marketing professionals.
4. Implement targeted training courses on AI-driven analytics.

Pairing 4: Innovating Customer Experience with Cutting-Edge Technologies

Big Skill: Foresight & Innovation – Identifying new technologies to enhance customer engagement.

Small Skill: Translating Vision into Action – Implementing structured customer experience improvements.

Teaching Technique: Customer journey innovation workshops.

Actions: Research technological advancements and create step-by-step initiatives for improving customer engagement.

Example:

1. Recognize the rise of augmented reality (AR) in e-commerce.
2. Set a vision for integrating AR shopping experiences.
3. Develop an implementation roadmap for AR product visualization.
4. Partner with tech vendors to roll out AR-enhanced shopping.

Pairing 5: Aligning Corporate Sustainability with Long-Term Strategic Goals

Big Skill: Foresight & Innovation – Integrating environmental, social, and governance (ESG) factors into business strategy.

Small Skill: Translating Vision into Action – Developing structured sustainability initiatives.

Teaching Technique: Sustainable business strategy development.

Actions: Establish sustainability objectives and create a roadmap for achieving measurable impact.

Example:

1. Forecast regulatory changes impacting sustainability reporting.
2. Develop a corporate vision for carbon neutrality by 2035.
3. Design a step-by-step emissions reduction strategy.
4. Implement a sustainability task force to track progress.

Pairing 6: Creating Agile Business Strategies for Market Volatility

Big Skill: Foresight & Innovation – Preparing for economic and geopolitical uncertainties.

Small Skill: Translating Vision into Action – Building adaptable business models.

Teaching Technique: Scenario planning and risk assessment.

Actions: Identify potential business risks, develop alternative strategies, and structure an adaptable implementation approach.

Example:

1. Anticipate supply chain disruptions due to global instability.
2. Develop a vision for diversifying suppliers and distribution channels.
3. Create contingency plans with alternative suppliers and routes.
4. Implement a risk monitoring dashboard to track vulnerabilities.

Pairing 7: Developing Data-Driven Innovation Strategies

Big Skill: Foresight & Innovation – Leveraging data to identify and predict industry shifts.

Small Skill: Translating Vision into Action – Structuring data-driven decision-making processes.

Teaching Technique: Data analytics-driven strategy sessions.

Actions: Use predictive analytics to guide long-term strategy and break down innovation goals into measurable actions.

Example:

1. Analyze consumer behavior data to anticipate demand shifts.
2. Set a vision for a data-driven product development approach.
3. Implement a data analytics team to track real-time market changes.
4. Develop a structured process for integrating data insights into decision-making.

Visionary Thinking – (Big Skill) Foresight & Innovation vs. (Small Skill) Crafting Persuasive Messaging

---

Pairing 1: Communicating Industry Transformation to Internal Stakeholders

Big Skill: Foresight & Innovation – Anticipating industry shifts and proactively shaping strategies.

Small Skill: Crafting Persuasive Messaging – Writing and delivering effective vision statements and presentations.

Teaching Technique: Executive vision communication training.

Actions: Train business leaders on how to craft compelling presentations that align teams with future industry trends.

Example:

1. Identify AI-driven automation as an emerging trend.
2. Develop a vision statement emphasizing the company's leadership in AI adoption.
3. Craft a presentation highlighting AI's impact on company operations.
4. Conduct an internal town hall to communicate this vision.

Pairing 2: Positioning a Company as an Industry Leader Through Thought Leadership

Big Skill: Foresight & Innovation – Shaping industry trends through innovative insights.

Small Skill: Crafting Persuasive Messaging – Writing compelling keynote speeches and white papers.

Teaching Technique: Thought leadership content development.

Actions: Guide executives in developing opinion pieces, reports, and speeches that establish authority in their industry.

Example:

1. Research the future of sustainable packaging.
2. Write a white paper advocating for biodegradable materials.
3. Develop a keynote speech for an industry conference.
4. Publish an article in a leading business journal on sustainability trends.

Pairing 3: Driving Stakeholder Buy-In for Strategic Innovation

Big Skill: Foresight & Innovation – Proactively shaping business strategies in anticipation of change.

Small Skill: Crafting Persuasive Messaging – Creating presentations that persuade investors and stakeholders.

Teaching Technique: Investor pitch training.

Actions: Develop an executive presentation that clearly outlines the rationale, risks, and benefits of an innovation initiative.

Example:

1. Identify cloud computing as the future of enterprise software.
2. Develop a vision statement emphasizing the company's transition to cloud-first solutions.
3. Craft an investor pitch showcasing market demand and growth potential.
4. Deliver the pitch to secure funding for cloud infrastructure expansion.

Pairing 4: Aligning Employees with a Future-Driven Vision Statement

Big Skill: Foresight & Innovation – Helping teams prepare for long-term industry shifts.

Small Skill: Crafting Persuasive Messaging – Writing and reinforcing vision-driven internal messaging.

Teaching Technique: Vision alignment workshops.

Actions: Develop and distribute messaging strategies that connect employees' roles to the company's future vision.

Example:

1. Forecast the rise of remote and hybrid work models.
2. Develop a vision statement promoting flexibility and digital transformation.
3. Deliver internal presentations emphasizing employee empowerment.
4. Launch an internal campaign reinforcing the company's commitment to innovation.

Pairing 5: Building Industry Partnerships with Future-Focused Messaging

Big Skill: Foresight & Innovation – Anticipating cross-industry collaborations and opportunities.

Small Skill: Crafting Persuasive Messaging – Creating compelling partnership proposals and presentations.

Teaching Technique: Strategic collaboration pitch development.

Actions: Help business leaders develop narratives that position their company as a valuable strategic partner.

Example:

1. Identify opportunities for fintech and healthcare integration.
2. Develop a vision-driven partnership proposal for digital health payments.
3. Create an executive pitch highlighting mutual benefits.
4. Present the proposal to potential partners in the healthcare sector.

Pairing 6: Securing Government and Regulatory Support for Emerging Innovations

Big Skill: Foresight & Innovation – Preparing for policy shifts and regulatory trends.

Small Skill: Crafting Persuasive Messaging – Writing and delivering advocacy statements.

Teaching Technique: Public policy and regulatory influence training.

Actions: Train executives to craft policy recommendations and communicate the societal benefits of their innovation.

Example:

1. Anticipate increased data privacy regulations.
2. Develop a corporate stance on ethical AI and data security.
3. Write a formal policy brief for legislators and regulators.
4. Deliver testimony at a government hearing on AI ethics.

Pairing 7: Launching a Future-Oriented Brand Messaging Campaign

Big Skill: Foresight & Innovation – Predicting shifts in consumer expectations and behaviors.

Small Skill: Crafting Persuasive Messaging – Designing marketing campaigns that reflect a forward-thinking vision.

Teaching Technique: Brand storytelling and message development.

Actions: Develop campaigns that position the company as a leader in industry innovation.

Example:

1. Recognize the demand for carbon-neutral products.
2. Develop a brand campaign highlighting the company's sustainability commitments.
3. Craft a series of video ads and blog posts reinforcing the vision.
4. Launch a social media initiative engaging consumers in eco-friendly choices.

Visionary Thinking – (Big Skill) Foresight & Innovation vs. (Small Skill) Stakeholder Alignment

---

Pairing 1: Aligning Leadership with Future Market Shifts

Big Skill: Foresight & Innovation – Anticipating industry shifts and proactively shaping strategies.

Small Skill: Stakeholder Alignment – Ensuring that different teams understand and commit to the vision.

Teaching Technique: Leadership alignment workshops.

Actions: Facilitate leadership discussions to ensure company executives understand market trends and align business strategies.

Example:

1. Identify the impact of automation on workforce structures.
2. Develop a leadership vision for reskilling employees.
3. Conduct executive roundtables to align on workforce transition plans.
4. Implement an internal communication strategy reinforcing leadership unity.

Pairing 2: Ensuring Cross-Departmental Buy-In for Strategic Initiatives

Big Skill: Foresight & Innovation – Driving forward-thinking strategies across departments.

Small Skill: Stakeholder Alignment – Creating cohesion in cross-functional teams.

Teaching Technique: Cross-departmental collaboration summits.

Actions: Host interactive summits where teams contribute to long-term strategy discussions and align their objectives.

Example:

1. Introduce an AI-driven customer service initiative.
2. Ensure marketing, IT, and operations teams align on implementation.
3. Assign each department specific objectives and integration points.
4. Conduct monthly alignment check-ins to measure progress.

Pairing 3: Gaining Employee Commitment to Innovation Strategies

Big Skill: Foresight & Innovation – Encouraging a company-wide culture of innovation.

Small Skill: Stakeholder Alignment – Ensuring employees understand their role in future strategies.

Teaching Technique: Vision-driven employee engagement programs.

Actions: Develop internal initiatives that reinforce employee contributions to the company's innovative vision.

Example:

1. Launch the "Innovation Champions" program.
2. Train employees on how industry shifts impact their roles.
3. Provide an internal platform for employees to propose innovation ideas.
4. Reward employees who contribute actionable ideas to strategic initiatives.

Pairing 4: Securing Investor Support for Future-Focused Business Models

Big Skill: Foresight & Innovation – Structuring businesses to attract long-term investment.

Small Skill: Stakeholder Alignment – Aligning investor expectations with company vision.

Teaching Technique: Investor relations and vision pitch development.

Actions: Develop investor presentations that clearly communicate the company's long-term vision and market positioning.

Example:

1. Predict the rise of sustainable manufacturing.
2. Develop a five-year roadmap for transitioning to eco-friendly production.
3. Craft a data-driven pitch for investors emphasizing ROI and market growth.
4. Deliver a presentation at an investor summit to secure funding.

Pairing 5: Aligning Suppliers and Partners with Future Business Strategies

Big Skill: Foresight & Innovation – Creating resilient supply chain and partnership models.

Small Skill: Stakeholder Alignment – Ensuring supply chain stakeholders commit to innovation.

Teaching Technique: Strategic partner engagement forums.

Actions: Establish long-term alignment strategies with key suppliers and business partners.

Example:

1. Forecast increased reliance on local supply chains.
2. Develop a vision for sourcing 70% of materials domestically.

3. Host supplier strategy meetings to align on shared growth opportunities.
4. Establish joint investment initiatives in domestic production facilities.

Pairing 6: Creating Industry Influence Through Visionary Messaging

Big Skill: Foresight & Innovation – Positioning a company as a thought leader in the industry.

Small Skill: Stakeholder Alignment – Gaining industry-wide consensus on innovation strategies.

Teaching Technique: Public relations and industry influence strategy.

Actions: Develop public messaging that encourages industry-wide adoption of the company's strategic vision.

Example:

1. Identify the impact of 5G on remote work infrastructure.
2. Publish a thought leadership white paper on future connectivity solutions.
3. Host an industry roundtable on digital transformation.
4. Align with other industry leaders to drive regulatory discussions.

Pairing 7: Integrating Customer Expectations into Future Strategy

Big Skill: Foresight & Innovation – Adapting business models based on changing consumer behavior.

Small Skill: Stakeholder Alignment – Ensuring customer needs shape innovation strategies.

Teaching Technique: Customer insight-driven strategy workshops.

Actions: Conduct market research and customer engagement programs to align future business strategies with consumer expectations.

Example:

1. Forecast the demand for personalized digital experiences.
2. Develop a strategic plan for AI-driven customer interactions.
3. Host consumer advisory panels to evaluate early-stage innovations.
4. Adjust product roadmaps based on real-time consumer feedback.

Example Teaching Strategies for Big vs. Small Thinking Skill Pairings of Tactical Execution at the Business Professional Level

Here are 63 unique teaching strategy examples for the big thinking and small thinking skill pairings, each described in detail to align with **Big Thinking Skills** (Concepts & Strategies) and **Small Thinking Skills** (Step-by-Step Execution) for the **Tactical Execution Skill** taught at the business professional level.

Tactical Execution

Big Thinking Skills (Concepts & Strategies)

- **Operationalizing Strategy:** Designing structured processes to move from idea to implementation.
- **Resource Allocation:** Understanding how to distribute time, talent, and capital effectively.
- **Risk Management:** Identifying potential roadblocks and developing contingency plans.

Small Thinking Skills (Step-by-Step Execution)

- **Project Planning & Task Management:** Using frameworks like OKRs, KPIs, and Gantt charts to track progress.
- **Delegation & Accountability:** Assigning tasks and ensuring follow-through.
- **Process Optimization:** Identifying inefficiencies and refining workflows for better performance.

Tactical Execution – (Big Skill) Operationalizing Strategy vs. (Small Skill) Project Planning & Task Management

---

Pairing 1: Converting Business Strategy into a Scalable Execution Plan

Big Skill: Operationalizing Strategy – Designing structured processes to move from idea to implementation.

Small Skill: Project Planning & Task Management – Using frameworks like OKRs, KPIs, and Gantt charts to track progress.

Teaching Technique: Strategic execution roadmap development.

Actions: Develop a framework that translates business strategy into a detailed execution plan with clear milestones and ownership.

Example:

1. Outline a corporate expansion strategy.
2. Use OKRs to set measurable objectives for new market entry.
3. Develop a Gantt chart to manage regional rollout timelines.
4. Assign KPIs to track growth and adjust execution as needed.

Pairing 2: Implementing Agile Project Management for Strategic Initiatives

Big Skill: Operationalizing Strategy – Aligning strategic goals with execution methods.

Small Skill: Project Planning & Task Management – Structuring execution plans using Agile methodologies.

Teaching Technique: Agile implementation and sprint planning.

Actions: Introduce Agile methodologies to drive flexible, iterative project execution while tracking KPIs.

Example:

1. Develop an AI-driven customer service chatbot.
2. Break down the project into Agile sprints using a Kanban board.
3. Use OKRs to measure chatbot adoption and user satisfaction.
4. Adjust development priorities based on real-time customer feedback.

Pairing 3: Scaling Operations Through Process Automation

Big Skill: Operationalizing Strategy – Structuring workflow improvements to enhance efficiency.

Small Skill: Project Planning & Task Management – Using KPIs to monitor automation impact.

Teaching Technique: Process optimization workshops.

Actions: Identify repetitive tasks, implement automation solutions, and track performance improvements.

Example:

1. Automate invoice processing in finance departments.
2. Implement workflow automation software and train employees.
3. Use KPIs to measure time and cost savings post-implementation.
4. Adjust workflows based on performance data and feedback.

Pairing 4: Driving Digital Transformation Through Structured Execution

Big Skill: Operationalizing Strategy – Executing technology-driven organizational changes.

Small Skill: Project Planning & Task Management – Structuring transformation projects with OKRs and roadmaps.

Teaching Technique: Digital transformation roadmap training.

Actions: Develop a digital adoption strategy with clear milestones and tracking metrics.

Example:

1. Transition a traditional retail business to an e-commerce model.
2. Use Gantt charts to schedule platform development and training.
3. Define KPIs for website traffic, conversion rates, and customer retention.
4. Adjust marketing and logistics based on real-time analytics.

Pairing 5: Managing Mergers & Acquisitions with Strategic Execution Plans

Big Skill: Operationalizing Strategy – Structuring large-scale business changes.

Small Skill: Project Planning & Task Management – Using structured project plans to track integration.

Teaching Technique: Post-merger integration frameworks.

Actions: Develop structured execution plans for culture integration, process alignment, and performance tracking.

Example:

1. Plan the merger of two technology firms.
2. Use Gantt charts to manage IT system consolidation.
3. Track employee engagement KPIs to ensure cultural alignment.
4. Monitor financial KPIs to measure post-merger revenue growth.

Pairing 6: Structuring a Global Expansion Plan

Big Skill: Operationalizing Strategy – Implementing international growth strategies.

Small Skill: Project Planning & Task Management – Using project frameworks to coordinate market entry.

Teaching Technique: Global expansion project planning.

Actions: Develop structured market entry plans with country-specific execution metrics.

Example:

1. Expand into the Asia-Pacific market.
2. Use OKRs to measure brand awareness and regulatory approvals.
3. Implement phased rollout schedules using Gantt charts.
4. Track sales performance KPIs for continuous market adaptation.

Pairing 7: Managing Enterprise Risk Through Strategic Execution Plans

Big Skill: Operationalizing Strategy – Aligning risk management with business goals.

Small Skill: Project Planning & Task Management – Using KPIs to track risk mitigation progress.

Teaching Technique: Enterprise risk management frameworks.

Actions: Develop structured execution plans for crisis preparedness and operational continuity.

Example:

1. Develop a cybersecurity risk mitigation plan.
2. Use risk assessment frameworks to prioritize vulnerabilities.
3. Assign KPIs to measure security incidents before and after mitigation.
4. Implement training and technology upgrades based on performance data.

Tactical Execution – (Big Skill) Operationalizing Strategy vs. (Small Skill) Delegation & Accountability

---

Pairing 1: Structuring a Product Launch Through Clear Task Delegation

Big Skill: Operationalizing Strategy – Designing structured processes to move from idea to implementation.

Small Skill: Delegation & Accountability – Assigning tasks and ensuring follow-through.

Teaching Technique: Project ownership framework.

Actions: Assign responsibilities across product development, marketing, and sales teams, ensuring accountability at each stage.

Example:

1. Define key milestones for the product launch.
2. Assign R&D to finalize product specifications.
3. Delegate marketing tasks such as campaign planning and outreach.
4. Track team progress through weekly check-ins and status reports.

Pairing 2: Implementing a Digital Transformation Strategy with Role-Based Responsibilities

Big Skill: Operationalizing Strategy – Managing digital transformation initiatives for efficiency and scalability.

Small Skill: Delegation & Accountability – Ensuring IT, operations, and leadership teams execute assigned tasks.

Teaching Technique: Role-based execution mapping.

Actions: Develop a structured digital transformation plan with assigned responsibilities and accountability checkpoints.

Example:

1. Establish cloud migration goals for IT teams.
2. Assign HR to train employees on new digital tools.
3. Delegate operations to align workflows with digital solutions.
4. Track implementation progress with biweekly stakeholder reviews.

Pairing 3: Aligning Sales Strategy Execution with Performance-Based Accountability

Big Skill: Operationalizing Strategy – Structuring sales expansion efforts with clear execution processes.

Small Skill: Delegation & Accountability – Assigning sales quotas and tracking team accountability.

Teaching Technique: Performance-based accountability system.

Actions: Develop sales execution plans where teams are responsible for meeting assigned targets.

Example:

1. Set quarterly revenue goals for sales teams.
2. Assign sales reps to target specific market segments.
3. Monitor individual and team performance against quotas.
4. Implement performance-based incentives for top achievers.

Pairing 4: Managing a Company-Wide Rebranding Initiative Through Delegated Workstreams

Big Skill: Operationalizing Strategy – Overseeing large-scale business branding transformations.

Small Skill: Delegation & Accountability – Assigning tasks across departments to ensure alignment.

Teaching Technique: Workstream-based project execution.

Actions: Establish a rebranding execution team with clear responsibilities.

Example:

1. Assign marketing to develop a new brand identity.
2. Task HR with internal brand training initiatives.
3. Delegate finance to oversee budget and vendor contracts.
4. Conduct monthly progress reviews to track implementation.

Pairing 5: Improving Customer Experience with Structured Service Initiatives

Big Skill: Operationalizing Strategy – Developing customer service improvements for business growth.

Small Skill: Delegation & Accountability – Assigning customer service enhancements to different teams.

Teaching Technique: Cross-functional accountability planning.

Actions: Implement service improvement initiatives with designated ownership per department.

Example:

1. Assign IT to develop a chatbot for quick support.
2. Task customer service to reduce response times by 30%.
3. Hold sales accountable for personalized customer follow-ups.
4. Monitor success through Net Promoter Score (NPS) tracking.

Pairing 6: Executing a Cost Reduction Strategy with Accountability at Multiple Levels

Big Skill: Operationalizing Strategy – Implementing cost-cutting initiatives without sacrificing quality.

Small Skill: Delegation & Accountability – Assigning financial efficiency responsibilities across departments.

Teaching Technique: Cost ownership structure.

Actions: Develop a company-wide expense reduction strategy with clear accountability metrics.

Example:

1. Assign procurement to renegotiate supplier contracts.
2. Task HR with optimizing employee benefit structures.
3. Delegate operations to reduce supply chain inefficiencies.
4. Track cost reduction impact through quarterly financial audits.

Pairing 7: Leading a Sustainability Initiative with Assigned Departmental Responsibilities

Big Skill: Operationalizing Strategy – Driving ESG (Environmental, Social, Governance) efforts within an organization.

Small Skill: Delegation & Accountability – Assigning sustainability objectives with tracking mechanisms.

Teaching Technique: Departmental ESG responsibility matrix.

Actions: Develop a sustainability roadmap with clear execution roles.

Example:

1. Assign R&D to explore eco-friendly product materials.
2. Task logistics with reducing carbon footprint in distribution.
3. Delegate marketing to develop sustainability awareness campaigns.
4. Track impact using sustainability key performance indicators.

Tactical Execution – (Big Skill) Operationalizing Strategy vs. (Small Skill) Process Optimization

---

Pairing 1: Streamlining Product Development Workflows

Big Skill: Operationalizing Strategy – Designing structured processes to move from idea to implementation.

Small Skill: Process Optimization – Identifying inefficiencies and refining workflows for better performance.

Teaching Technique: Lean product development framework.

Actions: Analyze existing product development cycles and implement lean methodologies to reduce waste and accelerate time to market.

Example:

1. Conduct a process audit to identify delays in R&D.
2. Implement agile methodologies to optimize prototype iterations.
3. Reduce approval bottlenecks by automating cross-functional sign-offs.
4. Track efficiency gains through time-to-market KPIs.

Pairing 2: Improving Cross-Departmental Communication in Strategy Execution

Big Skill: Operationalizing Strategy – Enhancing strategic coordination across departments.

Small Skill: Process Optimization – Refining collaboration methods to minimize miscommunication.

Teaching Technique: Workflow integration and collaboration mapping.

Actions: Design and implement streamlined communication channels to enhance cross-functional coordination.

Example:

1. Identify inefficiencies in interdepartmental handoffs.
2. Introduce a project management tool for real-time updates.
3. Standardize reporting structures to ensure alignment.
4. Measure success through reductions in project delays and miscommunications.

Pairing 3: Increasing Efficiency in Supply Chain Operations

Big Skill: Operationalizing Strategy – Optimizing logistics and vendor management.

Small Skill: Process Optimization – Reducing bottlenecks and delays in supply chain workflows.

Teaching Technique: Lean supply chain management training.

Actions: Identify inefficiencies in procurement, warehousing, and distribution and introduce workflow automation.

Example:

1. Map out existing supply chain workflows.
2. Automate inventory tracking to reduce stock shortages.
3. Implement just-in-time (JIT) manufacturing for cost savings.
4. Track lead time reductions and cost efficiency improvements.

Pairing 4: Enhancing Employee Onboarding and Training Efficiency

Big Skill: Operationalizing Strategy – Structuring workforce development and onboarding processes.

Small Skill: Process Optimization – Refining training workflows for faster employee ramp-up.

Teaching Technique: Digital onboarding and performance tracking.

Actions: Streamline training processes by implementing digital learning platforms and optimizing knowledge transfer.

Example:

1. Audit existing onboarding processes for inefficiencies.
2. Introduce e-learning modules to reduce instructor-led training time.
3. Automate onboarding checklists to track progress.
4. Measure onboarding efficiency through time-to-productivity KPIs.

Pairing 5: Reducing Costs Through Financial Process Optimization

Big Skill: Operationalizing Strategy – Improving financial workflow efficiency.

Small Skill: Process Optimization – Identifying and eliminating financial inefficiencies.

Teaching Technique: Finance process automation training.

Actions: Audit existing financial workflows and implement automation tools to enhance accuracy and speed.

Example:

1. Analyze invoicing and expense reporting processes.
2. Implement AI-driven accounting software to reduce manual data entry.
3. Streamline budget approvals through digital workflows.
4. Monitor improvements using financial accuracy and processing time metrics.

Pairing 6: Optimizing IT Infrastructure for Scalability

Big Skill: Operationalizing Strategy – Ensuring IT processes support business growth.

Small Skill: Process Optimization – Refining IT workflows to improve system performance.

Teaching Technique: IT workflow efficiency review.

Actions: Identify slow or redundant IT processes and implement cloud-based solutions for improved performance.

Example:

1. Assess IT ticketing system inefficiencies.
2. Implement an AI-driven IT helpdesk to manage repetitive queries.
3. Automate software deployment for faster system updates.
4. Measure success through IT response time and system uptime KPIs.

Pairing 7: Enhancing Customer Service Operations

Big Skill: Operationalizing Strategy – Structuring customer support for efficiency.

Small Skill: Process Optimization – Reducing resolution time and increasing service quality.

Teaching Technique: Customer service process redesign.

Actions: Audit service workflows, reduce manual inefficiencies, and implement automated ticketing and self-service solutions.

Example:

1. Identify bottlenecks in customer inquiry resolution.
2. Introduce AI chatbots to manage common queries.
3. Optimize call center workflows for reduced wait times.
4. Track service efficiency using average handling time (AHT) and customer satisfaction scores.

Tactical Execution – (Big Skill) Resource Allocation vs. (Small Skill) Project Planning & Task Management

---

Pairing 1: Optimizing Budget Allocation for a New Product Launch

Big Skill: Resource Allocation – Understanding how to distribute time, talent, and capital effectively.

Small Skill: Project Planning & Task Management – Using frameworks like OKRs, KPIs, and Gantt charts to track progress.

Teaching Technique: Budget-to-impact analysis.

Actions: Implement financial planning tools and performance tracking systems to ensure efficient capital allocation for a new product launch.

Example:

1. Use Gantt charts to map the product development timeline.
2. Allocate budget across R&D, marketing, and distribution channels.
3. Establish OKRs to measure product launch success.
4. Monitor financial efficiency through cost-per-customer acquisition KPIs.

Pairing 2: Managing Human Resource Allocation in a Consulting Firm

Big Skill: Resource Allocation – Strategically distributing workforce talent across projects.

Small Skill: Project Planning & Task Management – Tracking workload distribution and efficiency.

Teaching Technique: Workforce capacity modeling.

Actions: Implement project planning software to track consultant utilization rates and ensure even distribution of workload.

Example:

1. Develop a skills inventory to match employees with appropriate projects.
2. Use task management dashboards to monitor billable hours.
3. Adjust team assignments based on real-time project progress.
4. Optimize workload balance to prevent employee burnout.

Pairing 3: Structuring Capital Investments for Business Expansion

Big Skill: Resource Allocation – Allocating financial resources effectively for company growth.

Small Skill: Project Planning & Task Management – Tracking funding impact through financial metrics.

Teaching Technique: Capital deployment tracking.

Actions: Develop a funding roadmap using KPIs to measure the effectiveness of capital investments.

Example:

1. Use OKRs to define investment objectives.
2. Assign capital to R&D, technology upgrades, and hiring.
3. Track ROI through quarterly financial reports.
4. Adjust budget allocations based on market performance trends.

Pairing 4: Enhancing Team Efficiency in Agile Software Development

Big Skill: Resource Allocation – Distributing team efforts based on priority needs.

Small Skill: Project Planning & Task Management – Using Agile methodology to streamline workflow.

Teaching Technique: Agile resource sprint planning.

Actions: Implement an Agile framework to allocate resources dynamically based on evolving project needs.

Example:

1. Use a Gantt chart to outline sprint deliverables.
2. Allocate developer time based on task complexity.
3. Monitor velocity KPIs to optimize work distribution.
4. Adjust team assignments based on project backlog prioritization.

Pairing 5: Optimizing Marketing Budget Allocation for Maximum ROI

Big Skill: Resource Allocation – Maximizing marketing impact within budget constraints.

Small Skill: Project Planning & Task Management – Using data-driven tracking methods.

Teaching Technique: ROI-based marketing allocation.

Actions: Develop a marketing campaign resource plan based on customer acquisition cost and engagement analytics.

Example:

1. Set OKRs for lead generation and conversion rates.
2. Allocate budget across digital, print, and influencer marketing.
3. Track campaign performance using real-time KPIs.
4. Reallocate budget to high-performing channels.

Pairing 6: Managing Supply Chain Resources for Cost Efficiency

Big Skill: Resource Allocation – Optimizing inventory and logistics operations.

Small Skill: Project Planning & Task Management – Using supply chain KPIs to track efficiency.

Teaching Technique: Lean supply chain management.

Actions: Implement logistics tracking tools and performance analytics to optimize resource distribution.

Example:

1. Develop a Gantt chart for procurement timelines.
2. Allocate warehousing space based on demand forecasts.
3. Use KPIs to track supplier performance.
4. Adjust inventory levels to reduce storage costs.

Pairing 7: Aligning Talent Acquisition with Business Strategy

Big Skill: Resource Allocation – Allocating hiring resources to critical roles.

Small Skill: Project Planning & Task Management – Tracking hiring pipelines and onboarding success.

Teaching Technique: Workforce planning with talent acquisition KPIs.

Actions: Develop a hiring roadmap using OKRs to align with business growth needs.

Example:

1. Set recruitment goals based on workforce projections.
2. Allocate resources for employer branding and talent sourcing.
3. Track onboarding time-to-productivity KPIs.
4. Adjust hiring strategies based on attrition data.

Tactical Execution – (Big Skill) Resource Allocation vs. (Small Skill) Delegation & Accountability

---

Pairing 1: Optimizing Leadership Delegation in a Fast-Growing Startup

Big Skill: Resource Allocation – Understanding how to distribute time, talent, and capital effectively.

Small Skill: Delegation & Accountability – Assigning tasks and ensuring follow-through.

Teaching Technique: Role-based delegation training.

Actions: Train managers to distribute tasks based on employees' strengths and track accountability through performance metrics.

Example:

1. Assign project leads based on expertise in marketing, product development, and sales.
2. Use a task management system to monitor progress.
3. Hold weekly check-ins to ensure deliverables are met.
4. Adjust roles based on workload distribution and results.

Pairing 2: Enhancing Cross-Functional Collaboration in a Corporate Setting

Big Skill: Resource Allocation – Aligning resources for interdepartmental synergy.

Small Skill: Delegation & Accountability – Structuring accountability across teams.

Teaching Technique: Cross-functional delegation frameworks.

Actions: Implement a responsibility matrix (RACI) to clarify ownership and task dependencies.

Example:

1. Define roles for marketing, sales, and product teams in a campaign launch.
2. Assign task ownership using a RACI matrix.
3. Track progress through biweekly accountability meetings.
4. Adjust assignments based on efficiency and workload capacity.

Pairing 3: Managing Remote Team Delegation for Maximum Productivity

Big Skill: Resource Allocation – Distributing tasks in a remote work environment.

Small Skill: Delegation & Accountability – Ensuring remote employees meet deadlines.

Teaching Technique: Digital accountability tracking.

Actions: Use project management tools to assign, track, and review remote work progress.

Example:

1. Assign clear objectives with deadlines in Asana or Trello.
2. Use automated reminders to ensure follow-through.
3. Conduct virtual check-ins to review accountability metrics.
4. Adjust workload based on time zone challenges and team availability.

Pairing 4: Scaling Delegation in High-Growth Companies

Big Skill: Resource Allocation – Managing talent growth effectively.

Small Skill: Delegation & Accountability – Creating structured delegation processes.

Teaching Technique: Leadership delegation playbook.

Actions: Develop structured delegation templates to maintain accountability as teams expand.

Example:

1. Implement a standard operating procedure (SOP) for project delegation.
2. Assign team leaders with ownership over different business units.
3. Use KPI dashboards to measure team efficiency.
4. Adjust delegation plans based on performance data.

Pairing 5: Delegating High-Value Client Accounts in a Consulting Firm

Big Skill: Resource Allocation – Assigning talent to maximize client satisfaction.

Small Skill: Delegation & Accountability – Holding consultants accountable for client outcomes.

Teaching Technique: Client priority-based task allocation.

Actions: Assign high-value clients to senior consultants while mentoring junior staff for long-term capacity building.

Example:

1. Segment clients based on revenue contribution.
2. Assign lead consultants and establish clear deliverables.
3. Use performance dashboards to track client success metrics.
4. Adjust team assignments based on client feedback.

Pairing 6: Delegating Operational Tasks in a Manufacturing Company

Big Skill: Resource Allocation – Optimizing production line resource distribution.

Small Skill: Delegation & Accountability – Ensuring accountability in manufacturing workflows.

Teaching Technique: Lean operations delegation.

Actions: Implement a production tracking system with clearly defined roles and accountability metrics.

Example:

1. Assign production line workers specific quality control tasks.
2. Implement real-time monitoring dashboards.
3. Hold daily stand-ups to review task completion.
4. Adjust assignments based on bottleneck analysis.

Pairing 7: Delegating Responsibilities in a Crisis Management Plan

Big Skill: Resource Allocation – Strategically deploying resources during crises.

Small Skill: Delegation & Accountability – Assigning crisis response roles with accountability measures.

Teaching Technique: Emergency response delegation drills.

Actions: Develop a crisis response hierarchy with clear delegation and accountability measures.

Example:

1. Assign emergency response leads for IT security, communications, and operations.
2. Implement a communication protocol for real-time updates.
3. Conduct quarterly crisis response drills.
4. Adjust roles based on response effectiveness.

Tactical Execution – (Big Skill) Resource Allocation vs. (Small Skill) Process Optimization

---

Pairing 1: Optimizing Budget Allocation in Marketing Campaigns

Big Skill: Resource Allocation – Understanding how to distribute time, talent, and capital effectively.

Small Skill: Process Optimization – Identifying inefficiencies and refining workflows for better performance.

Teaching Technique: Cost-benefit analysis.

Actions: Implement performance tracking systems to ensure optimal budget allocation across different marketing channels.

Example:

1. Use A/B testing to assess ad performance.
2. Allocate more funds to high-performing channels.
3. Track ROI through customer acquisition cost (CAC) metrics.
4. Adjust campaign budgets based on real-time conversion data.

Pairing 2: Enhancing Workforce Productivity in Remote Teams

Big Skill: Resource Allocation – Ensuring talent is effectively deployed in a remote setting.

Small Skill: Process Optimization – Streamlining communication and workflow management.

Teaching Technique: Digital workflow automation.

Actions: Implement automated project tracking tools and asynchronous communication frameworks to enhance efficiency.

Example:

1. Use Slack for asynchronous team updates.
2. Implement Trello for task tracking and deadline automation.
3. Measure team output with time-tracking tools.
4. Optimize meeting schedules to reduce unnecessary calls.

Pairing 3: Improving Supply Chain Efficiency in Manufacturing

Big Skill: Resource Allocation – Ensuring efficient material and logistics allocation.

Small Skill: Process Optimization – Identifying bottlenecks and refining procurement strategies.

Teaching Technique: Lean Six Sigma analysis.

Actions: Conduct root cause analysis to eliminate supply chain inefficiencies.

Example:

1. Use real-time inventory tracking to prevent overstocking.
2. Streamline supplier relationships with just-in-time (JIT) procurement.
3. Optimize warehouse logistics using automation.
4. Reduce shipping costs by consolidating orders.

Pairing 4: Streamlining IT Resource Allocation in Digital Transformation

Big Skill: Resource Allocation – Distributing IT budgets and personnel efficiently.

Small Skill: Process Optimization – Refining IT deployment for optimal system performance.

Teaching Technique: Agile IT resource planning.

Actions: Implement DevOps best practices to enhance IT workflow efficiency.

Example:

1. Shift to cloud computing to reduce infrastructure costs.
2. Automate software updates and patch management.
3. Allocate cybersecurity resources based on risk assessments.
4. Measure system uptime and optimize server capacity.

Pairing 5: Optimizing Financial Resource Allocation in Corporate Budgeting

Big Skill: Resource Allocation – Managing corporate finances efficiently.

Small Skill: Process Optimization – Reducing financial waste and improving cost efficiency.

Teaching Technique: Zero-based budgeting.

Actions: Analyze department spending trends and reallocate budget based on operational needs.

Example:

1. Audit past expenses to identify unnecessary spending.
2. Adjust department budgets quarterly based on performance metrics.
3. Automate expense approvals for faster processing.
4. Optimize investment portfolios by reallocating underperforming assets.

Pairing 6: Increasing Customer Service Efficiency in Call Centers

Big Skill: Resource Allocation – Ensuring sufficient staffing and support tools.

Small Skill: Process Optimization – Reducing wait times and improving response rates.

Teaching Technique: Queue management optimization.

Actions: Use AI-driven analytics to balance workload and improve customer response times.

Example:

1. Implement AI chatbots for initial customer queries.
2. Allocate high-priority cases to senior agents.
3. Use predictive call routing to optimize agent efficiency.
4. Monitor service level agreements (SLAs) for continuous improvement.

Pairing 7: Refining Hiring and Onboarding Workflows in HR Management

Big Skill: Resource Allocation – Optimizing human capital investment.

Small Skill: Process Optimization – Improving recruitment and onboarding efficiency.

Teaching Technique: Data-driven HR analytics.

Actions: Leverage applicant tracking systems and structured onboarding programs to improve hiring efficiency.

Example:

1. Use AI-powered screening tools to filter candidates.
2. Automate onboarding with e-learning modules.
3. Implement feedback loops to refine training programs.
4. Reduce hiring cycles through data-driven recruitment strategies.

Tactical Execution – (Big Skill) Risk Management vs. (Small Skill) Project Planning & Task Management

---

Pairing 1: Identifying Financial Risks in Budget Planning

Big Skill: Risk Management – Identifying potential roadblocks and developing contingency plans.

Small Skill: Project Planning & Task Management – Using frameworks like OKRs, KPIs, and Gantt charts to track progress.

Teaching Technique: Scenario-based financial risk modeling.

Actions: Use financial forecasting models to assess potential budget overruns and establish mitigation strategies.

Example:

1. Identify cost overrun risks in the annual budget forecast.
2. Develop contingency plans for unexpected expenses.
3. Use Gantt charts to track spending vs. projections.
4. Adjust resource allocation based on real-time financial data.

Pairing 2: Mitigating Supply Chain Disruptions

Big Skill: Risk Management – Anticipating supplier or logistics failures.

Small Skill: Project Planning & Task Management – Tracking supply chain performance using KPIs.

Teaching Technique: Supply chain risk mapping.

Actions: Develop supplier contingency plans and monitor procurement risks using performance metrics.

Example:

1. Map key supplier dependencies using risk analysis tools.
2. Establish backup suppliers for high-risk materials.
3. Use KPIs to track supplier performance and reliability.
4. Adjust procurement strategy based on real-time supply chain disruptions.

Pairing 3: Managing Cybersecurity Risks in IT Projects

Big Skill: Risk Management – Identifying cybersecurity vulnerabilities and response plans.

Small Skill: Project Planning & Task Management – Tracking IT security initiatives using OKRs.

Teaching Technique: Cybersecurity risk simulation.

Actions: Implement cybersecurity OKRs and use dashboards to monitor potential vulnerabilities.

Example:

1. Set OKRs for system security enhancements.
2. Conduct quarterly vulnerability assessments.
3. Use KPIs to track cybersecurity incident resolution times.
4. Adjust security policies based on emerging threats.

Pairing 4: Reducing Operational Downtime in Manufacturing

Big Skill: Risk Management – Preventing delays and equipment failures.

Small Skill: Project Planning & Task Management – Scheduling preventive maintenance using Gantt charts.

Teaching Technique: Predictive maintenance modeling.

Actions: Implement machine learning-driven maintenance scheduling to reduce unexpected breakdowns.

Example:

1. Use IoT sensors to monitor machine performance.
2. Set up a preventive maintenance schedule in a Gantt chart.
3. Track equipment downtime and maintenance KPIs.
4. Adjust maintenance schedules based on real-time data.

Pairing 5: Addressing Legal & Compliance Risks in New Market Expansion

Big Skill: Risk Management – Identifying regulatory and legal challenges.

Small Skill: Project Planning & Task Management – Using OKRs to track compliance milestones.

Teaching Technique: Legal risk assessment workshops.

Actions: Establish compliance tracking dashboards to ensure adherence to international regulations.

Example:

1. Map key regulatory requirements for entering a new market.
2. Assign legal review tasks in a project management system.
3. Use KPIs to track compliance completion rates.
4. Adjust expansion strategies based on legal assessments.

Pairing 6: Ensuring Data Privacy & GDPR Compliance

Big Skill: Risk Management – Identifying data privacy risks and compliance issues.

Small Skill: Project Planning & Task Management – Using task management tools to monitor compliance efforts.

Teaching Technique: Data privacy task-force implementation.

Actions: Develop a data privacy risk matrix and track compliance initiatives with KPIs.

Example:

1. Assign data security-ownership roles in a Gantt chart.
2. Implement a data breach response plan with clear OKRs.
3. Use dashboards to track GDPR training completion rates.
4. Adjust privacy policies based on regulatory updates.

Pairing 7: Crisis Management Planning for Brand Reputation Risks

Big Skill: Risk Management – Preparing for PR crises and reputation management.

Small Skill: Project Planning & Task Management – Tracking crisis response actions with OKRs.

Teaching Technique: Crisis response drills.

Actions: Use crisis simulation exercises to prepare teams for brand reputation risks.

Example:

1. Develop a media response protocol with pre-approved messaging.
2. Assign PR roles and responsibilities in a task management tool.
3. Use KPIs to track brand sentiment before and after crises.
4. Adjust PR strategy based on public response analytics.

Tactical Execution – (Big Skill) Risk Management vs. (Small Skill) Delegation & Accountability

---

Pairing 1: Managing Crisis Response Teams

Big Skill: Risk Management – Identifying potential roadblocks and developing contingency plans.

Small Skill: Delegation & Accountability – Assigning tasks and ensuring follow-through.

Teaching Technique: Crisis simulation exercises.

Actions: Assign crisis management roles and conduct real-time response drills to evaluate preparedness.

Example:

1. Designate a crisis response leader for each department.
2. Use checklists to ensure key response actions are completed.
3. Conduct quarterly fire drills and cybersecurity breach simulations.
4. Review post-crisis debriefs to refine the response plan.

Pairing 2: Preventing Supply Chain Disruptions

Big Skill: Risk Management – Mitigating risks in vendor and logistics management.

Small Skill: Delegation & Accountability – Assigning supplier oversight responsibilities.

Teaching Technique: Vendor risk mapping.

Actions: Develop contingency plans by delegating vendor performance monitoring and contract compliance.

Example:

1. Assign procurement managers to track vendor reliability.
2. Establish KPIs for on-time delivery and contract adherence.
3. Implement a secondary supplier strategy for critical materials.
4. Monitor performance through quarterly vendor assessments.

Pairing 3: Ensuring Compliance with Industry Regulations

Big Skill: Risk Management – Managing legal and regulatory compliance risks.

Small Skill: Delegation & Accountability – Assigning roles for compliance tracking and audits.

Teaching Technique: Compliance accountability framework.

Actions: Distribute compliance responsibilities across legal, HR, and operations teams to track adherence.

Example:

1. Assign HR to track labor law compliance.
2. Delegate finance teams to ensure tax and financial reporting accuracy.
3. Conduct quarterly compliance audits with department heads.
4. Adjust policies based on regulatory updates.

Pairing 4: Reducing Financial Risk in Budget Planning

Big Skill: Risk Management – Identifying financial vulnerabilities in business operations.

Small Skill: Delegation & Accountability – Assigning budgetary oversight to key personnel.

Teaching Technique: Budget review and approval workflows.

Actions: Assign budget forecasting responsibilities and establish clear approval chains.

Example:

1. Delegate department heads to monitor and justify spending.
2. Assign finance teams to review and approve expenses.
3. Implement an automated budget tracking system with alerts.
4. Review quarterly financial risk assessments.

Pairing 5: Managing IT Security & Cyber Risk

Big Skill: Risk Management – Identifying cybersecurity threats and ensuring data protection.

Small Skill: Delegation & Accountability – Assigning IT security roles and responsibilities.

Teaching Technique: Cybersecurity incident response training.

Actions: Assign security teams to monitor risks, conduct penetration testing, and enforce best practices.

Example:

1. Delegate an IT security officer to oversee threat detection.
2. Assign network admins to conduct quarterly vulnerability scans.
3. Implement two-factor authentication enforcement.
4. Conduct phishing attack simulations to train employees.

Pairing 6: Reducing Workplace Safety Hazards

Big Skill: Risk Management – Identifying and mitigating workplace safety risks.

Small Skill: Delegation & Accountability – Assigning safety monitoring and compliance roles.

Teaching Technique: OSHA compliance training.

Actions: Appoint safety officers and assign routine hazard assessments to improve workplace safety.

Example:

1. Assign team leaders to conduct weekly safety inspections.
2. Implement a real-time incident reporting system.
3. Establish emergency response teams for fire and medical situations.
4. Provide monthly safety training to employees.

Pairing 7: Mitigating Employee Turnover Risk

Big Skill: Risk Management – Identifying talent retention challenges.

Small Skill: Delegation & Accountability – Assigning responsibility for employee engagement strategies.

Teaching Technique: HR retention action plans.

Actions: Assign HR and team leaders to track employee satisfaction and implement retention strategies.

Example:

1. Conduct stay interviews and pulse surveys.
2. Assign mentors for professional development support.
3. Implement work-life balance initiatives based on employee feedback.
4. Adjust hiring strategies based on turnover analysis.

Tactical Execution – (Big Skill) Risk Management vs. (Small Skill) Process Optimization

---

Pairing 1: Optimizing Financial Reporting to Reduce Errors

Big Skill: Risk Management – Identifying potential roadblocks and developing contingency plans.

Small Skill: Process Optimization – Identifying inefficiencies and refining workflows for better performance.

Teaching Technique: Automated financial workflow implementation.

Actions: Identify inefficiencies in financial reporting processes and implement automation tools to improve accuracy.

Example:

1. Audit the financial reporting workflow to find bottlenecks.
2. Implement automated data entry validation to reduce human error.
3. Establish contingency procedures for unexpected reporting discrepancies.
4. Reduce financial reporting time by 20% through process refinement.

Pairing 2: Improving Supply Chain Resilience

Big Skill: Risk Management – Mitigating supplier disruptions and delays.

Small Skill: Process Optimization – Streamlining procurement and logistics workflows.

Teaching Technique: Lean supply chain management.

Actions: Assess supply chain vulnerabilities and implement optimization techniques to improve efficiency.

Example:

1. Map the supply chain process and identify high-risk suppliers.
2. Implement just-in-time inventory management to reduce waste.

3. Develop contingency plans for critical supply shortages.
4. Optimize logistics routes to reduce delivery time by 15%.

Pairing 3: Enhancing Cybersecurity Risk Mitigation Strategies

Big Skill: Risk Management – Identifying cybersecurity vulnerabilities and threats.

Small Skill: Process Optimization – Refining security protocols for better performance.

Teaching Technique: Cybersecurity framework evaluation.

Actions: Analyze security policies and optimize procedures to minimize vulnerabilities.

Example:

1. Conduct a cybersecurity audit to detect weak points.
2. Implement a structured incident response workflow.
3. Automate system updates and vulnerability patching.
4. Reduce security breach risks by 30% with optimized protocols.

Pairing 4: Streamlining Compliance and Regulatory Processes

Big Skill: Risk Management – Ensuring regulatory compliance to avoid legal risks.

Small Skill: Process Optimization – Automating and refining compliance tracking systems.

Teaching Technique: Compliance risk assessment workshops.

Actions: Identify inefficiencies in compliance tracking and implement automation tools for efficiency.

Example:

1. Audit compliance workflows to identify redundant steps.
2. Automate regulatory reporting using compliance software.
3. Develop standardized checklists for audits and inspections.
4. Reduce compliance review time by 25% through workflow improvements.

Pairing 5: Reducing Employee Turnover Risk Through HR Process Optimization

Big Skill: Risk Management – Identifying risks related to workforce retention and engagement.

Small Skill: Process Optimization – Streamlining HR policies and talent management strategies.

Teaching Technique: HR process improvement framework.

Actions: Optimize hiring, onboarding, and performance management processes to enhance retention.

Example:

1. Assess the hiring process for delays and bottlenecks.
2. Implement structured onboarding workflows to improve employee integration.
3. Optimize performance review cycles to provide more actionable feedback.
4. Increase employee retention by 10% through process refinement.

Pairing 6: Optimizing Project Execution to Minimize Risks

Big Skill: Risk Management – Preventing project failures through structured planning.

Small Skill: Process Optimization – Refining task management workflows.

Teaching Technique: Agile project risk assessment.

Actions: Implement agile methodologies and streamline project management workflows to reduce risks.

Example:

1. Identify common project roadblocks through retrospective analysis.
2. Standardize project management workflows to reduce inefficiencies.
3. Implement daily stand-ups and sprint reviews to track progress.
4. Improve on-time project delivery rate by 15%.

Pairing 7: Reducing Operational Downtime in Manufacturing

Big Skill: Risk Management – Preventing unexpected production disruptions.

Small Skill: Process Optimization – Implementing predictive maintenance and workflow automation.

Teaching Technique: Lean manufacturing principles.

Actions: Optimize manufacturing workflows to minimize delays and reduce equipment failure risks.

Example:

1. Implement real-time machine monitoring to predict failures.
2. Automate maintenance scheduling to prevent breakdowns.
3. Optimize production line workflows for better efficiency.
4. Reduce downtime by 20% through proactive risk mitigation.

Example Teaching Strategies for Big vs. Small Thinking Skill Pairings of Continuous Improvement at the Business Professional Level

Here are 63 unique teaching strategy examples for the big thinking and small thinking skill pairings, each described in detail to align with **Big Thinking Skills** (Concepts & Strategies) and **Small Thinking Skills** (Step-by-Step Execution) for the **Continuous Improvement Skill** taught at the business professional level.

Continuous Improvement

Big Thinking Skills (Concepts & Strategies)

- **Iterative Thinking:** Viewing business processes as cycles of improvement rather than static plans.
- **Growth Mindset & Adaptability:** Embracing change and learning from failures.
- **Data-Driven Decision-Making:** Using metrics and feedback to refine strategies.

Small Thinking Skills (Step-by-Step Execution)

- **Implementing Feedback Loops:** Collecting and applying customer, employee, and market feedback.
- **Conducting Retrospectives:** Running structured reviews (e.g., post-mortems, lessons learned sessions).
- **Performance Monitoring:** Tracking trends using dashboards and key performance indicators.

Continuous Improvement – (Big Skill) Iterative Thinking vs. (Small Skill) Implementing Feedback Loops

---

Pairing 1: Enhancing Customer Service with Real-Time Feedback

Big Skill: Iterative Thinking – Viewing business processes as cycles of improvement rather than static plans.

Small Skill: Implementing Feedback Loops – Collecting and applying customer, employee, and market feedback.

Teaching Technique: Customer experience monitoring.

Actions: Set up real-time feedback loops through surveys and customer support interactions to refine service strategies.

Example:

1. Deploy post-call satisfaction surveys for customer support.
2. Use AI sentiment analysis to detect dissatisfaction patterns.
3. Train service teams based on common customer pain points.
4. Reduce complaint resolution time by 25% using iterative feedback.

Pairing 2: Refining Product Development Based on User Insights

Big Skill: Iterative Thinking – Using feedback-driven refinement in product design.

Small Skill: Implementing Feedback Loops – Gathering user input to enhance product features.

Teaching Technique: Agile development feedback cycles.

Actions: Integrate beta testing and customer insights into product iteration.

Example:

1. Conduct A/B testing for new app features.
2. Collect feedback from early adopters through user forums.
3. Adjust product roadmaps based on data-driven insights.
4. Increase feature adoption by 30% through iterative refinements.

Pairing 3: Improving Employee Performance through 360-Degree Feedback

Big Skill: Iterative Thinking – Continuously improving workforce effectiveness.

Small Skill: Implementing Feedback Loops – Creating structured review processes for skill development.

Teaching Technique: 360-degree feedback framework.

Actions: Use multi-source feedback to guide leadership and employee development programs.

Example:

1. Implement biannual peer and manager feedback assessments.
2. Use feedback data to create personalized development plans.
3. Adjust leadership training based on emerging employee concerns.
4. Boost employee engagement by 20% through continuous improvement initiatives.

Pairing 4: Optimizing Marketing Strategies Based on Campaign Performance

Big Skill: Iterative Thinking – Adapting marketing approaches through ongoing evaluation.

Small Skill: Implementing Feedback Loops – Monitoring audience engagement and adjusting strategies.

Teaching Technique: Performance analytics review.

Actions: Track and refine marketing campaigns based on engagement data.

Example:

1. Measure click-through rates and conversion metrics for ad campaigns.
2. Adjust messaging and targeting based on heatmap insights.
3. Continuously refine social media content based on audience reactions.
4. Improve ad conversion rates by 15% through rapid iterations.

Pairing 5: Enhancing Workplace Culture through Employee Feedback

Big Skill: Iterative Thinking – Continuously refining organizational culture.

Small Skill: Implementing Feedback Loops – Gathering and acting on employee satisfaction data.

Teaching Technique: Employee engagement pulse surveys.

Actions: Conduct regular surveys and implement changes based on workforce needs.

Example:

1. Collect biannual employee engagement survey data.
2. Identify trends in job satisfaction and work-life balance.
3. Adjust workplace policies based on recurring employee concerns.
4. Reduce voluntary turnover by 10% through targeted engagement strategies.

Pairing 6: Refining Sales Techniques Based on Customer Objections

Big Skill: Iterative Thinking – Continuously improving sales effectiveness.

Small Skill: Implementing Feedback Loops – Tracking and analyzing common customer concerns.

Teaching Technique: Objection-handling workshops.

Actions: Use sales data and customer feedback to refine selling strategies.

Example:

1. Gather feedback on common sales objections during calls.
2. Adjust sales scripts based on real-time objection trends.
3. Provide targeted training for sales teams to improve conversion rates.
4. Increase sales closure rates by 12% through feedback-driven coaching.

Pairing 7: Strengthening Product Quality Through Customer Complaints Analysis

Big Skill: Iterative Thinking – Using complaints as an opportunity for process refinement.

Small Skill: Implementing Feedback Loops – Monitoring complaint resolution trends for product improvement.

Teaching Technique: Root cause analysis sessions.

Actions: Analyze complaint patterns and refine production processes accordingly.

Example:

1. Identify top 5 recurring customer complaints in product defects.
2. Adjust manufacturing processes to address common product failures.
3. Implement proactive quality control measures based on insights.
4. Reduce defect-related returns by 18% through feedback-driven process updates.

Continuous Improvement – (Big Skill) Iterative Thinking vs. (Small Skill) Conducting Retrospectives

---

Pairing 1: Improving Product Development Cycles through Post-Mortem Reviews

Big Skill: Iterative Thinking – Viewing business processes as cycles of improvement rather than static plans.

Small Skill: Conducting Retrospectives – Running structured reviews (e.g., post-mortems, lessons learned sessions).

Teaching Technique: Agile post-mortem analysis.

Actions: Implement structured post-mortems at the end of each product launch to refine future development cycles.

Example:

1. Gather cross-functional teams to review product release successes and challenges.
2. Identify delays, miscommunications, and unforeseen technical challenges.
3. Document insights and refine future product roadmaps.
4. Reduce product development cycle time by 15% through iterative refinement.

Pairing 2: Enhancing Marketing Campaign Performance with Lessons Learned Reviews

Big Skill: Iterative Thinking – Continuously refining marketing strategies.

Small Skill: Conducting Retrospectives – Reviewing campaign effectiveness and audience engagement.

Teaching Technique: Marketing analytics debriefs.

Actions: Conduct structured reviews of marketing campaigns to optimize messaging and targeting.

Example:

1. Analyze engagement metrics (CTR, conversion rates, social shares).
2. Identify successful and underperforming campaign elements.
3. Adjust content, targeting, and budget allocation based on findings.
4. Improve campaign ROI by 20% through retrospective-driven refinements.

Pairing 3: Strengthening Customer Experience with Service Review Sessions

Big Skill: Iterative Thinking – Using continuous feedback to improve service quality.

Small Skill: Conducting Retrospectives – Analyzing customer complaints, support trends, and resolution effectiveness.

Teaching Technique: Customer service quality assessment.

Actions: Run retrospective meetings with customer support teams to address recurring service pain points.

Example:

1. Collect data from customer satisfaction surveys and complaint logs.
2. Identify common themes in customer dissatisfaction.
3. Adjust scripts, escalation procedures, and training protocols.
4. Reduce average resolution time by 25% through service workflow optimizations.

Pairing 4: Refining Sales Strategies Through Performance Evaluations

Big Skill: Iterative Thinking – Adapting sales strategies based on real-world performance.

Small Skill: Conducting Retrospectives – Analyzing sales conversions and objections for future improvements.

Teaching Technique: Sales debriefing workshops.

Actions: Hold structured sales review sessions to evaluate conversion rates and refine techniques.

Example:

1. Review common customer objections and pain points.
2. Identify successful pitch strategies from high-performing sales reps.
3. Adjust sales scripts and refine lead qualification criteria.
4. Improve sales conversion rates by 12% using retrospective-driven optimizations.

Pairing 5: Increasing Efficiency in Project Management through Sprint Retrospectives

Big Skill: Iterative Thinking – Treating projects as ongoing cycles of optimization.

Small Skill: Conducting Retrospectives – Reviewing project execution effectiveness to improve future workflows.

Teaching Technique: Agile sprint retrospectives.

Actions: Implement structured retrospective meetings at the end of each project phase to refine processes.

Example:

1. Hold biweekly sprint reviews in project teams.
2. Identify workflow inefficiencies and resource bottlenecks.
3. Adjust task allocation, timeline estimates, and communication strategies.
4. Increase project completion efficiency by 18% through iterative refinements.

Pairing 6: Enhancing IT System Upgrades with Risk-Based Retrospectives

Big Skill: Iterative Thinking – Refining IT deployment strategies for better efficiency.

Small Skill: Conducting Retrospectives – Analyzing system upgrade processes for optimization.

Teaching Technique: IT infrastructure post-implementation reviews.

Actions: Conduct post-upgrade retrospectives to identify deployment risks and inefficiencies.

Example:

1. Review system downtime and error logs post-upgrade.
2. Identify root causes of unexpected issues.
3. Adjust deployment strategies and rollback procedures.
4. Reduce downtime from system upgrades by 30% through refined processes.

Pairing 7: Improving Workplace Culture through Employee Feedback Retrospectives

Big Skill: Iterative Thinking – Continuously refining workplace engagement strategies.

Small Skill: Conducting Retrospectives – Reviewing employee feedback on workplace satisfaction and policies.

Teaching Technique: Employee engagement review forums.

Actions: Run quarterly retrospectives to assess and improve organizational culture.

Example:

1. Analyze trends in employee satisfaction surveys.
2. Identify areas of concern (workload balance, communication gaps, career development).

3. Adjust HR policies and team-building initiatives based on feedback.
4. Reduce voluntary turnover by 10% through data-driven culture improvements.

Continuous Improvement – (Big Skill) Iterative Thinking vs. (Small Skill) Performance Monitoring

---

Pairing 1: Enhancing Sales Performance Through KPI Dashboards

Big Skill: Iterative Thinking – Viewing business processes as cycles of improvement rather than static plans.

Small Skill: Performance Monitoring – Tracking trends using dashboards and key performance indicators.

Teaching Technique: Sales performance tracking.

Actions: Implement a sales dashboard to monitor and optimize sales cycles.

Example:

1. Use CRM software to track lead conversion rates.
2. Identify patterns in lost deals and winning sales tactics.
3. Adjust sales strategies based on real-time performance insights.
4. Increase quarterly sales by 15% through continuous data-driven adjustments.

Pairing 2: Improving Customer Support with Response Time Metrics

Big Skill: Iterative Thinking – Continuously refining customer service processes.

Small Skill: Performance Monitoring – Using dashboards to track response times and resolution rates.

Teaching Technique: Customer support analytics.

Actions: Monitor customer service KPIs to enhance responsiveness.

Example:

1. Track average response time and resolution speed for support tickets.
2. Identify peak times of high customer service demand.
3. Adjust staffing schedules based on data insights.
4. Reduce customer wait time by 25% by refining scheduling and training.

Pairing 3: Optimizing Marketing ROI with Campaign Performance Dashboards

Big Skill: Iterative Thinking – Adapting marketing strategies based on real-time insights.

Small Skill: Performance Monitoring – Tracking ad engagement, conversions, and cost-per-click data.

Teaching Technique: Marketing performance reviews.

Actions: Implement a marketing performance dashboard to track ROI.

Example:

1. Monitor cost-per-click, bounce rates, and customer acquisition cost.
2. Adjust ad placements and content based on performance trends.
3. Use A/B testing to refine messaging for higher engagement.
4. Increase ad campaign ROI by 20% through iterative refinements.

Pairing 4: Refining Employee Productivity with Workflow Dashboards

Big Skill: Iterative Thinking – Continuously improving workplace efficiency.

Small Skill: Performance Monitoring – Using dashboards to track individual and team performance.

Teaching Technique: Workforce analytics monitoring.

Actions: Use project management dashboards to track task completion rates and bottlenecks.

Example:

1. Identify delays in key project milestones.
2. Allocate resources based on real-time workload data.
3. Automate performance alerts to prevent workflow disruptions.
4. Improve project turnaround time by 18% through proactive monitoring.

Pairing 5: Enhancing Financial Planning with Real-Time Revenue Dashboards

Big Skill: Iterative Thinking – Using data to refine budgeting and financial forecasting.

Small Skill: Performance Monitoring – Tracking revenue trends and expense reports.

Teaching Technique: Financial performance tracking.

Actions: Implement dashboards to monitor financial KPIs in real time.

Example:

1. Analyze monthly revenue, profit margins, and expense breakdowns.
2. Identify cost-cutting opportunities based on financial trends.
3. Adjust pricing strategies based on revenue fluctuation insights.
4. Improve profit margins by 12% through continuous financial monitoring.

Pairing 6: Strengthening Supply Chain Efficiency with Logistics Dashboards

Big Skill: Iterative Thinking – Optimizing supply chain operations through data-driven decisions.

Small Skill: Performance Monitoring – Tracking supplier lead times and inventory levels.

Teaching Technique: Supply chain performance analysis.

Actions: Use dashboards to monitor logistics performance and optimize shipping processes.

Example:

1. Track order fulfillment rates and supplier reliability scores.
2. Identify bottlenecks causing delivery delays.
3. Adjust procurement strategies to maintain optimal stock levels.
4. Reduce shipping time by 20% through continuous supply chain monitoring.

Pairing 7: Enhancing Product Development with Feature Usage Analytics

Big Skill: Iterative Thinking – Refining products based on customer usage trends.

Small Skill: Performance Monitoring – Using dashboards to track feature adoption and retention rates.

Teaching Technique: Product performance review.

Actions: Monitor user engagement data to improve product design.

Example:

1. Analyze which software features have the highest engagement.
2. Identify underutilized features that require refinement or removal.
3. Adjust the product roadmap based on customer interaction patterns.
4. Increase feature adoption by 25% through data-driven development.

Continuous Improvement – (Big Skill) Growth Mindset & Adaptability vs. (Small Skill) Implementing Feedback Loops

---

Pairing 1: Improving Customer Satisfaction Through Continuous Feedback Analysis

Big Skill: Growth Mindset & Adaptability – Embracing change and learning from failures.

Small Skill: Implementing Feedback Loops – Collecting and applying customer, employee, and market feedback.

Teaching Technique: Customer experience enhancement.

Actions: Use customer feedback surveys to improve service offerings.

Example:

1. Gather customer satisfaction data through online surveys.
2. Identify recurring complaints and pain points.
3. Adjust service workflows and product offerings accordingly.
4. Increase Net Promoter Score (NPS) by 20% through iterative improvements.

Pairing 2: Strengthening Employee Engagement with Continuous Feedback Sessions

Big Skill: Growth Mindset & Adaptability – Encouraging a workplace culture that values change.

Small Skill: Implementing Feedback Loops – Collecting real-time employee feedback.

Teaching Technique: Open forum discussions.

Actions: Conduct regular employee feedback sessions to refine workplace policies.

Example:

1. Implement quarterly town hall meetings to collect employee insights.
2. Analyze trends in feedback on work-life balance and company culture.
3. Adjust HR policies and workplace flexibility based on input.
4. Improve employee engagement scores by 15% with data-driven policy changes.

Pairing 3: Enhancing Product Development with Beta Testing Feedback Loops

Big Skill: Growth Mindset & Adaptability – Using iterative product improvements.

Small Skill: Implementing Feedback Loops – Collecting insights from beta users.

Teaching Technique: User-driven product refinement.

Actions: Use beta testing groups to gather real-world product feedback.

Example:

1. Launch an early-access product version for select users.
2. Collect feedback on usability, features, and performance.
3. Refine product design based on user suggestions.
4. Reduce product launch defects by 30% through pre-launch improvements.

Pairing 4: Refining Marketing Strategies with Consumer Feedback Analytics

Big Skill: Growth Mindset & Adaptability – Adapting marketing approaches based on customer sentiment.

Small Skill: Implementing Feedback Loops – Monitoring consumer reactions to campaigns.

Teaching Technique: Social listening and sentiment analysis.

Actions: Adjust marketing campaigns based on audience feedback.

Example:

1. Use social media monitoring tools to analyze consumer sentiment.
2. Identify advertisement messaging that resonates best with target demographics.
3. Modify campaign visuals and messaging to align with preferences.
4. Increase ad engagement rates by 25% through audience-driven adjustments.

Pairing 5: Optimizing Business Operations Through Employee Process Feedback

Big Skill: Growth Mindset & Adaptability – Encouraging internal process improvements.

Small Skill: Implementing Feedback Loops – Using employee suggestions for workflow optimization.

Teaching Technique: Process improvement workshops.

Actions: Collect frontline employee insights on workflow inefficiencies.

Example:

1. Run monthly process feedback sessions with teams.
2. Identify bottlenecks and redundant steps in workflows.
3. Adjust operational procedures based on employee input.
4. Improve process efficiency by 20% through real-time feedback integration.

Pairing 6: Enhancing Training Programs with Learner Feedback Loops

Big Skill: Growth Mindset & Adaptability – Refining corporate learning strategies.

Small Skill: Implementing Feedback Loops – Collecting training effectiveness data.

Teaching Technique: Training program assessment surveys.

Actions: Use feedback from training participants to improve learning materials.

Example:

1. Gather feedback on course clarity and relevance from trainees.
2. Identify areas where employees struggle to apply learned skills.
3. Adjust course content and delivery methods based on feedback.
4. Increase training retention rates by 18% through continuous refinement.

Pairing 7: Improving Vendor Relationships with Supplier Performance Feedback

Big Skill: Growth Mindset & Adaptability – Adapting supplier management strategies.

Small Skill: Implementing Feedback Loops – Collecting performance data from vendors.

Teaching Technique: Supplier evaluation and improvement sessions.

Actions: Establish a structured vendor feedback program to improve supply chain efficiency.

Example:

1. Use vendor scorecards to track on-time deliveries and product quality.
2. Identify suppliers with consistent performance issues.
3. Adjust contract terms and alternative sourcing plans as needed.
4. Reduce supply chain disruptions by 22% through proactive vendor management.

Continuous Improvement – (Big Skill) Growth Mindset & Adaptability vs. (Small Skill) Conducting Retrospectives

---

Pairing 1: Refining Product Development Strategies Through Post-Launch Reviews

Big Skill: Growth Mindset & Adaptability – Embracing change and learning from failures.

Small Skill: Conducting Retrospectives – Running structured reviews (e.g., post-mortems, lessons learned sessions).

Teaching Technique: Post-product launch evaluation.

Actions: Conduct structured post-mortems after product launches to analyze successes and challenges.

Example:

1. Gather cross-functional teams to review customer feedback and sales data.
2. Identify technical issues, market misalignments, and process inefficiencies.
3. Adjust future product roadmaps based on retrospective insights.
4. Improve time-to-market efficiency by 15% through iterative adjustments.

Pairing 2: Improving Crisis Response Effectiveness with Incident Reviews

Big Skill: Growth Mindset & Adaptability – Learning from operational breakdowns.

Small Skill: Conducting Retrospectives – Analyzing crisis managing and refining future response plans.

Teaching Technique: Crisis post-mortem workshops.

Actions: Run retrospective reviews on crisis response strategies to strengthen future preparedness.

Example:

1. Evaluate company response to a cybersecurity breach.
2. Identify weaknesses in communication and containment efforts.

3. Develop updated response protocols based on lessons learned.
4. Reduce incident response time by 30% with refined crisis procedures.

Pairing 3: Optimizing Sales Processes Through Retrospective Analysis

Big Skill: Growth Mindset & Adaptability – Using feedback to refine sales strategies.

Small Skill: Conducting Retrospectives – Reviewing past sales cycles for effectiveness.

Teaching Technique: Sales team debrief sessions.

Actions: Conduct quarterly retrospectives to identify high-performing and underperforming sales tactics.

Example:

1. Analyze success rates of different sales pitches.
2. Identify obstacles that slowed deal closures.
3. Adjust training and sales materials based on feedback.
4. Increase conversion rates by 20% by eliminating ineffective strategies.

Pairing 4: Enhancing Employee Training Programs with Post-Training Evaluations

Big Skill: Growth Mindset & Adaptability – Refining corporate learning strategies.

Small Skill: Conducting Retrospectives – Running structured reviews on training effectiveness.

Teaching Technique: Training program post-mortems.

Actions: Gather feedback from training participants to improve instructional design.

Example:

1. Conduct post-training surveys to identify gaps in comprehension.
2. Review employee application of learned skills in real-world tasks.
3. Modify training materials to align better with business needs.
4. Increase training retention rates by 18% through continuous refinement.

Pairing 5: Strengthening Marketing Campaigns with Post-Campaign Analysis

Big Skill: Growth Mindset & Adaptability – Adjusting marketing strategies based on audience response.

Small Skill: Conducting Retrospectives – Reviewing campaign results for lessons learned.

Teaching Technique: Campaign post-mortem analysis.

Actions: Analyze key performance indicators (KPIs) after each campaign to refine marketing tactics.

Example:

1. Assess engagement rates, click-through rates, and conversion data.
2. Identify underperforming content and ad placements.
3. Adjust future messaging and targeting based on insights.
4. Improve ROI by 25% through iterative campaign optimization.

Pairing 6: Enhancing Customer Support with Service Review Sessions

Big Skill: Growth Mindset & Adaptability – Improving customer satisfaction through iterative learning.

Small Skill: Conducting Retrospectives – Reviewing service interactions for quality improvements.

Teaching Technique: Service case debriefs.

Actions: Review customer service cases to identify patterns in recurring complaints.

Example:

1. Analyze customer inquiries and complaint resolution efficiency.
2. Identify gaps in support training or resource availability.
3. Adjust support protocols based on customer pain points.
4. Increase customer satisfaction scores by 22% through enhanced service training.

Pairing 7: Refining Supply Chain Operations with Post-Delivery Analysis

Big Skill: Growth Mindset & Adaptability – Adapting logistics strategies based on past performance.

Small Skill: Conducting Retrospectives – Analyzing supply chain disruptions and inefficiencies.

Teaching Technique: Supply chain failure analysis.

Actions: Conduct periodic reviews of shipping delays and inventory management to optimize supply chain operations.

Example:

1. Track delivery failures and supplier inconsistencies.
2. Identify factors causing shipment delays.
3. Adjust sourcing and distribution strategies accordingly.
4. Reduce supply chain disruptions by 20% through process refinement.

Continuous Improvement – (Big Skill) Growth Mindset & Adaptability vs. (Small Skill) Performance Monitoring

Pairing 1: Enhancing Sales Forecasting Through KPI Tracking

Big Skill: Growth Mindset & Adaptability – Embracing change and learning from failures.

Small Skill: Performance Monitoring – Tracking trends using dashboards and key performance indicators.

Teaching Technique: Sales performance reviews.

Actions: Implement dashboards to monitor sales trends and adjust strategies based on insights.

Example:

1. Analyze weekly revenue trends and close rates.
2. Identify underperforming sales regions and adjust targets.
3. Implement new customer acquisition strategies.
4. Improve quarterly sales growth by 15% through data-driven adjustments.

Pairing 2: Refining Employee Productivity with Performance Dashboards

Big Skill: Growth Mindset & Adaptability – Adjusting workforce management based on data insights.

Small Skill: Performance Monitoring – Using real-time productivity tracking tools.

Teaching Technique: Workforce analytics.

Actions: Monitor employee performance metrics and optimize workloads accordingly.

Example:

1. Track project completion rates per employee.
2. Identify workflow bottlenecks affecting deadlines.
3. Adjust work distribution to improve efficiency.
4. Increase productivity by 20% through optimized task delegation.

Pairing 3: Optimizing Customer Service with KPI-Driven Adjustments

Big Skill: Growth Mindset & Adaptability – Adapting service strategies based on customer feedback.

Small Skill: Performance Monitoring – Analyzing customer support response times and satisfaction scores.

Teaching Technique: Customer service benchmarking

Actions: Use KPI dashboards to track customer support efficiency and implement process improvements.

Example:

1. Monitor response times for customer inquiries.
2. Identify peak hours where delays occur.
3. Adjust staffing schedules based on real-time data.
4. Reduce customer service resolution time by 25%.

Pairing 4: Improving Financial Decision-Making with Data-Driven Insights

Big Skill: Growth Mindset & Adaptability – Refining financial strategies through ongoing analysis.

Small Skill: Performance Monitoring – Using dashboards to track financial KPIs like revenue growth, expenses, and ROI.

Teaching Technique: Financial trend analysis.

Actions: Regularly assess financial KPIs to optimize budgeting and forecasting.

Example:

1. Monitor monthly cash flow and profit margins.
2. Identify underperforming expense categories.
3. Adjust spending to improve profitability.
4. Increase net revenue by 10% through strategic budget reallocations.

Pairing 5: Strengthening Marketing ROI Through KPI Analysis

Big Skill: Growth Mindset & Adaptability – Adjusting marketing campaigns based on real-time performance.

Small Skill: Performance Monitoring – Tracking engagement, conversion rates, and campaign success.

Teaching Technique: Marketing performance audits

Actions: Continuously evaluate digital marketing campaigns and refine messaging based on performance data.

Example:

1. Use A/B testing to compare ad performance.
2. Adjust target demographics based on engagement data.
3. Allocate more budget to high-performing campaigns.
4. Improve ad conversion rates by 30%.

Pairing 6: Enhancing Project Management Efficiency with KPI Dashboards

Big Skill: Growth Mindset & Adaptability – Refining project workflows based on measurable success indicators.

Small Skill: Performance Monitoring – Tracking task completion rates and milestone achievements.

Teaching Technique: Agile project tracking.

Actions: Use performance dashboards to assess project delays and optimize workflows.

Example:

1. Measure project completion times against deadlines.
2. Identify tasks consistently delayed across multiple projects.
3. Adjust team workflows to streamline execution.
4. Reduce project overruns by 25%.

Pairing 7: Driving Operational Excellence Through Real-Time Performance Monitoring

Big Skill: Growth Mindset & Adaptability – Using operational data to improve efficiency.

Small Skill: Performance Monitoring – Implementing dashboards to measure production output and quality.

Teaching Technique: Lean operations monitoring.

Actions: Track and analyze production data to eliminate inefficiencies and reduce waste.

Example:

1. Monitor manufacturing error rates per production cycle.
2. Identify equipment downtime patterns.
3. Implement predictive maintenance to prevent breakdowns.
4. Reduce defective product rates by 18%.

Continuous Improvement – (Big Skill) Data-Driven Decision-Making vs. (Small Skill) Implementing Feedback Loops

---

Pairing 1: Optimizing Customer Satisfaction Through Survey Analysis

Big Skill: Data-Driven Decision-Making – Using customer feedback metrics to enhance service strategies.

Small Skill: Implementing Feedback Loops – Collecting and applying survey responses to refine customer experience.

Teaching Technique: Customer feedback cycle.

Actions: Deploy surveys to measure customer satisfaction and make iterative service improvements.

Example:

1. Collect post-purchase feedback via Net Promoter Score (NPS) surveys.
2. Identify recurring service complaints.
3. Implement policy changes to resolve common issues.
4. Improve customer satisfaction by 20%.

Pairing 2: Enhancing Employee Engagement Through Continuous Feedback

Big Skill: Data-Driven Decision-Making – Utilizing workforce analytics to refine engagement strategies.

Small Skill: Implementing Feedback Loops – Regularly gathering employee sentiment data.

Teaching Technique: Employee pulse surveys.

Actions: Create feedback loops where employees provide insights on workplace satisfaction, and leadership adapts policies accordingly.

Example:

1. Conduct monthly anonymous employee engagement surveys.
2. Track trends in feedback related to leadership, workload, and work-life balance.
3. Adjust HR policies based on feedback.
4. Increase employee engagement scores by 15%.

Pairing 3: Improving Product Development with User Feedback Analysis

Big Skill: Data-Driven Decision-Making – Refining product design through iterative user testing.

Small Skill: Implementing Feedback Loops – Incorporating customer feedback into future product iterations.

Teaching Technique: Beta testing & iterative refinement.

Actions: Conduct usability testing and analyze customer feedback to optimize product features.

Example:

1. Release a beta version of a software product.
2. Gather user feedback on usability and performance.
3. Adjust UI/UX design based on user pain points.
4. Reduce user complaints by 30% after implementation.

Pairing 4: Optimizing Digital Marketing Strategies Using Real-Time Feedback

Big Skill: Data-Driven Decision-Making – Using engagement metrics to refine advertising approaches.

Small Skill: Implementing Feedback Loops – Adjusting content and ad targeting based on performance data.

Teaching Technique: A/B testing campaigns.

Actions: Monitor ad engagement and iterate on campaign elements based on audience feedback.

Example:

1. Run two ad variations with different headlines.
2. Analyze click-through rates (CTR) and customer feedback.
3. Adjust targeting strategy based on insights.
4. Increase CTR by 25% after refining messaging.

Pairing 5: Refining Sales Strategies with CRM Data and Customer Feedback

Big Skill: Data-Driven Decision-Making – Optimizing sales tactics using behavioral analytics.

Small Skill: Implementing Feedback Loops – Collecting real-time input from customers and sales teams.

Teaching Technique: Sales funnel analysis.

Actions: Use CRM tools to track sales performance and refine sales pitch strategies based on feedback.

Example:

1. Track customer objections in sales calls.
2. Identify common concerns that prevent conversions.
3. Adapt sales scripts to address objections proactively.
4. Increase close rates by 18%.

Pairing 6: Improving IT Support Through User Satisfaction Scores

Big Skill: Data-Driven Decision-Making – Enhancing IT operations based on user-reported issues.

Small Skill: Implementing Feedback Loops – Collecting system usability ratings and troubleshooting efficiency metrics.

Teaching Technique: IT helpdesk ticket analysis.

Actions: Track IT support response times and user satisfaction scores to optimize helpdesk workflows.

Example:

1. Gather feedback after IT tickets are resolved.
2. Identify trends in response time concerns.
3. Implement AI-driven ticket sorting for efficiency.
4. Reduce ticket resolution time by 40%.

Pairing 7: Optimizing Supply Chain Efficiency Through Vendor Feedback

Big Skill: Data-Driven Decision-Making – Enhancing logistics using supplier input and performance data.

Small Skill: Implementing Feedback Loops – Collecting feedback from vendors to streamline supply chain processes.

Teaching Technique: Vendor performance assessments.

Actions: Monitor supplier delivery accuracy and quality consistency to refine procurement decisions.

Example:

1. Rate supplier performance on delivery speed and quality.
2. Identify underperforming vendors.
3. Adjust contracts and seek alternative suppliers.
4. Improve supply chain efficiency by 22%.

Continuous Improvement – (Big Skill) Data-Driven Decision-Making vs. (Small Skill) Conducting Retrospectives

---

Pairing 1: Enhancing Project Delivery through Lessons Learned Reviews

Big Skill: Data-Driven Decision-Making – Using project performance data to refine delivery processes.

Small Skill: Conducting Retrospectives – Analyzing completed projects to identify successes and failures.

Teaching Technique: Post-mortem analysis.

Actions: Hold structured post-project reviews to extract key insights and improve workflows.

Example:

1. Gather project teams for a retrospective session.
2. Identify bottlenecks that delayed the timeline.
3. Implement process improvements for future projects.
4. Reduce project delays by 30%.

Pairing 2: Optimizing Sales Performance Using Win/Loss Analysis

Big Skill: Data-Driven Decision-Making – Refining sales strategies based on performance trends.

Small Skill: Conducting Retrospectives – Analyzing won and lost deals to extract patterns.

Teaching Technique: Sales performance debriefing.

Actions: Review past sales deals to determine factors contributing to successful and failed negotiations.

Example:

1. Analyze lost deals to identify common objections.
2. Adjust sales training to address identified challenges.
3. Track improvements in close rates over subsequent months.
4. Increase sales conversions by 15%.

Pairing 3: Improving Customer Service Through Call Quality Assessments

Big Skill: Data-Driven Decision-Making – Enhancing customer experience through service metrics.

Small Skill: Conducting Retrospectives – Reviewing past customer interactions to improve call handling.

Teaching Technique: Call review sessions.

Actions: Evaluate recorded customer service calls to identify best practices and areas for improvement.

Example:

1. Conduct weekly team retrospectives on call handling.
2. Identify common customer complaints.
3. Implement script refinements based on findings.
4. Increase customer satisfaction scores by 20%.

Pairing 4: Strengthening Cybersecurity with Incident Response Reviews

Big Skill: Data-Driven Decision-Making – Enhancing IT security by analyzing breach data.

Small Skill: Conducting Retrospectives – Reviewing past security incidents to improve response strategies.

Teaching Technique: Incident response retrospectives.

Actions: Analyze previous cybersecurity breaches to strengthen system defenses and response protocols.

Example:

1. Conduct a post-breach security audit.
2. Identify gaps in current security measures.
3. Implement AI-driven threat detection based on findings.
4. Reduce cybersecurity vulnerabilities by 40%.

Pairing 5: Increasing Manufacturing Efficiency Through Production Reviews

Big Skill: Data-Driven Decision-Making – Reducing waste and optimizing production processes.

Small Skill: Conducting Retrospectives – Reviewing past production runs to improve efficiency.

Teaching Technique: Lean manufacturing retrospectives.

Actions: Assess past production cycles to identify inefficiencies and refine workflows.

Example:

1. Hold monthly retrospective meetings with factory supervisors.
2. Identify machine downtime causes.
3. Optimize shift scheduling and preventative maintenance.
4. Improve production output by 25%.

Pairing 6: Boosting Employee Retention Through Exit Interview Analysis

Big Skill: Data-Driven Decision-Making – Using workforce analytics to enhance employee satisfaction.

Small Skill: Conducting Retrospectives – Reviewing exit interviews to uncover workplace trends.

Teaching Technique: HR retention analysis.

Actions: Analyze exit interviews to identify patterns leading to employee turnover.

Example:

1. Compile data from exit interviews over six months.
2. Identify common reasons for employee resignations.
3. Adjust company policies based on feedback.
4. Reduce turnover rates by 18%.

Pairing 7: Refining Marketing Campaigns with Post-Campaign Reviews

Big Skill: Data-Driven Decision-Making – Optimizing marketing tactics using campaign analytics.

Small Skill: Conducting Retrospectives – Reviewing past marketing efforts to enhance future strategies.

Teaching Technique: Post-campaign debriefs.

Actions: Evaluate digital campaign performance to refine messaging and targeting.

Example:

1. Analyze ad engagement rates and conversion trends.
2. Identify underperforming channels.
3. Adjust marketing strategy for better ROI.
4. Increase ad campaign effectiveness by 35%.

Continuous Improvement – (Big Skill) Data-Driven Decision-Making vs. (Small Skill) Performance Monitoring

---

Pairing 1: Enhancing Customer Satisfaction through Service Metrics

Big Skill: Data-Driven Decision-Making – Using customer feedback data to refine service strategies.

Small Skill: Performance Monitoring – Tracking customer experience metrics (CSAT, NPS).

Teaching Technique: Customer service KPI dashboard.

Actions: Implement dashboards to monitor real-time customer feedback and adjust service protocols accordingly.

Example:

1. Set up a Net Promoter Score (NPS) dashboard.
2. Track negative feedback trends.
3. Adjust training programs based on insights.
4. Increase NPS by 15% over six months.

Pairing 2: Optimizing Sales Strategy through KPI Analysis

Big Skill: Data-Driven Decision-Making – Refining sales tactics using performance trends.

Small Skill: Performance Monitoring – Measuring sales conversion rates, lead quality, and pipeline velocity.

Teaching Technique: Sales performance tracking.

Actions: Use CRM dashboards to analyze sales data and adjust sales outreach techniques.

Example:

1. Track lead response times.
2. Identify patterns in successful conversions.
3. Train sales teams on best practices based on trends.
4. Boost conversion rates by 10% in the next quarter.

Pairing 3: Reducing Supply Chain Delays with Logistics Data

Big Skill: Data-Driven Decision-Making – Using operational data to improve supply chain efficiency.

Small Skill: Performance Monitoring – Tracking logistics KPIs like on-time delivery and order accuracy.

Teaching Technique: Supply chain performance dashboard.

Actions: Monitor shipment delays and adjust vendor selection based on real-time logistics data.

Example:

1. Identify recurring delays with specific suppliers.
2. Shift sourcing strategy to mitigate risks.
3. Reduce delivery delays by 25% in six months.

Pairing 4: Improving Employee Productivity through Workload Tracking

Big Skill: Data Driven Decision-Making – Using workforce analytics to optimize productivity.

Small Skill: Performance Monitoring – Tracking task completion rates and employee efficiency metrics.

Teaching Technique: Workforce performance monitoring.

Actions: Use time-tracking tools to measure productivity and identify workload imbalances.

Example:

1. Identify departments struggling with workload distribution.
2. Reallocate tasks based on workload data.
3. Increase task completion efficiency by 20%.

Pairing 5: Enhancing Website Engagement through Data Analytics

Big Skill: Data-Driven Decision-Making – Using website traffic patterns to improve user experience.

Small Skill: Performance Monitoring – Tracking bounce rates, session duration, and conversion metrics.

Teaching Technique: Digital performance analysis.

Actions: Use Google Analytics to monitor website traffic and refine content strategy.

Example:

1. Identify pages with high bounce rates.
2. Adjust UI/UX design for better navigation.
3. Increase website retention rate by 30%.

Pairing 6: Increasing Manufacturing Efficiency with Real-Time Monitoring

Big Skill: Data-Driven Decision-Making – Using production data to reduce waste and optimize processes.

Small Skill: Performance Monitoring – Measuring cycle time, defect rates, and machine downtime.

Teaching Technique: Lean manufacturing analytics.

Actions: Set up real-time dashboards to monitor performance and preemptively address failures.

Example:

1. Track production line inefficiencies.
2. Implement predictive maintenance based on downtime patterns.
3. Reduce manufacturing defects by 18%.

Pairing 7: Boosting Financial Performance through Budget Trend Analysis

Big Skill: Data-Driven Decision-Making – Optimizing financial decision-making using historical trends.

Small Skill: Performance Monitoring – Tracking budget variance, revenue growth, and expense.

Teaching Technique: Financial dashboard optimization.

Actions: Use financial KPIs to adjust budget allocations and spending priorities.

Example:

1. Identify overspending areas.
2. Adjust departmental budgets based on ROI analysis.
3. Improve budget adherence by 25%.

Example Teaching Strategies for Big vs. Small Thinking Skill Pairings of Global Perspective at the Business Professional Level

Here are 63 unique teaching strategy examples for the big thinking and small thinking skill pairings, each described in detail to align with **Big Thinking Skills** (Concepts & Strategies) and **Small Thinking Skills** (Step-by-Step Execution) for the **Global Perspective Skill** taught at the business professional level.

Global Perspective

Big Thinking Skills (Concepts & Strategies)

- **Cross-Cultural Competence:** Understanding and respecting international business norms and cultural differences.
- **Global Market Analysis:** Evaluating economic, political, and technological trends in different regions.
- **International Strategy Development:** Adapting business models to different markets and regulatory environments.

Small Thinking Skills (Step-by-Step Execution)

- **Navigating International Business Regulations:** Applying compliance frameworks (e.g., GDPR, trade laws).
- **Global Communication Skills:** Using effective negotiation and collaboration techniques across cultures.
- **Monitoring Global Trends:** Keeping up with emerging market shifts and economic reports.

Global Perspective – (Big Skill) Cross-Cultural Competence vs. (Small Skill) Navigating International Business Regulations

---

Pairing 1: Understanding GDPR Compliance in Cross-Border Business

Big Skill: Cross-Cultural Competence – Adapting business practices to respect global privacy expectations.

Small Skill: Navigating International Business Regulations – Ensuring compliance with General Data Protection Regulation (GDPR).

Teaching Technique: Compliance case study analysis.

Actions: Analyze how GDPR affects multinational corporations operating in the EU and create a compliance strategy.

Example:

1. Research GDPR principles.
2. Assess how GDPR applies to a company's customer data practices.
3. Develop an internal compliance checklist.
4. Reduce data security risks by ensuring alignment with GDPR rules.

Pairing 2: Managing Trade Laws in International Supply Chains

Big Skill: Cross-Cultural Competence – Understanding import/export practices across different markets.

Small Skill: Navigating International Business Regulations – Complying with trade laws and tariffs.

Teaching Technique: Trade law simulation.

Actions: Role-play as a logistics manager handling import/export restrictions across different regions.

Example:

1. Identify key trade barriers for a product entering the EU, China, and the U.S.
2. Develop alternative sourcing plans based on regulations.
3. Optimize supply chain efficiency by reducing legal risks.

Pairing 3: Ensuring Anti-Bribery Compliance in International Deals

Big Skill: Cross-Cultural Competence – Navigating ethical business practices in different countries.

Small Skill: Navigating International Business Regulations – Understanding anti-bribery laws like the Foreign Corrupt Practices Act (FCPA).

Teaching Technique: Ethical decision-making workshop.

Actions: Discuss real-world bribery cases and develop a best-practices policy for avoiding corruption risks.

Example:

1. Compare U.S. FCPA rules with local bribery laws in emerging markets.
2. Design an anti-corruption policy for an international sales team.
3. Train executives on ethical risk assessment.

Pairing 4: Adapting to Data Localization Laws in Global Expansion

Big Skill: Cross-Cultural Competence – Understanding how different governments handle data security.

Small Skill: Navigating International Business Regulations – Complying with data residency requirements (e.g., China's Cybersecurity Law).

Teaching Technique: Comparative regulatory analysis.

Actions: Compare and contrast data localization laws and assess their impact on business strategy.

Example:

1. Identify countries with strict data residency laws.
2. Develop a localized data storage strategy.
3. Ensure compliance while maintaining operational efficiency.

Pairing 5: Managing International Tax Compliance for Multinational Operations

Big Skill: Cross-Cultural Competence – Understanding global taxation structures.

Small Skill: Navigating International Business Regulations – Adhering to OECD tax frameworks and local tax laws.

Teaching Technique: Tax compliance simulation.

Actions: Analyze corporate tax structures across different jurisdictions and develop a tax strategy.

Example:

1. Compare tax incentives in Ireland vs. Germany for business incorporation.
2. Develop a legal tax minimization strategy for a multinational firm.
3. Ensure compliance while optimizing tax efficiency.

Pairing 6: Understanding Global Labor Laws for International HR Strategy

Big Skill: Cross-Cultural Competence – Managing international employment standards.

Small Skill: Navigating International Business Regulations – Complying with labor laws (e.g., EU Working Time Directive).

Teaching Technique: Global HR compliance workshop.

Actions: Compare employment contract requirements across different countries.

Example:

1. Examine mandatory employee benefits in Japan vs. the U.S.
2. Develop a compliant international hiring process.

3. Reduce legal risks while maintaining fair labor practices.

Pairing 7: Handling Export Control Laws in High-Tech Industries

Big Skill: Cross-Cultural Competence – Understanding restricted technology trade policies.

Small Skill: Navigating International Business Regulations – Complying with ITAR and other export controls.

Teaching Technique: Compliance scenario role-play.

Actions: Simulate a scenario where a tech firm must determine whether a product can be exported under ITAR rules.

Example:

1. Assess if an AI software product falls under U.S. export control.
2. Determine the necessary licenses and restrictions.
3. Develop an internal compliance workflow.

Global Perspective – (Big Skill) Cross-Cultural Competence vs. (Small Skill) Global Communication Skills

---

Pairing 1: Navigating Cross-Cultural Business Negotiations

Big Skill: Cross-Cultural Competence – Understanding business etiquette in diverse global markets.

Small Skill: Global Communication Skills – Using culturally appropriate negotiation tactics.

Teaching Technique: Role-playing cross-cultural negotiation scenarios.

Actions: Simulate a business deal between Western and Asian executives to practice cultural negotiation techniques.

Example:

1. Role-play a contract negotiation between a U.S. and Japanese company.
2. Adjust communication styles based on cultural expectations (e.g., direct vs. indirect negotiation).
3. Identify key etiquette differences and their impact on deal outcomes.

Pairing 2: Adapting Communication Styles for Global Collaboration

Big Skill: Cross-Cultural Competence – Recognizing communication preferences in different cultures.

Small Skill: Global Communication Skills – Adjusting communication tone, formality, and structure.

Teaching Technique: Business email and meeting analysis.

Actions: Compare and contrast business email styles from different cultures and refine messages accordingly.

Example:

1. Analyze formal business emails from Germany vs. casual business emails from the U.S.
2. Adapt tone for international partners while maintaining professionalism.
3. Ensure clarity by avoiding culturally insensitive phrases.

Pairing 3: Leading Virtual International Teams

Big Skill: Cross-Cultural Competence – Managing distributed teams across different time zones and cultures.

Small Skill: Global Communication Skills – Establishing effective collaboration practices in remote settings.

Teaching Technique: Virtual team-building exercises.

Actions: Simulate a virtual team meeting with participants from multiple countries to practice inclusive leadership.

Example:

1. Use time-zone-friendly scheduling techniques.
2. Implement cross-cultural icebreakers to foster team cohesion.
3. Adjust communication tools to accommodate language barriers (e.g., closed captions for meetings).

Pairing 4: Conducting Cross-Cultural Client Presentations

Big Skill: Cross-Cultural Competence – Tailoring business presentations to international audiences.

Small Skill: Global Communication Skills – Adapting storytelling and persuasion techniques.

Teaching Technique: Business presentation adaptation exercise.

Actions: Modify a standard corporate pitch to suit different cultural preferences for engagement.

Example:

1. Adjust a sales pitch for an American vs. Chinese audience.
2. Incorporate culturally relevant storytelling techniques.
3. Use appropriate levels of data vs. emotional appeal based on the audience.

Pairing 5: Resolving Cross-Cultural Team Conflicts

Big Skill: Cross-Cultural Competence – Mediating misunderstandings arising from cultural differences.

Small Skill: Global Communication Skills – Applying culturally sensitive conflict resolution techniques.

Teaching Technique: Conflict resolution workshop.

Actions: Discuss real-world cases of cultural misunderstandings and develop solutions.

Example:

1. Analyze a case where direct feedback clashed with indirect communication styles.
2. Develop an action plan to mediate and realign expectations.
3. Role-play resolving the conflict using active listening strategies.

Pairing 6: Building Trust in International Business Relationships

Big Skill: Cross-Cultural Competence – Understanding how trust is built in different cultures.

Small Skill: Global Communication Skills – Using culturally appropriate relationship-building techniques.

Teaching Technique: Trust-building scenario analysis.

Actions: Compare trust-building methods in relationship-oriented vs. transaction-oriented cultures.

Example:

1. Compare long-term relationship-building norms in Latin America vs. fast-paced deal-making in the U.S.
2. Develop strategies to establish credibility with international clients.
3. Implement appropriate levels of personal rapport-building in negotiations.

Pairing 7: Using Professional Interpreters & Translation Tools Effectively

Big Skill: Cross-Cultural Competence – Ensuring accurate communication in multilingual business settings.

Small Skill: Global Communication Skills – Managing professional interpreters and translation software.

Teaching Technique: Business translation practice.

Actions: Practice using interpreters and AI-based translation tools in business discussions.

Example:

1. Conduct a bilingual meeting with live interpretation.
2. Analyze potential misinterpretations in machine translations.
3. Implement best practices for clear, culturally appropriate messaging.

Global Perspective – (Big Skill) Cross-Cultural Competence vs. (Small Skill) Monitoring Global Trends

---

Pairing 1: Analyzing Emerging Markets for Expansion

Big Skill: Cross-Cultural Competence – Understanding cultural nuances in new international markets.

Small Skill: Monitoring Global Trends – Tracking economic growth indicators in developing regions.

Teaching Technique: Market expansion case study.

Actions: Analyze recent shifts in emerging markets and assess their business implications.

Example:

1. Study the rise of fintech in Africa and its impact on banking accessibility.
2. Evaluate GDP growth, digital adoption, and regulatory trends.
3. Recommend strategies for companies to enter these markets effectively.

Pairing 2: Identifying Consumer Behavior Shifts in Global Markets

Big Skill: Cross-Cultural Competence – Recognizing how local values influence purchasing behavior.

Small Skill: Monitoring Global Trends – Examining global shifts in consumer preferences.

Teaching Technique: Consumer trend analysis.

Actions: Compare consumer preferences in different regions and predict future trends.

Example:

1. Analyze the increasing demand for sustainable products in Europe.
2. Compare brand loyalty differences between the U.S. and China.
3. Adjust marketing approaches to align with cultural purchasing behaviors.

Pairing 3: Understanding Political and Economic Risks in Global Business

Big Skill: Cross-Cultural Competence – Adapting to business risks influenced by political landscapes.

Small Skill: Monitoring Global Trends – Assessing geopolitical and economic risks worldwide.

Teaching Technique: Risk scenario planning.

Actions: Evaluate real-world geopolitical risks and develop risk mitigation strategies.

Example:

1. Analyze how Brexit affected supply chains in the EU.
2. Identify alternative supply routes in response to tariffs.
3. Recommend investment strategies for businesses in politically unstable regions.

Pairing 4: Adapting to Global Supply Chain Disruptions

Big Skill: Cross-Cultural Competence – Understanding regional variations in supply chain logistics.

Small Skill: Monitoring Global Trends – Tracking disruptions in global supply chains.

Teaching Technique: Global logistics simulation.

Actions: Develop contingency plans for businesses affected by supply chain bottlenecks.

Example:

1. Assess the impact of COVID-19 on semiconductor shortages.
2. Identify alternative sourcing strategies for manufacturers.
3. Recommend solutions for diversifying supplier networks.

Pairing 5: Predicting Future Workforce Trends Across Cultures

Big Skill: Cross-Cultural Competence – Navigating workforce dynamics in different countries.

Small Skill: Monitoring Global Trends – Analyzing employment trends and workforce automation.

Teaching Technique: Workforce trend forecasting.

Actions: Compare workforce shifts in different regions and predict skill demands.

Example:

1. Assess how AI adoption affects job opportunities in the U.S. vs. India.
2. Develop strategies for retraining workers in automation-heavy industries.
3. Recommend HR policies that accommodate cultural work expectations.

Pairing 6: Evaluating Cultural Influences on Digital Transformation

Big Skill: Cross-Cultural Competence – Understanding cultural attitudes toward technology adoption.

Small Skill: Monitoring Global Trends – Assessing digital transformation rates across regions.

Teaching Technique: Digital transformation analysis.

Actions: Compare tech adoption trends and recommend region-specific strategies.

Example:

1. Analyze how Japan vs. Brazil adapts to e-commerce.
2. Assess cybersecurity and data privacy concerns in different countries.
3. Recommend localization strategies for tech-driven businesses.

Pairing 7: Tracking Global Financial Market Fluctuations

Big Skill: Cross-Cultural Competence – Understanding how economic policies impact financial stability.

Small Skill: Monitoring Global Trends – Observing stock markets, interest rates, and inflation patterns.

Teaching Technique: Economic impact forecasting.

Actions: Evaluate financial trends and predict their impact on international business.

Example:

1. Analyze the effect of Federal Reserve rate hikes on emerging markets.
2. Predict how currency fluctuations impact multinational investments.
3. Recommend hedging strategies to mitigate exchange rate risks.

Global Perspective – (Big Skill) Global Market Analysis vs. (Small Skill) Navigating International Business Regulations

---

Pairing 1: Analyzing the Impact of Trade Tariffs on Global Business

Big Skill: Global Market Analysis – Evaluating how trade policies influence international markets.

Small Skill: Navigating International Business Regulations – Understanding tariffs and trade laws.

Teaching Technique: Trade regulation case study.

Actions: Assess recent tariff changes and their impact on industries.

Example:

1. Study the U.S.-China trade war and its effect on global supply chains.
2. Analyze tariff structures on key industries like technology and agriculture.
3. Develop risk mitigation strategies for companies affected by tariffs.

Pairing 2: Understanding GDPR Compliance in International Business

Big Skill: Global Market Analysis – Assessing how data protection laws affect market access.

Small Skill: Navigating International Business Regulations – Applying GDPR compliance requirements.

Teaching Technique: Data privacy workshop.

Actions: Examine GDPR's impact on multinational corporations and best compliance practices.

Example:

1. Analyze how GDPR affects digital marketing and customer data collection.
2. Compare compliance costs for businesses operating in the EU vs. the U.S.
3. Develop a checklist for companies to align with GDPR regulations.

Pairing 3: Evaluating Foreign Direct Investment (FDI) Regulations

Big Skill: Global Market Analysis – Understanding investment trends in different regions.

Small Skill: Navigating International Business Regulations – Learning investment compliance laws.

Teaching Technique: FDI policy analysis.

Actions: Analyze regional investment incentives and regulatory challenges.

Example:

1. Compare FDI policies in emerging markets vs. developed economies.
2. Study how countries like India and Brazil attract foreign investment.
3. Recommend strategies for businesses expanding into regulated markets.

Pairing 4: Understanding Tax Policies for Global Expansion

Big Skill: Global Market Analysis – Evaluating corporate tax trends and their impact.

Small Skill: Navigating International Business Regulations – Learning cross-border taxation laws.

Teaching Technique: International tax compliance training.

Actions: Compare corporate tax rates and regulations in different countries.

Example:

1. Study the impact of the OECD's global tax reforms on multinational corporations.
2. Evaluate tax incentives for companies operating in free trade zones.
3. Develop a compliance strategy to minimize tax risks in global markets.

Pairing 5: Analyzing Political Risk in International Business

Big Skill: Global Market Analysis – Assessing how political events shape global markets.

Small Skill: Navigating International Business Regulations – Understanding regulatory responses to political risk.

Teaching Technique: Political risk scenario planning.

Actions: Evaluate case studies of businesses affected by geopolitical risks.

Example:

1. Study how Brexit impacted trade laws between the UK and the EU.
2. Analyze how sanctions affect business operations in restricted markets.
3. Develop strategies for mitigating political risk in international expansion.

Pairing 6: Understanding Import-Export Regulations Across Regions

Big Skill: Global Market Analysis – Evaluating supply chain logistics in global trade.

Small Skill: Navigating International Business Regulations – Applying import/export compliance rules.

Teaching Technique: Supply chain compliance workshop.

Actions: Examine customs procedures and regulatory hurdles in different regions.

Example:

1. Compare import/export documentation requirements for the EU, U.S., and China.
2. Analyze how trade agreements like USMCA affect supply chain operations.
3. Develop a best-practice guide for businesses navigating international trade.

Pairing 7: Examining Digital Trade Regulations Across Countries

Big Skill: Global Market Analysis – Understanding the rise of digital economies.

Small Skill: Navigating International Business Regulations – Learning cross-border e-commerce laws.

Teaching Technique: Digital trade regulation case study.

Actions: Analyze the regulatory landscape for digital businesses across regions.

Example:

1. Study how China's cybersecurity laws affect international companies.
2. Compare digital taxation policies in the EU, U.S., and ASEAN countries.
3. Develop a compliance strategy for global e-commerce platforms.

Global Perspective – (Big Skill) Global Market Analysis vs. (Small Skill) Global Communication Skills

---

Pairing 1: Negotiating International Business Contracts

Big Skill: Global Market Analysis – Evaluating how political and economic policies shape international business agreements.

Small Skill: Global Communication Skills – Using effective negotiation strategies to close deals across cultures.

Teaching Technique: Role-playing international negotiations.

Actions: Simulate cross-border contract negotiations focusing on economic and legal considerations.

Example:

1. Participants role-play a negotiation between a U.S. tech firm and a European distributor.
2. They adjust their approach based on cultural expectations and regulatory policies.
3. Teams analyze how economic conditions impact contract terms and pricing.

Pairing 2: Understanding Communication Styles in Emerging Markets

Big Skill: Global Market Analysis – Evaluating technological trends in different regions.

Small Skill: Global Communication Skills – Adapting negotiation styles based on regional business etiquette.

Teaching Technique: Case study analysis of cultural communication differences.

Actions: Compare direct vs. indirect communication in business negotiations.

Example:

1. Study how Western companies adjust their approach when negotiating with Asian firms.
2. Examine how high-context cultures (Japan, China) differ from low-context cultures (U.S., Germany).
3. Develop region-specific negotiation strategies for multinational corporations.

Pairing 3: Managing Cross-Border Business Expansion

Big Skill: Global Market Analysis – Assessing economic conditions for market entry.

Small Skill: Global Communication Skills – Using diplomatic communication strategies to establish partnerships.

Teaching Technique: Market entry strategy simulation.

Actions: Develop and pitch expansion plans tailored to local business norms.

Example:

1. A company plans to expand operations to South America and must navigate language and business etiquette barriers.
2. Teams create strategies for collaborating with local partners while respecting cultural norms.
3. Participants simulate investor meetings with foreign stakeholders.

Pairing 4: Negotiating Trade Agreements Across Countries

Big Skill: Global Market Analysis – Evaluating trade policies affecting international commerce.

Small Skill: Global Communication Skills – Applying diplomatic skills to build consensus in trade negotiations.

Teaching Technique: International trade agreement workshop.

Actions: Analyze and draft a mock trade agreement between two countries.

Example:

1. Teams represent different governments negotiating import/export tariffs.
2. Participants apply active listening and persuasive negotiation tactics.
3. The goal is to reach a trade deal balancing national economic interests.

Pairing 5: Communicating Global Market Risks to Stakeholders

Big Skill: Global Market Analysis – Identifying financial, political, and economic risks.

Small Skill: Global Communication Skills – Delivering clear and concise risk reports to executives.

Teaching Technique: Business risk communication training.

Actions: Develop presentations on economic volatility and geopolitical risks.

Example:

1. Participants analyze risks of investing in different regions (e.g., inflation in Latin America, Brexit uncertainty).

2. They present their findings to an executive board, emphasizing clarity and cross-cultural awareness.
3. Feedback is given on improving communication style for diverse audiences.

Pairing 6: Handling Crisis Communication in International Markets

Big Skill: Global Market Analysis – Assessing political instability and crisis response strategies.

Small Skill: Global Communication Skills – Crafting culturally sensitive crisis messaging.

Teaching Technique: Crisis management role-play.

Actions: Create and deliver emergency responses for global business crises.

Example:

1. Teams develop a response plan for a company dealing with political unrest in a key market.
2. They draft press releases, stakeholder emails, and government negotiation strategies.
3. Participants present their plans to a panel simulating global business executives.

Pairing 7: Negotiating Technology Partnerships in Global Markets

Big Skill: Global Market Analysis – Evaluating technological innovation in different regions.

Small Skill: Global Communication Skills – Using strategic messaging to build cross-border tech collaborations.

Teaching Technique: Strategic partnership simulation.

Actions: Engage in simulated negotiations for tech collaboration deals.

Example:

1. Participants role-play executives negotiating a tech transfer agreement between a U.S. firm and an Asian semiconductor manufacturer.
2. They navigate legal, ethical, and market-specific concerns in their negotiations.
3. Teams analyze how technology trends shape global partnerships.

Global Perspective – (Big Skill) Global Market Analysis vs. (Small Skill) Monitoring Global Trends

---

Pairing 1: Predicting Economic Downturns and Market Recessions

Big Skill: Global Market Analysis – Assessing macroeconomic indicators to anticipate potential downturns.

Small Skill: Monitoring Global Trends – Tracking GDP growth, inflation rates, and unemployment figures.

Teaching Technique: Economic forecast simulation.

Actions: Analyze past economic crises and predict future downturns based on current data.

Example:

1. Teams review central bank reports and economic trends from multiple regions.
2. They use recession indicators (e.g., yield curve inversion, consumer confidence).
3. Participants present a risk mitigation plan for multinational businesses.

Pairing 2: Evaluating Political Risks in Global Investments

Big Skill: Global Market Analysis – Understanding how political instability affects foreign investments.

Small Skill: Monitoring Global Trends – Tracking regulatory changes, elections, and trade policies.

Teaching Technique: Political risk assessment exercise.

Actions: Assess the impact of global political shifts on investment decisions.

Example:

1. Teams analyze the effect of Brexit on European markets.
2. They identify industries most vulnerable to political disruptions.
3. Participants develop contingency plans for international firms.

Pairing 3: Understanding Trade Relations Between Major Economies

Big Skill: Global Market Analysis – Examining trade agreements and their business impact.

Small Skill: Monitoring Global Trends – Tracking tariff changes and trade disputes.

Teaching Technique: Trade policy case study.

Actions: Compare the effects of trade agreements such as NAFTA, USMCA, and RCEP.

Example:

1. Teams analyze the impact of U.S.-China trade tensions on tech exports.
2. They research tariff effects on supply chains and international prices.
3. Participants propose policy recommendations for affected industries.

Pairing 4: Identifying Emerging Market Opportunities

Big Skill: Global Market Analysis – Evaluating economic growth patterns in developing regions.

Small Skill: Monitoring Global Trends – Tracking foreign direct investment (FDI) and infrastructure developments.

Teaching Technique: Emerging markets opportunity mapping.

Actions: Identify top regions for investment based on economic indicators.

Example:

1. Teams research Africa's growing tech sector and rising consumer demand.
2. They examine foreign investment incentives in South America.
3. Participants present a business case for expansion into an emerging market.

Pairing 5: Analyzing Supply Chain Disruptions in Global Markets

Big Skill: Global Market Analysis – Evaluating logistics and production dependencies.

Small Skill: Monitoring Global Trends – Tracking shipping delays, raw material shortages, and geopolitical risks.

Teaching Technique: Supply chain risk modeling.

Actions: Assess supply chain vulnerabilities and develop mitigation strategies.

Example:

1. Teams analyze the effects of semiconductor shortages on global tech production.
2. They monitor the impact of trade sanctions on supply chains.
3. Participants develop alternative sourcing plans for companies.

Pairing 6: Forecasting Currency Exchange Rate Volatility

Big Skill: Global Market Analysis – Understanding how forex markets influence trade and investments.

Small Skill: Monitoring Global Trends – Tracking interest rates, inflation, and monetary policies.

Teaching Technique: Currency market analysis simulation.

Actions: Predict exchange rate fluctuations based on macroeconomic conditions.

Example:

1. Teams examine how the U.S. dollar strengthens or weakens based on Federal Reserve policies.
2. They analyze the impact of inflation on emerging market currencies.

3. Participants present strategies to hedge against forex volatility.

Pairing 7: Monitoring Technological Disruptions in Global Markets

Big Skill: Global Market Analysis – Evaluating the adoption of new technologies across industries.

Small Skill: Monitoring Global Trends – Tracking innovations such as AI, blockchain, and green energy.

Teaching Technique: Technology impact assessment.

Actions: Identify industries most affected by emerging tech trends.

Example:

1. Teams assess the impact of 5G rollout on telecom and digital services.
2. They research AI adoption trends in manufacturing and logistics.
3. Participants develop strategies for businesses to stay ahead of tech disruptions.

Global Perspective – (Big Skill) International Strategy Development vs. (Small Skill) Navigating International Business Regulations

---

Pairing 1: Expanding into the European Market Under GDPR Compliance

Big Skill: International Strategy Development – Adapting business expansion models to align with EU data protection laws.

Small Skill: Navigating International Business Regulations – Ensuring compliance with the General Data Protection Regulation (GDPR).

Teaching Technique: Regulatory impact analysis.

Actions: Assess the impact of GDPR on business operations and customer data management.

Example:

1. Teams analyze how GDPR affects e-commerce companies operating in Europe.
2. They design data storage policies that comply with GDPR restrictions.
3. Participants develop a compliance roadmap for entering the European market.

Pairing 2: Adjusting Supply Chain Strategy to Meet USMCA Trade Regulations

Big Skill: International Strategy Development – Aligning supply chain models with trade agreements in North America.

Small Skill: Navigating International Business Regulations – Adapting operations to comply with the United States-Mexico-Canada Agreement (USMCA).

Teaching Technique: Trade policy case study.

Actions: Compare supply chain models before and after USMCA implementation.

Example:

1. Teams analyze the impact of new tariff regulations on cross-border manufacturing.
2. They recommend alternative logistics strategies to reduce trade costs.
3. Participants develop a supply chain model that optimizes cost efficiency under USMCA.

Pairing 3: Localizing Financial Strategies for Asia's Varying Tax Regulations

Big Skill: International Strategy Development – Developing financial strategies for global tax compliance.

Small Skill: Navigating International Business Regulations – Adapting to tax policies in different Asian markets.

Teaching Technique: Comparative tax analysis.

Actions: Examine corporate tax structures across China, India, and Singapore.

Example:

1. Teams analyze how India's Goods and Services Tax (GST) affects multinational companies.
2. They compare corporate tax rates and incentives in Asian financial hubs.
3. Participants recommend tax-efficient business structures for market entry.

Pairing 4: Aligning Business Operations with Environmental Regulations

Big Skill: International Strategy Development – Ensuring sustainability compliance in different regions.

Small Skill: Navigating International Business Regulations – Adhering to global environmental laws (e.g., EU Green Deal, carbon taxes).

Teaching Technique: Sustainability compliance workshop.

Actions: Develop an international environmental compliance framework.

Example:

1. Teams analyze the impact of Europe's carbon border adjustment mechanism.
2. They assess how China's emission regulations affect manufacturing costs.
3. Participants design a sustainability strategy that aligns with global green laws.

Pairing 5: Adapting Employment Policies for Global Expansion

Big Skill: International Strategy Development – Customizing HR policies for compliance in new markets.

Small Skill: Navigating International Business Regulations – Understanding labor laws and employment rights worldwide.

Teaching Technique: Global HR policy simulation.

Actions: Develop compliant HR frameworks for hiring in different regions.

Example:

1. Teams research labor laws in France, where employee termination rules are strict.
2. They compare employment contract structures in North America vs. Europe.
3. Participants design an international recruitment and payroll strategy.

Pairing 6: Managing Intellectual Property Rights in International Markets

Big Skill: International Strategy Development – Protecting intellectual property (IP) across multiple jurisdictions.

Small Skill: Navigating International Business Regulations – Understanding copyright, patent, and trademark laws.

Teaching Technique: IP risk assessment.

Actions: Evaluate intellectual property protection strategies in key global markets.

Example:

1. Teams analyze how China's intellectual property laws impact technology companies.
2. They compare U.S. and EU trademark protections for consumer products.
3. Participants design an IP protection strategy for an international brand expansion.

Pairing 7: Developing an International E-Commerce Expansion Plan

Big Skill: International Strategy Development – Adapting digital business models to international regulations.

Small Skill: Navigating International Business Regulations – Complying with cross-border online sales laws.

Teaching Technique: E-commerce compliance case study.

Actions: Identify challenges and solutions for selling across different regulatory environments.

Example:

1. Teams analyze consumer data protection laws affecting online retail in Europe.
2. They examine VAT requirements for digital goods in international markets.

3. Participants create a cross-border e-commerce strategy that complies with global regulations.

Global Perspective – (Big Skill) International Strategy Development vs. (Small Skill) Global Communication Skills

---

Pairing 1: Negotiating Market Entry with Local Partners

Big Skill: International Strategy Development – Structuring entry strategies for new markets through local partnerships.

Small Skill: Global Communication Skills – Employing culturally aware negotiation tactics to secure partnerships.

Teaching Technique: Cross-cultural negotiation simulation.

Actions: Simulate negotiations for a joint venture in a foreign market, considering cultural norms and business etiquette.

Example:

1. Teams role-play negotiations between a U.S. tech company and a Japanese distributor.
2. They discuss indirect communication styles and decision-making hierarchies.
3. Participants adapt their negotiation strategy based on cultural insights.

Pairing 2: Adapting Brand Messaging for International Audiences

Big Skill: International Strategy Development – Customizing branding strategies for different cultural contexts.

Small Skill: Global Communication Skills – Refining brand messaging to resonate with diverse customer bases.

Teaching Technique: Cultural brand adaptation workshop.

Actions: Analyze marketing campaigns in different regions and modify messaging for cultural appropriateness.

Example:

1. Teams assess how McDonald's adapts its advertising in China vs. the U.S.
2. They adjust product names and slogans to fit cultural sensitivities.
3. Participants present a rebranded ad campaign for a specific international market.

Pairing 3: Managing Multinational Teams in a Global Expansion

Big Skill: International Strategy Development – Structuring effective multinational team collaboration strategies.

Small Skill: Global Communication Skills – Leading cross-border virtual teams with cultural awareness.

Teaching Technique: Virtual team leadership exercise.

Actions: Practice leading a remote team spanning multiple time zones and cultures.

Example:

1. Teams simulate a global product launch involving teams in Europe, Asia, and the U.S.
2. They adjust meeting times, communication tools, and work styles for inclusivity.
3. Participants resolve misunderstandings arising from cultural differences in leadership styles.

Pairing 4: Structuring Cross-Cultural Business Proposals

Big Skill: International Strategy Development – Developing expansion proposals that align with regional market needs.

Small Skill: Global Communication Skills – Writing and delivering persuasive proposals tailored to different business cultures.

Teaching Technique: Business proposal development.

Actions: Craft a market entry proposal customized for different cultural expectations.

Example:

1. Teams create a market expansion pitch for investors in Latin America vs. Germany.
2. They modify the proposal's tone, formality, and structure based on business norms.
3. Participants present their proposals and receive feedback from diverse stakeholders.

Pairing 5: Building Cross-Cultural Customer Engagement Strategies

Big Skill: International Strategy Development – Developing region-specific customer engagement models.

Small Skill: Global Communication Skills – Adapting communication styles for international customer service.

Teaching Technique: Multicultural customer service role-play.

Actions: Simulate interactions with customers from different regions and refine service approaches.

Example:

1. Teams manage customer inquiries from South Korea vs. France using culturally appropriate responses.
2. They practice adjusting formality levels and response times based on expectations.
3. Participants develop a global customer service training manual.

Pairing 6: Managing Crisis Communication in International Markets

Big Skill: International Strategy Development – Creating crisis response plans for different regions.

Small Skill: Global Communication Skills – Tailoring crisis messaging to diverse cultural expectations.

Teaching Technique: Crisis communication simulation.

Actions: Develop and deliver culturally appropriate crisis responses to real-world case studies.

Example:

1. Teams simulate a public relations crisis for a multinational brand.
2. They craft press releases and social media statements adjusted for each target country.
3. Participants compare crisis responses in collectivist vs. individualist cultures.

Pairing 7: Leading International Business Development Meetings

Big Skill: International Strategy Development – Expanding business opportunities through strategic alliances.

Small Skill: Global Communication Skills – Facilitating productive cross-cultural business discussions.

Teaching Technique: Business development role-play.

Actions: Conduct a simulated international business meeting with diverse stakeholders.

Example:

1. Teams role-play a merger discussion between a European and Middle Eastern company.
2. They practice adapting discussion structures to different business hierarchies.
3. Participants develop an agenda and communication strategy that balances directness and relationship-building.

Global Perspective – (Big Skill) International Strategy Development vs. (Small Skill) Monitoring Global Trends

---

Pairing 1: Adapting E-Commerce Strategies for Emerging Markets

Big Skill: International Strategy Development – Modifying e-commerce business models for varying regional demands.

Small Skill: Monitoring Global Trends – Analyzing digital payment trends and consumer behavior.

Teaching Technique: Market adaptation case study.

Actions: Study how global e-commerce platforms adjust to regional differences in consumer trust and infrastructure.

Example:

1. Teams analyze mobile payment adoption in China vs. credit card reliance in the U.S.
2. They assess how Alibaba and Amazon modify operations for different regulatory environments.
3. Participants develop an e-commerce launch plan for an emerging market.

Pairing 2: Expanding Renewable Energy Investments in Global Markets

Big Skill: International Strategy Development – Aligning investment strategies with regional sustainability policies.

Small Skill: Monitoring Global Trends – Tracking regulatory shifts in environmental and energy sectors.

Teaching Technique: Policy impact forecasting.

Actions: Assess policy changes in major economies and predict their effects on renewable investments.

Example:

1. Teams compare EU carbon reduction goals vs. U.S. energy policies.
2. They model investment risks based on changing subsidies and tax incentives.
3. Participants draft a strategic investment memo for renewable energy expansion.

Pairing 3: Navigating Global Supply Chain Disruptions

Big Skill: International Strategy Development – Developing resilient supply chain strategies across multiple markets.

Small Skill: Monitoring Global Trends – Tracking geopolitical risks and trade policy changes.

Teaching Technique: Supply chain risk assessment.

Actions: Use real-world supply chain disruptions to evaluate alternative sourcing and logistics strategies.

Example:

1. Teams assess the impact of tariffs on semiconductor supply chains.
2. They explore supplier diversification in response to labor shortages.
3. Participants create contingency plans for mitigating future disruptions.

Pairing 4: Adjusting Product Pricing for Currency Fluctuations

Big Skill: International Strategy Development – Adapting pricing models based on foreign exchange rates.

Small Skill: Monitoring Global Trends – Analyzing currency trends and inflation reports.

Teaching Technique: Dynamic pricing strategy simulation.

Actions: Use historical currency exchange data to model pricing adjustments in different markets.

Example:

1. Teams study how luxury brands adjust pricing based on currency strength.
2. They calculate the impact of inflation on manufacturing costs in different regions.
3. Participants develop a regional pricing strategy for a global brand.

Pairing 5: Adapting Banking and FinTech Models for Global Markets

Big Skill: International Strategy Development – Tailoring financial services to different regulatory and consumer landscapes.

Small Skill: Monitoring Global Trends – Tracking digital banking adoption and financial regulations.

Teaching Technique: FinTech trend analysis.

Actions: Compare financial technology regulations and adoption trends across multiple regions.

Example:

1. Teams examine mobile banking penetration in Africa vs. Europe.
2. They assess how data privacy laws affect cross-border financial transactions.
3. Participants design a market entry strategy for a FinTech startup.

Pairing 6: Global Expansion of Healthcare Services

Big Skill: International Strategy Development – Scaling healthcare models for diverse economic and regulatory conditions.

Small Skill: Monitoring Global Trends – Evaluating public health policy changes and emerging medical technologies.

Teaching Technique: Healthcare trend forecasting.

Actions: Study healthcare access trends and policy reforms to guide business strategy.

Example:

1. Teams analyze the effect of telemedicine adoption in developed vs. developing nations.
2. They explore the impact of government healthcare spending shifts.
3. Participants create a global expansion strategy for a medical technology company.

Pairing 7: Managing Global Workforce Trends in Remote Work

Big Skill: International Strategy Development – Creating flexible work policies for multinational teams.

Small Skill: Monitoring Global Trends – Tracking shifts in labor laws and remote work adoption.

Teaching Technique: Global workforce strategy planning.

Actions: Develop a remote workforce model based on regional labor laws and cultural norms.

Example:

1. Teams study remote work legislation in the U.S., EU, and Asia-Pacific.
2. They assess cultural attitudes toward hybrid work arrangements.
3. Participants create a policy framework for a multinational company adapting to global remote work trends.

References

Anderson, L. W., & Krathwohl, D. R. (Eds.). (2001). *A taxonomy for learning, teaching, and assessing: A revision of Bloom's Taxonomy of educational objectives*. Longman.

Argyris, C., & Schön, D. A. (1978). *Organizational learning: A theory of action perspective.* Addison-Wesley.

Aristotle. (c. 350 BCE). *Organon* (Trans. J. Barnes). Oxford University Press.

Arnold, R. D., & Wade, J. P. (2015). A definition of systems thinking. *Procedia Computer Science, 44*, 669-678.

Aquinas, T. (1274). *Summa theologica.* Christian Classics Ethereal Library.

Bacon, F. (1620). *Novum organum* (Trans. J. Spedding, R. L. Ellis, & D. D. Heath). Clarendon Press.

Baehr, J. (2013). *The inquiring mind: On intellectual virtues and virtue epistemology.* Oxford University Press.

Bloom, B. S. (1956). *Taxonomy of educational objectives, Handbook I: The cognitive domain.* David McKay.

Brickhouse, T. C., & Smith, N. D. (2000). *The philosophy of Socrates.* Westview Press.

Brown, P. C., Roediger, H. L., & McDaniel, M. A. (2014). *Make it stick: The science of successful learning.* Harvard University Press.

Carr, N. (2010). *The shallows: What the Internet is doing to our brains.* W.W. Norton.

Christensen, C. M. (1997). *The innovator's dilemma: When new technologies cause great firms to fail.* Harvard Business Review Press.

Collins, J. (2001). *Good to great: Why some companies make the leap... and others do not.* Harper Business.

Copleston, F. (1955). *Aquinas: An introduction to the life and work of the great medieval thinker.* Penguin Books.

Cottingham, J. (1992). *The Cambridge companion to Descartes.* Cambridge University Press.

De Bono, E. (1967). *The use of lateral thinking.* Penguin Books.

De Bono, E. (1985). *Six thinking hats.* Little, Brown and Company.

De Bono, E. (1992). *Serious creativity: Using the power of lateral thinking to create new ideas.* Harper Business.

Descartes, R. (1637). *Discourse on the method* (Trans. E. Haldane & G. R. T. Ross). Cambridge University Press.

Dewey, J. (1933). *How we think.* D.C. Heath.

Drucker, P. (1993). *The effective executive.* HarperCollins.

Duhigg, C. (2012). *The power of habit: Why we do what we do in life and business.* Random House.

Dunning, D. (2011). The Dunning-Kruger Effect: On being ignorant of one's own ignorance. *Advances in Experimental Social Psychology, 44*, 247-296.

Dweck, C. (2006). *Mindset: The new psychology of success.* Random House.

Eberle, B. (1996). *SCAMPER: Creative games and activities for imagination development* (2nd ed.). Prufrock Press.

Einstein, A. (1934). *The world as I see it.* Philosophical Library.

Eisenhardt, K. M., & Sull, D. N. (2001). Strategy as simple rules. *Harvard Business Review, 79*(1), 106-116.

Ennis, R. H. (2018). Critical thinking across the curriculum: A vision. *Inquiry: Critical Thinking Across the Disciplines, 33*(3), 1-8.

Epstein, R. M. (1999). Mindful practice. *JAMA, 282*(9), 833–839. https://doi.org/10.1001/jama.282.9.833

Ericsson, K. A. (2006). *The Cambridge handbook of expertise and expert performance.* Cambridge University Press.

Facione, P. A. (2011). *Think critically.* Pearson.

Facione, P. A. (2020). *Critical thinking: What it is and why it counts.* Insight Assessment.

Fine, G. (1999). *Plato 1: Metaphysics and epistemology*. Oxford University Press.

Fisher, R., Ury, W., & Patton, B. (2011). *Getting to yes: Negotiating agreement without giving in.* Penguin.

Flavell, J. H. (1963). *The developmental psychology of Jean Piaget.* D. Van Nostrand.

Flavell, J. H. (1979). Metacognition and cognitive monitoring. *American Psychologist, 34*(10), 906-911.

Gardner, S. (1999). *Kant and the critique of pure reason.* Routledge.

Gavetti, G., & Rivkin, J. W. (2007). On the origin of strategy: Action and cognition over time. *Organization Science, 18*(3), 420-439.

Gazzaley, A., & Rosen, L. (2016). *The distracted mind: Ancient brains in a high-tech world.* MIT Press.

Gigerenzer, G., & Brighton, H. (2009). Homo heuristicus: Why biased minds make better inferences. *Topics in Cognitive Science, 1*(1), 107–143. https://doi.org/10.1111/j.1756-8765.2008.01006.x

Goleman, D. (1995). *Emotional intelligence: Why it can matter more than IQ.* Bantam Books.

Goldman, E. F., Scott, A. R., & Follman, J. M. (2015). Organizational practices to develop strategic thinking. *Journal of Strategy and Management, 8*(2), 155-175.

Guilford, J. P. (1967). *The nature of human intelligence.* McGraw-Hill.

Haidt, J. (2012). *The righteous mind: Why good people are divided by politics and religion.* Pantheon.

Heath, C., & Heath, D. (2007). *Made to stick: Why some ideas survive, and others die.* Random House.

Heath, C., & Heath, D. (2013). *Decisive: How to make better choices in life and work.* Currency.

Hughes, R. L., & Beatty, K. C. (2005). *Becoming a strategic leader: Your role in your organization's enduring success.* Jossey-Bass.

James, W. (1890). *The principles of psychology.* Henry Holt.

Janis, I. (1972). *Victims of groupthink.* Houghton Mifflin.

Kahneman, D. (2011). *Thinking, Fast and slow.* Farrar, Straus, and Giroux.

Kant, I. (1781). *Critique of pure reason* (Trans. P. Guyer & A. Wood). Cambridge University Press.

Lafley, A. G., & Martin, R. (2013). *Playing to win: How strategy really works.* Harvard Business Review Press.

LeCun, Y., Bengio, Y., & Hinton, G. (2015). Deep learning. *Nature, 521*(7553), 436-444.

Liedtka, J. (1998). Strategic thinking: Can it be taught? *Long Range Planning, 31*(1), 120-129.

Maeda, J. (2006). *The laws of simplicity.* MIT Press.

Meadows, D. H. (2008). *Thinking in systems: A primer.* Chelsea Green Publishing.

Mill, J. S. (1863). *Utilitarianism.* Longmans, Green, and Co.

Mintzberg, H. (1994). *The rise and fall of strategic planning.* Free Press.

Misak, C. (2004). T*ruth and the end of inquiry: A Peircean account of truth.* Oxford University Press.

Newport, C. (2016). *Deep work: Rules for focused success in a distracted world.* Grand Central Publishing.

Newport, C. (2019). *Digital minimalism: Choosing a focused life in a noisy world.* Portfolio.

Nichols, T. (2017). *The death of expertise.* Oxford University Press.

Pariser, E. (2011). *The filter bubble: What the Internet is hiding from you.* Penguin.

Partnership for 21st Century Learning - P21. (2007). *Framework for 21st century learning.* https://www.battelleforkids.org/networks/p21

Paul, R., & Elder, L. (2019). *Critical thinking: Tools for taking charge of your learning and your life.* Pearson.

Peirce, C. S. (1877). The fixation of belief. *Popular Science Monthly*, *12*, 1-15.

Petrides, K. V., & Furnham, A. (2001). Trait emotional intelligence: Psychometric investigation with reference to established trait taxonomies. *European Journal of Personality, 15*(6), 425-448.

Piaget, J. (1936). *Origins of intelligence in the child.* Routledge & Kegan Paul.

Piaget, J. (1952). *The origins of intelligence in children.* Norton.

Plato. (c. 380 BCE). *The republic* (Trans. G. M. A. Grube). Hackett Publishing.

Roberts, R. C., & Wood, W. J. (2007). *Intellectual virtues: An essay in regulative epistemology.* Oxford University Press.

Robinson, K. (2011). *Out of our minds: The power of being creative.* Capstone.

Rumelt, R. (2011). *Good strategy, Bad strategy: The difference and why it matters.* Crown Business.

Runco, M. A., & Jaeger, G. J. (2012). The standard definition of creativity. *Creativity Research Journal, 24*(1), 92-96.

Russell, S., & Norvig, P. (2020). *Artificial intelligence: A modern approach* (4th ed.). Pearson.

Salovey, P., & Mayer, J. D. (1990). Emotional intelligence. *Imagination, Cognition and Personality, 9*(3), 185-211.

Sawyer, K. (2012). *Explaining creativity: The science of human innovation.* Oxford University Press.

Senge, P. M. (1990). *The fifth discipline: The art and practice of the learning organization.* Doubleday.

Senge, P. M. (2006). *The fifth discipline field book: Strategies and tools for building a learning organization.* Currency.

Schoemaker, P. J. (1995). Scenario planning: A tool for strategic thinking. *Sloan Management Review, 36*(2), 25-40.

Schön, D. A. (1983). *The reflective practitioner: How professionals think in action.* Basic Books.

Simon, H. A. (1996). T*he sciences of the artificial* (3rd ed.). MIT Press.

Skorupski, J. (1989). *John Stuart Mill*. Routledge.

Smith, R. (1997). *Aristotle: Prior analytics*. Hackett Publishing.

Sternberg, R. J. (1985). *Beyond IQ: A triarchic theory of human intelligence.* Cambridge University Press.

Sternberg, R. (1997). *Successful intelligence.* Plume.

Sweller, J. (1988). Cognitive load during problem solving: Effects on learning. *Cognitive Science, 12*(2), 257-285.

Toulmin, S. (1958). *The uses of argument.* Cambridge University Press.

Turing, A. (1950). Computing machinery and intelligence. *Mind, 59*(236), 433-460.

Tversky, A., & Kahneman, D. (1974). Judgment under uncertainty: Heuristics and biases. *Science, 185*(4157), 1124-1131.

Twenge, J. M. (2017). *iGen: Why today's super-connected kids are growing up less rebellious, more tolerant, less happy—and completely unprepared for adulthood—and what that means for the rest of us.* Atria Books.

Urbach, P. (1987). *Francis Bacon's philosophy of science: An account and a reappraisal.* Open Court.

Vlastos, G. (1991). *Socrates: Ironist and moral philosopher*. Cambridge University Press.

Vygotsky, L. S. (1978). *Mind in society: The development of higher psychological processes.* Harvard University Press.

Weston, A. (2018). *A rulebook for arguments.* Hackett Publishing.

Wolf, M. (2018). *Reader, come home: The reading brain in a digital world.* Harper.

Zagzebski, L. T. (1996). *Virtues of the mind: An inquiry into the nature of virtue and the ethical foundations of knowledge.* Cambridge University Press.

Appendix A: Five Synergistic Groupings of Thinking Skills

PCS is a structured, research-backed framework designed to guide cognitive development from foundational skills to advanced problem-solving and leadership. It addresses educational and workforce gaps by ensuring individuals develop a balanced set of cognitive skills that allow them to think critically, adapt flexibly, create innovatively, reason precisely, and collaborate effectively. By integrating theories from educational psychology, intelligence research, and real-world problem-solving, PCS serves as a comprehensive approach to fostering cognitive excellence across a lifetime.

PCS Development Stages vs. the Five Synergistic Thinking Skills

To ensure no confusion where PCS is mistakenly considered a "final solution," PCS avoids siloed learning by advancing cognitive development around five synergistic groupings of thinking skills:

1. **Integrative Thinking** – Includes systems thinking, critical thinking, and meta-thinking.
2. **Innovative Thinking** – Incorporates creative exploration, practical application, and habit formation.
3. **Reflective & Adaptive Thinking** – Comprises introspection, emotional intelligence, and adaptability.
4. **Precision Thinking** – Embodies logical analysis, evidence-based reasoning, and intellectual rigor.
5. **Collaborative & Contextual Thinking** – Encompasses teamwork, empathy, and contextual awareness.

Understanding the Development of the Five Synergistic Thinking Skills

While the PCS incorporates the learning strategies needed to develop the learning skills at each of the designated stages, additional training and application are necessary to fully obtain and master the Five Synergistic Thinking Skills. These advanced cognitive skills do not develop in isolation; rather, they require targeted training, real-world application, and continuous learning beyond standard education.

- The **PCS developmental stages provide the foundation** by strengthening cognitive abilities, but they do not independently cultivate fully developed synergistic thinking skills.
- The Five Synergistic Thinking Skills—**Integrative, Innovative, Reflective & Adaptive, Precision, and Collaborative & Contextual Thinking**—are higher-order

cognitive abilities that necessitate deliberate practice and exposure to complex problem-solving scenarios. For example:

  - **Precision Thinking** (logical analysis, evidence-based reasoning) may begin developing in grade school. However, without advanced training in data analysis, critical evaluation, or structured argumentation, it remains incomplete.
  - **Reflective & Adaptive Thinking** (emotional intelligence, adaptability) may be introduced in high school, but without leadership training, real-world experience, or crisis management exposure, it cannot be fully realized.

Thus, while the **PCS framework lays the groundwork**, mastery of the Five Synergistic Thinking Skills requires a combination of structured learning, professional development, and practical application.

The Role of PCS Developmental Stages in Five Synergistic Thinking Skills Acquisition

The **PCS developmental stages are essential** for building the foundational cognitive skills that eventually support the Five Synergistic Thinking Skills. These stages function as **progressive stepping-stones**, strengthening and refining fundamental cognitive abilities in preparation for advanced thinking.

- If a learner **skips** or does not complete certain PCS stages, they may experience gaps in their cognitive skillset that hinder their ability to develop synergistic thinking skills effectively. Examples:
  - A lack of structured reasoning skills from middle school can result in struggles with **Integrative Thinking** later on, as systems thinking and logical analysis were not reinforced.
  - If a professional has never been exposed to adaptability training or contextual awareness exercises, they will struggle to master **Collaborative & Contextual Thinking** in leadership roles.
- While alternative learning methods—such as **self-directed learning, accelerated training, or experiential learning**—can support skill development, they cannot replace the foundational cognitive abilities established through the PCS framework.

The Difference Between PCS Development Stages and the Five Synergistic Thinking Skills

- The Five Synergistic Thinking Skills require additional training beyond PCS development.
- The PCS developmental stages are essential for laying the foundation of cognitive development.
- Without completing the PCS stages, gaps in cognitive development may hinder full mastery of the Five Synergistic Thinking Skills.

Ultimately, developing these advanced cognitive abilities demands a combination of structured learning, real-world application, and ongoing professional growth. To develop the five synergistic groupings of thinking skills, we followed a structured logical analysis and synthesis approach. This involved breaking down the PCS framework, analyzing its alignment with cognitive skill development, and synthesizing the relationships between the stages and the synergistic groupings. Below is a detailed step-by-step reasoning that led to this conclusion:

Step 1: Establishing the Foundation of PCS (Previously Discussed in this Book)

The PCS framework is explicitly designed to foster cognitive development progressively across five stages:

- **Grade School** – Basic cognitive functions (recognition, logical reasoning, problem-solving).
- **Middle School** – Intermediate cognitive functions (critical thinking, creativity, systems thinking).
- **High School** – Advanced cognitive functions (abstract reasoning, strategic planning, reflection).
- **College/Technical School** – Integration of theory and practice (interdisciplinary problem-solving, adaptability).
- **Business Professionals** – Mastery of leadership and uncertainty management.

These stages are logically built upon one another, ensuring cognitive skills develop in a structured and increasingly complex manner.

Step 2: Identifying the Five Synergistic Groupings of Thinking Skills

The PCS framework leads to the development of cognitive abilities, which are then integrated into **five synergistic groupings**:

- **Integrative Thinking** – Combining critical, systems, and meta-thinking to address complex problems.
- **Innovative Thinking** – Blending creativity, practical application, and habit formation.
- **Reflective & Adaptive Thinking** – Enhancing introspection, emotional intelligence, and adaptability.
- **Precision Thinking** – Ensuring logical analysis, evidence-based reasoning, and intellectual rigor.
- **Collaborative & Contextual Thinking** – Strengthening teamwork, empathy, and contextual awareness.

Each of these **groupings represents a critical dimension of cognitive skill development** necessary for success at different educational and professional stages.

Step 3: Mapping PCS Stages to the Synergistic Groupings

Each stage of PCS contributes **incrementally** to the development of all five thinking skill groupings:

- **Grade School: Laying the Foundation**
  - **Integrative Thinking**: Early pattern recognition and problem-solving.
  - **Innovative Thinking**: Encouraging creativity and exploration.
  - **Reflective & Adaptive Thinking**: Developing early self-awareness and adaptability.
  - **Precision Thinking**: Learning logical reasoning and structured thinking.
  - **Collaborative & Contextual Thinking**: Basic teamwork and empathy.
- **Middle School: Strengthening Critical and Creative Thinking**
  - **Integrative Thinking**: Early systems thinking and deeper logical connections.
  - **Innovative Thinking**: Developing creative problem-solving strategies.
  - **Reflective & Adaptive Thinking**: Beginning to assess personal learning and emotional intelligence.
  - **Precision Thinking**: Applying structured reasoning and evidence-based analysis.
  - **Collaborative & Contextual Thinking**: Participating in teamwork and appreciating diverse viewpoints.
- **High School: Advancing Strategic and Abstract Thinking**
  - **Integrative Thinking**: Refining abstract reasoning and strategic planning.
  - **Innovative Thinking**: Applying creative solutions to structured problems.
  - **Reflective & Adaptive Thinking**: Developing independent learning and adaptability.
  - **Precision Thinking**: Using higher-order logic and structured argumentation.
  - **Collaborative & Contextual Thinking**: Understanding global and societal contexts in problem-solving.
- **College/Technical School: Applying Theory to Practice**
  - **Integrative Thinking**: Using interdisciplinary problem-solving in real-world scenarios.
  - **Innovative Thinking**: Bridging creativity with real-world applications.
  - **Reflective & Adaptive Thinking**: Adapting to diverse academic and professional challenges.
  - **Precision Thinking**: Engaging in intellectual rigor and critical analysis.
  - **Collaborative & Contextual Thinking**: Working in teams and navigating complex social dynamics.
- **Business Professionals: Mastery of Cognitive and Strategic Leadership**
  - **Integrative Thinking**: Executing high-level strategic decision-making.
  - **Innovative Thinking**: Driving innovation in organizational settings.

- **Reflective & Adaptive Thinking**: Leading through uncertainty and continuous learning.
- **Precision Thinking**: Making evidence-based decisions under pressure.
- **Collaborative & Contextual Thinking**: Navigating global business environments and leading diverse teams.

Step 4: Establishing the Logical Conclusion

Since each stage of PCS contributes progressively to all five thinking skill groupings, we can conclude that:

- The PCS framework ensures a structured, developmental pathway for cognitive skills.
- The five synergistic groupings of thinking skills are not developed in isolation but evolve progressively.
- Each PCS stage builds upon prior learning, leading to increasingly complex and refined thinking abilities.

Thus, the PCS instructional framework leads to the development of five synergistic thinking skills at every stage of learning and professional growth, ensuring a holistic and progressive cognitive development process.

Choosing the Five Synergistic Thinking Skills

The choice of the five synergistic thinking skills (**Integrative, Innovative, Reflective & Adaptive, Precision, and Collaborative & Contextual Thinking**) was not arbitrary. Instead, their selection was grounded in cognitive science, educational development theories, and the specific demands of real-world problem-solving and leadership. Below is a detailed explanation of how these particular five were identified and why they were chosen over alternative groupings.

Step 1: Establishing the Need for a Structured Thinking Skills Framework

The PCS framework was designed to develop **progressive cognitive abilities** that evolve from foundational thinking in grade school to mastery in professional settings. A structured approach to thinking skill development must:

- Encompass both **big thinking** (strategic vision) and **small thinking** (attention to detail).
- Provide **practical utility** in academic, professional, and leadership contexts.
- Enable **progressive skill-building** rather than isolated, non-integrative learning.

To ensure a balanced, holistic cognitive framework, the thinking skills must:

- **Cover a broad spectrum of mental functions** (not just one type, like analytical thinking).
- Be progressively built upon each other as one moves through the PCS stages.
- **Be applicable across disciplines** rather than being domain-specific.

Step 2: Examining Established Theories of Thinking and Learning

The selection of these five groupings aligns with **established cognitive and educational research**, including:

- **Bloom's Taxonomy of Cognitive Development**
  - Bloom's hierarchy moves from basic knowledge acquisition to higher-order thinking skills like evaluation and creation.
  - The PCS model follows a **progression** similar to Bloom's:
    - **Grade school** (knowledge, comprehension).
    - **Middle school** (application, analysis).
    - **High school** (synthesis, evaluation).
    - **College/Technical School & beyond** (creation, strategic decision-making).
  - Each stage aligns with Integrative, Innovative, Reflective, Precision, and Collaborative Thinking.
- **Sternberg's Triarchic Theory of Intelligence**
  - **Analytical intelligence** (logical, precision thinking) → Precision Thinking
  - **Creative intelligence** (idea generation, problem-solving) → Innovative Thinking
  - **Practical intelligence** (application of knowledge in real-world scenarios) → Integrative and Adaptive Thinking
  - This supports the need for skills that address not just analysis but also creativity, adaptability, and collaboration.
- **Divergent vs. Convergent Thinking** (Guilford, 1950s)
  - **Divergent Thinking** (generating ideas, flexibility) → Innovative Thinking
  - **Convergent Thinking** (logical reasoning, narrowing down) → Precision Thinking
  - Both must be included to create a balanced cognitive structure.
- **21st Century Skills Framework** (Partnership for 21st Century Learning - P21, 2007)
  - Identifies critical thinking, creativity, collaboration, communication, and adaptability as essential workplace skills.
  - These directly align with Integrative, Innovative, Collaborative, Reflective, and Precision Thinking.

Step 3: Justifying the Five Synergistic Thinking Skills Over Other Possible Choices

We could have chosen other groupings, such as:

- **"Logical Thinking, Problem-Solving, Creativity, Leadership, Adaptability"** or

- **"Abstract Thinking, Emotional Intelligence, Critical Thinking, Innovation, Communication"**

However, those are either too broad, too narrow, or fail to cover key aspects of cognitive development. Let us analyze why the five chosen groupings are the best fit:

- **Integrative Thinking (Systems, Critical, and Meta-Thinking)**
  - **Why?** Because real-world problems are interconnected and require the ability to see both details and the big picture.
  - **Alternative?** Just "Critical Thinking" would be too narrow; without systems thinking, critical thinking can lead to siloed problem-solving.
- **Innovative Thinking (Creative Exploration, Practical Application, Habit Formation)**
  - **Why?** Because thinking without execution is incomplete. True innovation requires both idea generation and practical application.
  - **Alternative?** Just "Creativity" would be too unstructured, and "Problem-Solving" alone does not emphasize originality.
- **Reflective & Adaptive Thinking (Self-Reflection, Emotional Intelligence, Adaptability)**
  - **Why?** Because successful cognitive growth requires the ability to learn from past experiences and adjust strategies when faced with uncertainty.
  - **Alternative?** Just "Adaptability" would not emphasize introspection, and just "Emotional Intelligence" would not cover cognitive flexibility
- **Precision Thinking (Logical Analysis, Evidence-Based Reasoning, Intellectual Rigor)**
  - **Why?** Because precise, evidence-based reasoning is crucial for decision-making, academic rigor, and problem-solving.
  - **Alternative?** Just "Logical Thinking" would not emphasize intellectual rigor and "Problem-Solving" would not focus on structured, evidence-based reasoning.
- **Collaborative & Contextual Thinking (Teamwork, Empathy, and Context Awareness)**
  - **Why?** Because learning and decision-making do not happen in isolation; understanding the social, cultural, and organizational context is key.
  - **Alternative?** Just "Communication" would not include empathy or strategic collaboration.

Step 4: Synthesizing the Conclusion

Since these five cognitive skill groupings:

- Align with established cognitive development frameworks (e.g., Bloom, Sternberg, P21),
- Ensure a progressive learning structure across all PCS stages,
- **Balance different dimensions of thinking** (logical, creative, strategic, and social), and
- Are directly applicable to academic, professional, and leadership settings,

We logically concluded that the PCS instructional framework is structured around these five thinking skill groupings (i.e., **Integrative, Innovative, Reflective & Adaptive, Precision, and Collaborative & Contextual Thinking**) to ensure a holistic, progressive, and practical cognitive development process and what is required for success across different life stages.

Conclusion

- The Five Synergistic Thinking Skills require additional training beyond PCS development.
- **The PCS developmental stages are essential** for laying the foundation of these skills.
- Without completing the PCS stages, gaps in cognitive development may hinder full mastery of the Five Synergistic Thinking Skills.

Ultimately, developing these advanced cognitive abilities demands a combination of structured learning, real-world application, and ongoing professional growth.

The Five Synergistic Thinking Skills and the Generalities of How They Are Taught/Learned

The five **synergistic groupings of thinking skills** are a framework for developing cognitive abilities in learners, typically aligned with **PCS (Progressive Cognitive Structuring)** and other cognitive development models. These thinking skills work together to enhance learning, problem-solving, and adaptability.

- **Focusing Skills** – Directing attention, filtering distractions, and prioritizing important details.
  - **Who Teaches It?** Parents, teachers, mentors, and structured learning programs.
  - **How It is Developed?** Activities like mindfulness exercises, puzzles, structured questioning, and practicing active listening.

- **Information Gathering Skills** – Observing, questioning, and remembering information.
  - **Who Teaches It?** Educators, librarians, researchers, and AI-powered learning tools.
  - **How It is Developed?** Inquiry-based learning, reading strategies, notetaking techniques, and interactive research projects.

- **Remembering Skills** – Storing, recalling, and using knowledge effectively.
    - **Who Teaches It?** Memory coaches, educators, and cognitive scientists.
    - **How It is Developed?** Mnemonics, spaced repetition, storytelling, and real-world applications.
- **Organizing Skills** – Structuring, categorizing, and sequencing information.
    - **Who Teaches It?** Teachers, business leaders, executive function coaches.
    - **How It is Developed?** Graphic organizers, project planning, outlining, and coding/programming exercises.
- **Analyzing & Generating Skills** – Comparing, contrasting, synthesizing, and creating new ideas.
    - **Who Teaches It?** Debate coaches, STEM educators, philosophers, and AI-enhanced learning platforms.
    - **How It is Developed?** Socratic questioning, case studies, simulations, and innovation-based challenges.

Teaching & Developing the Five Synergistic Groupings of Thinking Skills

PCS (Progressive Cognitive Structuring) avoids **siloed learning** by structuring cognitive development around five synergistic groupings of thinking skills. Each grouping interconnects with the others, ensuring a **holistic approach to learning and problem-solving.** Below is an overview of how these thinking skills can be taught and developed effectively.

---

1. Integrative Thinking – Systems Thinking, Critical Thinking, and Meta-Thinking

Who Teaches It?

- Educators, business strategists, philosophers, and interdisciplinary experts.

How It is Developed?

- **Systems Thinking**: Encourage mind mapping, flowcharts, and cause-effect analysis.
- **Critical Thinking**: Use Socratic questioning, case studies, and real-world problem-solving.
- **Meta-Thinking (Thinking About Thinking)**: Teach self-reflection techniques, cognitive biases, and reasoning strategies.

---

2. Innovative Thinking – Creative Exploration, Practical Application, and Habit Formation

Who Teaches It?

- Design thinkers, entrepreneurs, STEM educators, and innovation mentors.

How It is Developed?

- **Creative Exploration**: Use brainstorming, divergent thinking exercises, and artistic expression.
- **Practical Application**: Implement project-based learning, maker spaces, and prototype development.
- **Habit Formation**: Encourage discipline through goal setting, experimentation, and iteration cycles.

---

3. Reflective & Adaptive Thinking – Introspection, Emotional Intelligence, and Adaptability

Who Teaches It?

- Psychologists, leadership coaches, social-emotional learning (SEL) educators, and mentors.

How It is Developed?

- **Introspection**: Teach journaling, guided meditation, and self-assessment.
- **Emotional Intelligence**: Develop skills in empathy, active listening, and managing emotions.
- **Adaptability**: Use role-playing, situational analysis, and exposure to unpredictable challenges.

---

4. Precision Thinking – Logical Analysis, Evidence-Based Reasoning, and Intellectual Rigor

Who Teaches It?

- Mathematicians, scientists, lawyers, and data analysts.

How It is Developed?

- **Logical Analysis**: Teach deductive/inductive reasoning, argument mapping, and formal logic.
- **Evidence-Based Reasoning**: Encourage research methods, data literacy, and skepticism training.
- **Intellectual Rigor**: Use structured debates, error analysis, and rigorous evaluation of sources.

---

5. Collaborative & Contextual Thinking – Teamwork, empathy, and contextual awareness.

Who Teaches It?

- Team leaders, community organizers, historians, and cultural analysts.

How It is Developed?

- **Teamwork**: Encourage cooperative learning, group projects, and peer mentorship.
- **Empathy**: Foster perspective-taking, ethical discussions, and real-world simulations.
- **Contextual Awareness**: Teach historical thinking, cross-cultural literacy, and environmental scanning.

How to Obtain and Master These Five Synergistic Groupings of Thinking Skills

1. **Use Generative AI & Adaptive Learning**: AI tutors can provide personalized guidance on logic, creativity, and collaboration.
2. **Engage in Interdisciplinary Learning**: Encourage students to connect concepts across multiple fields.
3. **Develop Problem-Based & Inquiry-Based Learning**: Challenge students with open-ended questions and complex scenarios.
4. **Encourage Lifelong Learning & Reflection**: Promote journaling, mentorship, and exposure to diverse viewpoints.
5. **Simulate Real-World Decision-Making**: Use case studies, debates, and immersive problem-solving exercises.

Each **developmental stage** builds upon the previous one, refining cognitive abilities and deepening the learner's capacity for complex thought.

- **Grade School**: Foundational cognitive abilities such as pattern recognition, logical reasoning, and basic problem-solving are developed. These skills are essential for Precision Thinking and Innovative Thinking later.
- **Middle School**: Learners refine critical thinking, creativity, and early systems thinking, supporting Integrative Thinking and Precision Thinking as they explore problem-solving in broader contexts.
- **High School**: This stage enhances abstract reasoning, strategic foresight, and reflective thinking, laying the groundwork for Reflective & Adaptive Thinking and Integrative Thinking.
- **College/Technical School**: Students integrate theoretical knowledge with real-world problem-solving, cross-disciplinary thinking, and adaptability, strengthening Collaborative & Contextual Thinking and Innovative Thinking.
- **Business Professionals**: This stage focuses on high-level strategic decision-making, leadership, and managing uncertainty, leading to mastery in Integrative Thinking, Reflective & Adaptive Thinking, Collaborative & Contextual Thinking.

Skills Strengthened & Synergistic Thinking Skills Alignment

The following table directly maps **the skills strengthened at each PCS developmental stage** to their **corresponding synergistic thinking skills**, providing a clearer view of cognitive skill progression.

PCS Developmental Stage	Skills Strengthened	Corresponding Synergistic Thinking Skills
Grade School	Basic logical reasoning, structured problem-solving	Precision Thinking, Innovative Thinking
Middle School	Developing structured reasoning, expanding creative approaches	Integrative Thinking, Innovative Thinking, Precision Thinking
High School	Enhancing decision-making, refining strategic planning	Reflective & Adaptive Thinking, Integrative Thinking, Precision Thinking
College/Technical School	Interdisciplinary problem-solving, adapting knowledge in dynamic contexts	Integrative Thinking, Reflective & Adaptive Thinking, Innovative Thinking, Collaborative & Contextual Thinking
Business Professionals	Refining leadership, managing complex decision-making	Integrative Thinking, Reflective & Adaptive Thinking, Collaborative & Contextual Thinking

---

Specific Training Methods for Each Synergistic Thinking Skill

To fully develop and master the five synergistic thinking skills, targeted training beyond the PCS developmental stages is necessary. Below is a structured approach with specific training methods for each skill.

1. Integrative Thinking ***(Systems Thinking, Critical Thinking, Meta-Thinking)***

Goal: Ability to analyze complex systems, recognize patterns, and think critically about interconnected elements.

Training Methods:

- **Systems Thinking Exercises** – Use tools like causal loop diagrams, flowcharts, and network mapping to visualize relationships in complex problems.
- **Case Study Analysis** – Study real-world failures and successes (e.g., business case studies, historical events) to identify systemic causes and effects.
- **Interdisciplinary Learning** – Combine knowledge from different fields (e.g., economics, psychology, engineering, and philosophy) to develop holistic perspectives.
- **Socratic Questioning** – Engage in structured questioning to challenge assumptions and enhance deep thinking.

- **Simulation-Based Learning** – Use platforms like business simulations, war games, or policy-making scenarios to practice systemic decision-making.

Example Application:

- A business executive uses systems thinking to understand market dynamics and anticipate economic shifts.
- A scientist applies meta-thinking by analyzing how their own biases might affect experimental design.

---

2. Innovative Thinking *(Creative Exploration, Practical Application, Habit Formation)*

Goal: Foster creativity, develop practical innovations, and create habits for continuous improvement.

Training Methods:

- **Design Thinking Workshops** – Engage in hands-on innovation processes using methods like ideation, prototyping, and user feedback.
- **Divergent Thinking Exercises** – Use techniques like SCAMPER (Substitute, Combine, Adapt, Modify, Put to another use, Eliminate, Reverse) to explore alternative solutions.
- **Creative Constraints Challenges** – Solve problems with intentional limitations (e.g., "How can you build a house without using nails?").
- **Daily Creativity Practices** – Develop habits such as mind-mapping, journaling, or brainstorming one new idea per day.
- **Hackathons & Rapid Prototyping** – Participate in innovation challenges where solutions must be designed and evaluated under tight deadlines.

Example Application:

- A software engineer applies design thinking to develop a user-friendly app that solves an emerging problem.
- A teacher uses innovative thinking to create engaging lesson plans that adapt to different learning styles.

---

3. Reflective & Adaptive Thinking *(Introspection, Emotional Intelligence, Adaptability)*

Goal: Develop self-awareness, resilience, and the ability to adapt to changing environments.

Training Methods:

- **Self-Reflection & Journaling** – Regularly document thoughts, decisions, and lessons learned to develop introspective skills.
- **Emotional Intelligence Training** – Practice active listening, empathy exercises, and conflict resolution role-playing.

- **Exposure to Uncertainty** – Engage in improvisational activities, such as public speaking without preparation or navigating unpredictable scenarios.
- **Cognitive Reframing Techniques** – Learn to reframe negative experiences as learning opportunities using approaches like CBT (Cognitive Behavioral Therapy) strategies.
- **Cross-Cultural Immersion** – Adaptability increases when working in diverse environments or experiencing new cultures firsthand.

Example Application:

- A military veteran transitioning to civilian life uses cognitive reframing to adjust their leadership approach.
- A corporate leader applies adaptive thinking to change the company strategy during an economic downturn.

---

4. Precision Thinking *(Logical Analysis, Evidence-Based Reasoning, Intellectual Rigor)*

Goal: Develop structured, evidence-based reasoning and avoid cognitive biases.

Training Methods:

- **Formal Logic & Argumentation Training** – Learn deductive, inductive, and abductive reasoning through formal logic courses.
- **Data Literacy & Statistical Analysis** – Develop skills in interpreting data, identifying patterns, and recognizing flawed reasoning (e.g., correlation vs. causation).
- **Debate & Persuasive Argumentation** – Engage in structured debates where every claim must be supported with strong evidence.
- **Bias Detection & Cognitive Fallacy Exercises** – Study logical fallacies, cognitive biases, and heuristics that influence poor decision-making.
- **Peer Review & Critical Evaluation** – Practice reviewing research papers and identifying weak arguments or methodological flaws.

Example Application:

- A lawyer applies precision thinking to construct airtight legal arguments using precedent and evidence.
- A scientist ensures their hypotheses are rigorously evaluated and conclusions are logically sound.

---

5. Collaborative & Contextual Thinking (*Teamwork, Empathy, Contextual Awareness*)

Goal: Develop skills for working effectively in teams and understanding broader contextual influences.

Training Methods:

- **Perspective-Taking & Role-Reversal Exercises** – Engage in simulations where individuals must argue from opposing viewpoints.

- **High-Stakes Team Challenges** – Participate in escape rooms, military-style team-building exercises, or crisis management drills to enhance cooperation.
- **Cultural Competency Training** – Learn historical and cultural contexts to understand how beliefs and behaviors vary across groups.
- **Negotiation & Mediation Training** – Use structured negotiation role-plays to develop skills in diplomacy and conflict resolution.
- **Real-World Team Projects** – Engage in cross-functional teamwork (e.g., multidisciplinary research, business consulting projects).

Example Application:

- A diplomat applies contextual thinking to navigate complex international relations.
- A corporate leader uses collaborative thinking to resolve conflicts and build high-performing teams.

Final Thoughts: Mastering the Five Synergistic Thinking Skills

- The PCS developmental stages lay the groundwork, but true mastery of these thinking skills requires deliberate, ongoing training.
- These training methods ensure each thinking skill is actively developed, practiced, and applied in real-world settings.
- To accelerate progress, learners should engage in multiple training methods simultaneously, rather than relying on a single approach.

Made in United States
Orlando, FL
04 November 2026

90208242R00389